# Wetting Dynamics of Hydrophobic and Structured Surfaces

Jefferson Hotel, Richmond, Virginia, USA
12–14 April 2010

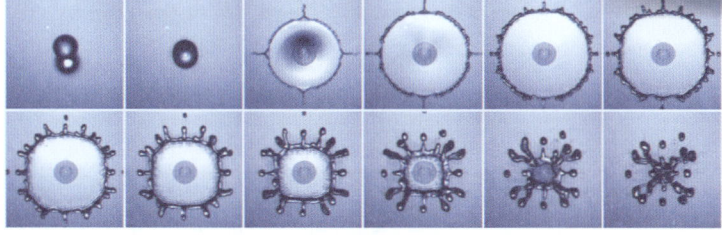

## FARADAY DISCUSSIONS
Volume 146, 2010

RSCPublishing

The Faraday Division of the Royal Society of Chemistry, previously the Faraday Society, founded in 1903 to promote the study of sciences lying between Chemistry, Physics and Biology.

**EDITORIAL STAFF**

**Editor**
Philip Earis

**Deputy editor**
Jane Hordern

**Senior Publishing editor**
Nicola Nugent

**Development editor**
Vibhuti Patel

**Publishing editors**
Helen Lunn, Anna Roffey

**Publishing assistant**
Kate Bandoo

**Publisher**
Niamh O'Connor

*Faraday Discussions* (Print ISSN 1359-6640, Electronic ISSN 1364-5498) is published 4 times a year by the Royal Society of Chemistry, Thomas Graham House, Science Park, Milton Road, Cambridge, UK CB4 0WF. Volume 146 ISBN-13: 978 1 84973 0563

2010 annual subscription price: print+electronic £622, US $1,160; electronic only £560, US $1,045. Customers in Canada will be subject to a surcharge to cover GST. Customers in the EU subscribing to the electronic version only will be charged VAT. All orders, with cheques made payable to the Royal Society of Chemistry, should be sent to RSC Distribution Services, c/o Portland Customer Services, Commerce Way, Colchester, Essex, UK CO2 8HP.
Tel +44 (0) 1206 226050;
E-mail sales@rscdistribution.org

If you take an institutional subscription to any RSC journal you are entitled to free, site-wide web access to that journal. You can arrange access *via* Internet Protocol (IP) address at www.rsc.org/ip. Customers should make payments by cheque in sterling payable on a UK clearing bank or in US dollars payable on a US clearing bank. Periodicals postage is paid at Rahway, NJ and at additional mailing offices. Airfreight and mailing in the USA by Mercury Airfreight International Ltd., 365 Blair Road, Avenel, NJ 07001, USA.

US Postmaster: send address changes to *Faraday Discussions*, c/o Mercury Airfreight International Ltd., 365 Blair Road, Avenel, NJ 07001. All despatches outside the UK by Consolidated Airfreight.

PRINTED IN THE UK

*Faraday Discussions* documents a long-established series of *Faraday Discussion* meetings which provide a unique international forum for the exchange of views and newly acquired results in developing areas of physical chemistry, biophysical chemistry and chemical physics.

**ORGANISING COMMITTEE, Volume 146**

**Chair**
Dr Hugo K Christenson (University of Leeds, UK) (Co-chair)
Professor Alenka Luzar (Virginia Commonwealth University, USA) (Co-chair)

Professor Bob Evans (University of Bristol, UK)
Dr Jim R Henderson (University of Leeds, UK)
Professor Pablo G Debenedetti (Princeton University, USA)
Professor John D Weeks (University of Maryland, USA)
Professor Michael L Klein (University of Pennsylvania, USA)

**FARADAY STANDING COMMITTEE ON CONFERENCES**

**Chair**
D E Heard (Leeds, UK)

W A Brown (UCL, UK)
I Hamley (Reading, UK)
J Hirst (Nottingham, UK)
A Mount (Edinburgh, UK)

© The Royal Society of Chemistry 2010. Apart from fair dealing for the purposes of research or private study, or criticism or review, as permitted under the Copyright, Designs and Patents Act 1988 and Related Rights Regulations 2003, this publication may only be reproduced, stored or transmitted, in any form or by any means, with the prior permission in writing of the Publishers or in the case of reprographic reproduction in accordance with the terms of licences issued by the Copyright Licensing Agency in the UK. US copyright law applicable to users in the USA. The Royal Society of Chemistry takes reasonable care in the preparation of this publication but does not accept liability for the consequences of any errors or omissions.

Royal Society of Chemistry: Registered Charity No. 207890.

∞The paper used in this publication meets the requirements of ANSI/NISO Z39.48-1992 (Permanence of Paper).

# Wetting Dynamics of Hydrophobic and Structured Surfaces

Faraday Discussions

www.rsc.org/faraday_d

A General Discussion on Wetting Dynamics of Hydrophobic and Structured Surfaces was held at Jefferson Hotel, Richmond, Virginia, USA on 12th, 13th and 14th April 2010.

*RSC Publishing is a not-for-profit publisher and a division of the Royal Society of Chemistry. Any surplus made is used to support charitable activities aimed at advancing the chemical sciences. Full details are available from www.rsc.org*

## CONTENTS

ISSN 1359-6640; ISBN 978-1-84973-056-3

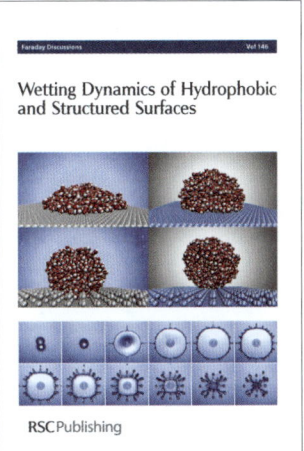

**Cover**
See Daub *et al.*, *Faraday Discuss.*, 2010, **146**, 67–77. Molecular picture (top): Nanoscale roughness renders hydrophilic surface more hydrophobic (left panel); Superhydrophobicity is achieved even when surface roughness is limited to nanoscale alone (right panel).

See Reyssat *et al.*, *Faraday Discuss.*, 2010, **146**, 19–33. Macroscopic picture (bottom): Top view of the impact of a water drop hitting a superhydrophobic solid decorated with a square lattice of micropillars. Interval between successive pictures is 1 ms.

Images reproduced by permission of Professors Alenka Luzar and David Quéré.

## PREFACE

9     **Preface**
        Alenka Luzar and Hugo K. Christenson

## INTRODUCTORY LECTURE

13    **Exploring nanoscale hydrophobic hydration**
        Peter J. Rossky

## PAPERS AND DISCUSSIONS

19    **Dynamical superhydrophobicity**
        Mathilde Reyssat, Denis Richard, Christophe Clanet and David Quéré

| | |
|---|---|
| 35 | **Superhydrophobic surfaces by hybrid raspberry-like particles**<br>Maria D'Acunzi, Lena Mammen, Maninderjit Singh, Xu Deng, Marcel Roth, Günter K. Auernhammer, Hans-Jürgen Butt and Doris Vollmer |
| 49 | **Microscopic shape and contact angle measurement at a superhydrophobic surface**<br>Helmut Rathgen and Frieder Mugele |
| 57 | **Transparent superhydrophobic and highly oleophobic coatings**<br>Liangliang Cao and Di Gao |
| 67 | **The influence of molecular-scale roughness on the surface spreading of an aqueous nanodrop**<br>Christopher D. Daub, Jihang Wang, Shobhit Kudesia, Dusan Bratko and Alenka Luzar |
| 79 | **General Discussion** |
| 103 | **Contact angle hysteresis: a different view and a trivial recipe for low hysteresis hydrophobic surfaces**<br>Joseph W. Krumpfer and Thomas J. McCarthy |
| 113 | **Amplification of electro-osmotic flows by wall slippage: direct measurements on OTS-surfaces**<br>Marie-Charlotte Audry, Agnès Piednoir, Pierre Joseph and Elisabeth Charlaix |
| 125 | **Electrowetting and droplet impalement experiments on superhydrophobic multiscale structures**<br>F. Lapierre, P. Brunet, Y. Coffinier, V. Thomy, R. Blossey and R. Boukherroub |
| 141 | **Macroscopically flat and smooth superhydrophobic surfaces: Heating induced wetting transitions up to the Leidenfrost temperature**<br>Guangming Liu and Vincent S. J. Craig |
| 153 | **Drop dynamics on hydrophobic and superhydrophobic surfaces**<br>B. M. Mognetti, H. Kusumaatmaja and J. M. Yeomans |
| 167 | **Dynamic mean field theory of condensation and evaporation processes for fluids in porous materials: Application to partial drying and drying**<br>J. R. Edison and P. A. Monson |
| 185 | **Molecular dynamics simulations of urea–water binary droplets on flat and pillared hydrophobic surfaces**<br>Takahiro Koishi, Kenji Yasuoka, Xiao Cheng Zeng and Shigenori Fujikawa |
| 195 | **General Discussion** |
| 217 | **First- and second-order wetting transitions at liquid–vapor interfaces**<br>K. Koga, J. O. Indekeu and B. Widom |
| 223 | **Hierarchical surfaces: an *in situ* investigation into nano and micro scale wettability**<br>Alex H. F. Wu, K. L. Cho, Irving I. Liaw, Grainne Moran, Nigel Kirby and Robert N. Lamb |
| 233 | **An experimental study of interactions between droplets and a nonwetting microfluidic capillary**<br>Geoff R. Willmott, Chiara Neto and Shaun C. Hendy |
| 247 | **Hydrophobic interactions in model enclosures from small to large length scales: non-additivity in explicit and implicit solvent models**<br>Lingle Wang, Richard A. Friesner and B. J. Berne |
| 263 | **Water reorientation, hydrogen-bond dynamics and 2D-IR spectroscopy next to an extended hydrophobic surface**<br>Guillaume Stirnemann, Peter J. Rossky, James T. Hynes and Damien Laage |
| 283 | **General Discussion** |

299     **The search for the hydrophobic force law**
        Malte U. Hammer, Travers H. Anderson, Aviel Chaimovich, M. Scott Shell and Jacob Israelachvili

309     **The effect of counterions on surfactant-hydrophobized surfaces**
        Gilad Silbert, Jacob Klein and Susan Perkin

325     **Hydrophobic forces in the wetting films of water formed on xanthate-coated gold surfaces**
        Lei Pan and Roe-Hoan Yoon

341     **Interfacial thermodynamics of confined water near molecularly rough surfaces**
        Jeetain Mittal and Gerhard Hummer

353     **Mapping hydrophobicity at the nanoscale: Applications to heterogeneous surfaces and proteins**
        Hari Acharya, Srivathsan Vembanur, Sumanth N. Jamadagni and Shekhar Garde

367     **General Discussion**

## CONCLUDING REMARKS

395     **Concluding remarks FD 146: Answers and questions**
        Frank H. Stillinger

## ADDITIONAL INFORMATION

403     **Poster titles**
409     **List of participants**
413     **Index of contributors**

# Preface

Alenka Luzar[*a] and Hugo K. Christenson[*b]

DOI: 10.1039/c005486k

Hydrophobic surfaces are ubiquitous in nature and in technology, and their importance has long been recognised. More recently, interest has grown in structured surfaces as a result of research into naturally occurring hydrophobic surfaces such as those of many plant leaves (the Lotus effect). It is now possible to engineer surfaces to show a range of properties related to, but not confined to, the traditional concepts of hydrophobicity. Non-wetting of a surface may thus be achieved not only by minimising the surface free energy, as with a classical hydrophobic surface like Teflon, but also *via* an appropriately tailored surface morphology. As a consequence, even low-energy liquids may dewet a surface and the term hydrophobicity is replaced by the more general term "lyophobicity". Theoretical interest in hydrophobic interactions and wetting has also been stimulated in several ways. Work on designing superhydrophobic surfaces has led to renewed interest in the theories of heterogeneous wetting due to Wenzel and Cassie–Baxter. Various surface reflectivity measurements have been interpreted as evidence for a layer (albeit thinner than the diameter of a water molecule) of depleted water density next to extended hydrophobic surfaces, and different techniques have been adopted to investigate the boundary conditions of flow next to both smooth and structured hydrophobic surfaces and the relationship to dewetting. Submicroscopic bubbles ("nanobubbles") have been discovered on many hydrophobic surfaces in water. Besides giving rise to a long-range attractive force that has been confused with a "hydrophobic attraction", they have raised the question of the importance of dissolved gas for wetting and surface properties in general.

The practical importance of *wetting dynamics* at hydrophobic and structured surfaces is considerable. The list of applications in many industrial and biological processes includes microfluidics, electrowetting and cell motility. The capillary driven motion of fluid through structures on a surface bears tremendous importance in the emerging field of nanofluidics and sensor development. The field of electrowetting continues to rapidly expand in applications ranging from lab-on-a-chip, liquid lenses and displays, to microelectronics. Surfaces showing significant drag reduction in liquids, as well as a decrease in turbulence at high flow rates may be constructed. This would lead to more efficient movement through liquid, *e.g.* of propellers, boats, ships and torpedos, *etc*. Surfaces engineered to be self-cleaning would reduce fouling and contamination, hence leading to longer working life. Rust-resistant surfaces, and anti-fog surfaces may be designed to prevent the growth of discrete droplets condensing from vapour. The potential applications of such surface engineering are numerous, and a common theme is greater efficiency in many industrial and domestic processes. Among more traditional areas of technology hydrophobicity is of key importance in mineral flotation, where efficient bubble attachment often requires surface modification through the use of additives. Most coal and petroleum products are hydrophobic and often render surfaces that come into contact with them hydrophobic. Because of their inherently low surface energy hydrophobic surfaces present a problem for many applications involving paints, coatings and

[a]Department of Chemistry, Virginia Commonwealth University, Richmond, VA, USA. E-mail: aluzar@vcu.edu
[b]School of Physics and Astronomy, University of Leeds, Leeds, LS2 9JT, UK. E-mail: h.k.christenson@leeds.ac.uk

adhesives, although this is also exploited in applications such as non-stick (paint resistant) surfaces.

A Faraday Discussion is the ideal forum for making progress towards a better understanding of these phenomena, and the published Discussion volume provides an invaluable reference to the current state of the field for a wide scientific community of physical chemists, biologists, engineers, materials scientists and nanotechnologists. We decided to act on this idea and the program started to take shape in late 2006, after invaluable input from Colin Bain, Pablo Debenedetti, Bob Evans, John Finney, Jim Henderson, Mike Klein, Peter Rossky and John Weeks. We wished to focus on nonpolar surfaces, with particular emphasis on dynamics and tunable wettability, as well as to emphasise the natural extensions to superhydrophobicity and surfaces with chemical and topological heterogeneities. In the late autumn of 2007 Colin Bain, Chairman of the Faraday Standing Committee sent us the wonderful news – that our meeting was finally on the official schedule for April 2010! We are grateful to the committee members who gave their time to sort through the large number of abstracts we received for contributed papers and took part in the final decision on selecting the best and the most appropriate ones for oral presentations.

It was an innovative event. Firstly, it was only the second time a Faraday Discussion had been held outside Europe (in North America) – the first one was at the University of Notre Dame, Indiana in 1963 (on radiation chemistry). The venue was the historic Jefferson Hotel, in Richmond, Virginia, which provided what we are sure, was an unsurpassed setting for a Faraday Discussion meeting. The location and the time of the year worked together to ensure a very successful meeting at the height of spring. On reflection, it was a good idea to bring Faraday Discussions back to US soil after 47 years. This way, we could raise the profile of the meeting and attract more delegates from the US. Secondly, it marked the introduction of a Faraday Discussion Graduate Research Seminar (FD-GRS) that we organized at Virginia Commonwealth University (VCU) over the preceding weekend. The "Concluding Remarks" by Frank Stillinger further describe the scope and success of FD-GRS, which ended with an outing to Maymont Park (see Fig. 1).

Questionnaire responses after the completion of FD146 demonstrated that all the students (the response rate was over 90%) felt much better prepared and less

**Fig. 1** Students and mentors in Maymont Park after completion of the FD-GRS. (Photograph: Alenka Luzar).

intimidated to engage in the FD format of critical and public "dissection" of papers. The published Discussion shows that graduate students posed 30% of the questions. The meeting was a great success, and the associated graduate research seminar provides a model that we hope many other Faraday Discussion meetings will seek to copy.

The subsequent Faraday Discussion meeting was well attended, with close to 130 delegates from over 15 countries, with the majority from the US, but quite a few from as far away as New Zealand, Australia and China. The photo of all (almost) the delegates was taken in the lobby of the Jefferson Hotel, on the magnificent Grand Staircase featured in the classic movie "Gone with the Wind" (Fig. 2).

We succeeded in bringing together communities that might not normally interact very closely, as they deal with length scales from the nanoscopic to the macroscopic. There was a perfect balance between experiment and theory. During the meeting we discussed recent breakthroughs in state-of-the-art techniques to control the behaviour of hydrophobic surfaces under specific conditions; *e.g.* thermal, optical, electrical, mechanical, chemical. Further, we discussed new developments to devise theoretical and simulation approaches to study nano-surfaces, which dominate nanoscale systems and are necessarily highly complex and heterogeneous. Two very eminent people in the field; Peter Rossky, and Frank Stillinger, delivered the introductory lecture of the Faraday Discussion Meeting and the concluding remarks, respectively. In between these there were twenty-two contributed and invited talks, all of which gave rise to interesting and lively debate. Indeed, on many occasions the discussions had to be cut short due to time constraints. The poster session was extremely well subscribed with over 70 contributions. "Flash" presentations gave the opportunity for all poster presenters to advertise their work using a single PowerPoint slide. Presenting research in a condensed way was a really good learning experience for young scientists. We congratulate all the

**Fig. 2** FD146 delegates taken on the historic Grand Staircase of the Jefferson Hotel. (Photograph: Alenka Luzar).

student participants who presented a rich variety of excellent work during the poster sessions.

Bob Evans, a member of the scientific committee, was standing in for his Bristol colleague and RSC-member Mike Ashfold who could not attend, and delivered an entertaining after banquet speech in his inimitable style. Bob referred to the history of previous FD meetings on related topics and gave his account of the two meetings that he attended, FD16 and FD20, both held in the freezing cold of an Oxford winter. The loving-cup ceremony passed without incident, except for the usual difficulty in explaining its conduct to inebriated delegates. It became much clearer to them when the Brazilian ensemble "Quatro na Bossa" started playing "Loving Cup" by Rolling Stones.

The feedback that we have received on FD-146 has been very positive. Numerous personal comments from delegates speak of how they enjoyed the meeting from a scientific and organisational point of view. Especially the US delegates, for many of whom this was the first Faraday Discussion, are now sold on this format! We expect that this Faraday Discussion and the published 146 volume will have significant impact on this exciting field in the years ahead, and we wish to thank all the contributing authors for their effort and support of our endeavour. Delegates also expressed their gratitude at being looked after so well and made to feel so welcome. This southern hospitality extended to a few "lucky" ones who had to prolong their stay at Jefferson because of Eyafjällajökull.

Our organisational achievements are due to many who worked tirelessly before, during and after the meetings. Firstly, we thank the entire staff of the VCU Chemistry Department, and especially Rose. Rose's energy was unsurpassed in overcoming all the administrative obstacles to deliver fellowships to students, and honoraria to mentors and poster judges. Shirley, from the College of Humanities and Sciences helped to ensure that the social events ran smoothly. Chemistry graduate students stretched their imagination to design the T-shirt logo (Kyler), and brought the professional Brazilian ensemble to the banquet (Fernando). RSC student stewards and members of Alenka's research group helped with the microphones, poster boards and computers, and at the registration desk. Meredith and Victor from the School of Engineering at VCU were responsible for ensuring that the FD-GRS meeting was held in a superb lecture hall with top-notch AVI technology and Victor diligently video-recorded the lectures. Mohammad could be seen tirelessly taking photos and movies during both meetings and social events. Ken Wynne had the brilliant idea of asking those with questions to queue up to overcome the shortage of microphones at FD146. We thank Fred Hawkridge, Sally Hunnicutt, and Scott Gronert for their welcoming speech at the reception for FD146, for introducing the young generation to Michael Faraday during the banquet, and for welcoming delegates in Maymont Park. Last, but not least, we are grateful to Morwenna, Anna and Helen of the RSC.

We acknowledge the Chemical Physics Division of the APS and the Division of Physical Chemistry of the ACS for their endorsement of FD146. We thank the following organisations for their generous financial support of FD-GRS and FD146: National Science Foundation (CHE-1016888), US Department of Energy (DE-FG02-10ER16152), American Chemical Society (Division of Colloid and Surface Science) and VCU. Without all the support, the inaugural FD-GRS, the overwhelming student participation in FD146, and all the social activities during *both* meetings, that contributed to a relaxing and engaging student/mentor atmosphere, would not have been possible.

The scientific discussion and educational commitment to the course of research in this field will continue in cyberspace. During the summer months we will be preparing a dedicated webpage (www.faraday.vcu.edu) that will serve as a continuing open-access discussion forum on this topic, along with providing the introductory lectures of the mentors, memorable photos, *etc*. We hope to ensure that the atmosphere established at the FD-GRS and carried over to FD146 will last much longer!

Hugo Christenson (Co-Chair)
Alenka Luzar (Co-Chair, Editor)

# Exploring nanoscale hydrophobic hydration

Peter J. Rossky*

*Received 29th April 2010, Accepted 29th April 2010*
DOI: 10.1039/c005270c

In this lecture, aspects of the hydration of hydrophobic interfaces that are emergent nanoscale properties of the interface chemical structure are discussed. General results inferred from systematic computational studies are emphasized, with a central theme focusing on the separate roles of surface topography and surface chemistry. The roles of surface curvature, polarity, and chemical heterogeneity, as well as the important role of solvent thermodynamic state are considered. The potential importance of understanding evolved natural biological interfaces on the same basis as model synthetic surfaces is pointed out, and progress in this direction is discussed.

## 1. Introduction

The phenomena to be discussed at the present *Faraday Discussion* 146 are remarkable. First and foremost in this regard is the fact that the wetting of an interface remains a topic of current and intense study on such a remarkably wide range of length and time scales. This range reflects the fact that the behavior observed initiates with molecular-scale phenomena, but that the emergent behavior of an interface is not simply predictable based on the properties of the hydration of small molecules alone. Indeed, both the chemical interactions occurring on a molecular level and the topography of a surface on a nano- to micro-scale are important determinants of the degree of surface hydrophobicity and of the dynamics of a liquid drop on such a surface. Correspondingly, studies addressing these phenomena on a variety of scales, *via* frameworks that range in detail from atomic coordinates to hydrodynamic fields, underlie progress toward fully understanding wetting broadly.

In this Introductory Lecture, I will address only one end of this scale, the nanoscale, and attempt to summarize what I believe to be the state of progress in understanding that scale. In particular, a considerable amount of effort has been invested in just the last several years in developing a *systematic* understanding of the relationship between the nanoscale chemical structure of interfaces and both the nanoscale hydration structure, on the one hand, and macroscopic measures of hydrophobicity, such as contact angle, on the other. Understanding the impact of molecular polarity, when coupled to interfacial molecular topography, is a particularly important element. An understanding of the role of chemical heterogeneity is also important, particularly considering that bio-macromolecular surfaces are both highly functional and heterogeneous.

The literature in this area is extensive and it is not possible to discuss or even cite all of the relevant work that underlies the current state of knowledge. The intent here is to capture certain major points that are relevant to nanoscale hydrophobic hydration and to point to particular work in the literature that illustrates these. A more in depth discussion can be found in the articles that are cited here.

*Department of Chemistry & Biochemistry and Institute for Computational Engineering and Sciences, University of Texas at Austin, Austin, TX, 78712-1165, USA. E-mail: rossky@mail.utexas.edu*

## 2. Generic effects of exclusion

This summary will first consider what I call exclusion surfaces, surfaces that, in the simplest case, have no chemical character but only present an impermeable interface to water. These are the ultimate non-polar, hydrophobic, surfaces and provide a useful theoretical construction and systematic starting point. The essential behavior of such surfaces is a function of the radius of curvature, with the hydrogen bonding network between water molecules fully capable of accommodating small radius, molecular scale, interfacial features (under ambient conditions) by restricting solvent orientations to those with hydrogen (H-)bonding directions either tangential to or pointing away from such surfaces, as in solid clathrate hydrates.[1] For features with characteristic radius larger than of the order of 1 nm, the H-bond strain induced in a tetrahedral network is too high, some H-bond rupture is unavoidable, and an inverted structure appears with a propensity for one H-bond direction to point inwards to the interface and three to point toward the bulk liquid.[2–5] This optimized organization leads with increasing radius of curvature to a decreasing solvent density in immediate contact with a simple exclusion surface.[6–8] In general, if one examines the dependence of solvation free energy on molecular size, one finds a cross-over with increasing size to a regime that is largely dependent on solute surface area. This phenomenon is, however, not limited to aqueous or even H-bonded liquids, but reflects more generally the stability of the liquid interface. The same behavior is seen for a simple Lennard-Jones liquid for thermodynamic states that lie near the coexistence curve, with the free energy of cavity formation dominated by enthalpic penalties.[6,9]

What has become clear as a result of a number of studies is, first, that this depression of density at the interface is not a general primary characteristic of hydrophobic hydration.[7] The density is sensitive to the existence of attractive forces exerted by the surface, and the observation of solvent "drying" at a single surface does not follow the chemistry in a simple way.[10] More recent studies indicate that the magnitude of solvent density *fluctuations*, or local solvent compressibility, is a more general indicator of hydrophobicity,[10–14] and it is one that increases with surface radius of curvature for exclusion surfaces. A reduction in the work to create a void in the vicinity of the surface (and a corresponding free energy driving force for non-polar moieties to preferably reside there, compared to the bulk solvent) correlates with this increased density fluctuation,[10] as one might expect. Correspondingly, those phenomena associated with enhanced compressibility are suppressed with increasing pressure, a result seen in several systems.[10,15] It is worth noting that simulations of similar quantities for the solvent surrounding proteins[16] yield an excess local solvent compressibility near protein apolar surfaces that is comparable to that found near model extended apolar surfaces.

## 3. Continuity of the scale of hydrophobicity

To chemists, the idea that the degree of hydrophobicity is a variable quality is a generally accepted concept, which has its origin in the variable atomic composition of molecular species and variations in their packing in materials. Computational studies of model systems have the ability to vary the various qualities of a surface completely independently of physical constraints, which can sometimes lead to additional insights. One example is a study carried out to understand the relationship between water contact angle, solvent structure, and surface polarity for a system where the nanoscale chemical homogeneity of the surface is constant.[17] The surface used corresponded to hydroxylated silica (quartz) and only the polarity of the model surface was varied, by scaling of surface atom partial charges. Having restricted the variations in the surface, it was clear that the contact angle tracked the solvent orientational preferences nearly ideally, with the surface that exhibited a "neutral" contact angle of 90° also being that surface exhibiting a neutral preference for any particular solvent orientation in proximity to the surface (in contrast to those described in Sec. 2). Perhaps of more interest, the orientational distributions observed at each level of polarity could be described by a superposition of those for the most polar and non-polar cases. This suggests that the two structures coexist in domains. This might be expected for an extended interface, based on the fact that the inverted structure with water H-bond loss, characteristic of an extended non-polar surface, is a result of optimizing the capacity of solvent to form water–water H-bonds, an inherently cooperative behavior.[18]

Given this correlation of contact angle with surface polarity, one might ask why fluorocarbon surfaces behave more hydrophobically than hydrocarbons.[19] The C–F bond is relatively more polar than is a C–H bond, suggesting the opposite at first guess. The answer recalls the discussion above about the role of dispersive van der Waals attractive forces. Fluorocarbons are relatively large compared to hydrocarbons and so the density of attractive centers on a self-assembled layer or in a polymeric film of fluorocarbon is reduced compared to hydrocarbons. At the same time, fluorine is relatively less polarizable (with correspondingly smaller dispersive forces) than one would expect of other atoms of similar size, with the overall result that fluorocarbon surfaces are the more hydrophobic. This explanation takes no account of changes in solvation structure. It has been appreciated for many years that the hydration structure of extended hydrophobic surfaces is relatively insensitive to mild polar surface interactions,[5] resulting from the very strong orientational preferences associated with satisfying solvent intermolecular H-bonding, already discussed. In fact, for weakly polar aprotic (non-network) solvents, fluorocarbon surfaces do appear to have enhanced attraction.[20]

## 4. Hydration at non-ambient conditions

Because water is essential for terrestrial biology, the variation of thermodynamic state away from ambient temperature and pressure is sometimes neglected in consideration of aqueous systems. Nevertheless, it is widely appreciated that proteins that are taken from mesophilic species, adapted to ambient conditions, can be denatured not only by heating but also by high pressures or by low temperatures.[21] The implication is that this is a result of changes in the strength of hydrophobic interactions. We have already touched on the impact of interfaces on local compressibility, *i.e.*, on the differential response to pressure of interfacial water compared to bulk water, and the resulting enhancement of hydrophobic interactions. In fact, studies show that it is the translational ordering of water that is most responsive to pressure while the orientational order, dictated by the strong H-bond network within water, is relatively refractory.[15] At a hydrophilic surface, where H-bonds also form between the surface and water, one finds that neither type of order is responsive to pressure.[15] The complementary perturbation associated with isobaric cooling to temperatures below ambient show remarkably similar solvent response to those seen in isothermal compression;[22] the solvent contact density with a hydrophobic surface is increased with decreasing temperature with little change in solvent orientational order at that surface, while hydrophilic surfaces show little structural change in solvation, beyond a sharpening of structural features. Hence, these model interfacial systems very clearly manifest the characteristics of decreasing hydrophobicity with increasing pressure and/or decreasing temperature, starting from ambient conditions.

## 5. Impact of chemical heterogeneity

For a heterogeneous patterned surface, hydration at the boundaries between hydrophilic and hydrophobic regions could well be expected to be of special significance. What is the resolution of such an atomic-scale boundary when seen from a solvent-mediated perspective? Correspondingly, one could ask whether the resolution that can be achieved in chemically patterning surfaces can be resolved in self-assembly of hydrophobic surfaces in solution. We find from the study of nanoscale patterned surfaces[23] that boundary lines defined by hydrophilic sites, in fact, are nearly fully resolved by the solvent density, except at very low hydrophilic site coverage. In fact, the presence of polar surface moieties at a boundary is found to be amplified by the solvent, with an excess hydration density at such boundaries compared to a uniform hydrophilic surface.[23] One consequence of this enhanced hydration is that sub-nanometre hydrophilic or hydrophobic patches (*e.g.*, involving only one or a few molecular surface sites) can become less distinct when resolved in terms of solvent density.[23,24]

Of course, since pressure and temperature blur the difference between hydrophobic and hydrophilic surfaces, the resolution of boundary lines is greatly reduced as pressure increases or temperature decreases.[22,23] This appears closely related to protein denaturation, already mentioned above.

## 6. Confinement between nanoscale hydrophobic surfaces

It is well established that, at least for macroscopic hydrophobic surfaces, there exists a critical separation below which the vapor phase is thermodynamically favored over the liquid, as reviewed, for example, in ref. 25. The argument is based simply on the increasing relative importance of the interfacial free energy of the fluid interfaces, when the volume of bulk fluid lying between these interfaces is decreasing. For nanoscale surfaces, it is clear that one can also get such a drying between surfaces, and that some minimum size surface is needed to observe the effect.[26] Nevertheless, the thermodynamic effects of finite size and the kinetic barriers to formation of vapor phases have not been developed fully.[25] Within the limitations of typically nanosecond timescales associated with conventional molecular dynamics, it is, however, of interest to explore the phase behavior of confined water and its response to temperature, pressure, and surface chemistry. For water in contact with and confined by a pair of model nanoscale hydrophobic (silica) plates, such studies have been carried out.[15,22,23,27] At ambient temperature, the results reveal a regime of inter-plate gaps ranging up to about 1 nm where water cavitation occurs between the plates at pressures where the bulk solvent remains mechanically stable. This regime of gaps $d$ decreases with increasing pressure, as expected, until the gap size for hypothetical cavitation is smaller than the physical size of one molecular layer of water (at about 0.15 GPa for SPC/E water[28] at 300 K), where macroscopic fluid concepts are untenable. Interestingly, at these high pressures, a number of additional phase transitions involving new *solid* phases occur for the confined condensed phase at 300 K,[27] in analogy to the formation of bilayer ice first observed between unstructured hydrophobic plates.[29] As temperature is reduced, the coexistence line in the gap-pressure ($d$, $P$) plane between the vapor and condensed (liquid or solid) phases moves increasingly toward the limit of bulk stability, so that the region in $d$ of vapor stability nearly vanishes at very low temperatures, while the region in $d$ manifesting a stable bilayer solid increases.[22]

## 7. Biological interfaces

Since proteins have evolved to form specific assemblies with each other, with other active biological molecules, and with metabolic substrates, it is of great interest to understand to what extent lessons learned about hydration of synthetic and natural surfaces can be profitably transferred between them. To this end, one model system, the dimer of the amphipathic polypeptide melittin, and its assembly in the crystallographic tetramer, has been examined in several simulation studies which address solvation structure and the impact of confinement.[18,24,30,31,32] Some years ago, the hydration structure of the melittin dimer was examined from the perspective of the impact of extended hydrophobic surface.[30] It was found that the solvent adjacent to a patch of "flat" hydrophobic surface did manifest the loss of H-bonding in water and the orientational structure seen in model extended interfaces. More recently, it was shown that if two dimers (fixed in their crystallographic geometries) were brought into close proximity along a vector that would bring them to the assembled crystal tetramer, a drying transition occurred, with solvent evacuating the gap between them in advance of dimer–dimer contact.[31] It was further shown that protruding hydrophobic groups on the dimers were a critical element in that process; these tended to cutoff the solvent gap from the bulk solvent prior to contact.

In order to examine this apparent role of topography in more detail and to provide a means to put synthetic and natural systems on a comparable footing, a study was carried out involving a "flat" melittin dimer surface.[32] To accomplish this, the protein interface was geometrically modified by shifting residues along the inter-dimer direction so that the contact interface between dimers became flat, while preserving the characteristic chemistry. The details are provided elsewhere.[32] A very small area of dewetting was observed at the smallest inter-dimer gap where water might still enter, and this occurred in the region of the gap proximal to the previously identified flat hydrophobic surface. Nevertheless, it was demonstrated that the surface would wet for gaps only an Angstrom larger, for moderately increased pressures, or upon substitution of a single polar amino acid for a non-polar one on the dimer interface. These observations suggested that the flat, natural chemistry, surface was only marginally hydrophobic. Direct comparison of the solvent $(d,P)$ phase diagrams, as well as solvent compressibilities, for the case of this natural chemistry interface with those for silica plates revealed a behavior for the protein-like case that was clearly intermediate between the synthetic hydrophobic and hydrophilic cases. The implication is that a prototypical hydrophobic protein surface, clearly involved in self-assembly, as evident in the crystal structure, need not be as hydrophobic as a simple Lennard-Jones (*e.g.*, alkyl) material in order to be functional.

## 8. Conclusions

The results just discussed suggest that nanoscale topography and chemical hydrophobicity can play equally important roles in natural system assembly and function. On a larger length scale, this is, of course, a guiding principle of superhydrophobic interface design, as the papers following in this Discussion demonstrate. Recent simulations probing the potential for drying in hypothetical gaps created at hydrophobic interfaces in multi-protein assemblies[33] show that even in systems selected as likely candidates, drying is not a common phenomenon. Earlier simulation results for hydration of the enzyme $\alpha$-chymotrypsin[34] showed both that the nominally hydrophobic binding pocket at the active site is occupied by water (as was already established by experiment) and that it exhibited substantial amphipathic character when examined at atomic resolution, despite its selectivity for binding non-polar amino acid residues. Extending studies of the relationship between nanoscale topography and hydrophobicity to a broad set of synthetic surfaces as well as proteins appears well justified. Such a program to establish specific relationships between topography and hydrophobicity in natural and model self-assembled systems may well yield a scheme which one could readily extend to complex self-organizing synthetic systems.

## Acknowledgements

The author's work described here is supported by a grant for collaborative research from the National Science Foundation Division of Chemistry (CHE-0910615), and is very gratefully acknowledged. Additional support from the R. A. Welch Foundation (F-0019) to PJR is also gratefully acknowledged. The contribution of my students and collaborators has been essential, with a special acknowledgement due to my co-authors Pablo Debenedetti and Nicolas Giovambattista.

## References

1. J. S. Loveday and R. J. Nelmes, *Phys. Chem. Chem. Phys.*, 2008, **10**, 937.
2. F. H. Stillinger, *J. Solution Chem.*, 1973, **2**, 141.
3. A. Luzar, Svetina and B. Zeks, *Chem. Phys. Lett.*, 1983, **96**, 485.
4. C. Y. Lee, J. A. McCammon and P. J. Rossky, *J. Chem. Phys.*, 1984, **80**, 4448; S. H. Lee and P. J. Rossky, *J. Chem. Phys.*, 1994, **100**, 3334.
5. G. M. Torrie, P. G. Kusalik and G. N. Patey, *J. Chem. Phys.*, 1988, **88**, 7826; G. M. Torrie and G. N. Patey, *J. Phys. Chem.*, 1993, **97**, 12909.
6. D. M. Huang and D. Chandler, *Phys. Rev. E: Stat. Phys., Plasmas, Fluids, Relat. Interdiscip. Top.*, 2000, **61**, 1501.
7. H. S. Ashbaugh and M. E. Paulaitis, *J. Am. Chem. Soc.*, 2001, **123**, 10721.
8. D. M. Huang and D. Chandler, *J. Phys. Chem. B*, 2002, **106**, 2047.
9. Henry S. Ashbaugh, *J. Chem. Phys.*, 2009, **130**, 204517.
10. R. Godawat, S. N. Jamadagni and S. Garde, *Proc. Natl. Acad. Sci. U. S. A.*, 2009, **106**, 15119.
11. S. Sarupria and S. Garde, *Phys. Rev. Lett.*, 2009, **103**, 037803.
12. J. Mittal and G. Hummer, *Proc. Natl. Acad. Sci. U. S. A.*, 2008, **105**, 20130.
13. A. P. Willard and D. Chandler, *J. Phys. Chem. B*, 2008, **112**, 6187.
14. A. J. Patel, P. Varilly and D. Chandler, *J. Phys. Chem. B*, 2010, **114**, 1632.
15. N. Giovambattista, P. J. Rossky and P. G. Debenedetti, *Phys. Rev. E: Stat., Nonlinear, Soft Matter Phys.*, 2006, **73**, 041604.
16. V. M. Dadarlat and C. B. Post, *Biophys. J.*, 2006, **91**, 4544.
17. N. Giovambattista, P. G. Debenedetti and P. J. Rossky, *J. Phys. Chem. B*, 2007, **111**, 9581.
18. Y. K. Cheng, W. S. Sheu and P. J. Rossky, *Biophys. J.*, 1999, **76**, 1734.
19. V. H. Dalvi and P. J. Rossky, Proc. Natl. Acad. Sci. USA (in press).
20. M. Graupe, M. Takenaga, K. Thomas, R. ColoradoJr and T. R. Lee, *J. Am. Chem. Soc.*, 1999, **121**, 3222.
21. L. Smeller, *Biochim. Biophys. Acta, Protein Struct. Mol. Enzymol.*, 2002, **1595**, 11; P. L. Privalov, *Crit. Rev. Biochem. Mol. Biol.*, 1990, **25**, 281.
22. N. Giovambattista, P. J. Rossky and P. G. Debenedetti, *J. Phys. Chem. B*, 2009, **113**, 13723.
23. N. Giovambattista, P. G. Debenedetti and P. J. Rossky, *J. Phys. Chem. C*, 2007, **111**, 1323.
24. Y. K. Cheng and P. J. Rossky, *Biopolymers*, 1999, **50**, 742.
25. K. Lum and A. Luzar, *Phys. Rev. E: Stat. Phys., Plasmas, Fluids, Relat. Interdiscip. Top.*, 1997, **56**, R6283; A. Luzar and K. Leung, *J. Chem. Phys.*, 2000, **113**, 5836; K. Leung and A. Luzar, *J. Chem. Phys.*, 2000, **113**, 5845.
26. T. Koishi, S. Yoo, K. Yasuoka, X. C. Zeng, T. Narumi, R. Susukita, A. Kawai, H. Furusawa, A. Suenaga, N. Okimoto, N. Futatsugi and T. Ebisuzaki, *Phys. Rev. Lett.*, 2004, **93**, 185701.
27. N. Giovambattista, P. J. Rossky and P. G. Debenedetti, *Phys. Rev. Lett.*, 2009, **102**, 050603.
28. H. J. C. Berendsen, J. R. Grigera and T. P. Straatsma, *J. Phys. Chem.*, 1987, **91**, 6269.
29. K. Koga, X. C. Zeng and H. Tanaka, *Phys. Rev. Lett.*, 1997, **79**, 5262.
30. Y. K. Cheng and P. J. Rossky, *Nature*, 1998, **392**, 696.
31. P. Liu, X. H. Huang, R. H. Zhou and B. J. Berne, *Nature*, 2005, **437**, 159.
32. N. Giovambattista, C. F. Lopez, P. J. Rossky and P. G. Debenedetti, *Proc. Natl. Acad. Sci. U. S. A.*, 2008, **105**, 2274.
33. L. Hua, X. H. Huang, P. Liu, R. H. Zhou and B. J. Berne, *J. Phys. Chem. B*, 2007, **111**, 9069.
34. C. Carey, Y. K. Cheng and P. J. Rossky, *Chem. Phys.*, 2000, **258**, 415.

PAPER

# Dynamical superhydrophobicity

Mathilde Reyssat,[†a] Denis Richard,[a] Christophe Clanet[ab] and David Quéré[ab]

Received 7th January 2010, Accepted 10th February 2010
DOI: 10.1039/c000410n

Superhydrophobicity is mainly remarkable for the special dynamical behaviours it generates: low adhesion, giant hydrodynamic slip, frictionless motion, rebounds after impacts. Here we discuss most of these properties. We first recall how contact angle hysteresis can be minimized in this state. Then, we show that a water drop first follows the Galilean law of free fall on an incline, before reaching a stationary state, for which we discuss the associated friction. Finally, the property of water repellency (that is, rebounds after impact) is presented. We describe in particular how the texture responsible for superhydrophobicity can also influence the figure of impact at a very large scale.

## 1. Introduction

When placed on a very hot plate, a drop of water levitates, owing to the formation of a vapour film that prevents the contact between the liquid and its substrate. This situation, often referred to as the Leidenfrost phenomenon,[1] is characterized by its remarkable mobility: suppressing the solid/liquid contact minimizes the viscous force associated with the drop motion, which is just resisted by the inertial friction arising from the presence of air around, as for a free-falling raindrop. The absence of a solid/liquid contact can be viewed as the ultimate hydrophobic state, where the "contact" angle reaches 180°, its maximum possible value.

At room temperature, water evaporation is much reduced but materials may approach the Leidenfrost limit if they are covered with a hydrophobic texture (typically at a scale of 100 nm to 10 µm), which acts as a spacer between the liquid and the solid (Fig. 1). The larger the quantity of air trapped in the textures, the larger the contact angle: water makes contact angles of the order of 160 to 175° on many natural[2] or artificial[3] materials of this kind. In addition, the hysteresis of contact angle is generally very low (about 5 to 10°) in this "fakir regime", since the drop can only pin on the few texture tops it contacts. As a consequence, liquid adhesion is highly reduced compared to usual materials.[4] Interestingly, special designs can also provide superoleophobicity, i.e. the possibility for a deposited oil to find an equilibrium position on the texture tops, leaving air trapped below.[5]

In a superhydrophobic situation, the liquids dynamics dramatically differs from what is known for usual materials, as demonstrated by three original effects: (i) A huge slip can be observed at the frontier between the liquid and the solid, when displaced relative to each other: the associated slip length can be more than 1000 times larger than on a hydrophobic flat material.[6] The magnitude of the slip depends on the design of the texture and on the pressure applied on the liquid. Research on slip has been particularly active for the last five years, and a recent comprehensive

[a]PMMH, ESPCI, 10 rue Vauquelin, 75005 Paris, France
[b]LADHYX, École Polytechnique, 91128 Palaiseau Cedex, France

† Current address: MMN, Gulliver, ESPCI, 10 rue Vauquelin, 75005 Paris, France.

**Fig. 1** A water drop deposited on a bed of hydrophobic micronails often stays at the top of the nails. This is directly observed in this photo, where the low density of nails (about 1%) together with the presence of air allows light to pass below the drop (of diameter 0.8 mm). The distance between the drop and its reflection is 24 μm, that is, twice the height of the micropillars used in this experiment. We discuss in this paper the dynamical properties generated by such a "fakir state".

review by Rothstein summarizes the main findings on this subject.[7] (ii) A second effect, first qualitatively reported by Leidenfrost who had to use (hot) spoons to trap drops,[1] is the remarkable mobility of pearl drops. On superhydrophobic solids, mobility arises from the conjunction of small hysteresis, which minimizes the force anchoring the liquid on its substrate, and low friction, as the drop moves.[8–11] Remarkably, very little was reported on the latter effect, and it is one of our primary goals here to discuss possible origins and magnitude for this friction. We also recall how contact angles and their hysteresis can be deduced from the density of the textures placed on the solid. (iii) The third spectacular effect generated by superhydrophobic materials is, literally, water repellency: when thrown on such materials, water bounces off, leaving the substrate dry after the rain.[12–15] We describe a few characteristics of these rebounds, and present new experiments where impact is accompanied by remarkable patterns directly related to the existence of a texture on the solid surface.

## 2. Residual drop adhesion

We first consider the question of pinning, for a drop of radius $R$ made of a liquid of surface tension $\gamma$ and density $\rho$, and deposited on an incline. We classically define the capillary length $\kappa^{-1}$ as $\sqrt{\gamma/\rho g}$, that is, 2.7 mm for water. We mainly consider drops smaller than $\kappa^{-1}$, i.e. whose behaviour is dictated by surface tension rather than by gravity. The reason why such drops do not move on inclines is often visible with a naked eye: the angle at the leading edge is larger than the one at the trailing edge, which generates a Laplace pressure difference opposing gravity.[16] This asymmetry is clearly visible in Fig. 2, where a fakir drop such as the one in Fig. 1 is tentatively displaced by a syringe contacting its top and moved laterally. Just before the drop starts moving, it is asymmetric with a leading angle $\theta_a$ significantly larger than the trailing angle $\theta_r$.

The detailed calculation of the force resisting the motion is not easy, because of the three-dimensional character of the drop geometry: the angle continuously decreases along the contact line, when following it from the front to the rear of the drop.[17] However, a simple approximation allows us to quantify the typical magnitude of the maximum sticking force $F$. Assuming that half the rear of the drop meets the solid with $\theta_r$, and that the other half meets the solid with $\theta_a$, we deduce that $F$ scales as $\gamma \ell (\cos\theta_r - \cos\theta_a)$, denoting $\ell$ as the radius of the contact

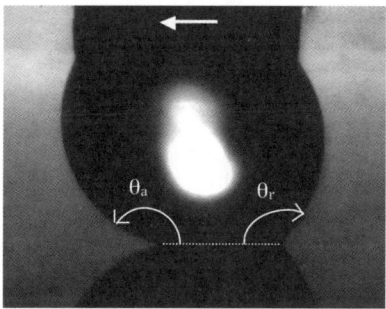

**Fig. 2** A drop resists the motion that we try to impose (with a syringe, the black body on the top, of diameter 1 mm) by pinning on the substrate, which clearly generates a difference of contact angle between the leading and the trailing edge. The arrow indicates the direction of the applied force. Here the drop is in a fakir state, on a substrate decorated with hydrophobic posts of height $\delta = 20$ µm, radius $b = 1.3$ µm and mutual distance $p = 7$ µm. The advancing and receding angles $\theta_a$ and $\theta_r$ are $159 \pm 2°$ and $126 \pm 2°$, respectively. The hysteresis $\Delta\theta = \theta_a - \theta_r \approx 30°$ is significant despite the small density of pillars ($\phi \approx 10\%$).

(taken as a circle, a good approximation in the highly-hydrophobic limit).[18] For ultra-hydrophobic situations ($\theta \approx 180°$), this contact is set by gravity,[8] but in most practical situations, it is simply given by the geometric relationship: $\ell \approx R\sin\theta$, where we define the mean contact angle $\theta$ as $(\theta_r + \theta_a)/2$. Since water repellent situations are characterized by both $\Delta\theta = \theta_a - \theta_r \ll \theta$ and $\varepsilon = \pi - \theta \ll 1$, we can expand the formula for $F$, which yields:

$$F \sim \gamma R \, \varepsilon^2 \, \Delta\theta \qquad (1)$$

Hence $F$ is smaller than a "usual" hysteresis force $\gamma R$ by a factor $\varepsilon^2\Delta\theta$, *i.e.* about 100 for $\varepsilon = \Delta\theta = 10°$. Only the drops verifying the inequality $F > \rho g R^3$ should remain pinned on vertical plates, that is, drops smaller than $\kappa^{-1}\varepsilon\Delta\theta^{1/2}$, instead of $\kappa^{-1}$ on usual solids. This implies droplets with a radius of typically 100 µm, much smaller than the millimetre-size drops sticking on our window panes. Note however that this average-based description might fail if the drop size becomes of the order of the texture characteristic length, in which case the drop obviously can sink and remain trapped in the texture.

Eqn (1) makes it clear that an efficient superhydrophobic surface (small $F$) combines a high contact angle ($\varepsilon \ll 1$) and a small hysteresis ($\Delta\theta \ll 1$). It is worth noticing that both these quantities can be fixed by a single parameter, often referred to as $\phi$, the proportion of solid in contact with the liquid. First, Cassie showed long ago that the contact angle on a superhydrophobic solid is an average between the angle $\theta_o$ on this solid, yet flat, and the angle on air, which is $180°$.[19] The average is calculated on the surface energies, that is, on the cosines of the angles, which yields in the limit we are considering here ($\varepsilon \ll 1$):

$$\varepsilon \approx [2(\cos\theta_o + 1) \, \phi]^{1/2} \qquad (2)$$

We often cannot observe trapped air, due to the small size of the textures and eqn (2) may be used to deduce the effective solid/liquid contact $\phi$ from the simple and direct measurement of $\varepsilon$.[20] For example, for $\varepsilon \approx 10°$, $\phi \approx \varepsilon^2$ is only 3% of the apparent contact between a drop and its substrate. Conversely, eqn (2) cannot always be used to predict the value of $\varepsilon$. On a disordered rough surface, we do not know *a priori* the value of $\phi$, which results from a minimization of surface energy in a complex energy landscape.[21] Even more sneakily, $\phi$ can be a function of the way the drop was deposited (gently, or brutally, *etc.*).[4] In any case, a model situation appears to be that of textures made of pillars or grooves, for which we select the value of $\phi$ (it is simply the density of texture), provided that the deposited liquid stays at the top of the texture.

In the pillar case, it is possible to calculate the value of the hysteresis $\Delta\theta$, i.e. to relate this macroscopic observation to the detail of contact line pinning on this well-defined texture. There is today no consensus on this question,[22] and we follow here the line proposed by Joanny and de Gennes, where the hysteresis $\gamma(\cos\theta_r - \cos\theta_a)$ is viewed as the energy (by unit area) stored by the contact line deformation on the defects.[23,24] The model assumes an ideal substrate decorated with a small density of defects. This condition is satisfied in the fakir picture, where the liquid sits on air (the most ideal substrate, indeed, with no kind of pinning), apart from a few pillars ($\phi \ll 1$). The contact line can be strongly pinned on the pillar edges, with a typical force per pillar (of radius $b$) scaling as $\gamma b$. We try here to quantify how a fakir drop resists motion. Since its leading edge hardly resists, we assume that hysteresis mainly arises from the pinning of the trailing edge. On each pillar, the line and thus the liquid surface deform as we try to displace the drop, as sketched in Fig. 3.

Surface energy is stored in the deformation, but the "tongues" that develop must remain in equilibrium with the rest of the drop. For a drop much larger than the defects, we take a condition of zero Laplace pressure everywhere. Therefore the curvature of a tongue must be zero, which is feasible provided its two radii of curvature have opposite signs. As learnt from wetting menisci on thin fibers, the corresponding shapes are catenoids of equation $y \sim b \cosh(x/b)$, taking $x$ as the axis along the deformation.[23] The deformation is larger than $b$, so that the former equation can be inverted: $x \sim b \ln(y/b)$, whose maximum is reached when the lateral deformation $y$ reaches (half) the distance $p$ between pillars. Then, we have $x \sim b \ln(p/b)$, typically a few times $b$, and the capillary force per pillar can be rewritten: $f \sim \gamma b \sim (\gamma/\ln(p/b)) x$.

The latter formula implies a classical Hookean elastic energy (varying as $x^2$), with a non-classical stiffness, $\gamma/\ln(p/b)$ instead of $\gamma$, the usual stiffness in surface elasticity. At the maximum deformation, the receding angle is reached and the surface energy can be written $\gamma b^2 \ln(p/b)$. Since there is one pillar per unit area $p^2$, and using the definition of $\phi$ ($\phi \sim b^2/p^2$), the surface energy per unit area stored at the maximum deformation scales as $\gamma\phi|\ln\phi|$. We identify this expression with the definition of the hysteresis $\gamma(\cos\theta_r - \cos\theta_a)$, and expand it at small $\Delta\theta = \theta_a - \theta_r$, from which we get:[18]

$$\Delta\theta \sim \phi^{1/2}|\ln\phi| \qquad (3)$$

where we used a simplified version of eqn (2) ($\varepsilon \sim \phi^{1/2}$).

As for the contact angle (eqn (2)), the hysteresis is found to be determined by the density of pillars $\phi$. And similarly, both dependencies are critical in this parameter (exponent ½ in both cases, plus a slowly diverging logarithm for the hysteresis, in agreement with recent experiments[18]). These behaviours are related to the choice of pillars as a texture. Rounded defects with no sharp edges, for example, minimize

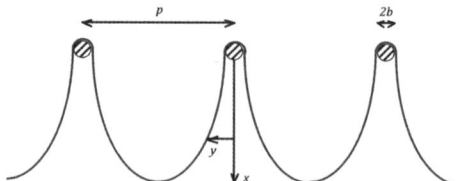

**Fig. 3** Top view of the trailing edge of a fakir drop, as we try to displace it in the $x$-direction. The drop resists the motion by pinning on the edges of the pillars; hence a surface energy stored in this deformation, as expressed by the macroscopically observed contact angle hysteresis.

the pinning of the contact line, which should make the hysteresis vanish, as indeed observed by Gao and McCarthy.[3] Conversely, the logarithmic term in eqn (3) explains why non-negligible residual adhesion is observed in the fakir state in the limit of dilute pillars.[18,25]

Note finally that the conjunction of eqn (2) and 3 allows us to specify how the adhesion force (eqn (1)) increases as a function of the pillar density. We find $F \sim \gamma R \; \phi^{3/2}|\ln\phi|$, indicating a strong reduction of adhesion (by a factor of $\phi^{3/2}|\ln\phi|$, *i.e.* about 25 for $\phi = 5\%$), compared to usual situations ($F \sim \gamma R$). This relationship also shows how drops can be separated according to their sizes, using a substrate of spatially-decreasing $\phi$. Then we expect the successive zones of the solid to stop gradually the drops, the largest ones going further on the incline.

## 3. Sliding pearls

A drop runs down a superhydrophobic incline (tilted by an angle $\alpha$ from the horizontal) if its weight $\rho g R^3 \sin\alpha$ exceeds the hysteretic force discussed in section 2.[16] Assuming this condition, we try here to understand the mobility of this drop. In the limit of zero wetting and at large viscosity, first, Mahadevan and Pomeau showed that the liquid should roll as a solid, and thus only dissipate energy by viscosity in the (tiny) Hertz zone where it contacts the solid.[8] In the limit of zero wetting ($\varepsilon = 0$), the size $\ell$ of the contact zone is fixed by the weight of the drop, which makes it increase as the square of its radius ($\ell \sim R^2 \kappa$). Hence the surface area $\ell^2$ of the contact grows as $R^4$, which generates a remarkable behaviour: the smaller the drops, the quicker they run down the hill![8,11]

Small viscous drops (of size $R < \varepsilon\kappa^{-1}$) on a superhydrophobic solid should behave differently. Then, the solid/liquid contact zone is the (residual) wetting contact $R\varepsilon$, larger in this limit than the gravitational contact $R^2\kappa$. At small Reynolds numbers, these drops also rotate, as proposed by Mahadevan and Pomeau.[8,9] Since viscosity $\eta$ only matters in the contact zone $R\varepsilon$, the viscous force scales as $(\eta V/R)(R\varepsilon)^2$, denoting $V$ as the descent velocity of the drop. $V$ is eventually fixed by a torque balance: $\eta V R\varepsilon^2 (R\varepsilon) \sim \rho g R^4 \sin\alpha$, which yields:

$$V \sim \frac{\rho g R^2}{\eta \varepsilon^3} \sin\alpha \qquad (4)$$

This formula appears to be a kind of Stokes velocity, yet increased by a factor of $1/\varepsilon^3$, typically of the order of 100! However, it should be emphasized that this regime is expected for small drops ($R < \varepsilon\kappa^{-1}$), yet large enough to move despite the residual contact angle hysteresis, that is, for solids with a very low hysteresis.

A much more common case is that of water, of low viscosity. After rain hits water repellent materials, drops quickly run on them, taking dust and contaminants, a cleaning behaviour often referred to as the lotus effect.[26] Here we discuss the dynamics of drops of low viscosity running down superhydrophobic inclines. Let us start by an experiment. To achieve long substrates, we simply glued lycopodium grains on aluminium plates. The static contact angle of water on this solid is $165 \pm 5°$ ($\varepsilon \approx 0.25$), and the contact angle hysteresis $\Delta\theta$ is $10 \pm 5°$. The total length of the plate is 1 m, allowing us not only to follow the beginning of the descent, but also to characterize the stationary state, once gravity balances the friction acting on the liquid. This state is reached after a run of typically 1 m. The drop radius is 2 mm, large enough to overcome the hysteresis force. We first focus on the first moments of the motion ($t < 0.2$ s), and plot in Fig. 4 the position $x$ of a millimetric water pearl deposited on the substrate tilted by $\alpha = 13°$, as a function of time $t$ (full circles). Data are deduced from a high-speed video recording shot at 2000 frames per second.

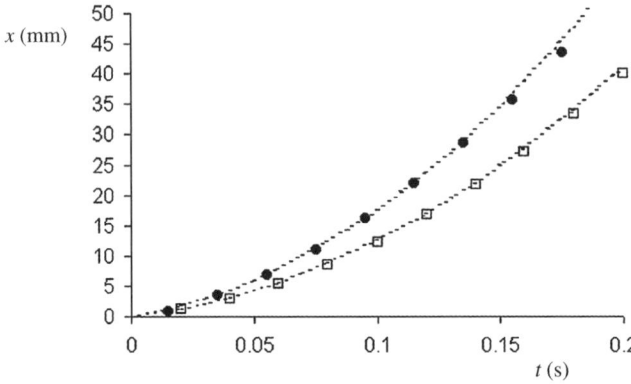

**Fig. 4** Position $x$ of a water drop on a super-hydrophobic incline tilted by $\alpha = 13°$, as a function of time $t$. The solid circles are the data, and the dotted line fitting them the law of free fall: $x = 1/2gt^2\sin\alpha$, with $g$ the gravity acceleration. In the same plot, we display the position of a steel marble running down the same incline (empty squares). Because of the rotation of the marble, the law of free fall writes $x = 5/14gt^2\sin\alpha$, represented in a dotted line and found to fit the data. The water drop thus moves faster than the solid marble.

At this small scale ($x < 5$ cm, typically the scale of a leaf), the drop is observed to constantly accelerate, in agreement with earlier observations by Nakajima et al.[10] The trajectory is even found to be fitted by the law of free fall, $x = 1/2gt^2\sin\alpha$, drawn with a dotted line in the same plot. Friction can be neglected during the first centimetres of the run, and the motion is purely slippery. This can be confirmed by including tracers inside the drop: the tracers are just translated as the drop moves, contrasting with viscous liquids for which rotation starts immediately.[9] Superhydrophobic surfaces thus evacuate water drops as quick as possible. A drop even moves faster than a steel marble (of the same size) running down the incline (open squares in Fig. 4). Then, we also observe a constant acceleration, but the speed is smaller, for a given position, because the marble rolls. Its moment of inertia being $2/5MR^2$, with $M$ its mass and $R$ its radius, the law of free fall for a rolling sphere writes $x = 5/14gt^2\sin\alpha$, indeed found in Fig. 4 to fit the trajectory of the marble.

The total force acting on the drop in this regime is the weight $Mg\sin\alpha$ minus the hysteresis $F$. Using eqn (1) implies that the descent can be described by an effective gravity scaling as $g(1 - \kappa^{-2}\varepsilon^2 \Delta\theta/R^2 \sin\alpha)$ instead of $g\sin\alpha$. The correction is negligible in the experiment of Fig. 4, for which the number $\kappa^{-2}\varepsilon^2 \Delta\theta/R^2\alpha$ is expected to be on the order of 5%. If we have $F \ll Mg\sin\alpha$, hysteresis slightly reduces the efficiency of the descent, but it preserves the ability of the drops to accelerate on centimetric distances. This allows these liquid pearls to reach large terminal velocities. In Fig. 5, we show how the data in Fig. 4 deviate at large scale ($x \approx 0.1$ to 1 m) from the law of free fall (still in dotted line). The drop velocity $V$ tends towards a constant, of the order of 1 m s$^{-1}$ in this case, despite the modest angle of inclination: drops of similar size on a vertical window pane typically move at 1 cm s$^{-1}$. The velocity $V$ here is comparable to the terminal speed of a drop in air (apart from the fact that gravity is reduced by a factor $\sin\alpha$), suggesting that air friction limits the drop speed.

In the terminal state, we also observe that tracers (which were translating in the acceleration regime) now have circular trajectories inside the drop, revealing a mixture of rolling and sliding in this regime. The combination of air friction and rolling dramatically affects the drop shape (Fig. 6). First, what was a sphere at a low speed (that is, in the acceleration regime described in Fig. 4) becomes elongated. At first glance, this is natural: a tear generally leaves a trace behind it; in other words, in partial wetting, there is a threshold velocity above which liquid is

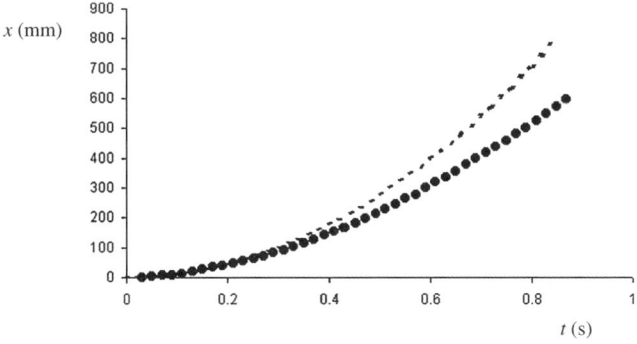

**Fig. 5** At a larger scale (1 m), the trajectory deviates from the law of free fall (in the dotted line) and the water drop reaches a terminal velocity $V$, here of approximately 1 m s$^{-1}$, 100 to 1000 times larger than on a window pane, or on a tilted plastic.

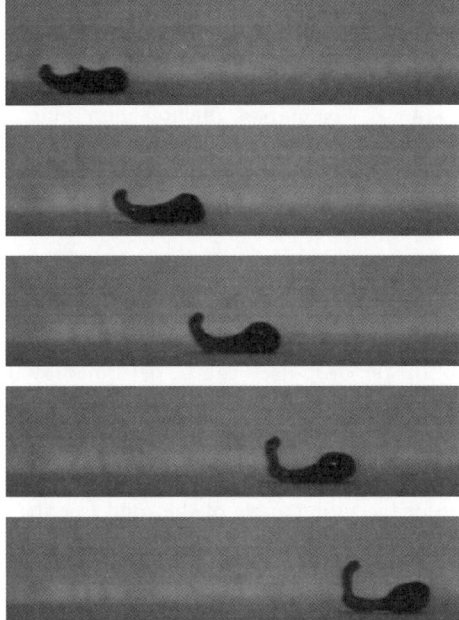

**Fig. 6** Successive snapshots of a water drop (initial radius $R = 2.5$ mm) running down a super-hydrophobic plate inclined by $\alpha = 60°$ (the high speed camera is tilted by the same angle, and shoots 18000 pictures per second). The picture width is 41 mm, the interval between snapshots 2.8 ms, and the drop speed $V = 2.5$ m s$^{-1}$. At such a high speed, a tail appears at the rear of the drop and the drop becomes a centimetre-long. Remarkably, this tail leaves the substrate, which remains dry behind the drop despite the speed.

abandoned behind the drop, because of the viscous friction at its trailing edge.[27] However, the liquid tail observed in Fig. 6 is original: it leaves the substrate, bends, and runs down at a velocity comparable to the drop speed $V$. This might arise from the centrifugation associated with the drop rotation: the Weber number, which compares the centrifugal force $\rho V^2/R$ to the capillary force $\gamma/R^2$ is on the order of 1 to 10 in this experiment. Since centrifugal force tends to expel liquid from the axis of rotation, *i.e.* perpendicularly to the direction of motion, the tail is indeed likely to bend. This effect is of practical interest: as a drawback of the quick motion, the drop should abandon a trace, leaving the solid partially wet (the tail then decaying in microdroplets because of the Rayleigh instability). The bending of the tail avoids this detrimental effect.

Despite the complexity of the shape of the tumbling drop, we can add a few simple thoughts about its terminal velocity. The "natural" Reynolds number $Re = \rho_a RV/\eta_a$ (with $\eta_a$ the air viscosity) being of the order 10 here, it would be "natural" to assume an inertial friction. Balancing such a force, of the form $\rho_a R^2 V^2$, with the drop weight, we immediately find the classical formula:

$$V \sim \left(\frac{\rho g R \sin\alpha}{\rho_a}\right)^{1/2} \quad (5)$$

Eqn (5) implies a descent velocity of a few meters per second, as observed experimentally. However, the drop shape does not seem to fit with this scenario: the drop elongates along the movement, which is characteristic of a viscous deformation. If inertia were dominant, the Bernoulli pressure would stretch the drop perpendicularly to the motion.[28] Our Reynolds number suggests that inertia dominates viscosity, but its construction might be too naive: drops are moving on a film of air, either trapped below the liquid, or possibly dynamically entrained by the motion itself. The large value of the static contact angle favours air entrainment,[29] generating a kind of dynamical Leidenfrost phenomenon and, subsequently, large slip.[30] If we assume a viscous friction of the order of $(\eta_a V/\delta)R^2$, where $\delta$ is the thickness of the air layer and $\eta_a$ the air viscosity, we get for the descent velocity:

$$V \sim \frac{\rho g R \delta \sin\alpha}{\eta_a} \quad (6)$$

which is also expected to be a few meters per second provided that $\delta$ is in the range of 1 to 10 μm. These estimates are of course quite rough: (i) the exact law for the dissipation depends on the detail of the flow of air, either inside the texture, or as it penetrates below the drop; (ii) the drop itself can be deformed by the flow (see Fig. 6), which impacts the surface area on which the viscous stress applies. However, the typical scale for the thickness $\delta$ should indeed be micrometric, whatever the nature (*i.e.* static or dynamical) of the film. We do need a series of precise experiments to test quantitatively the scaling laws expressed in the latter equations, for example as a function of the tilting angle of the solid (see the differences between eqn (5) and 6).

If the film of air is dynamical, we can go slightly further in the analysis. For a small drop, the pressure in the film is set by the Laplace pressure $\gamma/R$ and the length connecting the unperturbed drop to the film should scale as $(\delta R)^{1/2}$, as proposed by Landau and Levich. Following this line, the film thickness $\delta$ should result from a balance between viscous effects and capillary elasticity ($\eta_a V/\delta^2 \sim \gamma/(\delta^{1/2} R^{3/2})$), which yields the classical Landau–Levich scaling: $\delta \sim R (\eta_a V/\gamma)^{2/3}$ (or as $\kappa^{-1}(\eta_a V/\gamma)^{2/3}$ for bigger drops, for which the curvature at the front is set by the capillary length $\kappa^{-1}$). The capillary number $\eta_a V/\gamma$ is typically $10^{-3}$ in these experiments, which gives for the above scaling a thickness $\delta \sim 10$ μm for the air cushion. This thickness is clearly too small to be observed directly, as it can be for a Leidenfrost film ten times thicker. It is also worth noticing that $\delta$ can be comparable to the texture height, which might complicate the analysis.

## 4. Drop impact: from rebound to crystallographic splashing

The way superhydrophobic materials react to water drops impacts is probably the most well-known manifestation of water repellency. Provided that the capillary number $\eta V/\gamma$ is small enough (denoting $V$ as the impact velocity, and $\eta$ as the liquid viscosity), surface tension dominates viscous force, and a liquid drop can behave as a spring (Fig. 7). As it hits the solid, it converts a part of its kinetic energy into surface energy; after reaching a maximum size, the puddle (of diameter $D_M$ and thickness $h_m$) recoils and takes off. The contact time $\tau$ is the typical response time of this liquid spring, of mass $\rho R^3$ and stiffness $\gamma$, which yields:[12]

$$\tau \sim \left(\frac{\rho R^3}{\gamma}\right)^{1/2} \qquad (7)$$

As observed experimentally, this time is on the order of 1–20 ms for millimetre-size drops. The deformation is generally large ($D_M \gg R$, $h_m \ll R$), as in Fig. 7, which means by a Weber number ($We = \rho V^2 R/\gamma$) larger than unity. This condition is easily satisfied with water since it implies impact velocities larger than 30 cm s$^{-1}$. For $We \gg 1$, fragments form owing to the large drop deformation; this also makes the elasticity of the shock quite modest,[12] as seen in Fig. 7 where the drop reaches after take off a height of about one millimetre, while it was released from a height of 4 cm.

The differences between spreading and recoiling are generic, in the usual cases $Ca \ll 1$ and $We \gg 1$. A convenient representation is that of Fig. 8, where we report the diameter $D$ of the solid/liquid contact as a function of time, defining $t = 0$ as the time of contact, and $t = \tau$ as the moment of take off. Here a water drop of radius $R = 1.15$ mm hits at $V = 0.8$ m s$^{-1}$ a superhydrophobic solid, which yields $Ca \approx 10^{-3}$ and $We \approx 10$. The solid is decorated by a square lattice of hydrophobic posts of height 24 μm, diameter 2.7 μm and mutual distance 10 μm. The graph is found to be asymmetric: in a first stage, the drop spreads very quickly on the solid; the second step is longer: after reaching its maximum expansion $D_M$, the liquid

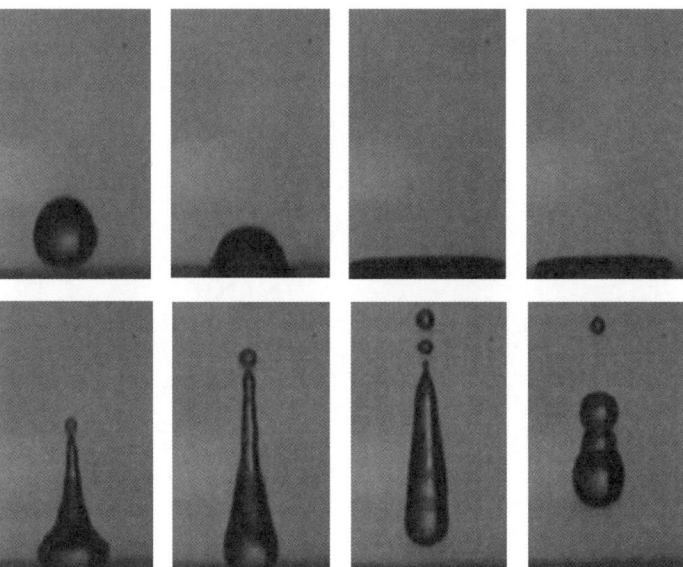

**Fig. 7** Millimetric water drop impacting a super-hydrophobic solid at high Weber number $We = \rho V^2 R/\gamma$, denoting $R$ as the drop radius, $V$ as the impact velocity, and $\rho$ and $\gamma$ as the density and surface tension of water. Here $We$ is 18, so that the drop gets highly elongated before taking off and emits droplets. The typical interval between successive pictures is 2 ms.

recoils at a constant dewetting velocity $V_d$ (here of approximately 0.4 m s$^{-1}$) before taking off for $t > \tau \approx 13$ ms.

In the first stage (inertial spreading), the behaviour is dictated by geometry and it expresses how a solid/liquid contact establishes as a spherical drop crashes on a planar surface. In the first moments, we deduce from the geometric contact between a sphere and a plane that the size $D$ of the contact should increase as $(RVt)^{1/2}$, where $Vt$ is the distance of penetration of the sphere in the plane. Slightly later, the drop reaches its maximum diameter $D_M$ and starts to recoil. The capillary force driving this stage depends on the detail of the contact: if a film of air comes between the substrate and the spreading water, then it is just $2\gamma$ (per unit length of the drop perimeter), as for a freely suspended film. It is likely that such an air cushion initially forms, but contrasting with the case of section 3, air is not constantly injected in this film, so that a static (wetting) contact may eventually set. Then, the retraction is accompanied by the creation of a new solid surface, and suppression of solid/liquid and liquid/vapour surfaces. It implies a balance of energy per unit area $\gamma_{SV} - \gamma_{SL} - \gamma$, where $\gamma_{SV}$ and $\gamma_{SL}$ are the effective solid/vapour and solid/liquid surface tensions in the fakir state (section 2). This yields a force per unit length $F \approx \gamma(1 - \cos\theta) \approx 2\gamma (1 - \varepsilon^2/4)$. Using eqn (2) (where we take $\cos\theta_o \approx 0$ for the sake of simplicity), we get: $F \approx 2\gamma (1 - \phi/2)$, which directly depends on the concentration $\phi$ of defects below the drop: the smaller $\phi$, the larger the force recoiling the drop.

At low capillary numbers, viscous effects can be ignored, and the main force resisting the motion is inertia. As proposed by Taylor, Culick and Brochard-Wyart (in the context of dewetting), a rim of mass $M$ forms at the receding edge of the puddle.[31] Newton's law expresses that $d(MV_d)/dt = F$; noting that $dM/dt = \rho h_m V_d$ (a relation again written per unit perimeter of the puddle), a solution of constant velocity is found, in agreement with the observation in Fig. 8:

$$V_d \sim \left(\frac{2\gamma}{\rho h_m}\right)^{1/2} \left(1 - \frac{\phi}{4}\right) \qquad (8)$$

An explicit formula for $V_d$ would require the knowledge of $h_m$, a quantity on which there is some debate. However, two remarks can be done. (i) The thickness

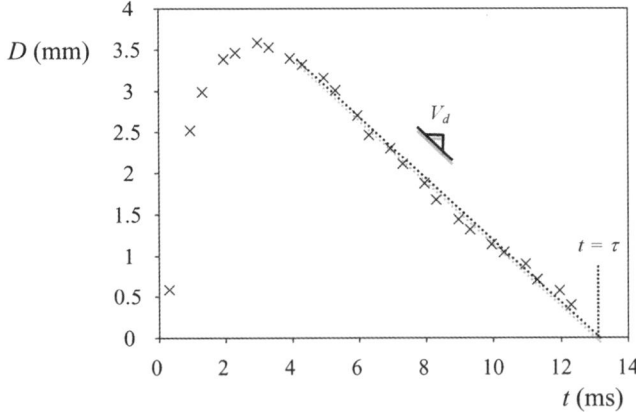

**Fig. 8** Diameter $D$ of the solid/liquid contact after a drop of radius $R = 1.15$ mm hits at $V = 0.8$ m s$^{-1}$ a superhydrophobic solid (here $We \approx 10$), consisting of a square lattice of hydrophobic posts of height $\delta = 24$ μm, radius $b = 1.3$ μm and mutual distance $p = 10$ μm. For $We \gg 1$, spreading and retraction are asymmetric: the recoiling stage is much longer than the spreading stage, and quite linear in time.

of the pancake can be directly deduced from the observation of $D_M$ (see Fig. 7 and 8), using volume conservation ($h_m D_M{}^2 \sim R^3$). From Fig. 8, for example, we deduce $h_m \approx 0.65$ mm, from which eqn (8) predicts $V_d \approx 45$ cm s$^{-1}$, slightly larger (yet of the same order of magnitude) than the value deduced from the figure ($V_d \approx 40$ cm s$^{-1}$). (ii) In the large deformation regime, the contact time is dominated by the recoiling step, that is: $\tau \sim D_M/V_d$. Using eqn (8) and volume conservation, we find, whatever the law for $h_m$ (!): $\tau \sim (\rho R^3/\gamma)^{1/2}(1 + \phi/4)$. The contact time in eqn (7) is corrected by a function of the density of defects, a correction in qualitative agreement with the recent experiments by Li, Ma and Lan.[32] More generally, an original program of research on the dewetting on water on superhydrophobic materials (for which dewetting should be ultra-fast, due to the quasi-absence of viscous resistance) remains to be conducted. We finally note that these different laws would be slightly modified (without changing our qualitative conclusions) by the logarithmic term discussed in eqn (3).

We finally investigate a remarkable behaviour which can happen as drops impact solids decorated with posts of moderate height $\delta$ ($\delta = 18$ µm), separated by a distance $p = 10$ µm much larger than the post radius $b = 1.4$ µm (hence we deduce for these surfaces $\phi = \pi b^2/p^2 \approx 0.06$). On such dilute pillars, impact can provoke the penetration of water inside the texture, which irreversibly pins a fraction of the drop, while the rest still bounces.[13] The simplest argument we can think of for understanding the threshold in impact velocity above which penetration takes place consists of balancing the dynamical pressure at impact $\rho V^2$ with the resisting Laplace pressure $\gamma/p$: hence, we find a velocity $(\gamma/\rho p)^{1/2}$, *i.e.* approximately 2.5 m s$^{-1}$. Fig. 9 shows a series of snapshots taken with a high-speed camera for an impact velocity just above this threshold ($V = 2.8$ m s$^{-1}$), and showing the figure of impact as seen from the top.

The impact figure exhibits original features: (i) At the place of impact, we can see a dark stain, which conveys the penetration of the drop inside the texture, where it gets irreversibly pinned. The size of the penetration zone is approximately the drop size, and it remains constant all along the sequence. (ii) A large sheet of liquid bounded by a thicker rim expands beyond this first region. Ahead of this "fakir" sheet, four fingers of liquid are emitted, in the main directions of the subjacent network of micropillars. This observation is close to what Xu *et al.* reported in similar situations, where the figure of impact was found to somehow conform to the symmetry of the underlying network.[33] Tsai and Lohse also recently did

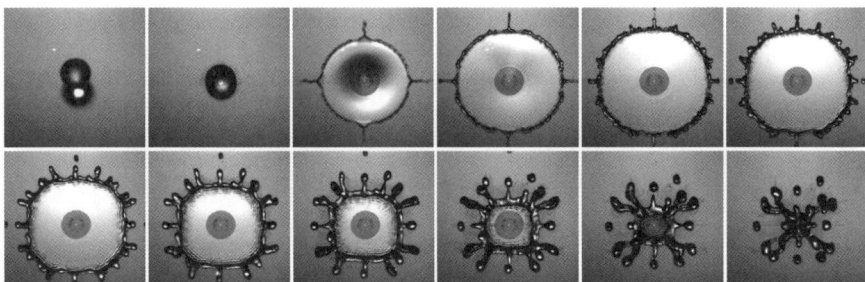

**Fig. 9** Top view of the impact of a water drop of radius $R = 1.9$ mm hitting at $V = 2.8$ m s$^{-1}$ a superhydrophobic solid decorated by a square lattice of hydrophobic posts of height $\delta = 18$ µm, radius $b = 1.4$ µm and mutual distance $p = 10$ µm. Three different effects can be observed: (i) the impact speed is large enough to provoke the penetration of liquid inside the texture, at the scale of the drop radius, as revealed by the darkening of the substrate at this place; (ii) as soon as the drop spreads inertially, four jets are emitted in the direction of the subjacent array of micropillars; (iii) in the recoiling stage, the presence of the posts induces an anisotropic dewetting, which transforms the circular drop in a quasi-square one. Interval between successive pictures: 1 ms.

comparable observations.[34] We measured the ejection speed $u$ of these jets using very high-speed cameras, and found approximately 10 m s$^{-1}$. (iii) The dewetting/recoiling stage has a remarkable characteristic, namely the square shape (also reflecting the subjacent network) transiently adopted by the drop (images 1 to 4 in the second row). This pattern is reminiscent from what can be observed as making hydraulic jumps on such surfaces – then polygons similarly form.[35]

If the impact velocity is slightly increased (same substrate as in Fig. 9, same drop, but impact at $V = 2.9$ m s$^{-1}$ instead of 2.8 m s$^{-1}$), an interesting variation is observed (see Fig. 10). Then, the peripheric jets also develop along the diagonals of the network, which generates an 8-fold figure of impact. The recoiling stage also seems to be characterized by the same symmetry. This phenomenon is observed for velocities up to 3.7 m s$^{-1}$. For still higher velocities, the impact figure becomes isotropic. Note that these figures tend to disappear gradually when decreasing the post heights; fixing this height and increasing the distance between posts broadens the window in velocity in which these "crystallographic" impacts are reported.

We do not claim to have any quantitative explanation to this phenomenon. But it seems clear that air plays a key role: the effect is observed only if the drop penetrates in the texture; then, the air present there is quickly dispelled. The volume of air brought in motion scales as $R^2\delta$, to which corresponds a flux of $R^2 V$. Conservation of this flux implies an ejection velocity for air $V_a$ (in the network of pillars) scaling as $RV/\delta$, which can initially be of the order of 100 m s$^{-1}$! In our interpretation, the flow of air follows preferential channels, in the main directions of the network (or possibly in the diagonals) of smaller hydrodynamic resistance. This anisotropic current drags the liquid above, which results in the development of liquid filaments in the directions of the flow of air.

The drag force acting on the liquid can be in this problem either inertial, or viscous. Considering the high velocity of air, we first assume inertia to be the dominant force. As the air comes out of the network of channels, it exerts on the edge of the water film a force $\rho_a V_a^2 hR$, where $h$ is the thickness of the liquid sheet entrained by the air. This force is mainly resisted by the liquid inertia $\rho u^2 hR$, denoting $u$ as the liquid velocity. Using the scaling for $V_a$, we deduce an injection velocity:

$$u \sim \left(\frac{\rho_a}{\rho}\right)^{1/2} \frac{R}{\delta} V \qquad (9)$$

We expect from eqn (9) a velocity $u$ on the order of 10 m s$^{-1}$, as observed experimentally. A natural criterion for observing the formation of filaments is $u > V$, which is found to be a condition on the design of the texture: $\delta < R (\rho_a/\rho)^{1/2}$, i.e. pillars smaller than approximately 30 μm. In addition, filaments can only exist if the drag force is large enough to overcome surface tension. If fingers develop on a width $L$, this criterion can be written: $\gamma L < \rho_a V_a^2 hL$, which can be rewritten: $V > V_c = (\gamma \delta^2/\rho_a hR^2)^{1/2}$, where $V_c$ is expected to be typically 1 m s$^{-1}$ for our parameters.

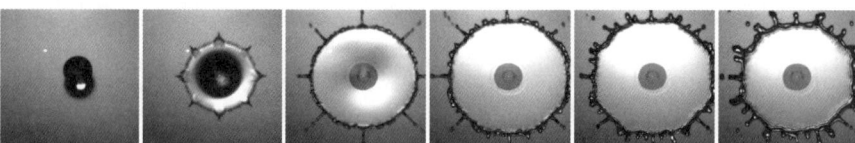

**Fig. 10** Top view of the impact of a water drop of radius $R = 1.9$ mm hitting at $V = 2.9$ m s$^{-1}$ the same superhydrophobic solid as described in Fig. 9. Eight jets are emitted, instead of four, and the drop becomes quasi-octogonal during the recoiling stage. Interval between successive pictures: 1 ms.

The viscous drag generated by the flow of air can also be significant. Considering only this drag for drawing the filament implies the force balance: $(\eta_a V_a/\delta)R^2 \sim \rho u^2 h R$, and thus an ejection velocity: $u \sim (\eta_a R^2 V/\rho h \delta^2)^{1/2}$. For our parameters, this velocity is expected to be approximately 1 m s$^{-1}$, smaller than above (which emphasizes that inertia might be the dominant force driving filamentation in our case). For small posts, however, viscous effects will dominate inertial ones. Then the criterion $u > V$ can be written: $\delta < R\,(\eta_a/\rho h V)^{1/2}$, which implies posts smaller than about 10 μm. Note however that very small posts would favour a viscous damping of the flow of air, thus suppressing the emission of filaments.

All these tentative ideas would need to be carefully confirmed (or not). It would be interesting to make more precise the phase diagram of these events, and in particular to understand the narrowness of the window of velocities where these crystallographic impacts are found. More generally, Fig. 9 and 10 show the possibility of (transient) coexistence between a trapped state (at the centre of the figure of impact) and a fakir state (the rest of the drop), which persists all along the sequence (so that the fakir part bounces off, while the rest of the drop remains pinned). The conditions for generating and preserving this coexistence also remain to be understood.

## 5. Conclusion

The aim of this paper was to put together the main and unique dynamical properties of superhydrophobic materials, when they are exposed to drops. Firstly, as noted by many authors, the contact angle hysteresis, which resists any drop motion, is minimized in this situation – and we presented a possible (quantitative) argument for understanding this fact for solids decorated by dilute pillars, a convenient model system. Secondly, the friction of water drops is dramatically reduced on these materials, compared to common solids. An original experiment was discussed, in order to stress both the very high velocities reached by the drops, and the dramatic changes in liquid shape induced by these fast motions. A few hints were given to understand the mobility of these drops, but clearly, a comprehensive experimental program remains to be conducted on this theme. Thirdly, we recalled a few features arising from drop impacts and rebounds. We also discussed how the figure of impact can reflect the presence of a texture at the solid surface, by presenting an original set of data on "crystallographic" impact, where the impact figure takes the symmetry of the underlying texture. There again, careful and complete experiments would be necessary (together with quantitative analysis) to fully capture the characteristics of these shocks.

## Acknowledgements

It is a pleasure to thank Tom Witten, L. Mahadevan and Howard Stone for stimulating discussions on this topic, and Elise Bourdin for her help in the impact experiments.

## References

1 J. G. Leidenfrost, *Int. J. Heat Mass Transfer*, 1966, **9**, 1153–1166.
2 T. Wagner, C. Neinhuis and W. Barthlott, *Acta Zool.*, 1996, **77**, 213–225; C. Neinhuis and W. Barthlott, *Ann. Bot.*, 1997, **79**, 667–677; X. Gao and L. Jiang, *Nature*, 2004, **432**, 36; B. Bhushan and Y. C. Jung, *Nanotechnology*, 2006, **17**, 2758–2772; E. Bormashenko, Y. Bormashenko, T. Stein, G. Whyman and E. Bormashenko, *J. Colloid Interface Sci.*, 2007, **311**, 212–216; X. F. Gao, X. Yan, X. Yao, L. Xu, K. Zhang, J. H. Zhang, B. Yang and L. Jiang, *Adv. Mater.*, 2007, **19**, 2213–2215.

3  T. Onda, S. Shibuichi, N. Satoh and K. Tsujii, *Langmuir*, 1996, **12**, 2125–2127; J. Bico, C. Marzolin and D. Quéré, *Europhys. Lett.*, 1999, **47**, 220–226; A. Nakajima, A. Fujishima, K. Hashimoto and T. Watanabe, *Adv. Mater.*, 1999, **11**, 1365–1368; D. Öner and T. J. McCarthy, *Langmuir*, 2000, **16**, 7777–7782; N. A. Patankar, *Langmuir*, 2004, **20**, 8209–8213; L. Gao and T. J. McCarthy, *J. Am. Chem. Soc.*, 2006, **128**, 9052–9053.
4  A. Lafuma and D. Quéré, *Nat. Mater.*, 2003, **2**, 457–460.
5  S. Herminghaus, *Europhys. Lett.*, 2000, **52**, 165–170; W. Chen, A. Y. Fadeev, M. C. Hsieh, D. Oner, J. Youngblood and T. J. McCarthy, *Langmuir*, 1999, **15**, 3395–3399; L. Cao, H. H. Hu and D. Gao, *Langmuir*, 2007, **23**, 4310–4314; A. Tuteja, W. Choi, M. L. Ma, J. M. Mabry, S. A. Mazzella, G. C. Rutledge, G. H. McKinley and R. E. Cohen, *Science*, 2007, **318**, 1618–1622; A. Ahuja, J. A. Taylor, V. Lifton, A. A. Sidorenko, T. R. Salamon, E. J. Lobaton, P. Kolodner and T. N. Krupenkin, *Langmuir*, 2008, **24**, 9–14.
6  C. Cottin-Bizonne, J. L. Barrat, L. Bocquet and E. Charlaix, *Nat. Mater.*, 2003, **2**, 238–240; J. Ou, B. Perot and J. P. Rothstein, *Phys. Fluids*, 2004, **16**, 4635–4643; J. Ou and J. P. Rothstein, *Phys. Fluids*, 2005, **17**, 103606; P. Joseph, C. Cottin-Bizonne, J. M. Benoit, C. Ybert, C. Journet, P. Tabeling and L. Bocquet, *Phys. Rev. Lett.*, 2006, **97**, 156104; P. Roach, G. McHale, C. R. Evans, N. J. Shirtcliffe and M. I. Newton, *Langmuir*, 2007, **23**, 9823–9830; A. Steinberger, C. Cottin-Bizonne, P. Kleimann and E. Charlaix, *Nat. Mater.*, 2007, **6**, 665–668; C. Lee, C. H. Choi and C. J. Kim, *Phys. Rev. Lett.*, 2008, **101**, 064501; F. Feuillebois, M. Z. Bazant and O. I. Vinogradova, *Phys. Rev. Lett.*, 2009, **102**, 026001.
7  J. P. Rothstein, *Annu. Rev. Fluid Mech.*, 2010, **42**, 89–109.
8  L. Mahadevan and Y. Pomeau, *Phys. Fluids*, 1999, **11**, 2449–2453.
9  D. Richard and D. Quéré, *Europhys. Lett.*, 1999, **48**, 286–291.
10 M. Miwa, A. Nakajima, A. Fujishima, K. Hashimoto and T. Watanabe, *Langmuir*, 2000, **16**, 5754–5760; M. Sakai, J. H. Song, N. Yoshida, S. Suzuki, Y. Kameshima and A. Nakajima, *Langmuir*, 2006, **22**, 4906–4909.
11 P. Aussillous and D. Quéré, *Nature*, 2001, **411**, 924–927.
12 D. Richard and D. Quéré, *Europhys. Lett.*, 2000, **50**, 769–775; D. Richard, C. Clanet and D. Quéré, *Nature*, 2002, **417**, 811–811; A. L. Biance, F. Chevy, C. Clanet, G. Lagubeau and D. Quéré, *J. Fluid Mech.*, 2006, **554**, 47–66.
13 M. Reyssat, A. Pépin, F. Marty, Y. Chen and D. Quéré, *Europhys. Lett.*, 2006, **74**, 306–312; D. Bartolo, F. Bouamrirene, E. Verneuil, A. Buguin, P. Silberzan and S. Moulinet, *Europhys. Lett.*, 2006, **74**, 299–305; P. C. Tsai, S. Pacheco, C. Pirat, L. Lefferts and D. Lohse, *Langmuir*, 2009, **25**, 12293–12298.
14 T. Deng, K. K. Varanasi, M. Hsu, N. Bhate, C. Keimel, J. Stein and M. Blohm, *Appl. Phys. Lett.*, 2009, **94**, 133109.
15 J. B. Boreyko and C. H. Chen, *Phys. Rev. Lett.*, 2009, **103**, 174502.
16 C. G. L. Furmidge, *J. Colloid Sci.*, 1962, **17**, 309–314.
17 E. B. Dussan and R. T. P. Chow, *J. Fluid Mech.*, 1983, **137**, 1–29.
18 M. Reyssat and D. Quéré, *J. Phys. Chem. B*, 2009, **113**, 3906–3909.
19 A. B. D. Cassie and S. Baxter, *Trans. Faraday Soc.*, 1944, **40**, 546–550; A. B. D. Cassie, *Discuss. Faraday Soc.*, 1948, **3**, 11–16.
20 L. Barbieri, E. Wagner and P. Hoffmann, *Langmuir*, 2007, **23**, 1723–1734.
21 P. S. Swain and R. Lipowsky, *Langmuir*, 1998, **14**, 6772–6780; G. Wolansky and A. Marmur, *Langmuir*, 1998, **14**, 5292–5297.
22 S. Brandon, A. Wachs and A. Marmur, *J. Colloid Interface Sci.*, 1997, **191**, 110–116; C. W. Extrand, *Langmuir*, 2002, **18**, 7991–7999; G. McHale, N. J. Shirtcliffe and M. I. Newton, *Langmuir*, 2004, **20**, 10146–10149; L. C. Gao and T. J. McCarthy, *Langmuir*, 2006, **22**, 6234–6237; H. Kusumaatmaja and J. M. Yeomans, *Langmuir*, 2007, **23**, 6019–6032; A de Simone, N. Grunewald and F. Otto, *Net. Heter. Med.*, 2007, **2**, 211–225; H. Kusumaatmaja and J. M. Yeomans, *Soft Matter*, 2009, **5**, 2704–2707.
23 J. F. Joanny and P. G. de Gennes, *J. Chem. Phys.*, 1984, **81**, 552–562.
24 Y. Pomeau and J. Vannimenus, *J. Colloid Interface Sci.*, 1985, **104**, 477–488.
25 K. Y. Yeh, L. J. Chen and J. Y. Chang, *Langmuir*, 2008, **24**, 245–251.
26 W. Barthlott and C. Neinhuis, *Planta*, 1997, **202**, 1–8.
27 T. Podgorski, J. M. Flesselles and L. Limat, *Phys. Rev. Lett.*, 2001, **87**, 036102.
28 E. Reyssat, F. Chevy, A. L. Biance, L. Petitjean and D. Quéré, *Europhys. Lett.*, 2007, **80**, 34005.
29 C. Duez, C. Ybert, C. Clanet and L. Bocquet, *Nat. Phys.*, 2007, **3**, 180–183.
30 O. I. Vinogradova, *Langmuir*, 1995, **11**, 2213–2220.

31 G. I. Taylor, *Proc. Roy. Soc.*, 1959, **A253**, 313–321; F. E. C. Culick, *J. Appl. Phys.*, 1960, **31**, 1128–1129; A. Buguin, L. Vovelle and F. Brochard-Wyart, *Phys. Rev. Lett.*, 1999, **83**, 1183–1186.
32 X. Li, X. Ma and Z. Lan, *Langmuir*, 2010, **26**, 4831–4838.
33 L. Xu, *Phys. Rev. E: Stat., Nonlinear, Soft Matter Phys.*, 2007, **75**, 056316; L. Xu, L. Barcos and S. R. Nagel, *Phys. Rev. E: Stat., Nonlinear, Soft Matter Phys.*, 2007, **76**, 066311.
34 P. C. Tsai and D. Lohse, personal communication (2010).
35 E. Dressaire, L. Courbin, J. Crest and H. A. Stone, *Phys. Rev. Lett.*, 2009, **102**, 194503.

# Superhydrophobic surfaces by hybrid raspberry-like particles

Maria D'Acunzi, Lena Mammen, Maninderjit Singh, Xu Deng, Marcel Roth, Günter K. Auernhammer, Hans-Jürgen Butt and Doris Vollmer*

*Received 8th December 2009, Accepted 29th January 2010*
DOI: 10.1039/b925676h

Surface roughness on different length scales is favourable for superhydrophobic behaviour of surfaces. Here we report (i) an improved synthesis for hybrid raspberry-like particles and (ii) a novel method to obtain superhydrophobic films of good mechanical stability. Polystyrene spheres with a diameter of 400 nm–1 μm are decorated with silica colloids < 100 nm in size, thus introducing surface asperities on a second length scale. To improve mechanical resistance, we then coated the polystyrene core and attached silica colloids with a smooth silica shell of 10 nm to 40 nm thickness. All three steps of this synthesis procedure can be sensitively tuned so that the average size and number of the silica colloids as well as the morphology of the resulting raspberry particles can be predicted. As the particles disperse in water, either monolayers can be prepared by dip coating or multilayers by drop casting. Although mechanically stable, the shells are porous enough to allow for leakage of molten or dissolved polystyrene from the core. In tetrahydrofuran vapour polystyrene bridges form between the particles that render the multilayer-film stable. Leaked polystyrene that masks some asperities can be removed by plasma cleaning. Surface roughness on larger scales can be tuned by the drying procedure. The films are hydrophobized by silanization with a semi-fluorinate silane.

## 1 Introduction

Over the last few years a great effort has been devoted to the fabrication of superhydrophobic surfaces[1–4] because of their self-cleaning properties. A water droplet on a superhydrophobic surface rolls off even at inclinations of only a few degrees while taking up contaminants encountered on its way. This property is appealing for many applications, such as coatings, paints, window panes, or stain resistant fabrics. It has been well known for years that superhydrophobicity is related to surface roughness.

The maximum contact angle reported for water on a smooth surface is about 120° (fluorinated surfaces).[5] In order to achieve higher contact angles, a rough surface is needed. Wenzel in 1936[6] and Cassie and Baxter in 1944[7] described two different superhydrophobic states that nowadays are referred to by the authors' respective names.

In the Wenzel state[6] the solvent surrounds the surface asperities and also wets the depressions in between, leading to an increase in actual contact area. Wenzel accounted for this by including a roughness factor $r$ in the Young equation, cos

*Max Planck Institute for Polymer Research, Ackermannweg 10, D-55128 Mainz, Germany. E-mail: vollmerd@mpip-mainz.mpg.de; Fax: +49 6131 379 110*

$\Theta^* = r \cos \Theta$, where $r$ relates the apparent, contact angle $\Theta^*$, with the contact angle of the flat surface, $\Theta$, and is given by the ratio of the actual and the projected surface areas. In the Wenzel state, the contact angle can be as high as 150° but the droplet does not roll off.

In the Cassie state,[7] a droplet is suspended on the tips of the asperities without filling the niches between them. Thus, the surface contact area is reduced as compared to flat surfaces. Partially the drop is in contact with air trapped in the niches. Cassie and Baxter assumed that the contact angle depends on the fractional areas of the asperities and air cushions. The contact angle $\Theta^*$ is an average between the contact angle on the surface $\Theta_s$ and on air $\Theta_v = 180°$. The Cassie equation reads: $\cos \Theta^* = -1 + \phi_s (1 + \cos \Theta)$, were $\phi_s$ is the fractional area of wetted solid substrate. The Cassie–Baxter equation applies for very rough or hydrophobic surfaces. In these cases the Wenzel equation predicts complete drying, which is not possible. The contact angle in the Cassie state can be higher than 150° and the droplet easily rolls off. Hence, droplets in the Cassie state exhibit self-cleaning properties.

Cassie and Wenzel models are based on thermodynamic considerations. They only predict one angle for a given material and surface topology. However, it is well known that real surfaces can show different values of contact angles[8] independent on the method of application of the drop on the surface, time scales, and local variations of surface topography and chemical nature (contact angle hysteresis).

McCarthy et al.[9] showed that it was not the area fractions underneath a droplet but the surface morphology close to the contact line which determined the contact angle and hysteresis.

The lotus leaf[10] is one of the most efficient examples of a superhydrophobic and self-cleaning surface. Its topography shows a surface roughness on the micrometre scale. This surface is coated with nano-sized asperities, which consist of an epicuticular wax with low surface tension. The combination of dual-size roughness and low surface tension material confers superhydrophobic properties to its surface.

Several methods have been proposed in order to make artificial superhydrophobic surfaces, such as lithography,[11–13] electrochemical deposition,[14,15] chemical deposition,[16,17] plasma etching,[18,19] sol–gel methods,[20] layer-by-layer (LBL) deposition,[21,22] and templation.[23] Some of these approaches involve colloids.[12,22] Recently, it has been shown that raspberry-like particles can be used for fabrication of superhydrophobic surfaces.[24,25] The term "raspberry-like particles" (short cut: raspberry particles) refers to the topography of their surfaces with nano-sized secondary spheres attached to a considerably larger primary particle. Mimicking the lotus leaf structure, the diameter of the primary particles is in the 1 μm range, while the smaller asperities are in the order of 20–100 nm. Ming et al.[24] developed a procedure to get superhydrophobic surfaces by covalently binding a monolayer of silica raspberry particles to an epoxy-based polymer matrix. The monolayer was partially embedded in the polymer matrix by a film forming reaction to improve the mechanical properties of the surface.

Lee et al.[25] prepared superhydrophobic surfaces with silica raspberry particles by a layer-by-layer technique. The raspberry-like nature of the silica particles was obtained by adsorption of smaller silica particles onto the surface of larger smooth ones. Although the surfaces survived prolonged immersion in water, the mechanical stability of these surfaces is poor, as attachment of the particles is achieved via physical adhesion instead of chemical forces.

Quian et al.[26] presented a preparation of superhydrophobic surfaces with raspberry-like particles made of a polystyrene core and small silica particles attached to its surface. Superhydrophobic surfaces are prepared by vertical lifting on glass substrates from particles dispersed in ethanol. After hydrophobization of the surface by silanization, the resulting samples were superhydrophobic.

So far, these techniques suffer from poor mechanical fixation of the colloids on the substrate. The particles can easily be removed from the substrate, causing a loss of its superhydrophobicity. In this work we introduce a novel, easy approach to form superhydrophobic surfaces with improved mechanical stability. This procedure is based on synthesis of hybrid raspberry particles made of a polystyrene core and a rough silica shell. As these particles are hydrophilic, multi-layers are prepared from water suspensions by water evaporation. Exposure to tetrahydrofuran (THF) vapour causes polystyrene to leak out of the cores and to form polymer bridges between the particles. Bridge formation results in surfaces with improved mechanical properties. Finally, the silica surface is hydrophobized by chemical vapour deposition of trichlorosilane.

## 2  Experimental

### Materials

Styrene was purchased from Acros Organics (99% extra pure, stabilized). The inhibitor was removed from the styrene by washing with 3 aliquots of 2 M sodium hydroxide solution and 3 aliquots of milli-Q water. The washed styrene was distilled under reduced pressure before use. All other chemicals and solvents were used without further purification: sodium hydroxide (WTL Laborbedarf GmbH, 99%), ammonium persulfate (Acros Organics, 98%), acrylic acid (Acros Organics, 99.5%, stabilized), absolute ethanol (Sigma Aldrich), tetrahydrofuran (THF) (Sigma Aldrich, 99.9%), poly(allylamine hydrochloride) (PAH) (Aldrich, average $M_w$ = 15,000), polyvinylpyrrolidone (PVP) (Fluka, K 90 $M_w$ = 360,000), ammonia (Fluka, 25% in water), tetraethoxysilane (TEOS) (Acros Organics, 98%), sodium chloride (Riedel-de Haën, 99.8%), (tridecafluoro-1,1,2,2-tetrahydrooctyl)-1-trichlorosilane (97%, Sigma Aldrich), and dioctyl sodium sulfosuccinate (AOT) (Fluka, > 99%). Milli-Q water was obtained from a Millipore purification system operating at 18.2 MΩ. The glass slides taken as a basis for the superhydrophobic surfaces were cleaned with a Hellmanex II solution (Hellma GmbH).

### Characterization

The particle morphology was characterized by Scanning Electron Microscopy (SEM, LEO 1530 Gemini, Oberkochen, Germany) and by Transmission Electron Microscopy (TEM, Tecnai F20, FEI; 200 kV). Furthermore, the films were imaged by SEM and 3D profile images of the surfaces are obtained by a μSurf® white-light confocal profilometer (Nanofocus AG, Germany). The effective surface charge of the particles was determined by zeta potential measurements (Delsa counter). The influence of polystyrene (PS) leakage on the contact angle was investigated on plasma-cleaned samples (Harrick, 200 W). Contact angle measurements were performed in the sessile drop configuration with a contact angle meter Dataphysics OCA35 (Data Physics Instruments GmbH, Germany). The tilting angle at which drops roll or slide off was measured at 5 positions after depositing a 5 μl water droplet on the surface, removing the needle and tilting the stage at a speed of 1.3°/s. Simultaneously the shape of the droplet was recorded. Increasing the speed by one order of magnitude or droplet volume by a factor of two did not change the tilting angle within experimental accuracy.

### Silanization

Silanization was performed at room temperature by putting the particle-coated substrate and a small glass vessel containing 1 ml of semi-fluorinated silane in a closed desiccator for 3 h. To increase the vapour pressure of silane we evacuated the desiccator for a few minutes and repeated this every half an hour. Afterwards the vessel containing the silane was removed from the desiccator and a vacuum was applied for one hour in order to remove unreacted silane residues.

**Table 1** Primary PS particles are obtained by soap-free emulsion polymerisation. The amount of acrylic acid is given relative to the amount of styrene in the reaction mixture. Batch E2E1 refers to seeded soap-free emulsion polymerisation. 38 ml of the reaction mixture E2 was transferred in a clean 500 ml flask and 300 ml of milli-Q water were added. The particle radius was measured by SEM images averaging over more than 100 particles

| Batch | Ammonium persulfate/g | NaCl/g | Acrylic acid % wt/wt | Styrene/g | Radius/nm |
|---|---|---|---|---|---|
| E1   | 0.11 | —    | 1   | 15.4 | 225 ± 8  |
| E2   | 0.11 | 0.2  | 0.6 | 25   | 334 ± 40 |
| E3   | 0.11 | 0.2  | 0.8 | 13.6 | 282 ± 16 |
| E4   | 0.11 | 0.2  | 1.6 | 13.6 | 290 ± 13 |
| E5   | 0.11 | 0.05 | 1.7 | 11.8 | 223 ± 9  |
| E2E1 | 0.4  | —    | 0.6 | 7.2  | 540 ± 30 |

### Synthesis of the polystyrene particles

If not stated otherwise, PS particles were synthesized by soap-free emulsion polymerization. In some cases these particles (radius 200–300 nm) were taken as seeds to increase their size towards 500 nm radius. Synthesis was performed in a 500 ml three-necked flask equipped with a condenser, a PTFE stirrer, and a gas inlet. 300 ml of Milli-Q water (300 ml) were bubbled with nitrogen for 20 min. Then ammonium persulfate and NaCl were added and the system was heated to 75 °C. A mixture of acrylic acid and styrene was added and the reaction was carried out at 75 °C. Depending on the composition (see Table 1) the particle radius ranged between 200 and 300 nm. To increase the particle radius up to 500 nm, part of the reaction mixture was transferred to a clean 500 ml flask for seeded soap-free emulsion polymerisation. 300 ml of milli-Q water were added and the apparatus was heated at 75 °C under a nitrogen flow. After adding ammonium persulfate, a mixture of acrylic acid and styrene was added dropwise within 1 h. After 1 day of stirring at 75 °C, particles were washed 6 times with fresh milli-Q water in subsequent centrifugation steps and in the end resuspended in ethanol (Tab. 2). When dispersed in distilled water at a concentration of 0.01 g L$^{-1}$, the resulting particles were negatively charged with a zeta potential of −55 ± 5 mV.

### Synthesis of composite particles

To obtain particles with dual scale roughness, we used PS particles as templates and coated these with small silica colloids. If not stated otherwise, the PS particles were dispersed in a solution of PVP in ethanol and stirred for 1 day prior to adding ammonia and a mixture of TEOS and ethanol for the Stöber synthesis.[27–29] The growth of small silica particles onto the surface of PS primary particles is governed by the condensation of TEOS catalyzed by ammonia and the relation between the amount of these reagents and the available primary particle surface. After 1 night, the composite particles were washed in fresh ethanol and then redispersed in water. Alternatively, the PVP coating was omitted (see Fig. 1).

**Table 2** Reaction conditions for the synthesis of composite particles according to the Stöber mechanism

|    | PS particles/g | PVP/ethanol/g ml$^{-1}$ | Ammonia 25%/ml | TEOS/ethanol/ml ml$^{-1}$ |
|---|---|---|---|---|
| S1 | 1.54  | 1.63/80   | 6.6  | 2/18 (in 4 steps in 5 h) |
| S2 | 0.05  | 0.013/5.8 | 0.44 | 0.16/0 |
| S3 | 0.153 | 0.163/8   | 0.66 | 0.5/1.5 |
| S4 | 0.153 | No PVP/8  | 0.66 | 0.5/1.5 |
| S5 | 0.156 | 0.04/8    | 1.32 | 0.5/0 |

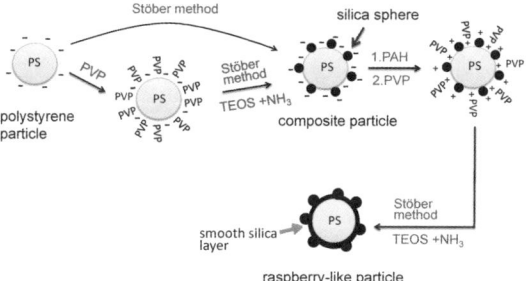

**Fig. 1** Sketch of the preparation procedure of raspberry particles.

## Synthesis of raspberry particles

Raspberry particles were obtained after coating the composite particles with a smooth silica shell (see Fig. 1). The reaction consists of two subsequent coatings of the composite particles with PAH and PVP.[29] To reverse their charge from negative to positive, 0.8 g of the composite particles were dispersed in 43 ml of water and added to a solution of 0.05 g PAH and 0.08 g NaCl. Subsequently the particles were coated with PVP. After 1 night under stirring in PAH aqueous solution, particles were washed in fresh water and in the end dispersed in a solution of PVP in ethanol (0.29 g PVP in 42 ml ethanol). After 30 min stirring, the particles were washed twice with fresh ethanol and dispersed in 60 ml ethanol. 4.8 ml ammonia and 0.6 g TEOS were added immediately. After 1 day the raspberry particles were washed several times with milli-Q water.

## Dip coating

A clean microscope glass slide (1 × 2 cm) was dipped into a water solution of PAH to reverse its charge from negative to positive. After rinsing twice with fresh water, the slide was immersed in a dispersion of raspberry particles in water (negatively charged). After drying, the sample surface was hydrophobized by silanization with trichloro(1$H$,1$H$,2$H$,2$H$-perfluorooctyl)silane, a semi-fluorinated silane.

## Solvent evaporation

To prepare superhydrophobic layers, a dispersion of raspberry particles in an aqueous solution (0.36 ml of a dispersion containing 2 g L$^{-1}$ up to 20 g L$^{-1}$) was filled in a circular sample holder of 1.5 cm in diameter with a glass bottom and removable Teflon walls. The particles sediment within about 2 h. Depending on temperature, complete evaporation of water takes 1–5 h. After the multi-layers had dried, the Teflon wall was removed. The glass slide covered with particles was exposed to THF vapor for 3 h and then put in vacuum for 3 h in order to remove residual THF. In the end, chemical vapor deposition of tridecafluoro-1,1,2,2-tetrahydrooctyl)-1-trichlorosilane on the films was performed in a vacuum chamber at room temperature.

## 3 Results and discussion

### 3.1 Raspberry particles

The synthesis of the hybrid raspberry particles is performed according to the scheme depicted in Fig. 1. In the following we describe a series of test experiments to analyze the influence of reaction conditions during the different steps of particle synthesis.

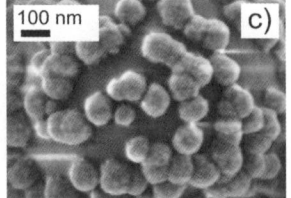

**Fig. 2** SEM images of composite particles using PS templates of different diameter: a) 400 nm sized PS particles (Table 1, E1). b) The 1000 nm sized PS (Table 1, E2E1). c) Higher magnification image of the surface of a particle shown in b). The Stöber reactions were performed according to procedures S5 (Table 2) and S1 (Table 2).

**Size of PS templates.** Synthesis of composite particles can be performed on PS particles of a wide range of sizes (see Fig. 2). Their overall diameter can be tuned by the size of the PS template. Both small (Fig. 2a) and large (Fig. 2b) PS templates are decorated by a large number of irregularly distributed silica nanoparticles with an almost uniform size of 100 ± 20 nm. According to the image taken at higher magnification (Fig. 2c) the silica colloids are not spherical but slightly elliptical. This hints that after nucleation or early attachment the growth of silica takes place predominantly on pre-existing silica instead of the PS surface. The size of the silica granules does not depend on the size of the PS template.

**Growth of silica colloids.** To investigate the growth process in more detail, we imaged aliquots of the reaction mixture without further purification after increasing the reaction times. The size of the colloids increased over the course of the reaction (Fig. 3). After only 20 min, 30 nm sized silica colloids appeared on the surface of the PS particles. After this initial fast growth, the silica grain size steadily increased until it approached 55 ± 5 nm after 3 h. Thereafter, the size remained almost constant. The growth process is limited by a lack of reaction material. From the images in Fig. 3 it cannot be decided whether silica particles nucleate on the template surface or new particles nucleated in the bulk continuously attach to it.

**Acrylic acid (AA) dependence.** The number of silica colloids attached to the surface of the PS particles depends on the amount of acrylic acid included in the primary particle synthesis (Fig. 4).

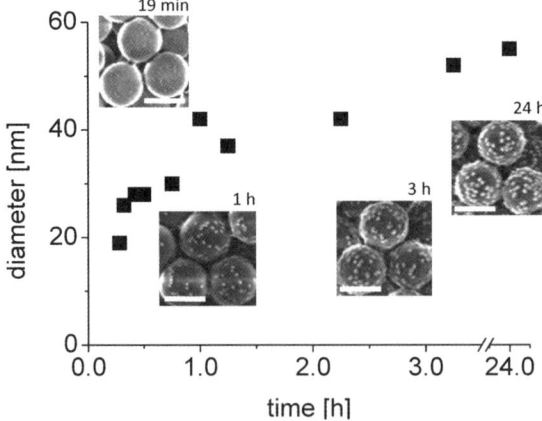

**Fig. 3** Dependence of the size of the silica colloids on reaction time. The PS particles were synthesized according to procedure E4 and the Stöber synthesis according to S2. Scale bar: 0.5μm.

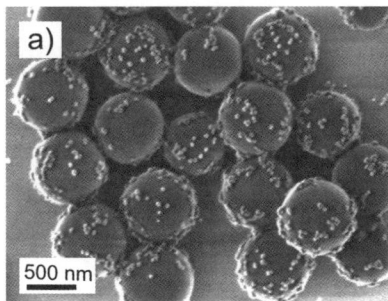
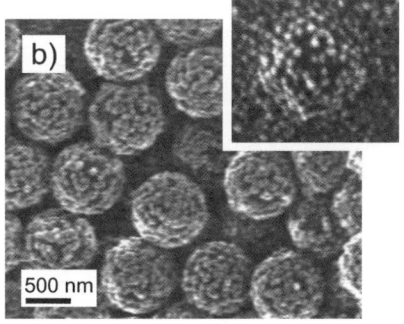

**Fig. 4** SEM images of composite particles. The PS templates contain different amounts of acrylic acid. a) Table 1, batch E3; b) Table 1, batch E4. The Stöber synthesis was performed according to procedure S2 (Table 2). Nucleation of silica particles occurs throughout the reaction mixture, indicated by abundant silica particles not attached to the PS surface in an unwashed sample (inset).

SEM images revealed that a large number of silica particles were attached to the surface of PS particles at all AA concentrations tested. However, at low AA concentrations the colloids detach during centrifugation and only a small number of silica colloids stick to the PS surface (Fig. 4a). Only at AA concentrations larger than ≈ 1% wt/wt the silica spheres are sufficiently strongly bonded to the surface that they remain attached even after several centrifugation steps (Fig. 4b). The particles formed in solution have almost the same size as those at the surface. We take this as a hint that either nucleation happens in the solution and the small particles afterwards bind to the large particles or that nucleation occurs both at the surface of the large particles and in solution at the same time (Fig. 4b, inset). Therefore, AA is responsible for binding silica particles to the PS surface.

**PVP dependence.** Why silica attaches to the PS surface is still unclear. Silica may bind directly to the PS surface due to electrostatic or hydrogen bond interaction between AA and silica. Higher AA concentrations may ensure better binding of silica by improving the PVP coating due to formation of hydrogen bonds between AA and PVP. The compatibility between PS and silica can be increased by coating PS with PVP.[29–32] To gain more insight on the influence of PVP on nucleation or attachment of silica on PS, we synthesized composite particles with (Fig. 5a) and without PVP (Fig. 5b). In both cases TEOS and ammonia were added to the reaction mixture without removing excess PVP. Whereas the diameter of silica colloids was $50 \pm 20$ nm with PVP coating, under otherwise identical conditions it increased

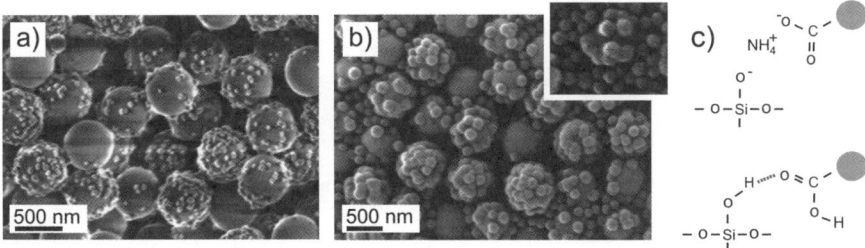

**Fig. 5** Composite particles synthesized using PS templates coated with (a: S3) or without (b: S4) PVP. The diameter of the silica colloids is $50 \pm 20$ nm (a) and $120 \pm 30$ nm (b), respectively. Inset: Silica particles also nucleate in solution, however, their size does not depend on whether they grow in solution or on the surface. Batch E5, Tab. 1. (c) Sketch: formation of hydrogen bonds between silica and the carboxylic groups of the PS particle.

up to 120 ± 30 nm if uncoated PS particles were used. To investigate whether silica particles detached we put a drop of the reaction mixture on top of an SEM substrate and let it dry. SEM images of uncleaned suspensions revealed that in the case of PVP coating the number of unattached silica colloids is much higher than without PVP coating (Fig. 5b, inset). This is consistent with conservation of the total volume of silica, which is determined by the amount of TEOS. In samples not cleaned after PVP coating, dissolved PVP induces disseminated silica nucleation in the solution. This explains the strongly increased number of silica particles. Conversely, without PVP coating, nucleation in the bulk fluid is strongly reduced and the colloids grow to a larger size. These (see Fig. 5) and further experiments varying the ammonia content show that the number of silica colloids bonded to the PS surface does not depend on previous PVP coating, *i.e.* PVP does not influence attachment of silica particles. Therefore AA is responsible for attachment of silica. This can be due to electrostatic interactions between negatively charged silica and carboxylate anions *via* ammonium cations (Fig. 5c, top) or due to the formation of hydrogen bonds between carboxylic groups (AA) and hydroxyl groups (silica) under the assumption that silica hardly dissociates in ethanol (Fig. 5c, bottom).

**Coating the composite particles with a smooth shell.** The term "raspberry particles" describes the final reaction product used to build up superhydrophobic films. These hybrid particles have a PS core and a completely closed silica shell with nanoscale roughness hardly different from that of the composite particles (Fig. 6a). The thickness of the shell can be tuned *via* the amount of TEOS. The thickness of the shell does not depend on the amount of acrylic acid used in the PS synthesis. To ensure complete silica coverage, we heated raspberry particles to 500 °C for 4 h in order to remove the PS core. The particles were still intact and their shape was unchanged (Fig. 6b). The thickness of occasionally found broken shells was estimated to be a few tens of nm (Fig. 6c). Coating the composite particles with a closed silica shell increases the rigidity of the particles. Smooth silica shells as thin as 30 nm already possess a Young modulus of ∼20 GPa.[29]

**PS leakage.** Compared to sintered silica, the shells are still porous. Hybrid raspberry particles with a shell thickness of a few tens of nm are still permeable to PS. Furthermore, some particles contain small holes or cracks, enhancing PS leakage. PS leakage can be visualized by transmission electron microscopy (TEM). For this purpose, a dilute dispersion of raspberry particles was dried on a grid (Fig. 7a). The particles appear black as they are impermeable to the electron beam. After exposing the sample to THF vapour for a few hours, PS leaks out of the particles, and PS bridges formed between neighbouring particles appear dark in TEM images (Fig. 7b). To investigate whether PS can be completely removed, aliquots of the same samples were dispersed in liquid THF and then washed several times in fresh THF. After dispersion in liquid THF, the difference in contrast

**Fig. 6** SEM images of raspberry particles before thermal treatment (a), hollow particles after annealing the particles at 500 °C for 4 h (b), and a broken empty shell (c). PS particles were synthesized *via* seeded emulsion polymerisation (E2E1) and the small silica colloids according to S1.

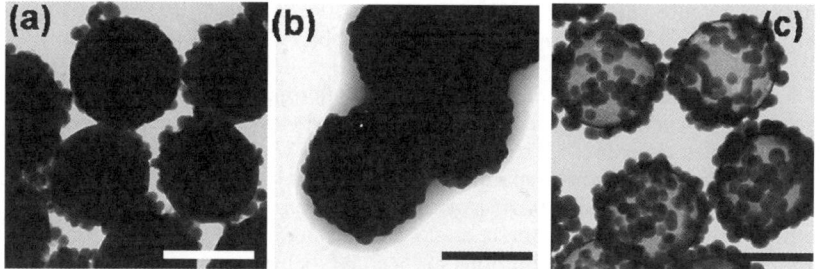

**Fig. 7** TEM image of pristine particles (a), particle after THF exposure (b), and after immersion in liquid THF (c). The time of vapor exposition required for bridging the particles by PS was typically 3 h. Scale bar: 1 μm.

compared to the original particles indicates that PS has been removed and hollow shells are obtained (Fig. 7c). The dark spots represent the silica colloids between which the thin shell can be discerned.

### 3.2 Preparation of superhydrophobic surfaces

To prepare superhydrophobic surfaces we used two different strategies: (i) dip coating or (ii) evaporation of the dispersant.

**Dip coating.** Monolayers of different degrees of surface coverage were prepared by dip coating (Fig. 8a). In these monolayers, the particles are randomly arranged, forming a homogeneous two-dimensional network. The large scale homogeneity of the surface coverage was verified by optical microscopy. The samples show a tilting angle of 10° ± 5°. Despite the low tilting angle and high reproducibility of the procedure, these surfaces suffer from poor mechanical resistance. Already after complete immersion of the surface into water for several minutes, particles detach from the substrate and the surface loses its superhydrophobicity.

To improve mechanical stability, we partially embedded the particles in a PS film. Therefore, the glass slides were dipped in a PAH solution (10 mg PAH and 16 mg sodium chloride in 10 ml of water), followed by successive dipping in a solution of uncrosslinked PS particles and water (17 g L$^{-1}$) and of raspberry particle and water (60 g L$^{-1}$). After each immersion the samples were rinsed twice intensively with fresh water. The "hybrid" multilayer obtained in this manner was heated to 170 °C for 1 h, *i.e.* well above the glass transition temperature ($T_g$) of PS. This procedure causes "molten" PS to spread on the glass surface and the raspberry particles to

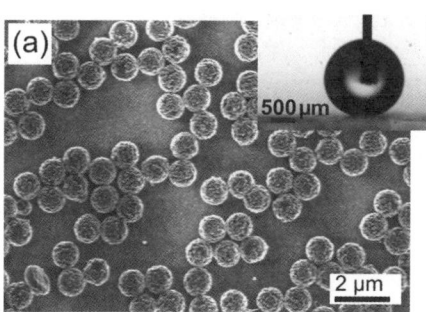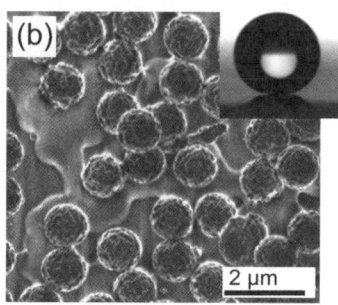

**Fig. 8** a) SEM image of a monolayer prepared by dip coating a PAH coated glass slide into a raspberry particle suspension. The surfaces show tilting angles of about 10°. b) SEM image of a monolayer of raspberry particles partially embedded in molten PS. These superhydrophobic surfaces show tilting angles below 10°. Insets: Water drops deposited on the surface.

partially sink into the polymer film (Fig. 8b). After hydrophobization with semi-fluorinated silane, the resulting immobilized raspberry particle films exhibit tilting angles below 10°.

The mechanical resistance of samples prepared in this way turned out to be much higher than in samples prepared with raspberry particles only. The raspberry particles did not detach even after repeated deposition of water drops at the same position. After one year from preparation, superhydrophobicity of such films is still unchanged. However, only 20–30% of the freshly prepared films were superhydrophobic. The others showed contact angles in the range of 140–150° and no roll-off of the water droplet upon tilting the surface. We believe that molten PS partially covers the resulting surface, inducing pinning sites and leveling out the (intended) surface roughness conferred by the particles.

**Solvent evaporation and THF leakage.** The samples prepared so far suffer from a loss of superhydrophobicity if particles detach from the surface. To circumvent this deficiency, we prepared films composed of 10 to 100 layers of raspberry particles, connected by PS bridges. To prepare thick films, a dispersion of raspberry particles was filled in a circular sample holder and the solvent was allowed to evaporate. Different evaporation temperatures were tested to evaluate the influence of evaporation times on film roughness and contact angle. The surfaces were exposed to THF vapour for 3 h, permitting PS to leak out of the pores in the silica shell. The surface topography consists of micro- and nano-asperities representing the PS and silica colloids, respectively (Fig. 9). At higher resolution (inset) PS bridges are visible between neighboring raspberry particles. These reduce detachment of particles from the film upon water contact or mechanical stress. Even if the topmost layer is removed, the layers underneath still ensure superhydrophobicity. After silanization most of these films showed tilting angles of a few degrees. However, in samples with high polystyrene leakage the contact angle hysteresis was higher than 10°.

**Plasma cleaning.** During THF exposure PS leaking out of the particles forms bridges between particles, *i.e.* PS fills some of the voids in the network and masks some asperities on the silica shells. If the amount of leaked-out PS is large, this affects the dual scale roughness and reduces the number of hydroxyl sites on the silica shells available for silane binding. Leaked-out PS accessible to water can be removed by plasma cleaning. Reactive species, *i.e.* ions and free radicals formed in the plasma induce surface crosslinking and degradation of PS. After 5 min of plasma treatment organic contaminants physically bonded to the surface are removed. Prolonging the treatment ensures that PS is indeed etched away. In

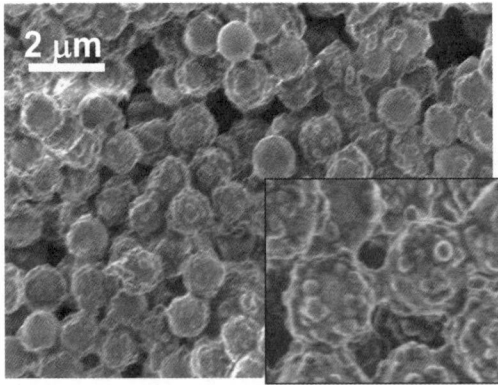

Fig. 9 Scanning electron microscopy (SEM) images of films prepared by water evaporation at 50°, THF exposure and subsequent silanization.

**Table 3** Water contact angles with their standard deviation. Contact angles were measured at positions corresponding to 3, 6, 9, and 12 of a clock and at the centre of samples

| Time of plasma cleaning/min | Contact angles before plasma cleaning/° | Contact angles after plasma cleaning/° |
|---|---|---|
| 5 | 150.5 ± 2.5 | 153 ± 2 |
| 45 | 151 ± 2.5 | 160 ± 4 |

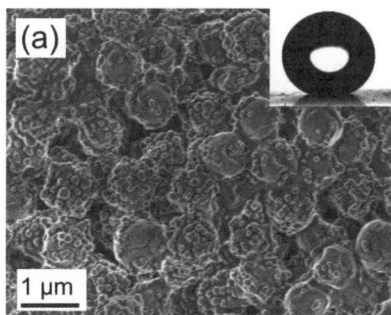
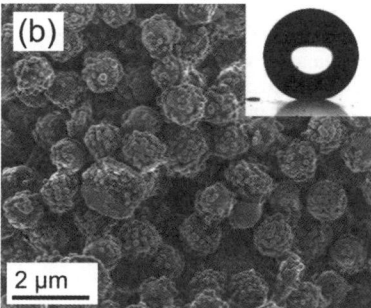

**Fig. 10** SEM images of films prepared from raspberry particles that were treated by argon plasma for 5 min (a) or 45 min (b) respectively. Before measuring the contact angle the films were silanized. Inset: Image of a water droplet deposited on the film.

silanized samples with identical initial contact angles (Table 3), 45 min of plasma cleaning significantly improved water contact angles as compared to plasma cleaning for 5 min. After longer plasma cleaning times part of the PS is removed from in between the silica colloids and between the raspberry particles (Fig. 10b). This increases the total surface area and thus the number of silane binding sites. Furthermore, after 45 mins of plasma cleaning the mechanical stability was increased due to partial crosslinking of residual PS.

**Drying procedure.** Formation of multilayers influences the roughness of the film, as given by the arrangement of raspberry particles during drying. Evaporation induced convection interacts with sedimentation of particles, causing them to sediment inhomogeneously. A three-dimensional view of the surface topography obtained with whitelight confocal microscopy (Fig. 11) shows that the surface consists of irregularly distributed "hills and valleys". To quantify the roughness we divided the image (typically 800 × 800 µm²) into small squares of 50 × 50 µm²

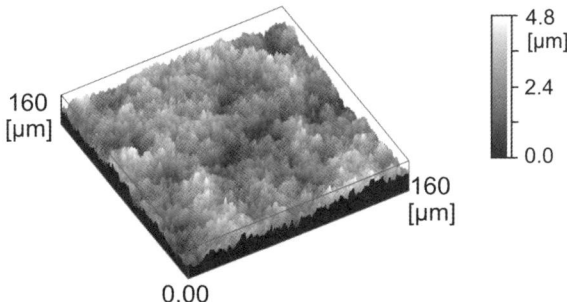

**Fig. 11** 3D profile over an area of 160 × 160 µm² of a raspberry film dried at 50 °C. Height is indicated by grayscale intensity.

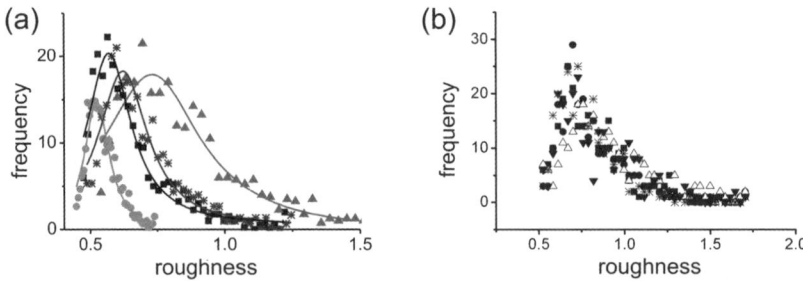

**Fig. 12** a) Dependence of the surface roughness on the drying temperature. The data show the arithmetic mean of the roughness determined by whitelight confocal microscopy and evaluated at different positions and on b) different samples. Typically, the roughness was measured at positions corresponding to 3, 6, 9, and 12 of a clock and at the centre of samples. To minimize large scale inhomogeneities or sample tilt the imaged areas of 800 × 800 μm² were divided into 256 small squares of 50 × 50 μm² each and the roughness was determined for each of these squares. The solid lines serve as guides to the eye. Evaporation temperature: circles: 20 °C, squares: 50 °C, stars: 70 °C, triangles: 90 °C. b) Reproducibility of roughness measurements of a film dried at 70 °C. The different symbols denote different samples and positions. 0.36 ml of a dispersion containing 10 g raspberry particles in 1 L of water was filled in a circular sample holder of 1.5 cm in diameter and the water was evaporated at different temperatures.

and determined the average roughness and its standard deviation. This procedure reduces the effect of inhomogeneous film thickness on the mm scale.

Evaporation time decreases and roughness increases with drying temperature, (Fig. 12). The increasing roughness is due to the increasing influence of advection of particles by the solvent. At room temperature the sedimentation time (∼2 h) is short compared to the water evaporation time (∼20 h). This reverses at high evaporation temperatures (see Table 4). After drying the sample at 90° even by eye the surface appears inhomogeneous (see Table 4). Measuring the roughness at different positions on the same sample or on different samples led to almost identical results, Fig. 12b.

As long as sedimentation is determined by gravity instead of evaporation, evaporation temperature hardly influences the tilting angle. However, in the case of high evaporation temperatures (90 °C) the tilting angle shows a strong dependency on position, varying between 4° and 40°.

**Table 4** Dependence of the tilting angle and roughness on temperature. The films were prepared by water evaporation, THF exposure and subsequent silanization. The contact angles were measured after annealing the silanized samples at 70° for 2 h to ensure complete evaporation of the remaining solvent

| $T/°C$ | Evaporation time/h | Tilting angles/° | Roughness | |
| --- | --- | --- | --- | --- |
| 20 | 20 | 6 | 0.52 ± 0.03 | |
| 50 | 5 | 10 | 0.56 ± 0.05 | |
| 70 | 2 | 9 | 0.62 ± 0.06 | |
| 90 | 0.5 | 15 | 0.73 ± 0.1 | |

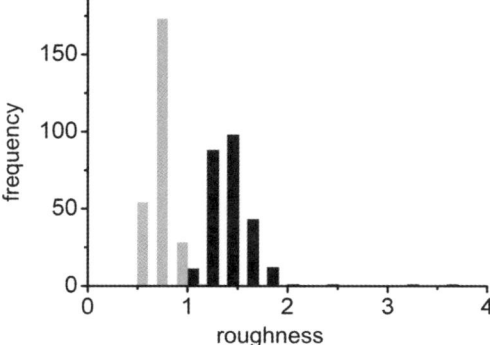

**Fig. 13** Comparison of the roughness for films prepared without AOT (grey) and with AOT (black).

The roughness is also influenced by interfacial tension between air and water. The interfacial tension can be reduced by adding a surfactant (AOT), where the concentration was taken to be well below the critical micellar concentration ($6 \times 10^{-4}$ g L$^{-1}$.). These samples also show increased roughness, Fig. 13. As the water–air interface passes the topmost particle layer during drying, the interfacial area increases. This excess interfacial energy is less if the particles are arranged in a less rough topography, *i.e.* in a better packing. With AOT, we obtained droplet slip at about a 5° tilting angle.

A 0.36 ml dispersion containing 2 g raspberry particles in 1 L of water was placed in a circular sample holder of 1.5 cm in diameter and the water was evaporated at different temperatures. AOT concentration: $6 \times 10^{-4}$ g L$^{-1}$.

## 4 Conclusions

In an approach for building superhydrophobic surfaces from hybrid raspberry-shaped particles, we studied the effect of the conditions during synthesis and film assembly on the surface roughness and on the contact angle towards water. The nano-sized roughness, conferred to the polystyrene particles by small silica spheres can be tuned by the reaction conditions. The amount of acrylic acid determines the number of silica particles attached to the polystyrene particles. A certain amount of acrylic acid is required to provide sufficient anchoring strength. The size of the colloids can be tuned by the amount of TEOS. Coating the particles with PVP prior to Stöber synthesis hardly changes the number of attached silica colloids, although coating may decrease their size. Covering these composite particles with a smooth silica shell not only enhances their mechanical stability, but also increases the number of binding sites for the fluorosilane used to hydrophobize the silica surface.

The silica surface of the raspberry particles is hard, chemically inert and can be easily modified. Since the particles disperse in water, mono- or multi-layers can be prepared on substrates by simple drop casting or water evaporation, even on surfaces with grooves and slots. This has the advantage that our suspension can easily take any form, *i.e.* it is not hindered by edges and can creep into small corners. Evaporation time can be decreased and surface roughness increased by drying well above room temperature. Within experimental accuracy both only slightly influence the tilting angle.

The mechanical stability of assembled raspberry particle films can be improved by polystyrene bridges. These offer the possibility of building thick particle layers. If the amount of leaked-out PS is large, this affects the dual scale roughness and reduces the number of hydroxyl sites on the silica shells available for silane binding. This reduces contact angles. Superhydrophobicity can be recovered by plasma cleaning and successive silanization.

Even after removing particles the surface morphology hardly changes due to its self-similarity. So far thick layers of superhydrophobic surfaces have not been reported yet. The preparation of multi-layers allows going one step further toward the design of "scratch resistant" superhydrophobic surfaces, *i.e.* surfaces that preserve the superhydrophobic behavior after scratching of the outer layer.

## Acknowledgements

We are grateful to G. Schäfer, and K. Kirchhoff for technical support. M.d.A., D.V. and G.K.A. acknowledge financial support by the DFG SPP 1273, and H.J.B. by the cluster of excellence Smart Interfaces Darmstadt.

## References

1 X. M. Li, D. Reinhoudt and M. Crego-Calama, *Chem. Soc. Rev.*, 2007, **36**, 1350–1368.
2 P. Roach, N. J. Shirtcliffe and M. I. Newton, *Soft Matter*, 2008, **4**, 224–240.
3 A. Lafuma and D. Quere, *Nat. Mater.*, 2003, **2**, 457–460.
4 D. Quere, *Rep. Prog. Phys.*, 2005, **68**, 2495–2532.
5 D. Quere, *Annu. Rev. Mater. Res.*, 2008, **38**, 71–99.
6 R. N. Wenzel, *Ind. Eng. Chem.*, 1936, **28**, 988–994.
7 A. B. D. Cassie and S. Baxter, *Trans. Faraday Soc.*, 1944, **40**, 0546–0550.
8 C. W. Extrand, *Langmuir*, 2002, **18**, 7991–7999.
9 L. C. Gao, T. J. McCarthy and X. Zhang, *Langmuir*, 2009, **25**, 14100–14104.
10 W. Barthlott and C. Neinhuis, *Planta*, 1997, **202**, 1–8.
11 D. Oner and T. J. McCarthy, *Langmuir*, 2000, **16**, 7777–7782.
12 J. Y. Shiu, C. W. Kuo, P. L. Chen and C. Y. Mou, *Chem. Mater.*, 2004, **16**, 561–564.
13 E. Martines, K. Seunarine, H. Morgan, N. Gadegaard, C. D. W. Wilkinson and M. O. Riehle, *Nano Lett.*, 2005, **5**, 2097–2103.
14 X. Zhang, F. Shi, X. Yu, H. Liu, Y. Fu, Z. Q. Wang, L. Jiang and X. Y. Li, *J. Am. Chem. Soc.*, 2004, **126**, 3064–3065.
15 F. Shi, Z. Q. Wang and X. Zhang, *Adv. Mater.*, 2005, **17**, 1005.
16 L. Huang, S. P. Lau, H. Y. Yang, E. S. P. Leong, S. F. Yu and S. Prawer, *J. Phys. Chem. B*, 2005, **109**, 7746–7748.
17 E. Hosono, S. Fujihara, I. Honma and H. S. Zhou, *J. Am. Chem. Soc.*, 2005, **127**, 13458–13459.
18 M. Morra, E. Occhiello and F. Garbassi, *Langmuir*, 1989, **5**, 872–876.
19 J. P. Youngblood and T. J. McCarthy, *Macromolecules*, 1999, **32**, 6800–6806.
20 K. Tadanaga, N. Katata and T. Minami, *J. Am. Ceram. Soc.*, 1997, **80**, 3213–3216.
21 L. Zhai, F. C. Cebeci, R. E. Cohen and M. F. Rubner, *Nano Lett.*, 2004, **4**, 1349–1353.
22 J. Bravo, L. Zhai, Z. Z. Wu, R. E. Cohen and M. F. Rubner, *Langmuir*, 2007, **23**, 7293–7298.
23 M. H. Sun, C. X. Luo, L. P. Xu, H. Ji, O. Y. Qi, D. P. Yu and Y. Chen, *Langmuir*, 2005, **21**, 8978–8981.
24 W. Ming, D. Wu, R. van Benthem and G. de With, *Nano Lett.*, 2005, **5**, 2298–2301.
25 H. J. Tsai and Y. L. Lee, *Langmuir*, 2007, **23**, 12687–12692.
26 Z. Qian, Z. C. Zhang, L. Y. Song and H. R. Liu, *J. Mater. Chem.*, 2009, **19**, 1297–1304.
27 N. A. M. Verhaegh and A. van Blaaderen, *Langmuir*, 1994, **10**, 1427–1438.
28 W. Stober, A. Fink and E. Bohn, *J. Colloid Interface Sci.*, 1968, **26**, 62.
29 L. Zhang, M. D'Acunzi, M. Kappl, G. K. Auernhammer, D. Vollmer, C. M. van Kats and A. van Blaaderen, *Langmuir*, 2009, **25**, 2711–2717.
30 J. N. Smith, J. Meadows and P. A. Williams, *Langmuir*, 1996, **12**, 3773–3778.
31 C. V. Graf, D.L.J., A. Imhof and A. van Blaaderen, *Langmuir*, 2003, **19**, 6693–6700.
32 I. W. Kellaway and N. M. Najib, *Int. J. Pharm.*, 1980, **6**, 285–294.

# Microscopic shape and contact angle measurement at a superhydrophobic surface

Helmut Rathgen[†]* and Frieder Mugele*

Received 9th December 2009, Accepted 25th January 2010
DOI: 10.1039/b925956b

We have studied the microscopic shape, contact angle and Laplace law behavior of the liquid–gas interfaces at a superhydrophobic surface. A superhydrophobic surface is immersed in water, and the radius of liquid gas menicsi that span between adjacent ridges of the surface texture is measured. The surface pattern consists of rectangular grooves, such that the sample is simultaneously an optical grating. The diffraction properties encode the shape of the menisci. The shape of the menisci is determined by measuring the intensity of several diffraction orders as a function of the incident angle, and fitting the data to numerical calculations of the diffraction. The uncertainty of the determined menisci deflections is a few nanometres. Observing the deflection as a function of externally controlled hydrostatic pressure, Laplace's law is probed for the menisci on the micrometre scale. The microscopic contact angle is determined by measuring the radius of the menisci prior to collapse. Close agreement with the macroscopic Young angle is found. A stability limit for the superhydrophobic-to-impregnated transition is given. The measurement is a microscopic analogue of 'bubble' and 'sessile drop' type methods.

The favorable properties of superhydrophobic[1–3] – or more generally: superlyophobic[4,5] – surfaces rely on the presence of liquid–gas interfaces spanning between adjacent ridges of the micropatterned surface, giving rise to low adhesion, self-cleaning[6] and drag reduction.[3] These favorable properties are lost when the superhydrophobic state collapses and the liquid invades the surface relief. The properties and the stability of the microscopic liquid menisci therefore determine the stability of the superhydrophobic state as a whole.[7–11] For generic patterns of rectangular grooves or pillars, standard models assume that the three-phase contact line is pinned along the edges of the texture.[9,12,13] When the pressure in the liquid increases, e.g. due to evaporation of the drop, or when a sufficient amount of energy is provided, e.g. by mechanical shaking, by electric fields, or by a finite impact velocity, the liquid micromenisci become unstable and the superhydrophobic state collapses. Despite considerable efforts[9,10,14] the mechanism of this transition is poorly understood. Moreover, recent experiments indicate that a quantitative characterization of the shape of the menisci is crucial for understanding the hydrodynamic slip on these composite surfaces.[3,15] Our understanding of superhydrophobic surfaces is thus limited by difficulties in quantitatively characterizing their properties on the level of the individual menisci.

*Physics of Complex Fluids, J.M. Burgers Centre of Fluid Dynamics and MESA+- and IMPACT-Institutes, University of Twente, The Netherlands. E-mail: helmut.rathgen@gmail.com; f.mugele@utwente.nl*

† Present address: 3. Physikalisches Institut, University of Stuttgart, Pfaffenwaldring 57, 70550, Stuttgart, Germany.

In this paper, we report quantitative measurements of the shape of liquid micromenisci at a model superhydrophobic surface consisting of one-dimensional hydrophobized rectangular grooves completely submerged under water. Extending our previously proposed optical diffraction technique[16] to variable angles of incidence, and evaluating a solution to the inverse diffraction problem through forward calculation, we measure the absolute deflection of the micromenisci with nanometre resolution as a function of an externally applied hydrostatic pressure. The menisci reversibly bend upward and downward obeying Laplace's law, up to a critical pressure, at which the menisci translate into the grooves, and the superhydrophobic state irreversibly collapses. The angle of the menisci with respect to the vertical faces of the grooves prior to collapse (which is here found in close agreement with Young's macroscopic angle) is the microscopic contact angle.

Experiments were performed using one-dimensional rectangular gratings with periodicity $T$ and groove width $w$ ranging from 5 to 12 μm and from 2 to 9 μm, respectively, with a depth of 6 μm, as shown in Fig. 1. All data presented in this work, (except Fig. 4c), correspond to $w = 5$ μm. The samples were fabricated by deep reactive ion etching (DRIE) into a silicon wafer and subsequently hydrophobized with a monolayer of $1H,1H,2H,2H$-perfluorodecyltrichlorosilane, leading to advancing and receding contact angles on the unpatterned surface adjacent to the grooves of 120° and 105°, respectively. The sample is mounted on the axis of a closed cylindrical glass container which is filled entirely with demineralized water. The hydrostatic pressure above the sample can be varied between −10 kPa and +25 kPa by adjusting the liquid level in a flexible tube that is connected to the container. The sample is illuminated with s-polarized light (wavelength $\lambda = 488$ nm) from an Ar-ion laser, and the diffracted light is detected with a photodiode. Multiple diffraction orders $m$ are observed at angles $\vartheta_m$ determined by the grating equation[17] $\sin\vartheta_m = \sin\vartheta + \frac{m\lambda}{nT}$, where $\vartheta$ is the incident angle (with respect to the surface normal) and $n$ is the refractive index of water. The photodiode and the glass container (including the sample) are mounted on individually controllable rotation stages, such that the intensity of individual diffraction orders can be recorded as a function of the incident angle.

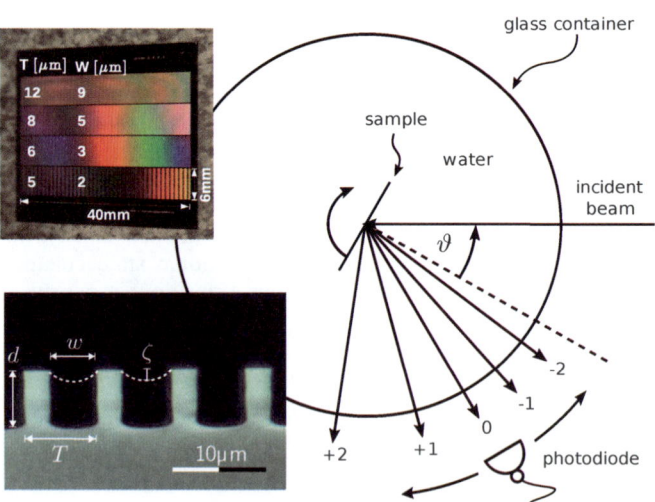

**Fig. 1** Diffraction setup with sample immersed in a water-filled glass container. Top inset: optical view of samples in air indicating periodicity $T$ and groove width $w$. Bottom inset: optical microscopy picture of a cleaved sample showing cross-sectional view of grooves. Dotted lines schematically indicate positions of the water menisci in the experiment.

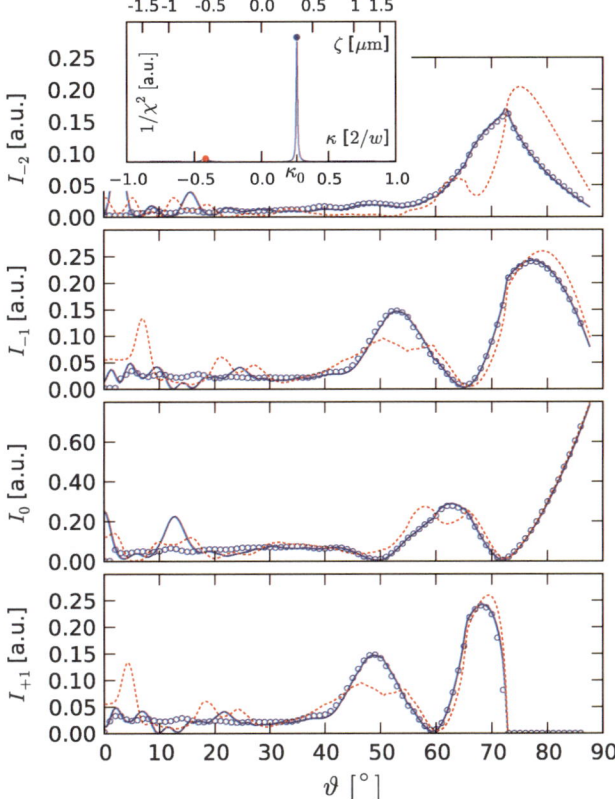

**Fig. 2** Diffracted intensity (normalized by incident intensity) vs. incident angle for $-2^{nd}$ to $+1^{st}$ diffraction order (from top to bottom). Open symbols: experimental data. Blue solid lines: best fit from optical model. Red dashed lines: diffraction curves for secondary maximum (red circle) in inset. Inset: inverse mean square deviation $1/\chi^2$ vs. model parameter $\zeta$ and $\kappa$, with Lorentzian peak at position $\zeta_0 = 338$ nm and FWHM $\Delta\zeta_0 = 7$ nm.

Fig. 2 shows a typical set of experimental data from the $-2^{nd}$ to the $+1^{st}$ diffraction order. The curves display two regimes. For small $\vartheta$ the diffracted intensities are small and almost constant. For incident angles beyond approximately the critical angle of total internal reflection at the water–air interface $\vartheta_T = 48.6°$, pronounced minima and maxima appear. This behavior is governed by the interference of elementary waves scattered from the Si surface and from the menisci, as shown schematically in Fig. 3, and therefore contains detailed information about the shape of

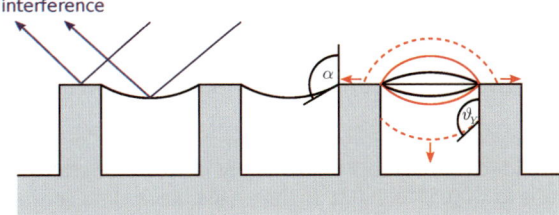

**Fig. 3** Schematic illustration of the main scattering mechanism (left) and the accessible range of meniscus configurations at variable pressure (right). See text for details.

the latter. (For $\vartheta < \vartheta_T$, little light is reflected from the water–air interfaces, and hence the interference pattern disappears.)

To obtain the shape of the menisci, we invert the diffraction data. The analysis proceeds as follows. We solve Maxwell's equations numerically by calculating the diffraction pattern with the so-called Rigorous Coupled Wave Analysis[18,19] (RCWA) for a set of trial surface configurations, namely circular arcs pinned at the ridges of the grooves with variable radius $r$, and a corresponding curvature $\kappa = 1/r$ (The geometric relation $\zeta = 1/\kappa \left(1 - \sqrt{1 - (w\kappa)^2/4}\right)$ relates $r$ and $\kappa$ to the deflection $\zeta$ of the apex of the menisci). The optimum shape is then found by minimizing the mean square deviation $\chi^2$ with the experimental data within this set. The solid lines in Fig. 2 and the sharpness of the Lorentzian peak of $1/\chi^2$ in the inset illustrate the accuracy of the analysis. For this specific data set, we obtain a meniscus deflection $\zeta = 338 \pm 7$ nm, corresponding to a curvature $\kappa = 0.265 \pm 0.0055$ $(w/2)^{-1}$. To demonstrate the uniqueness of the procedure, we also show for comparison the red dashed lines, which correspond to a secondary local maximum of $1/\chi^2$ for which the menisci are bent upward.

Following Laplace's law, this finite curvature implies a pressure jump $\Delta P_L = \sigma\kappa$ across the liquid–gas interface. Using the surface tension of water $\sigma = 73$ mN m$^{-1}$, we find a value of $\Delta P_L = 7446$ Pa. In mechanical equilibrium, this pressure jump is balanced by the externally applied hydrostatic pressure $P_h$. For the data in Fig. 2, the latter amounts to 7500 Pa, indeed in excellent agreement with the value $\Delta P_L$ determined from the curvature.

To analyze the behavior of the micromenisci in more detail, we increase the applied hydrostatic pressure in a step-wise fashion. For each pressure, we record the diffraction pattern as a function of the incident angle and determine the curvature of the menisci as described above. For a variation of $P_h$ from zero to +15 kPa, to −10 kPa, and back to zero, the peaks in the diffraction patterns shift in a reversible fashion (see Fig. 4a), corresponding to a reversible downward and upward bending of the menisci with positive and negative values of the curvature, respectively. Upon converting the measured curvatures into corresponding values of the Laplace Pressure, we find $\Delta P_L$ is in excellent agreement with $P_h$ over the entire range of pressures, as shown in Fig. 4b.

As we increase the hydrostatic pressure further, the distinct maxima and minima in the diffraction curves disappear irreversibly at a certain threshold pressure $P_c$ (red data in Fig. 4a). The loss of these features is accompanied by a change of the visual appearance of the sample from shiny and silvery to dark, as typically observed when the superhydrophobic state collapses to the impregnated one. The beautiful silvery appearance of the superhydrophobic state is due to total internal reflection at the liquid–gas interfaces, which is absent in the impregnated state. The transition to the impregnated state is also confirmed by the numerical data analysis. The full solid line in Fig. 4a calculated for completely filled grooves agrees with the measured data.

The collapse of the superhydrophobic state does not occur abruptly. In contrast to drops on superhydrophobic surfaces in ambient air, the gas in the present 'underwater' configuration is entrapped. When the superhydrophobic state collapses, it can only dissolve in the ambient liquid, which requires a finite time – typically a few minutes in the present experiments. All diffraction curves recorded after the completion of the transition are identical, irrespective of any further variation of $P_h$ (including negative pressure down to −10 kPa), confirming the irreversibility of the collapse.

To infer the critical microscopic condition for the collapse, we consider the configuration of the liquid–gas interfaces immediately prior to the transition. The highest stable curvature is $\kappa_c = 0.57 \pm 0.021$ $(w/2)^{-1}$. (see black data set in Fig. 4a.) Using elementary geometric relations, we find that this value corresponds to a critical angle $\alpha_c = 125 \pm 2°$ of the liquid–gas interfaces with respect to the vertical faces of the

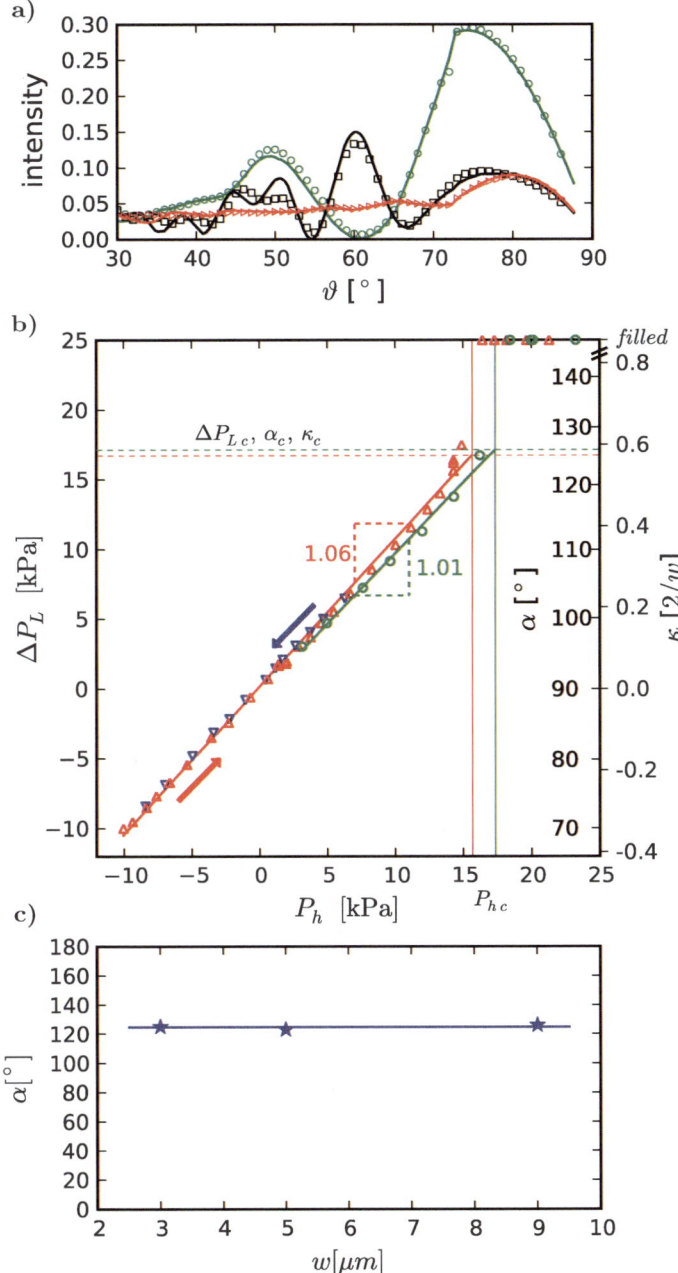

**Fig. 4** a) Diffraction curves (– 1$^{st}$ order) for several important meniscus configurations. Symbols: experimental data. Lines: best fit from optical model. Circles: moderate downward bending, $P_h$ = 6278 Pa. Squares: strong downward bending, $P_h$ = 14862 Pa, immediately before exceeding $P_c$. Triangles: filled grooves, after exceeding $P_c$, any pressure. b) Laplace pressure (left axis) and curvature (right axis) determined vs. applied hydrostatic pressure. $\alpha$ denotes the angle between the liquid–gas interface and the vertical faces of the surface (as defined in Fig. 3). The top boundary marks the filled state. Red and blue triangles: air-saturated water under ambient conditions, brown circles: degassed water at the coexistence pressure. c) Microscopic contact angle at which the superhydrophobic state collapses vs. groove width. Error bars are of the size of the symbols.

grooves. The value is in close agreement with the macroscopic advancing contact angle $\vartheta_a = 120°$ on the bare surface. The observed angle is independent of the groove width, as shown in Fig. 4c). Hence, our experiments represent a direct contact angle measurement, in analogy to the sessile drop or bubble method, on a length scale of a micrometre, and with a few nm absolute geometric resolution. The results are strong experimental evidence that (down to the micrometre scale) the *local* contact angle is identical to Young's macroscopic angle. Note that the experimentally determined local contact angle is slightly larger than Young's angle obtained from a macroscopic measurement. We tentatively attribute this to the inclination of the vertical faces of the grooves by a few degrees (due to underetching), as can also be seen in the cross section through the sample in Fig. 1c).

These results emphasize that the microscopic properties of the present generic superhydrophobic surface including the superhydrophobic-to-impregnated transition are governed by macroscopic principles of capillary theory. (1) Laplace's law describing the pressure drop across the liquid–vapor interface and (2) Young's law for the angle at the three-phase contact line. For the present case, the latter states that the liquid–gas interface can adopt any angle $\alpha$ between Young's angle with respect to the horizontal top surface and Young's angle with respect to the vertical face of the groove[25] (see Fig. 3). At zero pressure, the interface is flat and hence Young's condition is fulfilled. Upon increasing the pressure, the interface bends downwards and thereby $\alpha$ increases until it reaches Young's angle (or more precisely: the advancing contact angle $\vartheta_a$). From that moment on, the contact line can freely translate along the wall of the groove. The critical pressure is thus $P_c = -2\sigma \cos(\alpha_c)/w$. Since $P_c$ is the same for all menisci, the superhydrophobic state of the entire sample collapses at once. Note that this scenario is different from recent transition models for evaporating drops. In that case, different micromenisci depinning for different drop sizes (*i.e.* pressures) and the liquid is sometimes found to penetrate the topographic structure only partially.[9,10,13] Such finite size effects are absent in the present ideal system of a submerged superhydrophobic sample. Note that our experiments are direct experimental proof, that in contrast to previous opinion,[2,13,14,20] the superhydrophobic-to-impregnated transition is not governed by a comparison of the absolute energies of the two states. In the same spirit as the contact line depinning at high pressure, one also expects a depinning transition upon reducing the pressure. At a high enough negative pressure, $\alpha$ is expected to reach Young's angle with respect to the top surface and the contact line should move outwards (see Fig. 3). Experimentally, we do indeed observe this transition. It manifests itself by the spontaneous growth of macrosocopic bubbles on the sample surface, yet unlike the positive critical pressure for filling, the corresponding negative critical pressure is not well controlled and varies substantially from experiment to experiment. We tentatively attribute this to imperfections of the sample surface. Notwithstanding these uncertainties, with respect to the critical pressure $P_c$ for filling our experiments thus demonstrate that the stability of superhydrophobic surfaces at the level of individual cavities is governed by the same principles that govern pinning and depinning of liquid surfaces on macroscopic scales.[21]

The observations presented so far, are in agreement with the macroscopic laws of capillarity. This is remarkable in the sense that the same laws generally fail to correctly describe other properties of hydrophobic surfaces immersed into water, such as long-range hydrophobic forces and surface nanobubbles, and other phenomena, which have been observed on length scales up to about 1 μm.[22] The accurate validation of Laplace's law and Young's criterion demonstrate that such peculiarities are absent on the scale of a few micrometres.

Another important peculiarity of hydrophobic surfaces in water, is the importance of the presence or absence of dissolved gases, which controls amongst others the presence or absence of surface nanobubbles.[22] To investigate this aspect, we repeat our experiments using a closed container filled with degassed water as shown in Fig. 5. First we find that it is indeed possible to prepare such a sample in

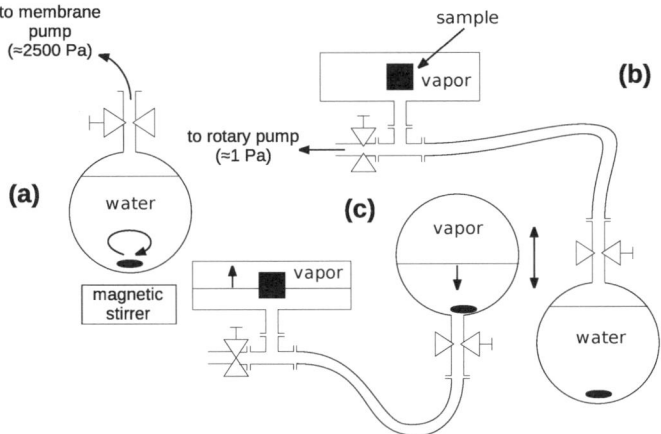

**Fig. 5** Setup and preparation procedure for measurements near liquid–vapor coexistence pressure (see text for details).

a superhydrophobic state with water vapor filling the cavities instead of air. The preparation procedure is as follows (see ref. 23 for details). a) Water is degassed overnight in a glass balloon using a magnetic stirrer and membrane pump. b) The glass balloon is connected to the measurement chamber through a stainless steel vacuum below and evacuated with a rotary pump. c) The glass balloon is lifted to a level higher than the measurement chamber, such that water flows from the balloon to the measurement chamber, whereby an amount of water in the glass balloon vaporizes and the vapor in the chamber condenses. The grooves of the sample contain vapor, while the liquid level rises above them.

We expect the pressure inside the cavities to be equal to the saturated vapor pressure $P_v = 2300$ Pa of water at room temperature. Upon varying the hydrostatic pressure by adjusting the level of the glass balloon, the menisci bend upward and downward exactly in the same manner as in the case of gas filled cavities (green symbols in Fig. 4b). Within the error bar, the critical pressure to induce filling is also the same as under ambient conditions. The transition remains equally irreversible as under ambient conditions. Even if the pressure of the water in the filled grooves is brought to a pressure very close to the pressure of the vapor ($P_w = P_v + P_h$, and $P_h$ is reduced to about 1 mm above the sample), the water in the grooves does not vaporize spontaneously to the superhydrophobic state. Note a difference between the air and the vapor case. For vapor the superhydrophobic-to-impregnated transition occurs faster – presumably because the water vapor condenses, rather than having to dissolve in the water.

Optical diffraction is thus a powerful tool for high resolution characterization of the superhydrophobic state. In particular, the technique also lends itself for incorporation into microfluidic channels in combination with particle image velocimetry measurements of hydrodynamic slip. In addition to such applications of superhydrophobicity, the present data also imply the feasibility of highly efficient tunable optical diffraction gratings.[24]

## Acknowledgements

We thank D. Lohse for access to his computing facilities, J. G. Bomer and C. Pirat for assistance with the sample preparation. This work was supported by the joint Micro- and Nanofluidics program of the Impact and MESA+ institutes at Twente University.

# References

1 D. Quéré, *Rep. Prog. Phys.*, 2005, **68**, 2495.
2 A. Lafuma and D. Quere, *Nat. Mater.*, 2003, **2**, 457.
3 A. Steinberger, C. Cottin-Bizonne, P. Kleimann and E. Charlaix, *Nat. Mater.*, 2007, **6**, 665.
4 A. Tuteja, W. Choi, M. Ma, J. M. Mabry, S. A. Mazzella, G. C. Rutledge, G. H. McKinley and R. E. Cohen, *Science*, 2007, **318**, 1618.
5 A. Ahuja, J. A. Taylor, V. Lifton, A. A. Sidorenko, T. R. Salamon, E. J. Lobaton, P. Kolodner and T. N. Krupenkin, *Langmuir*, 2008, **24**, 9.
6 A. Nakajima, K. Hashimoto, T. Watanabe, K. Takai, G. Yamauchi and A. Fujishima, *Langmuir*, 2000, **16**, 7044.
7 N. Patankar, *Langmuir*, 2003, **19**, 1249.
8 M. Reyssat, J. Yeomans and D. Quéré, *Europhys. Lett.*, 2008, **81**, 26006.
9 H. Kusumaatmaja, M. Blow, A. Dupuis and J. Yeomans, *Europhys. Lett.*, 2008, **81**, 36003.
10 S. Moulinet and D. Bartolo, *Eur. Phys. J. E*, 2007, **24**, 251.
11 M. Sbragaglia, A. Peters, C. Pirat, B. Borkent, R. Lammertink, M. Wessling and D. Lohse, *Phys. Rev. Lett.*, 2007, **99**, 156001.
12 J. Bico, C. Tordeux and D. Quéré, *Europhys. Lett.*, 2001, **55**, 214.
13 A. Dupuis and J. Yeomans, *Langmuir*, 2005, **21**, 2624.
14 N. Patankar, *Langmuir*, 2004, **20**, 7097.
15 A. Steinberger, C. Cottin-Bizonne, P. Kleimann and E. Charlaix, *Phys. Rev. Lett.*, 2008, **100**, 134501.
16 H. Rathgen, K. Sugiyama, C.-D. Ohl, D. Lohse and F. Mugele, *Phys. Rev. Lett.*, 2007, **99**, 214501.
17 M. Born and E. Wolf, *Principles of Optics* (Cambridge University Press, 1999), 7th ed.
18 M. G. Moharam, D. A. Pommet, E. B. Grann and T. K. Gaylord, *J. Opt. Soc. Am. A*, 1995, **12**, 1068.
19 H. Rathgen and H. Offerhaus, Optics Express 17, 4268 (2009), http://mrcwa.sourceforge.net/.
20 N. J. Shirtcliffe, G. McHale, M. I. Newton and C. C. Perry, *Langmuir*, 2005, **21**, 937.
21 J. Oliver, C. Huh and S. Mason, *J. Colloid Interface Sci.*, 1977, **59**, 568.
22 E. Meyer, K. Rosenberg and J. Israelachvili, *Proc. Natl. Acad. Sci. U. S. A.*, 2006, **103**, 15739.
23 H. Rathgen, Ph.D. thesis, University of Twente, Enschede (2008), http://doc.utwente.nl/60282/.
24 H. Rathgen, K. Sugiyama and F. Mugele, Frontiers in Optics, Optical Society of America, 2007, PDP_A1; http://www.opticsinfobase.org/abstract.cfm?URI=FiO-20070PDP_A1.
25 This formulation of Young's law is equivalent to stating that the angle at the three-phase contact line is always Young's angle and the radius of curvature of the edge is small compared to all other geometric dimensions.

# Transparent superhydrophobic and highly oleophobic coatings

Liangliang Cao and Di Gao*

*Received 18th February 2010, Accepted 22nd February 2010*
DOI: 10.1039/c003392h

We report a facile process for fabrication of transparent superhydrophobic and highly oleophobic surfaces through assembly of silica nanoparticles and sacrificial polystyrene nanoparticles. The silica and polystyrene nanoparticles are first deposited by a layer-by-layer assembly technique. The polystyrene nanoparticles are then removed by calcination, which leaves a porous network of silica nanoparticles. The cavities created by the sacrificial polystyrene particles form overhang structures on the surfaces. Modified with a fluorocarbon molecule, such surfaces are superhydrophobic and transparent. They also repel liquids with low surface tensions, such as hexadecane, due to the overhang structures that prevent liquids from getting into the air pockets even though the intrinsic contact angles of these liquids are less than 90°.

## Introduction

Superhydrophobic surfaces are often found in nature, such as on plant leaves,[1,2] water strider legs,[3] and butterfly wings.[4,5] Liquid water on these surfaces form beads with a contact angle of greater than 150° and drip off rapidly if the surface is slightly tilted. The extraordinary water-repellency of natural superhydrophobic surfaces has stimulated extensive research interest in understanding the fundamental mechanisms underlying the behavior of liquid on such surfaces[6–8] as well as in fabricating artificial superhydrophobic surfaces for diverse practical applications.[9–18] It is generally recognized that the superhydrophobicity, either natural or artificial, is a result of the interplay between the surface chemical composition and the surface texture with a two-tier roughness in micrometre and nanometre scales, respectively, for each tier. Because the surface textures with the two-tier roughness may induce significant light scattering, fabrication of transparent superhydrophobic surfaces has been a difficult task. By optimizing and manipulating surface topographies through delicate approaches, *e.g.* by layer-by-layer assembly,[19,20] using eutectic liquids,[21] and depositing silicone nanofilaments,[22] several groups have successfully fabricated transparent superhydrophobic surfaces, which are promising for many applications where the substrate material needs to be rendered superhydrophobic without significantly changing its optical properties. However, these transparent superhydrophobic surfaces are vulnerable to oil contamination—they repel water but not oil, and contamination of liquids with a low surface tension such as vegetable oil can easily remove their superhydrophobicity. Therefore, to fully explore the potential applications of transparent superhydrophobic surfaces, it is desirable to make them repel not only water but also oil and liquids with a low surface tension.

*Department of Chemical and Petroleum Engineering, University of Pittsburgh, Pittsburgh, Pennsylvania, 15261, USA. E-mail: gaod@pitt.edu; Fax: +(412) 624-9763; Tel: +(412) 624-8488*

Fabricating superhydrophobic surfaces that are able to repel oil in the same manner as they repel water has been challenging, because this typically requires the solid surface to possess a sufficiently low surface free energy so that its intrinsic oil contact angle is greater than 90°, which in practice has been exceptionally difficult if possible at all. Recently, researchers have shown that by fabricating "overhang" or "re-entrant" structures on the surface, the requirement on the extremely low surface free energy may be relaxed, and surfaces that are both super water-repellent and super oil-repellent have been demonstrated.[16,18] However, these coatings are not transparent, and we are unaware of published work to make transparent coatings with super repellencies to both water and oil. In this paper, we report a facile process for fabrication of transparent superhydrophobic and highly oleophobic surfaces by first depositing silica nanoparticles and sacrificial polystyrene nanoparticles through a layer-by-layer assembly technique and then remove the sacrificial polystyrene particles by calcination.

## Experimental

### Materials

Poly(sodium 4-styrenesulfonate) (PSS, $M_w = 70,000$), poly(diallyldimethylammonium chloride) (PDDA, 20 w.t.%), TM-40 colloidal silica (40 w.t.% suspension in water, ca. 20 nm diameter silica particles), chloroform (anhydrous), hexadecane (anhydrous) and isooctane were obtained from Sigma-Aldrich. Negatively charged polystyrene nanoparticles (8 w.t.% solid, 60 nm in diameter) were obtained from Interfacial Dynamics Corp. Isopropanol was obtained from Fisher Scientific. (Tridecafluoro-1,1,2,2-tetrahydrooctyl) trichlorosilane ($n$-$C_6F_{13}CH_2CH_2SiCl_3$, abbreviated as FTS, >95%) was purchased from Gelest Inc. De-ionized (DI) water was prepared using the Milli-Q (Millipore) system.

### Substrate preparation

Before the coating process, the substrates were cleaned by immersion in piranha solution (3 : 1 mixture of 98 w.t.% $H_2SO_4$ and 30 w.t.% $H_2O_2$; piranha solutions may result in explosion or skin burns if not handled with extreme caution) for 15 min, thoroughly rinsed with DI water, and dried with a nitrogen flow. This process rendered the freshly cleaned substrates negatively charged in solution.

### Layer-by-layer assembly of nanoparticles

During the assembly, the substrates were repetitively immersed into cationic and anionic aqueous solutions. In each cycle, the substrates were first immersed into cationic solutions, followed by rinsing with DI water and drying with a nitrogen flow. Then, the substrates were immersed into anionic solutions, followed by the same rinsing and drying steps. One bilayer was formed upon completion of each cycle. The assembly process started with deposition of five bilayers of PDDA and PSS, [PDDA/PSS]$_5$, before deposition of nanoparticles. A few bilayers of polymers, such as [PDDA/PSS]$_5$, have been shown to promote the adhesion of the film to the substrate during the subsequent layer-by-layer assembly of nanoparticles.[19,20] For deposition of [PDDA/PSS]$_5$, PDDA was used to make the cationic aqueous solution (2 mg mL$^{-1}$, pH 4), and PSS was used to make the anionic aqueous solution (2 mg mL$^{-1}$, pH 4). After five bilayers of PDDA and PSS were deposited, deposition of body layers started by switching the anionic solution from the PSS solution to an aqueous solution (pH 9.0) containing two kinds of nanoparticles with different sizes (60 nm polystyrene + 20 nm $SiO_2$, ca. 0.04 wt% each), while the PDDA aqueous solution was still used as the cationic solution. To investigate the effect of the number of the body bilayers on the wettability, samples with 5, 10, 15, 20 and 25 bilayers of nanoparticles, [PDDA/(60 nm polystyrene + 20 nm $SiO_2$)]$_x$ (where $x$

denotes the number of body bilayers), respectively, were prepared. For each coating of [PDDA/(60 nm polystyrene + 20 nm $SiO_2$)]$_x$, two sets of samples were prepared and compared. One set was used without removing the polystyrene particles while the other set was calcined as described below to remove the polystyrene particles before it was functionalized with FTS.

### Calcination of the coatings

One set of the [PDDA/(60 nm polystyrene + 20 nm $SiO_2$)]$_x$ samples was calcined at 550 °C for 4 h, during which both the polystyrene particles and polyelectrolytes were removed from the coatings and a nanoporous network of silica nanoparticles was formed.

### Surface functionalization of the coatings

Both sets of the [PDDA/(60 nm polystyrene + 20 nm $SiO_2$)]$_x$ samples were finally functionalized with FTS. The samples, either directly after layer-by-layer assembly or after calcination, were treated with ultraviolet ozone (UVO) (Jelight Inc.) for 5 min. Then, they were immersed into a coating solution, made by dissolving 0.5 mM FTS into a 4 : 1 (v/v) mixture of hexadecane and chloroform. After 15 min, the samples were thoroughly rinsed by isooctane, isopropanol, and DI water, sequentially.

### Characterization

The contact angle was measured using a VCA-OPTIMA drop shape analysis system (AST Products, Inc.) with a computer-controlled liquid dispensing system. Droplets of liquids in a volume of 5 μl were used to measure the static contact angle. The advancing and receding angles were recorded during expansion and contraction of the droplets induced by placing a needle in the liquid droplets and continuously dispensing and withdrawing liquid through the needle. All of the tests were performed under normal laboratory ambient conditions (20 °C and 40% relative humidity). Each contact angle measurement was repeated three times at different places of the sample, and the average value was reported. Scanning electron microscopy (SEM) images were taken by a Philips XL-30 field emission SEM setup. Transmittance was measured by a UV-Vis-NIR microspectrophotometer (Craic QDI 2010).

## Results and discussion

The process for preparing the coatings is schematically shown in Fig. 1 and detailed in the experimental section. It mainly consists of two steps: (i) layer-by-layer assembly of silica nanoparticles and sacrificial polystyrene nanoparticles and (ii) removal of sacrificial polystyrene nanoparticles by calcination. The layer-by-layer assembly process employed here is similar to previously published methods,[19,20] which are based on deposition of positively and negatively charged particles, polymers, or particle-polymer complexes alternately in sequential cycles. In our process, five bilayers of PDDA and PSS, [PDDA/PSS]$_5$, were first deposited as adhesion layers. Then, samples with 5, 10, 15, 20 and 25, bilayers of nanoparticles, [PDDA/(60 nm polystyrene + 20 nm $SiO_2$)]$_x$ (where $x$ denotes the number of body bilayers), respectively, deposited on top of the adhesion layers were prepared. By virtue of electrostatic interactions involved in this process, organic–inorganic nanocomposite coatings consisting of different numbers of body layers were readily achieved. As expected, the thickness and the roughness of the coatings monotonically increased in a well controlled manner with increasing numbers of bilayers. This feature of the layer-by-layer assembly technique provides us with the capability to gradually tune the surface roughness and study its effect on the wettability and transparency

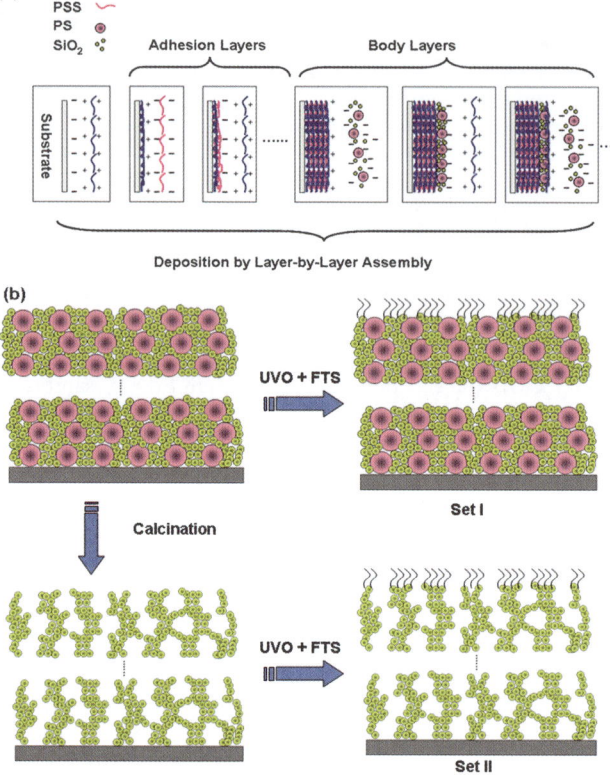

**Fig. 1** Schematic process for preparing the coatings. (a) Layer-by-layer assembly of 20 nm silica (SiO$_2$) nanoparticles and 60 nm sacrificial polystyrene (PS) nanoparticles. Adhesion layers are deposited before the deposition of body layers. (b) The assembled nanocomposite coating is treated by UVO and functionalized by FTS either directly after the assembly or after removal of the sacrificial PS particles through calcination.

of the coatings, and is particularly advantageous when both superhydrophobicity and transparency are desired.

For each coating of [PDDA/(60 nm polystyrene + 20 nm SiO$_2$)]$_x$, two sets of samples were prepared and compared. One set was treated by UV ozone and functionalized by FTS without removing the sacrificial polystyrene particles, while the other set was calcined to remove the sacrificial polystyrene particles before it was functionalized with FTS. Fig. 2 shows the SEM images of the coatings with 25 bilayers before and after removing the polystyrene particles. As shown in Fig. 2a, after the layer-by-layer assembly process, the smaller-sized silica particles (*ca.* 20 nm) crowd around the bigger-sized polystyrene particles (*ca.* 60 nm). After the calcination process, the bigger sacrificial polystyrene particles are removed (Fig. 2b) while leaving behind cavities with a similar size and shape, resulting in a porous network of silica nanoparticles.

After functionalizing both sets of samples with FTS, the hydrophobicity of these coatings is examined. Fig. 3a and 3b present the advancing and receding angles of water droplets on these two sets of coatings as a function of the number of the bilayers. On the first set of coatings which are prepared without removing the sacrificial polystyrene particles, the advancing angles are all in excess of 150°. However, large contact angle hysteresis (the difference between the advancing and the receding angles) of more than 90° is observed on all these samples and it increases with

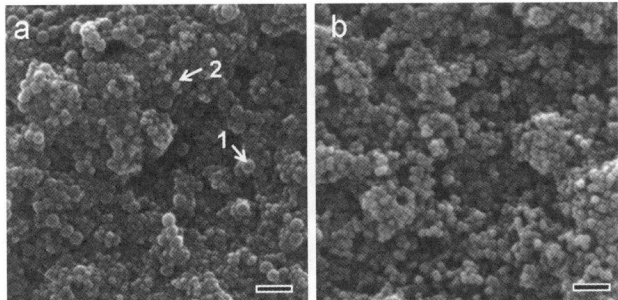

**Fig. 2** SEM images of the coatings. (a) and (b) are SEM images of the coatings with 25 bilayers before and after removing the sacrificial polystyrene particles, respectively. Labeled in (a) are a representative 60 nm polystyrene particle (labeled by "1") and a representative 20 nm silica particle (labeled by "2"). The polystyrene particles disappear in (b) after calcination. The scale bars are 200 nm.

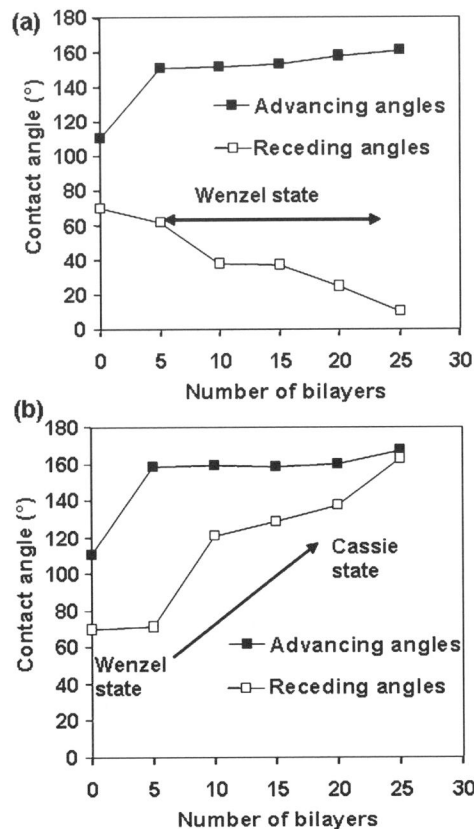

**Fig. 3** Advancing and receding contact angles as a function of the number of the bilayers on (a) the first set of coatings that are prepared without removing the sacrificial polystyrene particles and (b) the second set of coatings that are prepared after removing the sacrificial polystyrene particles.

increasing the number of the bilayers. The combination of large advancing angles and large hysteresis indicates that water droplets on these surfaces are most likely in a Wenzel state. In Wenzel state, water is in complete contact with the rough solid surface, and the apparent contact angle on the rough surface ($\theta_{rough}$) can be correlated to the intrinsic contact angle of the solid surface ($\theta_{flat}$) by the following equation:[23]

$$\cos\theta_{rough} = r\cos\theta_{flat} \qquad (1)$$

where $r$ is the roughness factor defined as the ratio of the actual surface area to the projection area. Since $\theta_{flat}$ of the FTS-grafted surface is $\sim 110°$, according to eqn (1), $\theta_{rough}$ will increase as $r$ increases. Therefore, as the number of the bilayers increases and thus $r$ increases, the advancing contact angle of the coatings increases. On the other hand, water droplets in Wenzel state on a rough surface are known to exhibit a large hysteresis, which is also consistent with our result.

During the wettability test on the second set of the coatings which are essentially porous networks of silica nanoparticles after removing the polystyrene particles, a transition in the wetting regime is clearly observed as the number of bilayers increases. As seen in Fig. 3b, the advancing contact angles on this set of samples are all greater than 150°, but the hysteresis decreases significantly from around 87° for 5 bilayers to about 4° for 25 bilayers. When the number of bilayers is small, both the advancing contact angle and the hysteresis are large, which is similar to what has been observed on the first set of samples. Therefore, the system is in the Wenzel wetting regime. However, as the number of bilayers increases to 25, both a large advancing contact angle and small hysteresis are observed, which indicates that the system is in the Cassie wetting regime. When the number of bilayers is between 5 and 25, water droplets on the coatings gradually transit from the Wenzel state to the Cassie state.

In the Cassie state, liquid is in contact with a composite surface of solid and air, and forms droplets (known as fakir droplets). The apparent contact angle, $\theta_{rough}$, in this case, has been correlated to $\theta_{flat}$ by the Cassie–Baxter equation:[6]

$$\cos\theta_{rough} = \phi_S \cos\theta_{flat} - (1 - \phi_S) \qquad (2)$$

where $\phi_S$ is the fraction of solid surface area contacting the liquid. By setting eqn (1) and (2) equal to each other, one can solve $\theta_{flat}$, which corresponds to a critical angle, $\theta_c$, determined by eqn (3):

$$\cos\theta_c = -\frac{1 - \phi_S}{r - \phi_S} \qquad (3)$$

Because $\phi_S < 1$ and $r > 1$, $\theta_c$ is greater than 90°. When $\theta_c < \theta_{flat}$, the Cassie state is thermodynamically more favorable; otherwise, the Wenzel state is preferred. We believe that a decrease in $\theta_c$ is responsible for the transition of the system from Wenzel state to Cassie state as the number of bilayers increases. Because the roughness of the coating and thus $r$ increases with increasing the number of bilayers, $\theta_c$ decreases according to eqn (3). $\theta_{flat}$, on the other hand, is determined by the surface chemical composition of the coating, which in this case is the tethered FTS molecule. This means that $\theta_{flat}$ does not change significantly with the number of bilayers. Therefore, it is very likely that as the number of bilayers increase, $\theta_c$ will change from a value that is originally greater than $\theta_{flat}$ to a value that is smaller than $\theta_{flat}$. Correspondingly, the system transits from the Wenzel to the Cassie wetting regimes.

The difference between the completely two different states that water is in on the two sets of coatings, before and after the sacrificial polystyrene particles are removed, respectively, may also be explained by the decrease in $\theta_c$. As presented in Fig. 1b, removal of polystyrene particles will create cavities between the network of silica particles and apparently will increase the actual solid surface area and hence $r$. This, in turn, will result in a decrease in $\theta_c$. Because the difference in $\theta_{flat}$ between the two sets of these coatings is very small, $\theta_c$ could decrease from a value that is greater than $\theta_{flat}$ when the polystyrene particles are not removed, to a value that is smaller than $\theta_{flat}$ after the polystyrene particles are removed. We believe that this decrease of $\theta_c$ accompanying the removal of the polystyrene particles induces the Wenzel-to-Cassie wetting regime transition.

In addition to the much larger $r$ which induces the wetting regime transition of water on the coatings, removal of sacrificial polystyrene particles creates an important surface texture on the coating surfaces, known as "overhang"[17,18] or "re-entrant"[16] structures. Such surface textures are able to induce a metastable Cassie state when liquid is in contact with the rough solid surface even when $\theta_{flat}$ is less than 90° and certainly less than $\theta_c$. The application of this mechanism to our system results in a coating that repels both water and oil with low surface tensions such as hexadecane (surface tension ca. 27.5 mN m$^{-1}$). Fig. 4a and 4b presents photographs

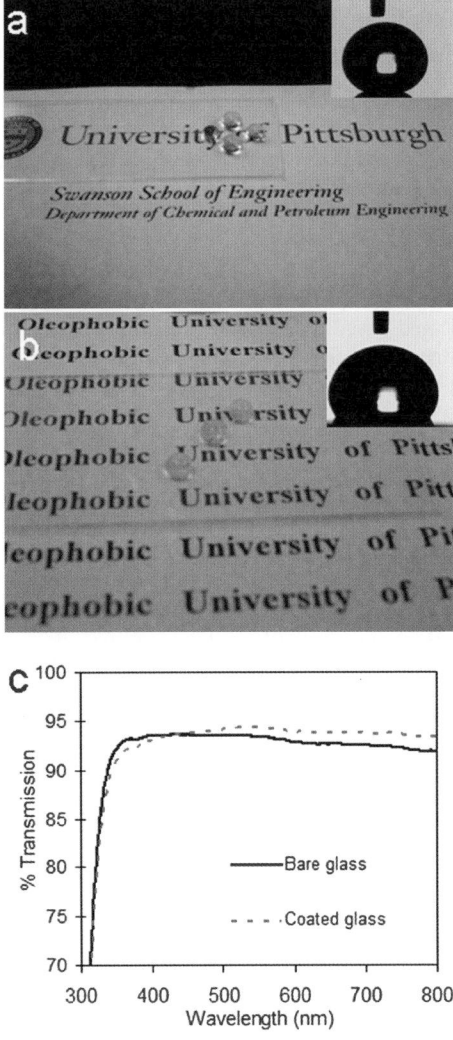

**Fig. 4** Photographs of a coated glass slide with droplets of (a) water and (b) hexadecane to demonstrate the superhydrophobicity, high oil-repellency, and transparency of the coating. Insets in (a) and (b) are images taken during the contact angle measurement. (c) Transmission spectrum of the glass slide before and after the coating process. The glass slide is coated with 25 bilayers, and is calcined and functionalized with FTS.

of a glass slide coated with 25 bilayers of polystyrene and silica nanoparticles, followed by calcination and functionalization with FTS. Both water and hexadecane form small beads when they are in contact with the coating. The contact angle of water is in excess of 160° with a small hysteresis of less than 5°. The contact angle of hexadecane on the coating is in excess of 140°, although the $\theta_{flat}$ of the FTS-grafted surface for hexadecane is only about 70°, which is much less than 90°. We believe that this high oil repellency of the coating is induced by the overhang structures on the surface which is formed during the removal of the sacrificial polystyrene particles.

Remarkably, the superhydrophobic and highly oil-repellent coating fabricated by this approach is also highly transparent. Because the polystyrene particles employed in this process are opaque, the transparency of the nanocomposite coatings fabricated by the layer-by-layer assembly process gradually decreases with increasing the number of the deposited bilayers during the assembly process. This effect becomes more apparent as the number of the bilayers exceeds 15, when the glass slide starts turning translucent. However, after the polystyrene particles are removed in the calcination step, the transparency of the glass slide recovers. This is evident from the photographs shown in Fig. 4a and 4b where the glass slide has been coated with 25 bilayers. Comparison of the transmission spectra of the glass slide before and after the coating process (shown in Fig. 4c) indicates that no significant decrease in the transparency of the glass slide is observed after it is coated, which further proves the high transparency of the coating.

## Conclusions

We have demonstrated that a transparent superhydrophobic and highly oleophobic coating can be fabricated based on a porous network of nanoparticles. The porous network can be constructed by first assembling an organic–inorganic nanocomposite and then removing the organic moieties by calcination. One example of this approach that we have shown in this paper is to first construct a silica-polystyrene nanocomposite by using a layer-by-layer assembly technique and then remove the sacrificial polystyrene particles by calcination. The resulting nanoporous structure is highly transparent and can be made both superhydrophobic and highly oleophobic after functionalization with a fluorocarbon molecule. We believe that removal of the polystyrene particles from the nanocomposite forms overhang structures on the surface of the nanoporous structure, which induces the high oil-repellency of the coating even though the intrinsic oil contact angle of the solid surface is less than 90°. Although transparent superhydrophobic coatings have been demonstrated using the layer-by-layer assembly technique previously,[19,20] to the best of our knowledge, this is the first time the fabrication of transparent coatings that are both superhydrophobic and highly oleophobic by using the layer-by-layer assembly technique has been demonstrated. In addition, a key innovation in our process is the employment of sacrificial organic particles, which not only form overhang structures for inducing the oil-repellency but also saves the final step of adding a top layer of smaller particles which has been previously employed in the literature[19,20] for making transparent superhydrophobic coatings.

## Acknowledgements

This work was supported by the National Science Foundation CMMI grant #0626045 and the University of Pittsburgh Mascaro Sustainability Innovation Center.

## References

1 C. Neinhuis and W. Barthlott, *Ann. Bot.*, 1997, **79**, 667–677.
2 W. Barthlott and C. Neinhuis, *Planta*, 1997, **202**, 1–8.

3 X. F. Gao and L. Jiang, *Nature*, 2004, **432**, 36–36.
4 L. Barbieri, E. Wagner and P. Hoffmann, *Langmuir*, 2007, **23**, 1723–1734.
5 W. Lee, M. K. Jin, W. C. Yoo and J. K. Lee, *Langmuir*, 2004, **20**, 7665–7669.
6 A. B. D. Cassie and S. Baxter, *Trans. Faraday Soc.*, 1944, **40**, 546–551.
7 S. Herminghaus, *Europhys. Lett.*, 2000, **52**, 165–170.
8 A. Lafuma and D. Quéré, *Nat. Mater.*, 2003, **2**, 457–460.
9 R. Blossey, *Nat. Mater.*, 2003, **2**, 301–306.
10 H. Y. Erbil, A. L. Demirel, Y. Avci and O. Mert, *Science*, 2003, **299**, 1377–1380.
11 T. Onda, S. Shibuichi, N. Sotah and K. Tsujii, *Langmuir*, 1996, **12**, 2125–2127.
12 T. Sun, L. Feng, X. Gao and L. Jiang, *Acc. Chem. Res.*, 2005, **38**, 644–652.
13 L. Gao and T. J. McCarthy, *J. Am. Chem. Soc.*, 2006, **128**, 9052–9053.
14 Y. Xiu, L. Zhu, D. W. Hess and C. P. Wong, *Nano Lett.*, 2007, **7**, 3388–3393.
15 A. Nakajima, K. Hashimoto and T. Watanabe, *Monatsh. Chem.*, 2001, **132**, 31–41.
16 A. Tuteja, W. Choi, M. Ma, J. M. Mabry, S. A. Mazzella, G. C. Rutledge, G. H. McKinley and R. E. Cohen, *Science*, 2007, **318**, 1618–1622.
17 L. Cao, T. Price, M. Weiss and D. Gao, *Langmuir*, 2008, **24**, 1640–1643.
18 L. Cao, H. Hu and D. Gao, *Langmuir*, 2007, **23**, 4310–4314.
19 J. Bravo, L. Zhai, Z. Wu, R. E. Cohen and M. F. Rubner, *Langmuir*, 2007, **23**, 7293–7298.
20 Y. Li, F. Liu and J. Sun, *Chem. Commun.*, 2009, 2730–2732.
21 Y. H. Xiu, F. Xiao, D. W. Hess and C. P. Wong, *Thin Solid Films*, 2009, **517**, 1610–1615.
22 G. R. J. Artus, S. Jung, J. Zimmermann, H.-P. Gautschi, K. Marquardt and S. Seeger, *Adv. Mater.*, 2006, **18**, 2758–2762.
23 R. N. Wenzel, *Ind. Eng. Chem.*, 1936, **28**, 988–994.

# The influence of molecular-scale roughness on the surface spreading of an aqueous nanodrop

Christopher D. Daub,* Jihang Wang, Shobhit Kudesia, Dusan Bratko* and Alenka Luzar*

*Received 22nd December 2009, Accepted 26th January 2010*
DOI: 10.1039/b927061m

We examine the effect of nanoscale roughness on spreading and surface mobility of water nanodroplets. Using molecular dynamics, we consider model surfaces with sub-nanoscale asperities at varied surface coverage and with different distribution patterns. We test materials that are hydrophobic, and those that are hydrophilic in the absence of surface corrugations. Interestingly, on *both* types of surfaces, the introduction of surface asperities gives rise to a *sharp increase* in the apparent contact angle. The Cassie–Baxter equation is obeyed approximately on hydrophobic substrates, however, the *increase* in the contact angle on a hydrophilic surface differs qualitatively from the behavior on macroscopically rough surfaces described by the Wenzel equation. On the hydrophobic substrate, the superhydrophobic state with the maximal contact angle of 180 degrees is reached when the asperity coverage falls below 25%, suggesting that superhydrophobicity can also be achieved by the nanoscale roughness of a macroscopically smooth material. We further examine the effect of surface roughness on droplet mobility on the substrate. The apparent diffusion constant shows a dramatic slow down of the nanodroplet translation even for asperity coverage in the range of 1% for a hydrophilic surface, while droplets on corrugated hydrophobic surfaces retain the ability to flow around the asperities. In contrast, for smooth surfaces we find that the drop mobility on the hydrophilic surface *exceeds* that on the hydrophobic one.

## I. Introduction

Two models are often applied to provide practical estimates for contact angles of macroscopic sessile liquid drops on corrugated surfaces. When the liquid in a drop can wet the surface, the Wenzel picture[1] predicts that, on the corrugated surface the added interfacial area causes the contact angle to differ from the Young angle on the smooth surface $\theta_Y$:

$$\cos\theta_c = R\cos\theta_Y \tag{1}$$

Here, parameter $R$ is the ratio of the surface area on the corrugated surface to the projected area on the smooth surface. When the surface is chemically or geometrically heterogeneous, the model of Cassie and Baxter[2] assumes that the cosine of the contact angle $\theta_c$ of the drop on a mixed surface can be approximated by a linear combination of the cosines of contact angles of the components of the surface:

---

*Department of Chemistry, Virginia Commonwealth University, Richmond, VA, USA. E-mail: cdaub@vcu.edu; dnb@berkeley.edu; aluzar@vcu.edu*

$$\cos\theta_c = \sum_i f_i R_i \cos\theta_{i,Y}$$

where $f_i$, $\theta_{i,Y}$ and $R_i$ are the projected surface fractions, contact angles, and roughness parameters for homogeneous surfaces made up of each component $i$ of the surface. Useful estimates of $\cos\theta_c$ are expected presuming the average properties of the substrate surface are representative of those of the surface under the three-phase contact line.[3] The original Cassie–Baxter equation considered the scenario where one of the surface components was air, or vapor, trapped between chemically homogeneous surface asperities.[2] Contact angle on smooth air pocket surface was assumed to be 180°. The second component comprised wetted asperity tips with contact angle $\theta_a$ and roughness (the ratio of wetted area and its projection on the surface plane) $R_a \sim 1$. Assuming $R_a \sim 1$ ($\theta_a = \theta_{a,Y}$), the contact angle can be given by a simplified expression,

$$\cos\theta_c = f_a(\cos\theta_a + 1) - 1 \qquad (3)$$

where $f_a$ and $(1 - f_a)$ are the surface fractions of the surface covered by exposed asperity tips and air, respectively. Equation (2) can also be applied to describe a partially wetted surface when $R_a$ exceeds unity but does not depend on $f_a$, provided $R_a$ is already absorbed in $\cos\theta_a$ according to Eq. (1).

Traditionally the two regimes described by Eqs. (1) and (3) have been thought to apply to different surface interactions between liquid and solid. For lyophobic interactions, the surface does not wet and the Cassie–Baxter equation, Eq. (3) should predict $\theta_c$, while for lyophilic interactions the Wenzel equation, Eq. (1) should apply.

Both of these models were developed to describe surfaces with macro- or mesoscopic corrugations, and even in this regime the details of the geometry of surface corrugations can be important, with the interpretation of these details also leading to some controversy. For example, whether the patches of each component lie below the bulk of the drop or under the contact line could be important.[4–8] Another effect not accounted for by these equations is the possibility of kinetic barriers, which may force the drop to remain in a metastable state.[9] At the nanoscale where atomic and molecular details become significant, we might expect new influences on the microscopic analogue of the contact angle.[10,11] Clearly, the concepts of surface tension and contact angle are not rigorous when drop dimensions and the length scale of surface texture become comparable to the range of molecular interactions. Nonetheless, for nanodrops exceeding a few tens of Angstroms (above Tolman length in water) the contact angle has been generally regarded as a useful measure of surface wettability in this regime.[12] A few theoretical studies of surfaces with nano-sized roughness have been reported, both by simulation,[13–15] and with analytical work[16,17] but for the most part these studies have focused on corrugated surfaces where the length scale of the roughness is sufficiently large to allow water molecules to coat the rough surface, if this is energetically favorable. Our study focuses on nanoscale surface roughness over a wide range of surface-pattern length scales, so that we can explore the transition between surfaces with low asperity coverage to high-coverage regimes where the space between asperities becomes too tight for hydrogen-bonded water molecules to penetrate into the rough surface. We have come across one study of water droplets on randomly roughened hydrophilic and hydrophobic surfaces,[14] which makes the important point that the typical molecule–molecule separation is much larger than the atomic separation in a typical substrate (see for example Fig. 24 in Ref. 14), and hence the Cassie-like state may be preferred in cases where the length scale of the surface roughness is very small. Other than this, however, the transition from nanoscopic to mesoscopic roughness has not been much addressed.

To study the transition between surfaces with nano- and mesoscopic inter-asperity spacings, we have completed simulations of sessile water drops on a variety of atomically corrugated surfaces, constructed by adding partially occupied atomic layers on top of a regular, (111) graphite-like model surface. For high fractions of these additional atoms, our findings are in qualitative agreement with the Cassie–Baxter equation, but deviate noticeably, with the contact angles being larger than those predicted by the equation. As the coverage lowers, the water drop can penetrate down into these partial layers, and the contact angle lowers. However, we never observe Wenzel behavior, as the contact angle is always *higher* than that on the original, molecularly smooth surface.

We also report interesting data on the lateral mobility of water drops on corrugated surfaces. Koishi and coworkers have studied the vertical mobility of water drops on surfaces with very deep corrugations,[13] but not lateral motions. Lundgren and coworkers have demonstrated that water molecules on mixed surfaces will flow to the hydrophilic regions,[15] but have not quantified these observations. In this study we have quantified the motion of water nanodrops on corrugated surfaces by computing the diffusion constants of the water nanodrops on nano-corrugated surfaces. These calculations reveal that water droplets become "pinned" on hydrophilic surfaces with only a few asperities under each water nanodrop, while droplets on corrugated hydrophobic surfaces retain the ability to flow around the asperities. In contrast, for smooth surfaces we find that the nanodrop motion due to thermal fluctuations is faster on the hydrophilic surface than on the hydrophobic one.

## II. Models and methods

We simulate sessile water drops, consisting of 2000 molecules of SPC/E[18] water. Additional simulations with 4000 molecules showed no significant finite-size effects. We used a recent version of the LAMMPS code[19] to do classical molecular dynamics (MD) simulations in the NVT ensemble, using a Nosé-Hoover thermostat to maintain a temperature of 300 K. The surfaces are modeled on the (111) face of graphite, as we[20] and others[21] have done in previous work on sessile water drops. The simulation box is a rectangular prism measuring 117.9 × 119.1 × 200 Å$^3$, and periodic boundary conditions are implemented throughout by using Ewald sums. This procedure removes the need to use a cut-off for intermolecular potentials without dramatically increasing the computation cost. The contact angles are also computed in the same fashion as in our previous work,[20] by modeling the edge of the drop as a circular section and extrapolating down to the surface of the substrate.

Clearly, for nanodroplets the range of intermolecular potentials is not negligible compared to other characteristic length scales, notably the droplet size. This renders the choice of the drop/substrate contact plane, the position of the three-phase coexistence line, and the resulting value of the contact angle itself slightly arbitrary. Specifically, the droplet contour that determines the contact angle must be extrapolated from a few water layers height down to the contact plane as described in detail in our previous work[20] and by others.[21,22] To validate this established simulation procedure for contact angle calculations, we have performed direct comparisons between simulated contact angles on smooth surfaces with those calculated from wetting surface free energies[23,24] for laterally infinite surfaces,[25] determined by applying the pressure tensor technique[26] on identical but fully wetted surfaces. These *thermodynamic* calculations do not depend on any assumption regarding the precise location of the interface, or on a specific technique for contact angle sampling from the droplet shape. Despite small finite-size effects[21] reported in previous nanodrop calculations, a general agreement between the average contact angles of nanodroplets and thermodynamic calculations of wetting surface free energies, $\Delta \gamma = \gamma_{sl} - \gamma_{sv} \cong \gamma_{lv}\cos\theta_c$, was nonetheless established. Here $\gamma_{\alpha\beta}$ are the interfacial free energies between solid ($s$), liquid ($l$) or vapor ($v$) phases. Both types of calculations pertain to equilibrium situations, which are by definition unable to confirm the presence or lack of hysteresis effects consistently observed in macroscopic experiments.

By tuning the Lennard-Jones interaction between the surface atoms and the oxygen atoms of water molecules in the drop, behaviors ranging from extremely

**Table 1** Lennard-Jones parameters and contact angles $\theta_c$ for atomically smooth surfaces. The surface lattice is that of the (111) face of graphite. Contact angles are determined from several 400 ps MD simulations, and quoted errors are one standard deviation in the mean

| $\sigma_{CO}$/Å | $\varepsilon_{CO}$/kJ mol$^{-1}$ | $\theta_c$/° |
|---|---|---|
| 3.190 | 0.50 | 59 ± 1 |
| 3.190 | 0.25 | 118 ± 1 |
| 3.230 | 0.165 | 139 ± 1 |
| 3.190 | 0.117 | 153 ± 1 |

hydrophobic to extremely hydrophilic may be captured. In Table 1 we show all of the examples of Lennard-Jones parameters and observed $\theta_c$ on atomically smooth, homogeneous surfaces that we have studied in this work. The first two models are the same as cases 19 and 17, respectively, from Werder et al.[21] Our contact angles are somewhat lower due to the omission of a pair-potential cut-off applied in Ref. 21.

Systems were equilibrated for at least 400 ps before contact angles were computed, with equilibration extended up to 1.2 ns for cases where additional time was required for contact angles to stabilize. Each determination of the contact angle was derived from one 400 ps trajectory, and was repeated several times until sufficient data to report a reasonable mean contact angle was obtained.

We construct nanoscale corrugated surfaces by addition of carbon atoms into two partially occupied layers on top of a regular, two-layer (111) graphite surface. These atoms are located at the same positions they would be located if additional, full atomic layers were added. Examples of the surfaces we have used are depicted in a simplified representation in Fig. 1, along with representative snapshots from some of the simulations. Identical corrugation topologies, illustrated in the top section of Fig. 1 were considered for both the hydrophobic and hydrophilic materials. Since the droplets on corrugated hydrophobic surfaces consistently displayed Cassie type behavior, a single graph (graph 3 in the bottom row) corresponds to this type of the surface and the remaining snapshots illustrate droplet spreading on model hydrophilic substrates over the whole range of asperity coverage.

## III. Results and discussion

### Contact angles

In Fig. 1 we display some snapshots of simulated water drops on varied corrugated surfaces. In Fig. 2 we show the contact angles obtained on corrugated surfaces. We have used two different models for the substrate atoms in our study, one hydrophilic (first row in Table 1) and one hydrophobic (second row in Table 1). At high coverage, we observe that $\theta_c$ rises as the density of additional atoms reduces, in accordance with the predictions of the Cassie–Baxter equation. However, our observed contact angles deviate from the Cassie–Baxter equation, being more hydrophobic. In fact, in the case of hydrophobic water–graphite interactions, below an areal coverage of 1/6 (shown in Fig. 1) we start to observe superhydrophobic[9,27] behavior. Here, the apparent contact angle $\theta_c$ reaches or even exceeds 180°, as the droplet contour, hanging over asperities, no longer intersects the asperity height. To ensure that this discrepancy is not a by-product of our choice of surface construction, we have verified that when we replace the vacancies by extremely hydrophobic atoms (third row in Table 1), the Cassie–Baxter equation is an excellent predictor of $\theta_c$.

In nature, superhydrophobicity often results from a two-scale roughness, the macroscopic roughness combined with finer roughness at the μm scale. Our models confirm that superhydrophobicity can also be achieved by nanoscale roughening of a macroscopically smooth material.

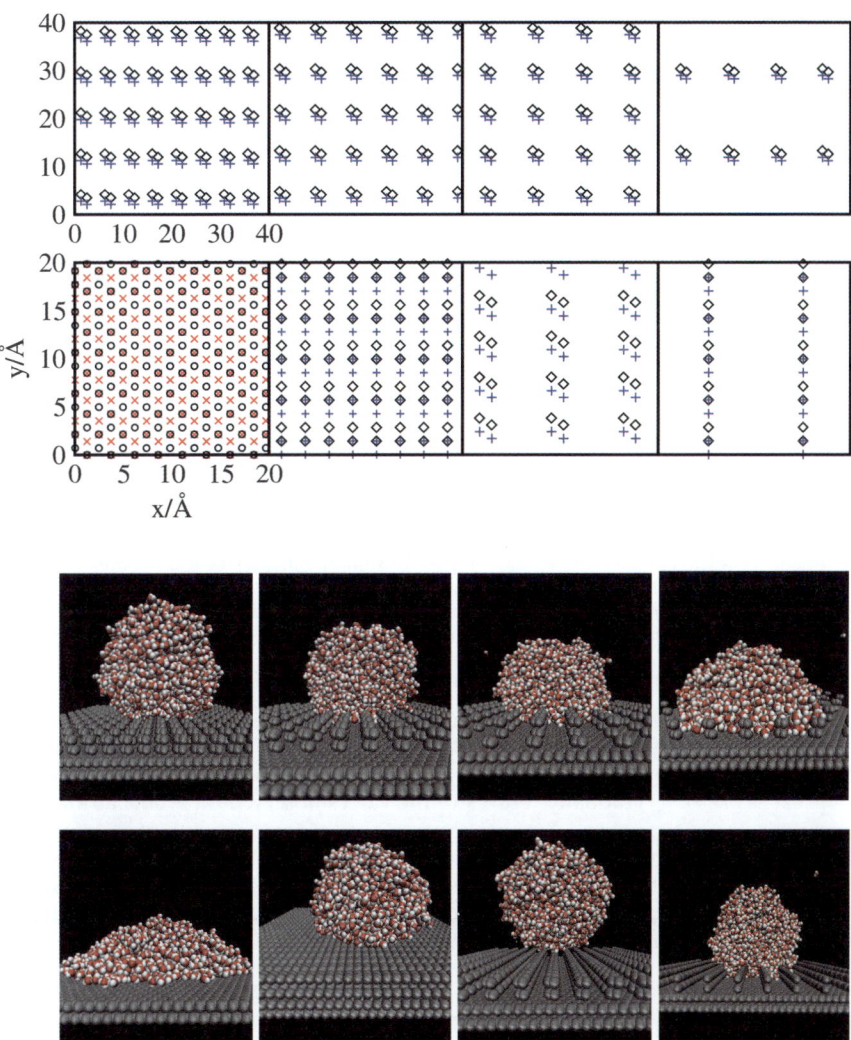

**Fig. 1** *Top*: Simplified representation of surface topologies under study. Lower row, L to R: 2-layer (111) surface with no asperities, 1/2 coverage by asperities, 1/6 coverage, 1/8 coverage with rows of asperities along the *y*-axis. Upper row, L to R (note different scale): 1/8 coverage, 1/12 coverage, 1/16 coverage, 1/32 coverage. Red ×'s represent the first layer, black circles the second, black diamonds the third, blue +'s the fourth layer. The first two layers are shown only for the regular (111) surface. *Bottom*: Simulation snapshots of sessile water drops, on the same surfaces as in the top part of the figure. All snapshots except the 1/6 coverage surface, which is hydrophobic (Row 2 of Table 1) are for hydrophilic surface atoms (Row 1 of Table 1).

As the surface coverage is reduced below ≈ 10%, we observe that the contact angles begin to lower. The reason for this lowering is that, when the spacing between asperities reaches ≈ 10 Å, *i.e.* the size of a few water molecules, some of the water molecules can start to penetrate into the top two layers. This contact angle dependence on the asperity coverage, $f_a$, on a hydrophobic substrate (top curve in Fig. 2) is qualitatively identical to the experimental dependence on PF$_3$-coated silicon surface covered by ∼5 μm wide, cylindrical asperities, plotted as a function of spacing factor $S_f = 2\pi^{-1/2} f_a^{1/2}$ (Fig. 3a of Bhushan and coworkers in Ref. 28). The abrupt decline in the contact angle at low coverage is similar to that expected

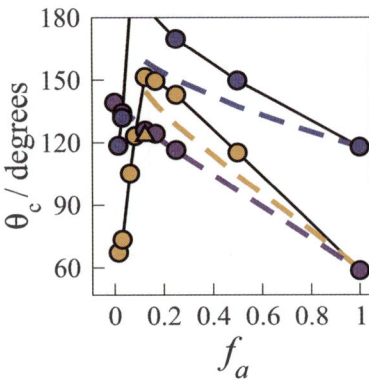

**Fig. 2** Contact angles as a function of the surface coverage, $f_a$, by the asperities, determined from several 400 ps MD simulations (4 × 400 ps for smooth surfaces, up to 12 × 400 ps near the threshold for water penetration). Orange circles: surfaces with the hydrophilic potential, contact angle on a smooth surface, $\theta_1$, equal to 59°; Orange triangle: the surface constructed with rows of asperities along the $y$-axis; Blue circles: surfaces with hydrophobic potential, $\theta_1 = 118°$; Purple circles: mixed surfaces with hydrophobic (2) and hydrophilic (1) components with characteristic contact angles on smooth surfaces of each component $\theta_1 = 59°$, $\theta_2 = 139°$. Solid lines: guide to the eye. Dashed lines: predictions from the Cassie–Baxter equation, in respective colors. Error bars (one standard deviation in the mean) are not shown since they are typically smaller than the symbol size.

upon transition to the Wenzel regime, however, our observed contact angles are never reduced below those observed on the smooth surfaces, in contradiction of the Wenzel equation prediction for hydrophilic substrates. As the surface coverage approaches zero, $\theta_c$ gradually returns to the values observed on atomically smooth surfaces.

Comparing the results at the two different patterns of 1/8 coverages (see Fig. 1) demonstrates the effect that different surface topologies can have on the observed contact angles. Here, the fact that one of the surfaces allows water penetration (orange triangle in Fig. 2), while the other does not, causes a difference in the contact angle of about 25°.

Our finding that the Cassie–Baxter equation provides a good description of the contact angle dependence on the smooth but chemically heterogeneous surface agrees with the related modeling studies[14,15,29] where as long as the size of the different domains was well below the size of the drop, the Cassie–Baxter equation was found to hold true. Deviations from the Cassie–Baxter prediction for the rough surfaces are more difficult to explain. One possibility is that there could be kinetic effects, which cause the drop to remain in a metastable state with an increased contact angle, without being able to cross the kinetic barrier to the state where quantitatively Cassie–Baxter behavior would manifest. On macroscopically rough hydrophilic surfaces, a metastable Cassie–Baxter regime has been predicted[30] and demonstrated experimentally[31] for special geometry of the asperities. In the present study, for nanodroplets on a macroscopically smooth surface with nanoscale asperities, especially near the threshold for water penetration, it did take up to 1 ns for the contact angles to slowly lower until they stabilized at their final values. Therefore if a kinetic barrier is the cause for the deviation from the Cassie–Baxter predictions, it must have a timescale much greater than 1 ns, which seems less likely for the small nanodroplets considered. Further, explaining the deviation from Cassie–Baxter behavior in terms of contact angle hysteresis would not conform with the observation that hysteresis primarily occurs on hydrophilic surfaces, and not hydrophobic ones.[14]

Another factor that could contribute to the increase in contact angles on both types of corrugated surfaces is relatively weak attraction of water molecules to nanoscale asperities. When the roughness occurs on the length-scale comparable to the range of attractive forces between the substrate and water molecules, the attraction of water molecules to an asperity is appreciably weaker than the attraction to an equally distant extended surface.

A major reason why we never see the Wenzel state would appear to be the very shallow depth of our rough surfaces. This is consistent with a DFT-based analytical study which demonstrated that the Wenzel state cannot be produced when the surface roughness is very shallow.[17] Indeed, contact angles agreeing with the Wenzel prediction have been produced by another MD study where the depth of the surfaces was much more pronounced than ours.[15] However, we note that when the depth of the roughness is close to the dimensions of the water drop, the location of the drop/substrate contact plane becomes ambiguous.

### Interfacial hydrogen bonding

The number of interfacial hydrogen bonds has a direct impact on the interfacial free energy of the water–surface interface, which in turn affects the contact angles *via* Young's equation, $\cos\theta_c = (\gamma_{sv} - \gamma_{sl})/\gamma_{lv}$. We have computed the average number of hydrogen bonds for interfacial water molecules and plot these as a function of the surface fraction of the partial surface layers in Fig. 3. Two water molecules are presumed to share a hydrogen bond when oxygen–oxygen distance $r_{OO} < 3.5$ Å, the distance $r_{OH} < 2.45$ Å, and the angle between the $r_{OO}$ and $r_{OH}$ vectors is $< 30°$.[32] One of the waters participating in the bond must be within $\sigma_{CO}$ of one of the surface "carbons". Fig. 3 shows that as the surface fraction decreases from full coverage, the number of interfacial hydrogen bonds decreases, in agreement with other results showing the same trend as the water–surface interaction becomes more hydrophobic.[33]

However, as the surface coverage lowers towards the threshold for water penetration, the number of interfacial hydrogen bonds rises and reaches a maximum for surface coverage $\approx 0.02$ to $0.05$. This can be attributed to the fact that at coverages in this range the water molecules may form bonds around the asperities in analogy to

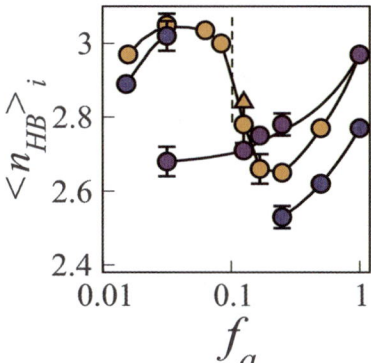

**Fig. 3** The number of hydrogen bonds formed by interfacial water molecules as a function of the surface fraction of filled atomic positions in the top two layers of graphite. Interfacial water molecules reside within the range of the repulsive part of the water-surface potential, *i.e.* $< \sigma_{CO}$, from any surface atom. The vertical dashed line roughly indicates where the water molecules can start to penetrate into the partially filled surface layers. The color symbols have the same meaning as in Fig. 2. Black lines are guides to the eye. There are no data for the hydrophobic surface (blue circles) within the interval $0.08 < f_a < 0.25$ since a water drop detaches from the surface in this regime (see the range in Fig. 1 with apparent contact angle exceeding 180°). Indicated error bars are one standard deviation of results from individual 400 ps trajectories (not shown when smaller than the symbol size).

hydration of small solutes,[34,35] hence raising the number of bonds formed above the maximum of three for a planar interface.[36,37]

**Droplet translation**

Because of thermal fluctuations, the droplets move randomly along the surface, showing quite different mobilities for different substrate types. To quantify these findings, we have determined apparent diffusion constants $D_{x,y}$ in directions parallel to the interface (Fig. 4). Each of these diffusion constants has been derived from mean-square displacements of the drop's center of mass observed in about ten 400 ps trajectories. Despite some statistical uncertainty, they represent useful measures of the mobility of the water drops. We see that the mobility is high in all cases where water molecules cannot penetrate into the partially filled surface layers. However, as the water molecules penetrate the top layers, for the hydrophilic substrate we see that the mobility of the water drops lowers dramatically. In this case, the asperities pin the water drops in essentially one place, at least over the characteristic timescale of our simulation. Remarkably, this pinning is evident even at the lowest coverage we simulated, where only $O(1)$ asperity is located under the water drop. We note also the interesting case where water molecules may freely penetrate only along one lateral axis, due to the presence of a "wall" of surface atoms (the last graph in the top section in Fig. 1). Here the diffusion constant parallel to the rows of aligned asperities remains high, but the drop cannot diffuse across the asperities in the perpendicular direction. If the underlying water–surface interaction is hydrophobic, on the other hand, the presence of the asperities does not have the same impact on the mobility of the water drop. Apparently in this case the water

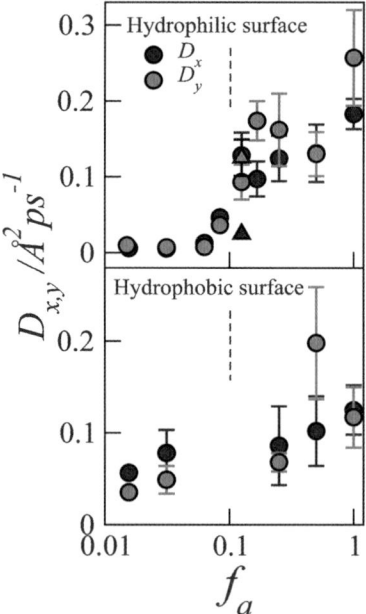

**Fig. 4** Apparent diffusion constants $D_{x,y}$ of sessile nanodrops as a function of the surface coverage, $f_a$, for: Top graph; hydrophilic Lennard-Jones interactions and Bottom graph; hydrophobic Lennard-Jones interactions (first and second row in Table 1). The diffusion constants were determined from the slope of the mean-squared displacement in $x$ and $y$ directions at times between 30 and 40 ps. The triangles represent the surface constructed with rows of asperities along the $y$-axis. Error bars are one standard deviation in the mean (not shown when smaller than the symbol size).

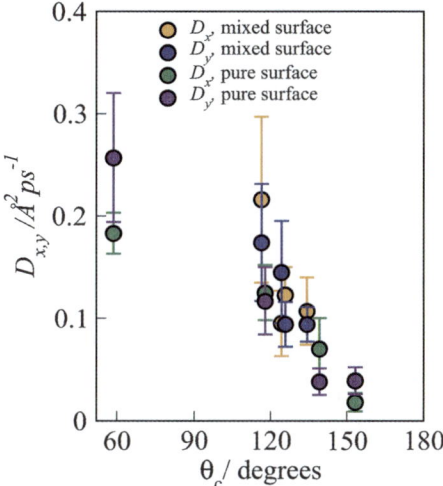

**Fig. 5** Diffusion constants $D_{x,y}$ of sessile water drops as a function of the contact angle on a smooth atomic surface. Lennard-Jones parameters for each pure surface's water–carbon interactions are listed in Table 1, while the mixed surface is the same as the purple circles in Fig. 2 and 3. Vertical error bars are one standard deviation in the mean, while the uncertainty in the contact angle is ±1°.

molecules remain able to flow around the asperities even when they can penetrate into the partially vacant surface layers.

Another interesting finding has to do with the mobility of water drops on an atomically smooth surface without any partial surface layers. These results are shown in Fig. 5. In this case, it appears that the water drops move faster on a hydrophilic surface than a hydrophobic one. For a fluctuation-driven process like the droplet motion on a flat surface, this apparently contradictory result may be explained in terms of stronger surface–water attraction in the case of the hydrophilic surface, which may allow larger random fluctuations in the contour position of the water drop. In the present scenario, thermal fluctuations contribute to the motion of the drop as a whole. Note that thermal motion is the sole cause of the droplet diffusion in our system. If additional forces were at play, *e.g.* due to gravity on a sloping surface or due to wind on a lotus leaf,[9,16,38] the drops would be much more mobile. The effect of surface texture on the droplet mobility under a systematic external force[39] would be different as well.

## IV. Conclusions

The majority of prior computational studies of wetting on corrugated surfaces considered mesoscopic length scales, where dynamical issues dominate whether superhydrophobic Cassie–Baxter states or fully wetted Wenzel-like states are observed. For instance, Koishi and coworkers[13] show that if a water drop impacts on a hydrophobic pillared surface with a spacing of 12.3 Å with sufficient velocity, it will overcome an energy barrier between the Cassie–Baxter state and the Wenzel state where it can penetrate between the pillars. What is unique about our study is that we probe very small length scales, where water molecules are prevented from penetrating between asperities by not only energetic or dynamical considerations, but also sterically. Thus, we probe the lowest length scales where a corrugated surface can be completely wetted, regardless of whether the substrate material is hydrophobic or hydrophilic.

We emphasize two new findings pertinent to nanoscale roughness:

First, we demonstrate that roughness on this length scale renders hydrophilic surfaces more hydrophobic, a trend opposite to observations on macroscopically rough materials. Second, we show that a superhydrophobic surface with a contact angle of 180° can also be achieved when surface roughness is limited to the nanoscale alone.

The increase of the apparent contact angle of an intrinsically hydrophobic material ($\theta_c = 118°$) in the presence of surface vacancies qualitatively conforms to, but exceeds the effect predicted by the Cassie–Baxter equation. Upon further decrease in the coverage, the contact angle gradually decreases towards the value characteristic of the original, smooth surface as the water droplet surrounds the asperities.

Our results for the hydrophilic substrate (contact angle of the asperity-free surface below 90°), contradict the prediction of the conventional Wenzel relation according to which the roughness should enhance the surface hydrophilicity. In fact, the qualitative structure of the drop-surface system differs little on either surface; the only difference is that on the hydrophobic surface, an "ultrahydrophobic" state with the maximal contact angle $\theta_c = 180°$ is observed when the surface coverage lies within the window between ~8–25%. Rough interpolation of our results would suggest that this state could be achieved *via* atomic corrugation of surfaces with intrinsic contact angles exceeding 90°, the traditional boundary dividing hydrophilic and hydrophobic surfaces. When the vacancies are replaced by extremely hydrophobic atoms, the behavior predicted by the Cassie–Baxter equation is observed.

The upward deviations from macroscopic predictions of the contact angle can arise from several effects. First we note reduced water/surface attraction when the roughness occurs on the length-scale comparable to the range of intermolecular forces. We hypothesize metastable vapor pockets trapped between hydrophilic asperities can also contribute to the observed contact angle increase although the Wenzel equation suggests a reduction in contact angle if the substrate below the drop is in a fully wetted state. We demonstrate that the reduction in hydrogen bonding represents a major contribution to the increase in surface free energy and the contact angle on the rough surfaces. While complete (100%) surface coverage by the asperities renders surface properties close to those of the original (zero coverage) smooth surface, we observe the maximal contact angle, and minimal hydrogen bonding among interfacial water molecules in the neighborhood of 15 ± 5% coverage, corresponding to an inter-asperity separation of two or three water molecules.

In contrast to the static contact angle measurements, the dynamics of droplet mobility is much more affected by the corrugations on the hydrophilic substrate compared to the hydrophobic one. The mean squared displacement of the droplet's center of mass (associated with thermal motion and droplet shape fluctuations) as a function of time shows a strong pinning effect by hydrophilic asperities. Lateral diffusion of the droplet as a whole is relatively fast on asperity-free surfaces (and, equivalently, on surfaces with 100% coverage), with the droplet being *more* mobile when the surface is hydrophilic. The droplet mobility is generally reduced in the presence of surface corrugations, however, the effect is much stronger on hydrophilic surfaces, where we observe a dramatic slow down of the droplet translation for asperity coverage as low as 1%. In the case of a nanodrop comprising 2000–4000 water molecules, this coverage corresponds to $O(1)$ asperities under the droplet base. A minute surface decoration proves sufficient to change the droplet mobility by an order of magnitude or more.

## Acknowledgements

This work was supported by the National Science Foundation through awards CHE-0718724 (to A.L.) and CBET-0432625 (D.B.). We acknowledge the support by the National Science Foundation through TeraGrid resources (CHE090108). We thank Jim Henderson for pointing out the proper use for Eqs. (1–3).

# References

1. R. N. Wenzel, *Ind. Eng. Chem.*, 1936, **28**, 988.
2. A. B. D. Cassie and S. Baxter, *Trans. Faraday Soc.*, 1944, **40**, 546.
3. L. C. Gao and T. J. McCarthy, *Langmuir*, 2009, **25**, 7249.
4. G. McHale, *Langmuir*, 2007, **23**, 8200.
5. L. C. Gao and T. J. McCarthy, *Langmuir*, 2007, **23**, 13243.
6. L. C. Gao and T. J. McCarthy, *Langmuir*, 2007, **23**, 3762.
7. P. S. Swain and R. Lipowsky, *Langmuir*, 1998, **14**, 6772.
8. A. Marmur and E. Bittoun, *Langmuir*, 2009, **25**, 1277.
9. C. Dorrer and J. Ruhe, *Soft Matter*, 2009, **5**, 51.
10. J. Hautman and M. L. Klein, *Phys. Rev. Lett.*, 1991, **67**, 1763.
11. W. Mar, J. Hautman and M. L. Klein, *Comput. Mater. Sci.*, 1995, **3**, 481.
12. B. Widom, Personal communication.
13. T. Koishi, K. Yasuoka, S. Fujikawa, T. Ebisuzaki and X. C. Zeng, *Proc. Natl. Acad. Sci. U. S. A.*, 2009, **106**, 8435.
14. C. Yang, U. Tartaglino and B. N. J. Persson, *Eur. Phys. J. E*, 2008, **25**, 139.
15. M. Lundgren, N. L. Allan and T. Cosgrove, *Langmuir*, 2007, **23**, 1187.
16. G. Carbone and L. Mangialardi, *Eur. Phys. J. E*, 2005, **16**, 67.
17. G. O. Berim and E. Ruckenstein, *J. Chem. Phys.*, 2008, **129**, 014708.
18. H. J. C. Berendsen, J. R. Grigera and T. P. Straatsma, *J. Phys. Chem.*, 1987, **91**, 6269.
19. S. Plimpton, *J. Comput. Phys.*, 1995, **117**, 1.
20. C. D. Daub, D. Bratko, K. Leung and A. Luzar, *J. Phys. Chem. C*, 2007, **111**, 505.
21. T. Werder, J. H. Walther, R. L. Jaffe, T. Halicioglu and P. Koumoutsakos, *J. Phys. Chem. B*, 2003, **107**, 1345.
22. R. L. Jaffe, P. Gonnet, T. Werder, J. H. Walther and P. Koumoutsakos, *Mol. Simul.*, 2004, **30**, 205.
23. D. Bratko, C. D. Daub and A. Luzar, *Phys. Chem. Chem. Phys.*, 2008, **10**, 6807.
24. D. Bratko, C. D. Daub and A. Luzar, *Faraday Discuss.*, 2009, **141**, 55.
25. D. Bratko, R. A. Curtis, H. W. Blanch and J. M. Prausnitz, *J. Chem. Phys.*, 2001, **115**, 3873.
26. D. Bratko, C. D. Daub, K. Leung and A. Luzar, *J. Am. Chem. Soc.*, 2007, **129**, 2504.
27. M. Nosonovsky; B. Bhushan In *12th International Conference on Vibrations at Surfaces* Erice, ITALY, 2007; Vol. 20.
28. B. Bhushan, M. Nosonovsky and Y. C. Jung, *J. R. Soc. Interface*, 2007, **4**, 643.
29. J. Wang, D. Bratko and A. Luzar, Manuscript in preparation, 2010.
30. J. Wang and D. Chen, *Langmuir*, 2008, **24**, 10174.
31. J. Wang, F. Liu, H. Chen and D. Chen, *Appl. Phys. Lett.*, 2009, **95**, 084104.
32. A. Luzar, *J. Chem. Phys.*, 2000, **113**, 10663.
33. C. Y. Lee, J. A. McCammon and P. J. Rossky, *J. Chem. Phys.*, 1984, **80**, 4448.
34. F. H. Stillinger, *J. Solution Chem.*, 1973, **2**, 141.
35. D. Chandler, *Nature*, 2005, **437**, 640.
36. A. Luzar, S. Svetina and B. Zeks, *Chem. Phys. Lett.*, 1983, **96**, 485.
37. A. Luzar, S. Svetina and B. Zeks, *J. Chem. Phys.*, 1985, **82**, 5146.
38. D. Quere, *Rep. Prog. Phys.*, 2005, **68**, 2495.
39. J. D. Halverson, C. Maldarelli, A. Couzis and J. Koplik, *J. Chem. Phys.*, 2008, **129**, 164708.

# General Discussion

**Professor Luzar** opened the discussion of the papers by Professor Quéré and Professor Butt: Let me open the discussion with a question to both speakers, Professors Quéré and Butt:

Is there any experimental evidence showing that a superhydrophobic surface can also be achieved when the surface roughness is limited to nanoscale alone? Our experiments in silico (paper 6[1] in this volume) suggest such a scenario.

1 Christopher D. Daub, Jihang Wang, Shobhit Kudesia, Dusan Bratko and Alenka Luzar, *Faraday Discuss.*, 2010, **146**, DOI: 10.1039/b927061m.

**Professor Butt** responded: To my knowledge plants always use at least two length scales to achieve superhydrophobicity. They have optimized their surface structure and I guess there is a good reason for the two length scales. This, however, is only an indication because plants might have to fulfil different boundary conditions. With his arrays of pillars David Quéré has already made a superhydrophobic surface with only one length scale.

**Professor Quéré** replied: Taking a given pattern for a microstructured surface, decreasing the size of this pattern and measuring how the contact angle and hysteresis vary is indeed a very interesting program of research. Superhydrophobicity should resist thereductionn of scale, but long range forces (such as van der Waals) might modify the "macroscopic" Wenzel/Cassie views. On a more practical perspective, we might also anticipate in this limit a better mechanical robustness, a property of obvious interest.

**Mr Wang** opened the discussion of the paper by Professor Rossky by communicating: I think there are two conditions which lead to the depletion layer at large length scale hard sphere solutes: (1) the physical condition is close to liquid vapor coexistence, (2) there is no attractive interactions between the hard sphere and water. So only in the hard sphere case do we see the depletion layer, while in any real systems there is no depletion layer. In both cases, large density fluctuations around the solute are similar to the liquid–vapor interface. The connections you made between the hydrogen bonding orientation at the solute water interface is special for water. So it would be better if you could point out what is common for all liquids (like LJ fluid) and what is special for water.

**Professor Rossky** communicated in reply: I largely agree with what you say, if one interprets the depletion layer to mean drying and one is in the vicinity of the liquid–vapor coexistence curve. I would not go as far as to say that there are no real systems which dry, but there is no evidence that it is at all common. As I discussed in my talk, and hope that I have adequately conveyed in my written lecture, I believe that the behavior of water and other fluids with respect to density depletion and fluctuation behavior are not distinguished, when the fluid states are thermodynamically similar, as others have shown. What distinguishes water is the location of the coexistence curve compared to ambient conditions and the widely variable interfacial free energy that is physically accessible. These are the result of the underlying molecular features; solvent hydrogen bonding with itself and with a surface are intimately connected with the density response. If one is interested in interpreting the origin of the thermodynamic response or attempting to modify it *via* chemical means, then one needs to explore the detailed structure and how it correlates with the properties of interest.

**Professor McCarthy** commented: The silica–water interface is a chemically complex system and the silica + water − silicic acid equilibrium is involved. Water hydrates silica and silica dissolves in water. The textbook by R. K. Iler,[1] discusses this in detail.

1 R. K. Iler, The Chemistry of Silica: Solubility, Polymerization, Colloid and Surface Properties and Biochemistry, John Wiley & Sons, New York, 1979.

**Professor Rossky** responded: I appreciate the comments on "real" silica chemistry, and Iler's monograph is cited in our publications. Our work necessarily uses an idealized, but not unphysical, interface in order to develop systematic understanding of the impact of different types of interface chemistry on water behavior.

**Professor J. Klein** addressed Professor Rossky and Professor Garde: Early in your talk you showed data on compressibility of hydration shells, obtained by simulations. How do these values compare with real behaviour, obtained for example directly by compressing hydrated ions between two solid surfaces[1] in the surface force balance?

1 U. Raviv and J. Klein, 'Fluidity of bound hydration layers', *Science*, 2002, **297**, 1540–1543.

**Professor Rossky** answered: Perhaps I don't fully understand the question, but the quantity we, and others—including Professor Garde in his Discussion paper—have examined is the change in interfacial density as a function of external bulk pressure or, in the same vein, the magnitude of the density fluctuations in the interfacial region at a given bulk pressure. In the article you cite, the resistance to compression of the gap is measured, at fixed ambient pressure, so that I don't see a direct connection between this measurement and ours. However, if one compares experimental results to the simulated volumetric response for model water solutions of small molecules and proteins, they do appear to be semi-quantitatively in agreement. Hence, I believe that the important conclusion from simulations—which is a qualitative one—that the compressibility of water is significantly increased near nanoscale hydrophobic surfaces and essentially unperturbed at hydrophilic surfaces is virtually certain to be correct.

**Professor Garde** answered: One of the key results of our *Faraday*[1] and *PNAS*[2] papers is that in the vicinity of hydrophobic surfaces, water displays enhanced density fluctuations and correspondingly higher compressibility. Experimental measurements of local density fluctuations at solid–liquid interfaces using scattering techniques are challenging and have not been reported to our knowledge. In light of this, surface force balance (SFB) experiments could potentially play an important role. We understand that the use of mica surfaces introduces ions in the system making the interfaces charged. What is needed is a similar experiment using purely hydrophobic/hydrocarbon chemistries, which could in future be extended to include a range of homogeneous chemistries from hydrophobic to hydrophilic, as well as to mixed chemistries. We look forward to such experimental data to complement our simulation and theory work.

1 Hari Acharya, Srivathsan Ranganathan, Sumanth N. Jamadagni and Shekhar Garde, Faraday Discuss., 2010, **146**, DOI: 10.1039/b927019a.
2 Rahul Godawat, Sumanth N. Jamadagni and Shekhar Garde, *Proc. Natl. Acad. Sci. USA*, 2009, **106**(36), 15119–15124.

**Dr Patankar** communicated: It was commented to the effect that hydrophobicity or hydrophilicity are relative rather than absolute definitions. I agree with this in general since the meaning of what is implied by these terms could be problem specific. In the case of roughness modulated wetting behavior the critical value of

90° contact angle represents a reasonable boundary that separates some specific trends. *E.g.* roughness amplifies the apparent hydrophilicity of a substrate made from a hydrophilic material whereas it amplifies the apparent hydrophobicity of a substrate made from a hydrophobic material. Thus, the trends of contact angles are typically opposite on either side of the traditionally considered critical value of the contact angle.

**Professor Rossky** communicated in reply: That is quite an interesting point, and I agree that in specific contexts there are natural boundaries. A contact angle of 90° also defines the critical contact angle for the appearance of a dewetting transition within a finite gap for a liquid confined between macroscopic surfaces. My intent was more focused on recognizing that there are degrees of hydrophobicity than on the contrast between hydrophobic and hydrophilic.

**Mrs Seyed-Yazdi** addressed Professor Quéré: (a) Does the figure of impact depend on the pattern of surface network of the pinning area or the pattern of the whole area ultimately wetted by the drop?
(b) What is the threshold density of pillars that is sufficient to see the figure of impact and what is the relation between the impact speed and the threshold density of pillars?
(c) How would the mechanism of the impact change if the surface is replaced by a hydrophilic one?

**Professor Quéré** answered: All these excellent questions underline the necessity of a more systematic study on this topic. We observed that "crystallographic" impacts disappear when the pattern of the surface becomes random, instead of ordered. For denser patterns, much larger impact velocities are needed to make the drop penetrate in the array of pillars (a condition for observing the ejection of anisotropic jets), but high impact velocities also "kill" the anisotropic jets, so that it is not obvious that crystallographic impacts occur in this limit. More generally, we need to make a complete "phase diagram" to fully characterize the phenomenon.

**Professor Yeomans** commented: Please can you comment on the relative importance of friction due to velocity gradients within the drop and friction due to contact line motion? Is the ratio different for hydrophobic and superhydrophobic geometries?

**Professor Quéré** replied: I would be very happy to be able to answer these questions. This would mean that I would know what is the line friction in a very hydrophobic limit (and even in a hydrophobic one), which is not understood today (to the best of my knowledge). Jacco Snoejjer in Twente is working in this direction. It would also mean that I could describe the velocity gradient inside the drop, at the high Reynolds number (of the order of 1000) they run down. Preliminary experiments with tracers show that velocity gradients can be ignored during the first few centimetres of the run (pure slip), while much more complex flow profile (including a mixture of rotation and translation) are observed later.

**Professor Gao** asked Professor Butt: What holds the raspberry-like particles together in the coating? Is there any chemical binder used to improve the adhesion of the film to the substrate? Has any experiment been conducted to test the scratch-resistance of the coating?

**Professor Butt** answered: The particles are held together and to the substrate by capillary bridges of polystyrene. The scratch-resistance has been tested only qualitatively, by rinsing, blowing, wiping the surface with a piece of paper, or by scratching it with finger nails. Before polystyrene leakage the layers were fragile and the particles are easily displaced. After polystyrene leakage the layers are much more stable and scratch resistant.

**Professor Debenedetti** asked Professor Quéré: How much is known about the sensitivity of the phenomenology you showed (*e.g.*, behavior upon impact) to the composition of the droplet, as one goes from water to aqueous mixtures?

**Professor Quéré** responded: To the best of my knowledge, there is mainly one study where people tried to understand how the contact angle varies as a function of the liquid surface tension, on textured hydrophobic surfaces (as Zisman did long ago on planar surfaces). Amirfazli and his collaborators looked at the effect of surfactants on wetting of superhydrophobic materials, and found that superhydrophobicity surprisingly "resists" the addition of surfactants, probably because of a specific adsorption. This is not observed for more classical mixtures (water/alcohol) for which the substrate becomes superhydrophilic when the contact angle (defined on a planar surface of the same chemical composition) becomes of the order of 90° (in agreement with the Wenzel model). We did a few (unpublished) experiments, where we observed that a drop containing surfactants bounces while it spreads if gently deposited, which we interpreted as resulting from the low contact time, compared to the surfactant adsorption time: then, the drop behaves (roughly) as if it were made of pure water.

**Professor Varanasi** remarked: Most practical engineering surfaces have random roughness. My question is with respect to such surfaces:
 1. Can you please comment on how one could approach wetting hysteresis and impact phenomena on surfaces with random roughness?
 2. If the length scales of the random roughness are very small (in comparison with the drop size), would we still expect some symmetry in droplet impact (similar to the crystallographic impact)?

**Professor Quéré** answered: How to connect model experiments and theory with the reality of random surfaces is one of the main challenges in the field. We know that both the degree of wetting and its hysteresis are related to the solid/liquid contact, but we do not how to model this contact on a random texture! Impacts are even more complicated, but I would tend to say that both reduction in size and randomness should oppose the formation of symmetric jets. We partially proved it by looking at the impact on random pillars—no anisotropic jets were observed; as for small length scales, the increase of friction and the role of surface tension should both contribute to suppress these special jets.

**Professor Weeks** asked Professor Butt: You mention that the lotus leaf presents a superhydrophobic and self-cleaning surface with roughness on both a micrometre scale along with nanoscale asperities. Your raspberry particles also exhibit dual-scale roughness. To make contact with the large amount of theoretical work along these lines, is it possible experimentally to produce a system with only nanoscale roughness with controllable spacing and depth of the asperities? What features of dual-scale superhydrophobicity would you expect to change with only nanoscale roughness?

**Professor Butt** responded: With current technology such as focused ion beam and electron beam lithography it is possible to construct surfaces which are structured on the 10–100 nm and the 1–10 μm length scales. Thus such an experiment should be possible and allow a direct comparison with theory. I think the dual (or multi) scale roughness leads to an increase in contact angle because it again reduces the contact area between the liquid and the solid surface.

**Mr Wu** asked:

1. Were you able to perform any measurements that quantify the mechanical stability you mentioned in the paper? Such as nanoindentation?

2. THF (and other solvents, as mentioned during the meeting) were used in the vapour phase to promote migration and leaching of polymer chains. From experience, THF that are embedded into the polymer chains of polystyrene cannot be removed from pure vacuum alone. Thus, have you conducted any spectroscopic or analytical studies to ensure the complete removal of solvent molecules from the polymer matrix? Reason being, the increase in plasticity caused by solvent incorporation is exponential even at extremely low concentrations. Unless substantial removal is achieved, the mechanical stability of the film would be compromised.

**Professor Butt** answered: We prepared free-standing films and performed first force-elongation measurements. Nanoindentation turned out to be difficult due to the large number of holes but measurements are under way. It is, however, too early to report reliable numbers. Qualitative tests such as rinsing with water, blowing, scratching with a piece of paper, or with the finger nails showed that the layers were stable. I agree that always trace amounts of solvent are left in the polymer. We have not applied spectroscopic techniques to detect trace amounts of solvent. In my experience with spin cast films dried under similar conditions the mechanical properties of the polystyrene bridges should not be significantly different from the pure polystyrene.

**Mr Wu** addressed Professor Quéré: Since the surfaces presented were typically generated using photolithography, specifically, using deep reactive ion etching, would you expect that the observed expulsion jets be affected by the geometry of the patterned substrate?

Perhaps, parameters such as the angle at the side-wall of each pillar, each pillar's placement with respect to one another (*i.e.* packing density) and shape of each pillar would govern the jet's expulsion?

**Professor Quéré** replied: Indeed, the correspondence between pattern and jet symmetries implies that the jet symmetry should be affected by the pattern geometry. It is less obvious to claim that the detail of the pillar shape could also affect the jets expulsion, since it should affect very little the air flow in the array of pillars.

**Professor Bratko** commented: The droplet bounce on a superhydrophobic surface is typically presumed to involve only Cassie-like states with none or only minute penetration of water between the substrate asperities. Upon increasing the droplet momentum at collision, partial penetration appears conceivable. Can a threshold extent of penetration preventing the droplet rebounce be predicted from surface structure and wetting thermodynamics?

**Professor Quéré** responded: You are right. We discuss briefly in the paper the transition of impalement that occurs as the impact velocity increases. It is a function of the pattern (as proved by early experiments by Bartolo, and by Reyssat), but the detail of the transition remains to be fully described. Conversely, this transition can be used to build a criterion of "robustness" for superhydrophobic surfaces, since it quantifies the ability of water to resist pinning (and thus to bounce).

**Dr Chunder** addressed Professor Butt: Under the section 'Solvent evaporation and THF leakage' I was wondering if the polystyrene (PS) leaked out of the silica shell and completely filled up the nanopores of the raspberry like particles, then they would lose the nanoscale roughness and may not show superhydrophobicity. How do you control this phenomena, *i.e.* how much PS would leak out and how do you get rid of it so that it can not fill the nanopores.

Did the group investigate long-term stability of the raspberry particle film under water? How stable is the fluorocarbon that they used to lower the surface energy of the film surface, under water? If a surface is superhydrophobic and is immersed under water, how long it will maintain this property, if the height of the water pillar is not very high so as to wet the surface (if the radius of curvature of the meniscus of the water droplet ($R$) inside the roughness groove at the air–water interface is lower than the typical scale roughness of the surface), then only a surface should get wet under water, otherwise not, if it is immersed under shallow water. For example, for a depth of 1 m water pillar, $R$ is about 10 μm, so it is large compared with a typical roughness scale of 100 nm–1 μm, so the surface will not get wet).

**Professor Butt** answered: The surfaces mechanically stabilized by polystyrene bridges remained superhydrophobic after immersion into water for a couple of hours. They lost their superhydrophobicity after immersion into water for a few days. This might be caused by partial removal of the binding sites of the fluorosilanes. However, they remain superhydrophobic for more than a year (the age of our oldest samples) if they are regularly exposed to water droplets ("simulation of rain").

**Dr Sarupria** addressed Professor Quéré: Can you comment on how the flexibility of the posts affects the behavior of the droplets in the case of (a) fabricated surfaces (b) biosurfaces such as the lotus leaf or water striders.

**Professor Quéré** answered: (1) The flexibility of posts can be used to prevent the contact between water and the "bottom" surface lying at the foot of pillars. This was discussed theoretically by Herminghaus, but I think that experiments remain to be done to quantify how post flexibility can prevent liquid penetration inside a texture. This phenomenon seems particularly relevant for water striders whose legs are decorated by long hairs, allowing them to jump on water. I do not think that it plays a role for lotus whose texture motif is much more compact (sugar-loaf micromountains), and thus much less flexible. (2) It is indeed important to stress how difficult it is to measure contact angles when they are high. The (small) effect of the drop weight becomes in this limit of the same order as the wetting deformation, for example, which can generate large errors. There are specific methods to measure small contact angles, we would need similarly special methods in the opposite limit.

**Dr Sarupria** addressed Professor Butt: What are the difficulties in measuring contact angles experimentally and what are the cautionary points one should keep in mind while interpreting the contact angle measurements?

**Professor Butt** replied: On superhydrophobic surfaces it was difficult to adjust the position of the needle to the center of the droplet. The drop always had a tendency to move to the side of the needle and it became asymmetric. Therefore we regard the sliding angle as most reliable.

**Professor Mugele** asked Professor Quéré: Dynamic contact angle "hysteresis" for flat surfaces are commonly associated with the divergence of the viscous dissipation in the vicinity of moving contact lines. For superhydrophobic surfaces, dissipation is mostly considered in terms of quasi-static arguments of breaking capillary bridges. It is important to point out that these are two completely different mechanisms. Yet, drop motion on superhydrophobic surfaces does also involve flow, in particular in the case of impacting drops, which may also lead to a contribution of the more classical dissipation known due to diverging viscous dissipation. Do you have an idea how important viscous dissipation is as compared to the quasi-static breakage of capillary bridges for drop motion on superhydrophobic surfaces?

**Professor Quéré** replied: If the impact velocity is not large (which avoids strong vibrations of the drop, and thus energy transfer), drops bounce off superhydrophobic surfaces with a remarkable elasticity. This proves that dissipation is dramatically different from what we usually know, and that the line contribution is somehow negligible—maybe for geometrical reasons, or it maybe because there is no line at all in a dynamic experiment (where dynamical air films might form). However, if you reduce more and more the drop size (which means: increase more and more surface effects), you observe that the rebound disappears, which we are tempted to interpret as resulting from hysteresis (*i.e.* the formation of capillary tails). A quantitative analysis remains to be built on this phenomena.

**Mr Kang** asked: First of all, it is a very interesting and inspiring study.

I have two questions—one about the impalement of drops to the patterned surface and the other about the bouncing of drops.

(1) When you showed the movie of the impacting water drop with different speeds, it seems to me that the shape and/or the size of the impalement changes as a function of the drop impact speed.

Could you comment on this, please?

(2) In addition, how would the drop bouncing change if the surface is flexible or can be deformed by impact?

Thank you.

**Professor Quéré** replied: Thanks for asking these questions:

(1) Above a threshold velocity, drops indeed impale in surfaces textured with micropillars and the threshold depends on the nature (height, distance, *etc.*) of the pillars. If filmed from above, the impalement is made visible by the darkening of the substrate at the place where the drop penetrates. The impalement pattern indeed depends on the impact speed: at small velocities, the pattern magnifies the pillar array (with a square shape, if the pillars are on a square lattice), despite the difference in size (the pattern has roughly the millimetre-size of the drop, while the mutual distance between pillars is a few micrometres); at large velocity, the pattern becomes circular. The transition square/circle remains to be understood.

(2) Drop bouncing should be impacted by the flexibility of the substrate, because it depends on the way the drop kinetic energy is stored as it contacts the substrate. Deforming a flexible substrate is a way to store this energy, so that flexibility might generally favour the drop rebound. This phenomenon should be particularly relevant in the context of leaves.

**Dr Patankar** asked Professor Butt: The design of roughness-based superhydrophobic surfaces, that are characterized by large contact angles and low hysteresis, has relied upon or assumed the presence of air pockets in roughness grooves. This is reasonable in open air applications where air is available from the surrounding environment. Now consider applications such as low drag superhydrophobic surfaces for liquid flow, either in enclosed fluidic networks or on immersed bodies. In these cases air is not available from the ambient. Even if air is initially present, it can diffuse out of the roughness grooves depending on the degree of saturation of air in the liquid. Thus, the surfaces may not be robust against wetting transition. A proposed way around it is to design surface roughness such that the vapor of the liquid itself is stabilized in the roughness grooves by way of capillary evaporation.[1,2] It is predicted that roughness geometries on the order of hundreds of nanometres and smaller could stabilize the vapor phase at standard conditions. Superhydrophobic surfaces based on raspberry-like particles have geometric parameters in the same range. Is it possible to experimentally verify whether these surfaces do indeed retain the Cassie–Baxter state by stabilizing the vapor, of the liquid itself, in the roughness grooves even if there is no air available?

1 N. A. Patankar, *Soft Matter*, 2010, **6**, 1613–1620.
2 N. A. Patankar, *Langmuir*, 2010, **26**(11), 8783–8786.

**Professor Butt** replied: I suggest measuring the flow through a capillary coated with the superhydrophobic layer at different pressures and different pressure differences.

**Mr Pacheco Benito** asked: What do you think is the role of the silica nano-shells regarding the enhancement of the hydrophobicity? Is there a minimum layer thickness necessary in order to obtain superhydrophobic surfaces also when silica nano-shells are used?

**Professor Butt** answered: The silica nano-shells serve as anchoring sites for the fluorosilane. Therefore they are essential for hydrophobicity. We did not encounter a minimum layer thickness although a closed shell requires a shell thickness above 15 nm.

**Mr Boreyko** asked Professor Quéré: For the water droplets moving down the superhydrophobic decline, did you observe a critical Weber number at which point the drops consistently began to elongate and form a tail? Do these tails begin to form exactly when the drop transitions from the purely sliding regime to the sliding/rolling regime, or do rolling drops remain spherical at first?

**Professor Quéré** replied: This question is difficult, and we will try to answer it precisely in our current experimental program on these dynamical behaviours. (1) I would tend to think that it is not a critical Weber number that triggers the elongation of the drop, because (generally) inertia distorts a falling drop in the direction perpendicular to its direction (as observed with big raindrops). So the tail rather betrays a viscous effect, which implies that it is rather a critical capillary number which should be invoked here. (2) Rolling drops can remain spherical if they are slow enough, which we observed with liquid marbles (drops decorated by hydrophobic grains). Increasing the marble velocity leads to spectacular deformations, which are not of a tail-type: centrifugation can indeed deform the drop, which is captured by a kind of Weber number—here you are absolutely correct to refer to this number!

**Professor C. Chen** asked: What is the mechanism of two-tier roughness? Could you comment on the following theory: the nanoscale roughness is primarily enhancing the robustness of the Cassie state, while the microscale roughness is primarily reducing contact angle hysteresis.

**Professor Butt** responded: It is certainly intriguing to relate the different length scales to different effects. I do not know what to expect. But the question might be answered experimentally by making surfaces with different nano- and microscale structures and measuring contact angle hysteresis and robustness of the Cassie state independently. The robustness of the Cassie state could be measured by pressing drops of different sizes and with different pressures from a capillary to a superhydrophobic layer.

**Professor Quéré** responded: Many papers were published on this topic, but I am happy that you ask this question: it means that, like me, you were not fully convinced by what we read in these papers! As discussed with Alenka Luzar earlier, and as you say, a nanoscale roughness should yield a robust Cassie state. So why would we need a larger scale? I am not proud to answer your question by the same question, but this is where we are!

**Professor McCarthy** addressed Professor Butt and Professor Quéré: Do you see a contact line pinning (and lower receding contact angles) in some surfaces? Are defects the cause of the "tails" observed in rapidly moving drops?

**Professor Butt** responded: We mostly found contact line pinning and lower receding angle in samples with a very high degree of polystyrene leakage. The polystyrene is less hydrophobic than fluorinated silica and its leakage levels out the roughness of the surface. If the exposure to solvent was adjusted properly such a high polystyrene leakage could be avoided.

**Professor Quéré** replied: We commonly observed that water receding on patterned surfaces (such as it does when recoiling in a rebound experiment, before taking off) leaves a shining surface, which slowly vanishes as evaporation takes place. This is indeed related to line pinning, which produces tiny drops at the top of each pillar (hence the shining surface), as proposed in your own work. This seems to be quite different from the formation of large-scale tails for rapid drops, for which the global flow of water seems to matter. However, we often observed that a tail can appear when a drop crosses a defect on which it can pin, which elongates it, and eventually makes it leave a thin rivulet behind it—a different behaviour from what is reported in our paper here, where the tail follows the drop without being left behind, neither as a collection of tiny drops (first part of my answer) nor as a macroscopic rivulet (second part).

**Dr Patel** opened the discussion of the paper by Dr Daub: Fig. 2 of the paper indicates that a mixed surface (results shown in purple) with hydrophobic particles occupying the crevices between hydrophilic asperities, leads to contact angles that are in agreement with the Cassie–Baxter equation (eqn. 1). If we are to replace the hydrophobic particles (with a contact angle of 139°) by hard spheres (contact angle of 180°), it is reasonable to expect that (and by your own admission), the resultant surface would once again obey the Cassie–Baxter equation. (eqn. 2 and shown as dashed lines in the figure). Now consider the case of the structured hydrophilic surface that has been obtained by removing the hard spheres from the crevices of the corresponding mixed surface. (results shown in yellow.) This corresponds to a relaxation of constraints on the water molecules. They can now choose to occupy the crevices that were occupied by the hard spheres in the mixed surface if they so desire, or not if they don't. According to the second law of thermodynamics, when a constraint is relaxed, as in the case of going from a mixed surface to a hydrophilic one with asperities, the free energy of the system must go down. In other words, the energetics of the hydrophilic surface with asperities and hence its contact angle must be *lower* than that of the corresponding mixed surface. This same argument applies for a hydrophobic surface with asperities as well. (results shown in blue.) Since the reported contact angles for structured surfaces and asperity fractions greater than 0.2 (yellow and blue symbols) are all *higher* than contact angles expected for the corresponding mixed surfaces (yellow and blue dashed lines), it implies that the reported contact angles are metastable and not equilibrium contact angles.

Do you agree with the above?

**Dr Daub** replied: The free energy argument you make is clearly valid provided the observed geometric contact angles of nanodroplets and Young contact angles are identical.[1] Several factors may be responsible for the difference in the behaviors of observed *contact angle* on molecularly mixed and corrugated surfaces. First, the introduction of corrugations introduces some ambiguity about the height of the solid/liquid interface to be used in contact angle calculations, especially at lower coverage. In contrast, this height is well defined on mixed but smooth substrates at all mixing ratios. Second, a similar uncertainty applies to surface fractions on a substrate with molecular-scale grooves and asperities, but not to chemically mixed but geometrically homogeneous surfaces. Third, an extended surface would exert greater attraction on the water drop than small asperities. For a specified range of C–O pair potential[2] we can suppose that each water molecule at the interface must interact with carbon atoms of at least this range in order to recover the same findings as were observed on the smooth surface, *i.e.* asperities should have an area of $\sim 60$–$70$ Å$^2$. Note that all of our surface atoms have essentially the same size $\sigma \sim 3.2$ Å, so even the very hydrophobic atoms cannot be considered to be hard spheres.

1 Ingebrigsten and Toxvaerd, *J. Phys. Chem.*, 2007, **111**, 8518.
2 R. L. Jaffe, P. Gonnet, T. Werder, J. H. Walther and P. Koumoutsakos, *Mol. Simul.*, 2003, **30**, 205.

**Dr Henderson** made a general comment: I have counted nine papers in this Discussion which refer to the Wenzel, Cassie–Baxter or Cassie formulation of Young's equation for a drop placed on a superhydrophobic surface. This subject has a history of special relevance to *Faraday Discussions* and I am wondering if all participants are aware of the unified picture that emerged in 1948. I am concerned that a key paper in *Discussions of the Faraday Society* No. 3, by Shuttleworth and Bailey[1] may not be sufficiently noted. This seminal work emphasizes the significance of geometry and in particular introduces what is now called the filling transition between a Wenzel state and a state of complete filling (by vapour in the case of drying). The concept of the superhydrophobic state of adsorption was introduced as early as 1930 by N. K. Adam.[2] Adam's definition can be seen in Fig. 1. In 1936 Wenzel[3] derived the equivalent of Young's equation on the assumption that there were no filled regions,

$$\cos \theta = r_u \cos\theta_Y. \quad \text{(W)}$$

In 1944 Cassie and Baxter[4] generalized Wenzel's result to adsorption on a porous substrate, which is equivalent to removing the portions of the substrate denoted by $r_f$ and replacing by gas,

$$\cos \theta = r_u f \cos\theta_Y - (1 - f). \quad \text{(CB)}$$

In the CB equation the quantity $f$ denotes the fractional projected planar area under the solid regions; *i.e.* $r_u f$ is the actual area of solid–liquid interface. In the 1948 *Discussion of the Faraday Society* No. 3, the opening paper is given by N. K. Adam. He notes that Wenzel's picture supports his requirement that roughening a surface made of hydrophobic material ($\theta_Y > \pi/2$) increases the observed contact angle $\theta$. Adam also notes the extension by Cassie and Baxter. Cassie is the next speaker[5] and he generalizes the CB equation to situations like Fig. 1, where the final term in (CB) is kept general like the first term on the right side. Some current workers refer to the Cassie equation as if there were no Wenzel roughness factors present on the right side. This is definitely not what Cassie had in mind and in

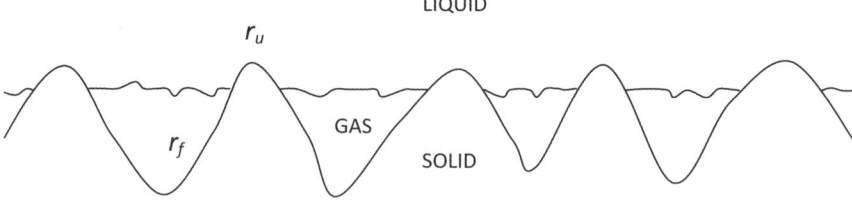

**Fig. 1** Adam's concept of superhydrophobicity. A portion of a drop of liquid placed on a rough surface is shown sitting on pockets of gas (vapour) that fill the bottoms of the valleys. The drop is therefore sitting largely on gas and will have a contact angle $\theta$ much closer to 180° than for a planar substrate of the same material (Young's contact angle $\theta_Y < \theta$). In specific geometries one is only required to distinguish the Wenzel roughness of the unfilled region of the substrate $r_u$ (see text).

fact CB consider a system of cylinders (representing fibres) where they explicitly calculate the Wenzel roughness factor required. Nevertheless, Cassie's equation is applicable to planar chemically patterned substrates, and to filling up to the tops of flat posts, where all the roughness factors happen to be unity. Below I shall refer to this special case as a Cassie state (since so many recent workers have adopted this language). I will use Cassie–Baxter–Adam (CBA) to denote the general case of Cassie's law, as would be required to analyse states like Fig. 1. The paper following Cassie, by Shuttleworth and Bailey,[1] is important because of its emphasis on geometry and because of the filling transition which in some geometries is a genuine interfacial phase transition between a Wenzel state and a CBA state. Here, it is perhaps worth briefly reviewing the derivation of generalized Young's equations like (W), (CB) and (CBA). Firstly, one is doing classical capillarity at a fixed volume of excess liquid, so that one may assume that an adsorbed drop is a spherical cap (as required to satisfy Laplace's force balance). Then one imagines an infinitesimal change of contact line radius (from one spherical cap to another). In this process the change in solid–liquid (SL), solid–vapour (SV) and liquid–vapour (LV) interfacial areas are related by $\cos\theta \, dA_{SL} = dA_{LV} = -\cos\theta \, dA_{SV}$. Outside the drop the entire substrate is filled by gas and so the filled regions of the substrate are unaffected by the above process (cancel out of the resulting generalized Young's equation). If we denote the actual solid–liquid interfacial free energy by $\tilde{\gamma}_{SL}$, then by reference to Fig. 1 we can see that $\tilde{\gamma}_{SL} = r_u f \gamma_{SL} + r_f (1-f)\gamma_{SV} + (1-f)\gamma_{LV}$, where the absence of a tilde denotes the planar values (defining $\cos\theta_Y$). Similarly, $\tilde{\gamma}_{SV} = [r_u f + r_f(1-f)]\gamma_{SV}$. Setting the work done at fixed temperature, chemical potentials and volume to zero (local mechanical equilibrium) requires $0 = (\tilde{\gamma}_{SL} - \tilde{\gamma}_{SV})dA_{SL} + \gamma_{LV}dA_{LV} = dA_{SL}[r_u f(\gamma_{SL} - \gamma_{SV}) + (1 - f + \cos\theta)\gamma_{LV}]$. Inserting Young's equation for the planar unfilled material and dividing through by the liquid–vapour surface tension leads immediately to the CB equation above (really the CBA equation when applied to Fig. 1 rather than a porous substrate). In general, of course, we always obtain

$$\cos\theta = \frac{\tilde{\gamma}_{SV} - \tilde{\gamma}_{SL}}{\gamma_{LV}}. \quad \text{(CBA)}$$

This derivation emphasizes that the only relevant physics is that concerned with the nature of the unfilled region sensed by the three-phase contact line of the adsorbed drop. This is local mechanical equilibrium, rather than total thermodynamic equilibrium. The filling transition is, by definition, the state where the system has the same free energy as if the liquid–vapour interface had lifted entirely off the substrate (a macroscopic film of adsorbed vapour between solid and liquid—known as complete drying). From (CB) and a wide variety of generalizations (since the filled regions are irrelevant to the derivation above) the filling transition is defined by the condition $r_u \cos\theta_Y = -1$. In the case of wetting ($\theta_Y < \pi/2$) the physics is determined by the solid–vapour interfacial free energy $\tilde{\gamma}_{SV}$ (equivalent to placing a bubble on a substrate immersed in liquid) and an identical derivation implies filling by liquid occurs at $r_u \cos\theta_Y = 1$.

Shuttleworth and Bailey[1] note the relevance of geometry (especially the parts of the surface with the largest roughness) but wisely choose to illustrate the consequences for a particular geometry in which the roughness is invariant with filling; namely, linear triangular wedges. Here the filling transition is a genuine interfacial phase transition between a Wenzel state and complete filling and occurs when Young's contact angle at the wedge walls allows for a planar liquid–vapour interface across the wedge. Thus, if one were able to decrease the wedge angle (making the grooves more acute) until reaching the critical angle for which the roughness (ratio of wedge surface area to projected area) satisfies the filling condition then the wedge will fill either by vapour or liquid. For a very acute wedge the Wenzel state can only be maintained by choosing a material for which Young's contact angle is precisely 90°. In this sense, the crossover between hydrophilicity and hydrophobicity can be said to occur at $\theta_Y = \pi/2$.

Geometry determines the position of this transition, but in some situations the molecular interactions will play a role in determining the order of the transition. As long as the roughness either increases or remains constant as the substrate fills, then the filling transition will be complete. However, if the roughness decreases with filling, then the filling condition will no longer be a genuine free-energy minimum; e.g. extreme superhydrophobicity will be pre-empted by a Cassie state for which $\cos\theta > -1$. Thus, from Shuttleworth and Bailey we learn that substrates fall into two distinct classes. An example of the unfavourable geometry is an array of posts with flat tops. Imagine Fig. 1 redrawn with partial filling up to some height below the tops of the posts. If this height satisfied the filling condition it would have the same free energy as complete drying, but the Cassie state (filling to the top of the flat posts) would be the thermodynamic minimum. Perhaps the best way to visualize this is to start with the drop in the Cassie state; now the roughness of the unfilled region (the flat tops) is unity and one cannot fulfil the filling transition condition. If one then pushes the drop partially into the posts then the roughness of the unfilled region steadily increases (the drop is now seeing a rough substrate). For typical numbers the state of complete filling is soon reached, but if one calculates $\tilde{\gamma}_{SL}$ it is easy to see that it is not a minimum; the Cassie state is the lowest free energy. Both Wenzel and Cassie were industrial scientists studying the water repellent properties of textiles. Thus, the historical synthesis described above was also applied to attempts at understanding contact angle hysteresis (the difference between soaking and extracting water from a textile). There is much to be found in these seminal papers. Recently, theoretical physicists such as A. O. Parry have paid special attention to the geometry of adsorption. When models are defined to possess a single class of geometry (roughness) continuous filling transitions have geometry-dependent critical exponents and associated fluctuation phenomena. Note that it is not necessary to have a drop of excess liquid on the surface to study filling/drying, since in a well-controlled system (or simplified model) all the physical information is contained in the solid–liquid part of the substrate (e.g. consider a desorption isotherm under liquid). In this context it is interesting to speculate on whether or not a true drying transition does exist on substrates whose asperities show an increase in roughness as filling proceeds. Lifshitz theory (dispersion forces) would continue to favour a Cassie state where the liquid–vapour interface remains pinned to the top of the sharp summits, but this mean-field interfacial approach does not contain the physics of capillary-wave fluctuations and it may be possible to show that the entropic cost of suppressing capillary waves is enough to allow the interface to detach altogether from the substrate.

1 R. Shuttleworth and G. L. J. Bailey, *Discuss. Faraday Soc.*, 1948, **3**, 16.
2 N. K. Adam, The Physics and Chemistry of Surfaces, Clarendon Press, Oxford, 1930, pp. 180–181.
3 R. N. Wenzel, *Ind. Eng. Chem.*, 1936, **28**, 988.
4 A. B. D. Cassie and S. Baxter, *Trans. Farad. Soc.*, 1944, **40**, 546.
5 A. B. D. Cassie, *Discuss. Faraday Soc.*, 1948, **3**, 11.

**Professor Choi** asked:
1. I wonder what the incident beam size (*i.e.* diameter) was in the experiment. Does beam size affect the measurement? Then how?
2. What is the spatial resolution of the diffractive technique? In other words, what is the minimum lateral size of the liquid/air meniscus (*e.g.* minimum groove width, $w$) that can be detected? If it is mainly limited by the given wavelength of the incident beam, what is the relationship between them?

**Professor Mugele** answered:
1. The beam diameter is about 1 mm. Thus it covers about 100 periods. Qualitatively, reducing the number of unit cells to scatter from broadens the diffraction peaks as in any diffraction experiments. We have not analyzed this in detail, so

I cannot answer how this effect will quantitatively affect the resolution of the measurement.

2. There are two aspects to "lateral resolution" of the technique. One is the number of unit cells that should be covered, the other is indeed the size of the smallest feature within each unit cell. The first one relates to the structure factor, the second one to the form factor in conventional X-ray scattering. Regarding the second issue, you need higher Fourier components to resolve the shape of the meniscus to some reasonable degree of accuracy. Again, we did not perform a quantitative analysis; practically, the deflection of a 1 micron meniscus is difficult to measure accurately with 500 nm light.

**Professor Butt** opened the discussion of the paper by Professor Gao: Layer-by-layer deposition usually leads to smooth surfaces. I would expect the roughness to be not higher than the diameter of the particles. Where do you think the roughness is coming from?

**Professor Gao** answered: During the deposition of the body layers, the negatively charged layer consists of a mixture of nanoparticles with two different sizes (60 nm polystyrene + 20 nm silica). The random packing of these particles in different sizes resulted in the roughness.

**Ms Mishchenko** addressed Dr Daub: You mentioned simulation of perfectly smooth surfaces and that mobility is higher on smooth hydrophilic surfaces than on hydrophobic ones. Do you think this may be an actual effect? Could this ever be measured experimentally?

**Dr Daub** replied: This result is counterintuitive, however, it is supported by the work of others, who have observed that the diffusion constant of water confined between planar surfaces of varied polarity reaches a maximum at a degree of polarity/hydrophilicity equivalent to a contact angle of ~ 30°.[1]

1 S. R. V. Castrillón, N. Giovambattista, I. A. Aksay, P. G. Debenedetti, *J. Phys. Chem. B*, 2009, **113**, 1438.

**Mr Epstein** addressed Professor Mugele: You described diffraction-based contact angle measurements for a 1-D system of grooves. Could you comment on any challenges that might arise for more complex symmetries, *e.g.*, a hexagonal array of posts, when the meniscus has multiple radii of curvature?

**Professor Mugele** replied: Experimentally this is absolutely no problem. In fact, the first paper that we published in this context was on the diffraction from a hydrophobized array of cylindrical holes immersed in water.[1] For a general surface with arrays of posts, the symmetry of the problem is very complex. From an experimental perspective, this does not pose substantial problems, yet, in terms of analysis more efficient numerical schemes involving various approximations would have to be devised to achieve a quantitative description. This is challenging, but definitely possible.

1 H. Rathgen, Kazuyasu Sugiyama, Claus-Dieter Ohl, Detlef Lohse, and Frieder Mugele, *Phys. Rev. Lett.*, 2007, **99**, 214501.

**Mr Wu** addressed Professor Gao: In your paper, you mentioned that once the spacers are removed, it will leave behind pores within the silica nanoparticle network. Have you conducted any cross-sectional TEM studies to confirm the existence of these pores?

The justification is that, at such elevated temperatures over a long period of time, the silica nanoparticles can undergo localized collapse and sintering. While it is intuitive to think that the silica network will not be affected by the thermal treatment, we have found from experience, such systems are highly sensitive to heating rate and duration.

Have you conducted any roughness measurements on your surface? It would be worthwhile to find out the thickness of the coating as well as the roughness to justify the origin of the observed transparency.

Since fluorocarbons are considered as a separate phase in comparison to aqueous and hydrocarbons, would you comment on the affect of substituting the fluorocarbon species with a common hydrocarbon on oleophobicity? The reason being, it would be worthwhile to see whether the overhang structure is really crucial in oleophobic behaviour for your specific coating, or is it simply a combination of roughness and fluorocarbon chemistry.

**Professor Gao** answered: We have not conducted cross-sectional TEM studies to confirm the existence of the pores. However, the porous structure of the film after removal of the sacrificial PS particles can be observed by SEM. The RMS roughness of the coating increased as a function of the number of bilayers. For the superhydrophobic and highly oleophobic films we demonstrated with 25 bilayers, the RMS roughness measured by AFM was in the range of hundreds of nanometres.

We did try to use SAM coatings of hydrocarbons instead of fluorocarbon molecules in our experiments, but the hydrocarbon molecules do not possess a low enough surface free energy to make the film oleophobic, even with the overhang structures.

**Professor Lamb** communicated: Is it correct to assume that the high film transparency is a result of lowered silica nanoparticle density, which was achieved through thermal degradation of the polymer spacers? If so, this would result in no significant change in the film's refractive index, and thus, this should give the film high transparency over a wider wavelength (*i.e.* beyond visible light).

**Professor Gao** replied: The film transparency significantly increases after the polystyrene particles are removed by calcination. It is very likely that this is due to the low density of silica nanoparticles and the highly porous structure of the film. We have not yet measured the film refractive index. We have only measured the film transparency of 320–800 nm wavelength light. The transparency for light beyond 800 nm wavelength has not been measured.

**Professor Law** addressed Dr Daub: In your computer simulation of nanometre sized droplets on flat and structured surfaces, do you need to account for the line tension of the droplet at the three phase solid–liquid–vapor boundary line?

**Dr Daub** answered: It has been observed that at the nanoscale, line tension manifests as a small finite-size effect, which causes nanoscale droplets to have slightly increased contact angles.[1] In our simulations, we have done trials with droplets of 4000 and 8000 water molecules and have not noted any significant changes with the increased size. It is possible that our use of Ewald summation rather than a cut-off reduces the finite-size effects.

1 T. Werder, J. H. Walther, R. L. Jaffe, T. Halicioglu and P. Koumoutsakos, *J. Phys. Chem. B*, 2003, **107**, 1345.

**Dr Sarupria** asked: Can you comment on how the droplet behavior will change if you make your asperities mobile, something similar to billiard balls or dumbbells rolling on the surface or like marbles rolling on a flat surface?

**Dr Daub** answered: This is an interesting question, but not one we have addressed. If the asperities were mobile and hydrophobic, they would likely be excluded from the liquid–surface interface. On hydrophilic surfaces, on the other hand, mobile asperities could facilitate water penetration in and wetting of the grooves.

**Professor Patankar** remarked: This question is about the finding in the paper that "for smooth surfaces...the drop mobility on hydrophilic surfaces exceeds that on the hydrophobic one". It is known that liquid slip (or the mobility of liquid molecules) next to a solid wall depends, among other things, on the effective potential experienced by the first layer of liquid molecules from the wall atoms. This potential is dominantly dependent on the liquid-wall interactions which include Lennard-Jones potential. The effective potential experienced by the first layer of liquid molecules is not constant with respect to the $x$–$y$ plane parallel to the wall even if the wall atoms are placed without any "roughness". This is due to the finite size of the atoms. If the effective potential in the $x$–$y$ plane has shallow valleys then the liquid mobility is high because the liquid molecules do not have strongly preferred locations along the $x$–$y$ plane.[1] Thus, even if a solid is hydrophilic but the wall potential experienced by the first layer of liquid molecules has shallow valleys with respect to the $x$–$y$ plane, compared to another hydrophobic solid, then the relative liquid mobility could be high. Perhaps you have verified this, but is this a possible reason underlying your results? One way to check this is to plot the density profile of the first layer of liquid molecules next to the "smooth" hydrophilic and hydrophobic surfaces. The case that has the less undulated density profile will be the one with the less undulatory wall potential which could cause higher mobility of liquid molecules.

1 H.-Y. Hsu and N. A. Patankar, *J. Fluid Mech.*, 2010, **645**, 59–80 and references therein.

**Dr Daub** replied: We have previously made some plots of the density profile of water drops near surfaces of varying hydrophilicity, but have only looked at the variation perpendicular to the surface. Here there is an increase in the discreteness of layers as the hydrophilicity is increased. We would therefore expect any variation parallel to the drop to follow the same trend, *i.e.* enhanced structure on hydrophilic surfaces. In any case, we expect this variation to be rather small near atomically smooth surfaces.

**Professor Patankar** addressed Professor Mugele: The Laplace pressure of a sessile drop causes wetting transition of smaller drops. Dynamic effects can also lead to high pressure at the base of a drop and induce transition. One example is a new proposition that transition is caused by water hammer effects.[1] This effect is hypothesized to be active even when a drop is supposedly "gently" deposited on rough surfaces (see the poster by Kwon *et al.*, Deceleration driven wetting transition during "gentle" drop deposition, in this volume), thus leading to wetting transition of larger drops. This observation is opposite to that caused by Laplace pressure of a sessile drop where smaller drops exhibit wetting transition. Can your techniques be used to directly verify dynamically driven transitions such as the water hammer? If not, what innovations or techniques would make such observations a reality or what are the challenges?

1 Tao Deng, Kripa K. Varanasi, Ming Hsu, Nitin Bhate, Chris Keimel, Judith Stein, and Margaret Blohm, *Appl. Phys. Lett.*, 2009, **94**, 133109.

**Professor Mugele** responded: In principle our diffraction technique has the advantage of being very fast. In fact, we measured in our paper[1] the response of arrays of micromenisci to ultrasound fields at frequencies up to MHz. (In terms of the measurement technology, even hundreds of MHz are not a problem.) Being a diffraction technique, though, our approach requires that the system is laterally homogeneous over a few tens of periods. Dynamic transitions such as the proposed water hammer probably involve lateral variations of the shape of adjacent menisci on much smaller scales, which would make the diffraction technique fail. In contrast, high speed video imaging with imaging rates of 100.000 fps using transparent substrates also allow for a rather good time resolution while simultaneously offering high lateral resolution. Despite the poor resolution regarding the vertical deflection, this approach probably provides a good idea of the general phenomenology and may allow for discriminating between competing transition scenarios.

1 H. Rathgen, Kazuyasu Sugiyama, Claus-Dieter Ohl, Detlef Lohse, and Frieder Mugele *Phys. Rev. Lett.*, 2007, **99**, 214501.

**Professor Debenedetti** addressed Dr Daub:
1. Have you checked whether you are at steady-state with respect to evaporation? In our computational studies of similar systems, we saw a non-negligible number of water molecules in the vapor phase. This could be a source of center-of-mass fluctuations in the drop.
2. Have you studied the structure of the water layer in contact with the surface, to see whether differences in water structure near hydrophobic *vs.* hydrophilic surfaces correlate with the diffusivity trends that you show in Figure 5 in your paper?

**Dr Daub** replied: To answer the first question, all we can say is that we equilibrated simulations for upwards of 400 ps before gathering data, sufficient to reach the steady-state with respect to evaporation as well as with any fluctuations in the interfaces. The number of molecules in the vapour phase is close to the thermodynamic prediction at the given temperature and droplet size; relocation of vapour molecules can contribute to droplet movement, however, the contribution should not be significant given only a minute fraction of molecules exist in the gas phase. To answer the second part of your question, we have not studied the details of the water hydration structure near the interface, other than *via* the interfacial hydrogen bonding. The number of interfacial hydrogen bonds near the smooth hydrophobic surfaces is in line with your own group's results.[1,2] Also in agreement with your results, the number increases near hydrophilic surfaces, although the presence of water-surface hydrogen bonds in your study leads to quantitative differences.

1 S. R. V. Castrillon, N. Giovambattista, I. A. Aksay, P. G. Debenedetti, *J. Phys. Chem. B*, 2009, **113**, 1438.
2 S. R. V. Castrillon, N. Giovambattista, I. A. Aksay, P. G. Debenedetti, *J. Phys. Chem. B Lett.*, 2009, **113**, 7973 and Supporting Material.

**Dr Hummer** commented: First, the sharply peaked contact angle in Figure 2 (in your paper) as a function of the surface coverage by asperities $f_a$ suggests a crossover between states typical of high and low coverage. For conditions near the cusp (where $f_a \sim 0.1$; see also Fig. 3 in your paper), do you see co-existence between different "states" of the droplet, possibly with dynamic transitions between penetrating and non-penetrating states? In our simulations[1] of water near a rough surface, we see such "transitions" locally.

Second, which physical and simulation factors affect the center of mass motion of the droplet that is used to determine its apparent diffusion constant? Does the choice of thermostat play a role by transferring momentum between the particle system and a "heat bath"? And, in light of Professor Debenedetti's comment, could differing degrees of evaporation explain some of the observed differences in the apparent diffusion?

1 Jeetain Mittal and Gerhard Hummer, *Faraday Discuss.*, 2010, **146**, DOI: 10.1039/b925913a.

**Dr Daub** replied:
(1) We did rather long simulations to see eventual coexistence between the penetration regime and the "fakir" state, but we never observed this. Perhaps more careful tuning of the coverage near $f_a \sim 0.1$ could reveal such coexistence.

(2) The primary mechanism behind the center-of-mass motion in our simulation is perpetual shape fluctuation of the drop. This fluctuation involves lateral motions of the molecules in the solvation layer, a process sensitive to water/surface interaction. Even if increased in number, loose water molecules in the vapor phase would still represent a very small fraction of the total number of water molecules in the system (O(10) water molecules in our simulation box of volume ~ $2.8 \times 10^3$ nm$^3$ at room temperature). It is unlikely that the exchange between the two phases could affect the drop dynamics significantly or that the dynamics of this mechanism would differentiate between hydrophilic and hydrophobic surfaces.

**Mr Stirnemann** asked: The faster translation diffusion of the drop center of mass that you measure for increasingly hydrophilic surfaces (Fig. 5 in your paper) is very intriguing. Is this faster translational dynamics of the drop center reflected by faster translational dynamics of the interfacial water molecules? Do the thermal fluctuations in the drop shape change with the surface hydrophilicity?

**Dr Daub** communicated in reply: Yes, we observe enhanced random fluctuations when the smooth surface is hydrophilic. With regards to your first question, please see our reply to Gerhard Hummer, and the article from the Debenedetti group,[1] where they observe similar trends as we do for translational diffusion of water molecules near surfaces of different hydrophilicities.

1 Santiago Romero-Vargas Castrillón, Nicols Giovambattista, Ilhan A. Aksay and Pablo G. Debenedetti, *J. Phys. Chem. B*, 2009, **113**, 1438.

**Professor McCarthy** said: You conclude that an atomically smooth surface would become hydrophobic. Wouldn't there be water vapor present near a drop? Wouldn't it adsorb (or condense)?

**Dr Daub** replied: Atomically smooth surfaces may be either hydrophilic or hydrophobic, depending on the chemical composition of the surface. Our droplets are in equilibrium with vapour, so we see a small number of water molecules evaporate and generally rejoin the droplet sometime later. In some tests, we have simulated a dense vapour of water molecules and observed the condensation on the surface. We observe many small droplets condensing on the surface. Presumably, these droplets would coalesce into one large droplet after sufficiently long simulation. We also see a small number of water molecules (O(10)) adsorb in the grooves of the corrugated, hydrophilic surfaces. Their presence could have some small effect on a water droplet above these adsorbed molecules, likely causing a small reduction in the contact angle, but the change would not be significant in any case.

**Dr Jamadagni** asked: I had one comment and one question.

**Comment**: I would like to bring what I feel is an under appreciated point to the attention of researchers employing simulations to study "nanodrops" — the effect of cutoffs on the measured contact angle. When a surface is present on one side of the interface, there is effectively a much longer ranged 9-3 potential instead of the typical 12-6 Lennard Jones potential. Thus conventionally used cutoffs of 1.0-1.2 nm may not be sufficient. The problem is especially important when simulating hydrophilic surfaces. This issue was highlighted in a recent publication.[1]
See Fig. 5, C. Sendner, D. Horinek, L. Bocquet, R. Netz, *Langmuir*, 2009, **25**, 10768.

**Question**: I find the result that the diffusivity of drops on hydrophilic surfaces to be more than that on hydrophobic surfaces difficult to rationalize. The diffusivity was calculated by fitting the mean square displacement data from a 30–40 ps time interval. Fitting over such a short time interval is probably not correct since the droplet center of mass motion arises from fluctuations in its shape rather than any global motion. A minimum time interval for such fitting would of the order of the time required for the drop to diffuse by its diameter. This might be rather large for droplets comprising a few thousand water molecules, especially on hydrophilic surfaces, but might be the physically correct thing to do (a few hundred picoseconds to 1–2 nanoseconds).

**Dr Daub** answered: Our comment to the comment:

Jaffe et al.[1] systematically examined the impact of truncation and while they identify a noticeable effect, they reparameterize the CO water potential to offset the cut-off effect. We use their parameters in our study. We have not noticed any systematic variation in the contact angle with larger drops (4000 and 8000 water molecules). Instead of the rather short-ranged cut-off of Coulombic interactions used in Jaffe et al. (smooth cut-off interpolated between 9 and 10 Å pair separation),[1] we have used the Ewald summation. For situations we studied, this seems to greatly reduce the importance of finite-size effects, or equivalently, the magnitude of the line tension at the three-phase interface line. See also our reply to Prof. Errington.

Our reply to the question:

I agree that quantitative determination will require much longer simulations. However, the italic trend of faster diffusion on hydrophilic *vs.* hydrophobic surfaces is quite clear in runs we have been able to make. Increased droplet motion on hydrophilic surfaces conforms with faster diffusion of individual molecules in the first solvation layer observed by the Debenedetti group.[2] In their work the variation of lateral diffusion constant with surface hydrophilicity is non-monotonic and peaks at a contact angle of about 30°. A reduction in contact angle from 108° to 59° in their study leads to about a four times faster lateral diffusion.

1 R. L. Jaffe, P. Gonnet, T. Werder, J. H. Walther, P. Koumoutsakos, *Mol. Simul.*, 2004, **30**(4), 205.
2 Santiago Romero-Vargas Castrillón, Nicols Giovambattista, Ilhan A. Aksay and Pablo G. Debenedetti, *J. Phys. Chem. B*, 2009, **113**, 1438.

**Professor Errington** remarked: I have two questions — one addressing the shape of the droplet and the second dealing with the nanodroplet technique for measuring contact angles.

First, in a recent paper by Yong and Zhang,[1] which focused on the wetting properties of Lennard-Jones particles on substrates with one-dimensional grooves, it was reported that anisotropic droplets form on the surface. Did you observe anisotropic droplets in your work?

Second, there has been a bit of controversy regarding the validity of the nanodroplet approach. MacDowell et al.[2] and Ingebrigtsen and Toxvaerd[3] have reported that the nanodroplet technique can give results that disagree with a pressure tensor or free energy based approach. The disagreement seems to be significant when the contact angle is less than 90°. In addition, Werder et al.[4] have reported substantial finite-size effects with the nanodroplet approach. In your work, you report that you do not observe such finite-size effects. Could you suggest a reason for this difference? What was different about your approach that prevented finite size from playing a role?

1 Yong and Zhang, *Langmuir*, 2009.
2 MacDowell et al., *Colloids Surf., A*, 2002.
3 Ingebrigsten and Toxvaerd, *J. Phys. Chem. C*, 2007, **111**, 8518.

**Dr Daub** replied: Some small anisotropies are observed in the droplet shape; it may look somewhat "square", for example, in the case of low surface fraction of asperities on the hydrophilic surface. We did not, however, attempt to quantify any difference in contact angles or droplet shape in different directions, which in any case would have been quite small in all of our systems unless an external field was applied.[1] In the Yong and Zhang paper, they observe a large anisotropy only in the case of rather deep grooves. Our case with a one-dimensional groove would correspond with a rather small surface roughness factor, where they only observe small anisotropy (*ca.* 5° difference in contact angle parallel and perpendicular to the surface). In addition, they study a mercury/copper interface, which is likely to behave much differently than a water/carbon interface.

As you mention, the deviations of geometric contact angles discussed by Ingebrigtsen and Toxvaerd[2] become significant for contact angles below ~ 90°, while in all but three out of seventeen systems we study are above this empirical threshold. The Werder *et al.* study[3] used a very short cut-off for the Coulombic contribution to the pair interactions. In our trial calculations, the dependence of the contact angle on the size of the droplet appeared greatly reduced by the use of Ewald sums for the Coulombic contribution to the energy. Also, the Ewald results agreed with those we obtained in non-periodic calculations when performed without any cut-off of water pair potentials. See also our response to Dr Jamadagni.

1 C. D. Daub, D. Bratko, K. Leung, A. Luzar, *J. Phys. Chem. C Lett.*, 2007, **111**, 505.
2 Ingebrigtsen and Toxvaerd, *J. Phys. Chem. C.*, 2007, **111**, 8518.
3 T. Werder, J. H. Walther, R. L. Jaffe, T. Halicioglu, P. Koumoutsakos, *J. Phys. Chem. B*, 2003, **107**, 1345.

**Professor Corrales** commented: You indicate in your data that there is an asymmetry in the droplet velocity, consistent with the asymmetry of your surface structure with rows along one direction and ruts along the other. Why is it that there isn't some favored development of an asymmetric shape? A statistical mechanical argument would venture to state that there will be a higher occupation along the energy component with more states available.

In your manuscript you indicate that the simulations are sampled within the NVT ensemble. Did you test to see if you were truly at equilibrium by running the simulation within the NVE ensemble? Sampling is typically done within the NVE ensemble. Could this be why you don't see an asymmetric spreading of the drop?

**Dr Daub** replied: Some small asymmetry can be seen in the spreading of the drop, as it accommodates the underlying surface structure, but this asymmetry seems to be confined to the first or second molecular layers nearest the surface. The droplet shape remains circular to a good approximation. However, the dynamics of the droplet's motion is dependent on the behaviour of the surface layers, so that asymmetry in the capability of these surface molecules to move in different directions affects the overall diffusion of the droplet.

The relative fluctuation in the energy of the droplet is small (<0.1% or so) and test runs at NVE conditions validated equilibration. Therefore we do not expect our results to vary much if a different ensemble was used.

**Dr Vollmer** addressed Professor Gao: The authors determined a contact angle of hexadecane in excess of 140°. How quickly does this value change over the course of time? How reproducible is the contact angle if it is measured several times at the same spot?

**Professor Gao** responded: The measurement was typically finished within 30 s. We did not observe significant change in contact angles during this period. Each contact angle measurement was repeated three times at different places of the sample, and the average value was reported. If the contact angle is measured several times at the same spot, the standard deviation is typically less than 5°.

**Professor C. Chen** asked Dr Daub: Instead of varying the solid–liquid contact fraction $f$, can you vary the actual length scale of the surface texture? How do you bridge the length scale of MD simulation to the 100 nm scale that seems to be magic in promoting robust superhydrophobicity?

**Dr Daub** responded: 100 nm is far too large for atomistic simulations with today's computers. Some form of coarse-grained simulation, for example that studied by Julia Yeomans, would be a way to explore this size regime, and they have many interesting results in this area.

**Dr Patel** said: The paper by Acharya *et al.*[1] shows that a small hydrophilic patch on a hydrophobic surface makes the surface locally hydrophilic to an extent that is disproportionate to the area of the hydrophilic patch. You have shown (Fig. 2) that even a small fraction of asperities makes the surface more hydrophobic. Would you expect a hydrophobic surface to become more or less hydrophobic if it had a small fraction of hydrophilic asperities?

1 Hari Acharya, Srivathsan Ranganathan, Sumanth N. Jamadagni and Shekhar Garde, *Faraday Discuss.*, 2010, **146**, DOI: 10.1039/b927019a.

**Dr Daub** responded: This would depend on the size of the asperities, depth of the asperities, their hydrophilicity, *etc*. A small fraction of moderately hydrophilic asperities might increase the contact angle of a rigid hydrophobic surface such as we study, as long as the water is prevented from penetrating the top layers.

**Mr Stirnemann** communicated: The surfaces in your study are rendered more hydrophilic by tuning the water-surface Lennard-Jones interaction, while keeping the surface uncharged. Can the conclusions obtained with these surfaces be generalized to the more chemically relevant case of hydrophilic surfaces including charged groups and pinning sites, where the water structure is probably very different?

**Dr Daub** communicated in reply: We would expect that the main effect of changing the surface in this way would be on the first layer or two of water molecules. Introducing surface components able to form hydrogen bonds could affect the threshold for penetration by the water, likely up towards higher coverage. The mobility of the water drop might also be affected. A sufficiently highly charged surface group could possibly play the same pinning role as an asperity does on the hydrophilic surface.

**Mr Kang** said: I appreciate your interesting study.
During your talk, you mentioned that the surface should be rigid to keep superhydrophobicity. Is this requirement applicable regardless the relative size of the drop with respective to the size of the roughness?
If so, could you comment on what sets the criteria for the rigid surface?

**Dr Daub** answered: There is no indication that the size of the drop will appreciably change these conclusions. For one thing, we have done some trials with drops of 4000 and 8000 water molecules and have not noticed any change in the measured contact angles. The reason rigid corrugations are more effective is simple. In our study, the crucial requirement for poor wettability is that water be prevented from penetrating into the very small grooves or holes in our surface. If the surface is too soft, and inherently hydrophilic, it is likely that water will interact with the surface and alter the interfacial geometry such that water will be able to penetrate.

**Dr Christenson** addressed Professor Mugele: Would it be possible to construct a surface similar to the one described except with wedge-shaped grooves? In that case one would expect to be able to apply very high pressures as the interface curvature could be made very small down at the bottom of the wedge. The measurements should also be reversible as cavitation would occur in the wedge on lowering the pressure.

**Professor Mugele** responded: Yes, this would be possible. The numerical method used to analyze the diffraction data is developed for the design for the optimization of optical diffraction gratings, which often include triangular and even asymmetric triangular (blazed) surface topographies. Additional technical challenges would arise because the positions of the contact lines would have to be included as additional free parameters. Moreover, when the menisci approach the bottom of the grooves upon increasing the pressure, they will become laterally rather small. Both effects will probably lead to a somewhat reduced resolution compared to our present experiments. Yet, fundamentally such an experiment would be possible and is in fact an interesting perspective — for the physical reasons that you mention.

**Professor Evans** asked: Professor Mugele has shown the results of beautiful experimental measurements of the shape of liquid–gas interfaces at hydrophobic surfaces. In particular he investigates the validity or accuracy of the macroscopic Laplace law at micron length scales. At first sight the accuracy of this law at such small scales is surprising. It reminds me of related observations on the validity of the macroscopic Kelvin equation for capillary condensation in a model pore. In a partial wetting situation this equation, which describes the transition between 'gas' and 'liquid' states in a narrow capillary, remains accurate down to capillary radii, or wall spacings in a slit pore, of 10–15 molecular diameters, *i.e.* a few nanometres. See, for example, R Evans *et al.*[1]

Why these macroscopic capillarity equations remain quantitatively accurate down to such small sizes remains somewhat puzzling.

Robert Evans, Umberto Marini Bettolo Marconi and Pedro Tarazona, *J. Chem. Soc., Faraday Trans.* 2, 1986, **82**, 1763.

**Professor Mugele** responded: I agree completely. It is indeed a remarkable observation that the macroscopic laws of continuum physics hold down to extremely small scales. Even dynamically (*e.g.* with respect to viscous dissipation) most liquids behave bulk-like down to a few nanometres of thickness. In that sense, our present experiments are not a record in "smallest scale continuum behavior" — yet they probably are an example of a particularly accurate verification of macroscopic laws down to the submicrometer scale.

**Dr Chunder** asked Dr Daub: You mentioned in one of your slides presented here that the superhydrophobic surface, you were discussing, required rigid and rough surface at small scale. And also you mentioned that increasing roughness of a hydrophilic surface may render superhydrophobicity. What type of material are you talking about? I know mica which is atomically flat, what will happen if the roughness of the mica surface is increased? Superhydrophobic surfaces can be made up of several flexible substrates too, but probably you were talking about very small scales like, scale less than an Ångstrom.

**Dr Daub** replied: The length scale where penetration starts to occur is about the size of a few water molecules, *i.e.* a few Ångstroms. So the surface must be sufficiently rigid to maintain corrugation at this length scale. An experimental analogue of our study is probably not feasible with current technology, but a possibility would be some sort of metal surface precisely etched at a resolution of only 2 or 3 atoms. Another intriguing possibility is a very cold surface, with the droplet composed of a liquid noble gas.

**Professor Yeomans** commented: Experiments and simulations have been carried out for ridge geometries with mesoscopic length scales.[1] Significant hysteresis was seen. Drops placed gently on the surface spread preferentially along the posts because pinning on the post edges impeded perpendicular motion. Drops 'stamped' onto the surface and then released, however, had a final state elongated perpendicular to the ridges, again because it was easier for them to retreat along, than perpendicular to, the ridges.

It would be interesting to compare the microscopic and the mesoscopic results and try to understand the cross over between the two length scales.

1 R. J. Vrancken, C. W. M. Bastiaansen, and J. M. Yeomans, *Langmuir*, 2008, **24**, 7299.

**Professor Bratko** asked Professor Mugele: In the case of sharp asperity edges, is there any indication for the contact angle of the meniscus at the very edge of the groove to differ from the contact angle observed on vertical walls in a partially filled groove?

**Professor Mugele** answered: Unfortunately, we are not able to address this question. We have no means to directly measure the contact angle on the side walls.

**Professor Garde** remarked: You referred to Laplace pressure playing a role in your system. Laplace pressure is expected to be quite high, several hundred bars, inside small (nanoscopic) droplets. Are there any experimental probes of local pressure inside such droplets?

**Professor Mugele** answered: I am only aware of indirect measurements. In the present paper, the characteristic size of the structures is of the order of a few micrometres. This leads to typical pressures of the order of a few hundred millibars. For non-wetting porous media and for nano-channels it is possible to measure the intrusion pressure. This is a standard way to determine pore sizes based on the application of the Laplace pressure on the nano-scale. The Ducker group[1] determined the pressure in surface nanobubbles based on the absorption of IR light by vibrational resonances of $CO_2$. Their measurements indicate that the pressure in the gas phase of these structures is approximately 1 bar. I am not aware of any methods to determine directly the pressure in a single nano-drop.

1 Xue H. Zhang, Abbas Khan, and William A. Ducker, *Phys. Rev. Lett.*, 2007, **98**, 136101.

**Professor Jiang** addressed Dr Daub:

1 Contact angle definition of superhydrophobic state : contact angle > 150° and sliding angle <10° (or rotation angle).
2 Definition of hydrophilic state : contact angle <65°.

**Dr Daub** replied: So the question is about the definition of "superhydrophobic" or "hydrophilic". We feel that these definitions are really arbitrary and relative; in the context of our study, "superhydrophobic" means a very large contact angle near 180°, and "hydrophilic" means a moderately low contact angle, less than 90°. Other definitions are of course valid in other contexts.

**Professor Butt** asked: A simple estimation using continuum theory and steady state indicates that a water drop of 10 nm radius should evaporate within 10–100 ns. Does evaporation occur in you simulations and might it somehow influence the results?

**Dr Daub** answered: Our nanodroplets do not evaporate on the timescale of our simulations (a few ns). The droplet is in equilibrium with its own vapour, so that when occasional water molecules evaporate, they can rejoin the droplet at a later time. As we noted in the answers to Prof. Debenedetti's and Dr Hummer's questions, the number of vapour water molecules is too low. We do not believe our contact angle results would be significantly changed with larger droplets. See also replies to Prof. Errington and Dr Jamadagni.

**Professor Giovambattista** asked: This question is related to the reported diffusion of water droplets on the different surfaces studied. It is difficult to understand why such droplets are indeed displacing. Is the velocity of the droplet's center of mass fixed to zero at the beginning of the simulation? If not, this could explain the droplets motion.

**Dr Daub** responded: The nonzero velocities associated with diffusive motion of the drop corresponded to pre-equilibrated samples where surface friction would have quenched any ballistic motion of the drop even if present at the outset. Thermal motion causes some fluctuation in the position of such a small droplet and there is no reason to set it to zero at an arbitrarily chosen start of the sampling process. We note somewhat reduced diffusion in limited simulations of larger-sized droplets.

**Mr Vanzo** addressed Dr Daub: In your model system the water droplet is surrounded by a very low density gas, making it more similar to a (real) experiment performed in a vacuum. This feature practically rules out any water vapor adsorption on the structured surfaces, *e.g.* single water molecules penetrating into the surface grooves. In the case of larger vapor pressure of water molecules do you expect these adsorption phenomena might modify the droplet spreading behaviour and induce the transition from Cassie to Wenzel state? Could you comment on that?

**Dr Daub** responded: We recently did some trial runs where we introduced a high-density vapour into a simulation box empty except for the graphite surface and watched it as it interacted with the surface. When the surface was corrugated and hydrophilic, we observed only a small number of water molecules (*ca.* 10) adsorbing inside the grooves. So it seems that this phenomenon would not cause a major change in the wetting behaviour. Any small change would likely be towards slightly lower contact angles, as the adsorbed water molecules act like hydrophilic surface sites.

# Contact angle hysteresis: a different view and a trivial recipe for low hysteresis hydrophobic surfaces

Joseph W. Krumpfer and Thomas J. McCarthy*

*Received 27th November 2009, Accepted 22nd January 2010*
DOI: 10.1039/b925045j

Contact angle hysteresis is addressed from two perspectives. The first is an analysis of the events that occur during motion of droplets on superhydrophobic surfaces. Hysteresis is discussed in terms of receding contact line pinning and the tensile failure of capillary bridges. The sign of the curvature of the solid surface is implicated as playing a key role. The second is the report of a new method to prepare smooth low hysteresis surfaces. The thermal treatment of oxygen plasma-cleaned silicon wafers with trimethylsilyl-terminated linear poly(dimethylsiloxane) (PDMS - commercial silicone oils) in disposable glass vessels is described. This treatment renders silicon/silica surfaces that contain covalently attached PDMS chains. The grafted layers of nanometre scale thickness are liquid-like (rotationally dynamic at room temperature), decrease activation barriers for contact line motion and minimize water contact angle hysteresis. This simple method requires neither sophisticated techniques nor substantial laboratory skills to perform.

## Introduction

Contact angle hysteresis complicates much of the history of wetting (liquids interacting with solid surfaces). Thomas Young[1] initiated modern thought on and scientific research concerning contact angle and wetting in 1804 assuming that there was no hysteresis, but rather "an appropriate angle of contact" for every liquid/solid pair. In retrospect it is apparent that Young could not have carried out many contact angle measurements and maintained his belief in this faulty assumption. Bartell,[2] who measured thousands of contact angles and always observed hysteresis, assumed prior to 1932 *"that either an advancing or a receding angle would, within a short time, so adjust itself as to give finally a definite equilibrium angle which would be the same whether approached from the advancing or the receding angle."* Subsequently he discounted this assumption and reported, *"We have since obtained good evidence that advancing angles and receding angles may each exist as definite, but different, equilibrium angles."* Wenzel[3] in 1936 and Cassie and Baxter[4] in 1944 developed theories that have been very influential on others' perspectives of wetting. Their theories do not address hysteresis, although at least Cassie was well aware of it. A section of his 1948 *Faraday Discussion* presentation[5] is titled "The Problem with Receding Contact Angles." These theories have had an arguably destructive influence[6,7] on wetting research and surface science education in the subsequent decades. Modern theoretical treatments of wetting describe numerous differently defined contact angles,[8] but not the experimentally accessible advancing and receding

*Polymer Science and Engineering Department, University of Massachusetts, Amherst, MA, 01003, USA. E-mail: tmccarthy@polysci.umass.edu*

contact angles that determine hysteresis and both shear and tensile hydrophobicity. Contact angle hysteresis has been neglected, tacitly ignored and generally considered linked to defects in solid surfaces.

## Overview

We write this paper with two goals in particular: First, in preparation for a discussion on contact angle hysteresis, we present a new perspective of water drop motion on superhydrophobic surfaces. This perspective is a consolidation of views that we have presented in one 2008,[9] two 2006,[10,11] and two 1999[12,13] papers. This perspective could be gleaned from these manuscripts, but its formulation would be difficult. Second, we describe a simple and reproducible method (a trivial recipe) for preparing smooth surfaces with low hysteresis.

We refer to a 2006 paper[10] titled "*Contact Angle Hysteresis Explained*" and make only brief remarks here that summarize more detailed explanations made in the referenced paper. When a liquid drop moves on a solid surface the 3-phase contact line around the entire drop perimeter continuously redefines itself[14] through advancing and receding events that may be concerted, synchronous or sequential. The mechanisms of these events might be described as rolling or sliding (as extremes) and combinations/hybrids of these extremes could be operative around the contact line of a given moving drop. Some contact line sections could advance and recede back and forth between two different metastable states at greater rates, in microns per millisecond, than the macroscopic drop motion, in millimetres per second (note that the units are equivalent). Advancing and receding events are not generally the reverse of one another[12] and thus will almost always have different barriers. This view leads to the expectation that most surfaces should exhibit contact angle hysteresis - even if they are not dirty, rough or chemically heterogeneous. This is neither a common view nor one that a student of wetting would be taught from any textbook or almost any of the literature. Surface scientists have generally learned to view hysteresis as a "fault" that imperfect surfaces exhibit. Most have not considered that one dimensional intersections of homogeneous (at some length scale) gas, solid and liquid (3D) materials can be heterogeneous, that molecular (bond length scale) heterogeneities in solids[15] as well as the molecular volume and structure of liquids[16] can contribute to hysteresis or that contact lines do not have radial symmetry. Fig. 1a shows a cross section of a sessile drop perpendicular to the solid/liquid/vapor contact line (shown as a point). Moving the contact line to the left (advancing) will involve different barriers than moving it to the right (receding).

From a practical (experimental) perspective, the task of preparing surfaces that exhibit negligible or low hysteresis is not a small one. There is ample literature to refer to, but filtering reproducible methods and legitimate contact angle data from the mass of information is nearly impossible. The recent attention to "superhydrophobity" has complicated this task.[17] Some rough surfaces exhibit little or no

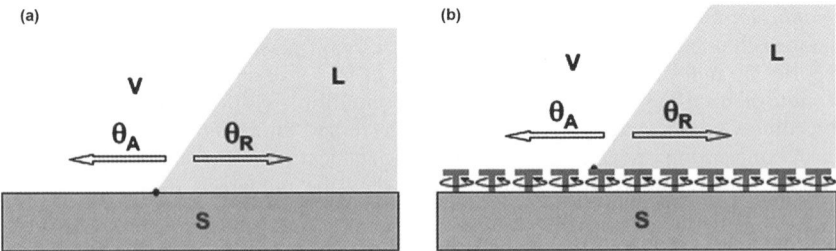

**Fig. 1** Cross sectional views of contact lines at solid/liquid/vapor interfaces on rigid (a) and liquid-like (b) surfaces.

hysteresis while others exhibit extremely high hysteresis. These surfaces have been described as "slippy superhydrophobic" and "sticky superhydrophobic" as well as other adjectives.[9] This terminology has been criticized[9] and we discuss these issues below. Our group has reported preparations of smooth surfaces that exhibit negligible hysteresis,[13,18,19] but these involve techniques of air-sensitive organic synthesis and silane monolayer preparative chemistry. These methods are inherently irreproducible due to "holes," that are reagent size but larger than the probe fluid, which occur as the result of random covalent attachment. Defects and heterogeneities (even at molecular length scales, for example an ethyl group rather than a methyl group[15]) contribute to hysteresis and the contact line will locate and focus on these defects during advancing and receding contact angle measurements. Our strategy for minimizing hysteresis has been to prepare randomly covalently attached monolayers that are not "close packed," but in which the attached molecules are free to rotate (exhibiting liquid-like or disordered plastic crystalline behavior - Fig. 1b), heal defects in the surface, and lower activation barriers to the level that the contact line is dynamic at room temperature.

## Contact angle hysteresis on superhydrophobic surfaces

Droplets roll on superhydrophobic surfaces containing topographical features that minimize contact between liquid and solid. This rolling motion and the ability of rolling droplets of water to pick up debris from the solid surface is termed the "lotus effect." Surfaces containing "posts" have been used to model this effect and we have reported[20] water contact angle data for surfaces containing posts of different size, shape, spacing and surface chemistry (covalently attached alkyl, perfluoroalkyl and silicone groups). A scanning electron micrograph of one of these surfaces is shown in Fig. 2a. This surface was prepared[20] using standard photolithography techniques and contains staggered rhombus-shaped posts that are smooth on their tops

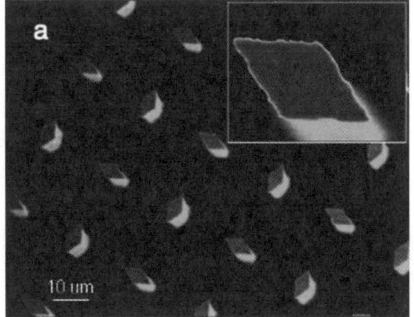

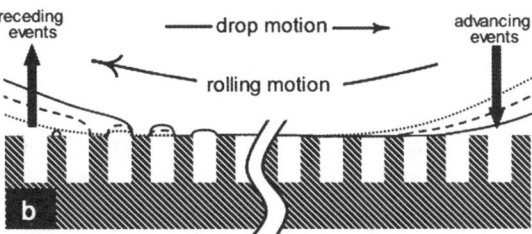

**Fig. 2** (a) SEM of a superhydrophobic modified silicon surface containing staggered rhombus posts (8 × 4 × 40 μm). (b) Advancing and receding events that occur at the contact lines of a rolling water drop on a surface such as the one shown in (a).

and are all the same height. This surface, when modified with dimethyldichlorosilane,[11] exhibits water contact angles of $\theta_A/\theta_R = 176°/156°$. The smooth surfaces of the post tops exhibit contact angles of $\theta_A/\theta_R = 104°/103°$.

Drops of water roll easily on this surface when it is tilted a few degrees from the horizontal (the minimum tilt angle depends on the drop volume), but are pinned when the surface is horizontal. The 20° hysteresis requires the rolling drop on a tilted surface to distort from a section of a sphere to a shape that exhibits a 176° contact angle (advancing) at the downhill edge and a 156° contact angle (receding) at the uphill-most section of the contact line. Fig. 2b describes the advancing and receding events that must occur at the contact lines. The advancing contact line does not move, but a new one forms as the liquid–vapor surface descends onto the next posts to be wet. The advancing contact angle of the post tops is 104° and the macroscopic drop contact angle is 176° so the drop spontaneously wets the post tops (no activation barrier). The receding events at the rolling drop contact line involve tensile hydrophobicity.[9] The rolling drop with the macroscopic receding contact angle, $\theta_R = 156°$, cannot recede across the post tops ($\theta_R = 103°$) and must disjoin from the posts being dewet in a near vertical (tensile) manner. This must involve the cohesive failure of a capillary bridge (Fig. 3a) and the formation of small sessile droplets on the recently dewet post tops. We have not observed these small droplets, but have not yet looked for them. We have studied capillary bridge rupture between smooth hydrophobic surfaces and small droplets are retained by the dewet surfaces. Fig. 3b shows frames from a movie of a capillary bridge rupturing as two smooth hydrophobic surfaces are separated. A small droplet of water is apparent on the upper surface.

We have shown[11] that introducing a second level, submicron topography to the post tops relieves the receding contact line pinning and eliminates hysteresis. We expect, however, that this pinning could also be eliminated by introducing positive curvature to the post tops. Surfaces with posts containing pyramidal, conical or

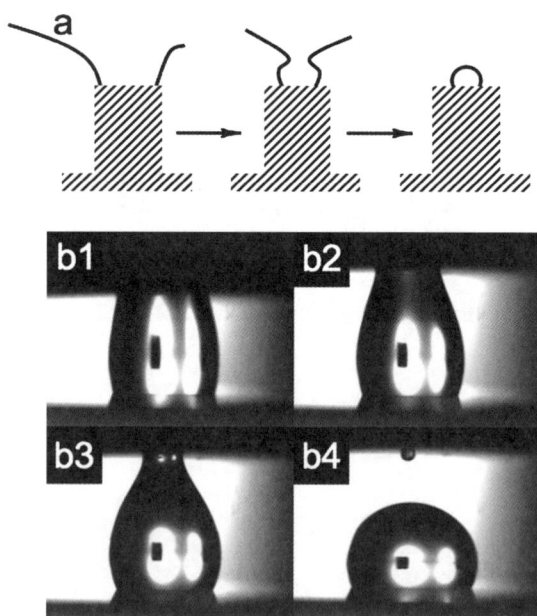

**Fig. 3** (a) Depiction of a capillary bridge rupturing during a receding event at the contact line. (b) Selected frames from a movie of a capillary bridge rupturing as two smooth hydrophobic surfaces are separated.

spherical caps should exhibit higher receding contact angles. Surfaces with pyramidal, conical or spherical depressions should exhibit lower receding angles and greater pinning (Fig. 4). This intuitively obvious prediction is based essentially only on the simple contact line arguments[21] made by Bartell in 1953 to explain differences in hysteresis on surfaces containing pyramid-shaped asperities. We have not prepared surfaces with the constructs shown in Fig. 4, but include these to stimulate discussion at the meeting. We note that pairs of surfaces with positive and negative curvature post caps would exhibit an identical liquid–solid contact area (have the same Cassie & Baxter $f_1$ and $f_2$ values), but very different receding contact angles. In Fig. 5 we show two surfaces with constant mean curvature that should exhibit direction-dependent receding contact angle pinning and a surface with a constantly changing curvature (French curve) that should exhibit and could test curvature-dependent tensile hydrophobicity. This figure is drawn in 2D, but is meant to represent a curved ribbon-like surface.

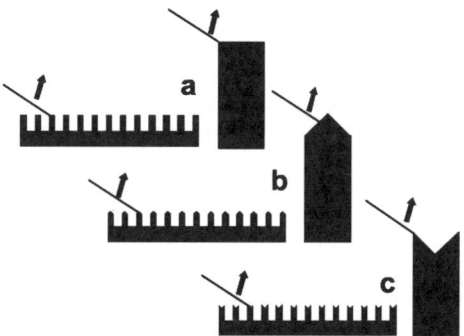

**Fig. 4** Compared with flat post tops (a), surfaces with posts containing pyramidal, conical or spherical caps (b) should exhibit higher receding contact angles and reduced receding contact line pinning. Surfaces with pyramidal, conical or spherical depressions (c) should exhibit lower receding angles and greater pinning.

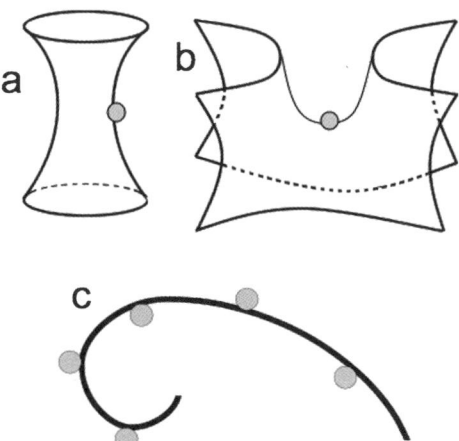

**Fig. 5** Two surfaces (a and b) with constant mean curvature, that should exhibit direction-dependent receding contact angle pinning, and a surface (c) with constantly changing curvature (French curve) that should exhibit curvature-dependent tensile hydrophobicity and direction-dependent shear hydrophobicity.

## Smooth surfaces without contact angle hysteresis

As mentioned above, preparing non-hysteretic surfaces is a synthetic challenge that the literature does not readily aid. Two publications that stand out as having the potential to impede the task of preparing smooth low hysteresis surfaces are (1) a 1994 *Langmuir* paper,[22] entitled *"Silanization of Solid Substrates: A Step toward Reproducibility,"* that claims to prepare surfaces with "less than 1°" hysteresis and (2) a 2005 *J. Am. Chem. Soc.* paper,[23] entitled *"Mixed Self-Assembled Monolayers of Alkanethiolates on Ultrasmooth Gold Do Not Exhibit Contact-Angle Hysteresis."* Another paper[24] that demonstrates the authors' significant insight into wetting, reports hysteresis values of 1–2° using hexadecane as a probe fluid and that hysteresis using water "is larger." In a subsequent paper[25] by the same authors, the water contact angle data were reported and hysteresis was indeed higher than 1–2°: reported values ranged from 23° to 61°.

We report here a procedure to prepare low hysteresis hydrophobic solid surfaces that uses a similar strategy (monolayer flexibility) as the one described above, but is improved over previous ones. This method neither produces surfaces that have lower hysteresis nor is it more reproducible in skilled hands than those we reported in the past,[13,18,19] however it does not involve air-sensitive reactive silanes, requires no cleaning of glassware, is not dependent on air humidity or the water content of solvents and does not require the skills of an experienced chemist. The procedure is simple: A silicon wafer sample[26] is cleaned in a commercial plasma cleaner,[27] placed in a just-opened commercial screw cap vial,[28] and wet with as-received trimethylsilyl-terminated linear poly(dimethylsiloxane) (PDMS).[29] The cap is replaced and the vial is placed in an oven at 100 °C for 24 h. After the vial cools to room temperature, the sample is rinsed with copious solvent[30] and allowed to dry.

Tables 1 and 2 show water contact angle[31] data for silicon wafer surfaces prepared using two different commercial PDMS samples with reported molecular weights of 2,000 and 9,400 g mol$^{-1}$. 12 individual reactions were carried out with PDMS$^{2\text{ K}}$ and 10 with PDMS$^{9\text{ K}}$. Each of the reported $\theta_A$ and $\theta_R$ values are the averages of 18 measurements made at 3 locations on each sample. Individual measurements were recorded to the nearest degree and the values to one decimal place reported in the tables are averages of the 18 measured values. The averages and standard deviation values of 216 measurements for surfaces prepared with PDMS$^{2\text{ K}}$ are $\theta_A/\theta_R = 104.0° \pm 0.8°/102.4° \pm 1.4°$. The 180 measurements on PDMS$^{9\text{ K}}$-derived surfaces averaged $\theta_A/\theta_R = 105.6° \pm 0.7°/104.8° \pm 0.9°$.

These easily prepared surface-modified silicon wafers exhibit water contact angle behavior that is indistinguishable from numerous other surfaces that we have

**Table 1** Water contact angles of PDMS$^{2\text{ K}}$-derived silicon wafers

| Sample | $\theta_A$ (°) | $\theta_R$ (°) |
| --- | --- | --- |
| PDMS$^{2\text{ K}}$-A | 104.7 | 102.3 |
| PDMS$^{2\text{ K}}$-B | 105.8 | 105.0 |
| PDMS$^{2\text{ K}}$-C | 104.2 | 103.2 |
| PDMS$^{2\text{ K}}$-D | 104.3 | 103.7 |
| PDMS$^{2\text{ K}}$-E | 103.7 | 103.2 |
| PDMS$^{2\text{ K}}$-F | 104.2 | 100.2 |
| PDMS$^{2\text{ K}}$-G | 104.0 | 102.2 |
| PDMS$^{2\text{ K}}$-H | 104.2 | 102.7 |
| PDMS$^{2\text{ K}}$-I | 103.3 | 99.8 |
| PDMS$^{2\text{ K}}$-J | 103.7 | 102.0 |
| PDMS$^{2\text{ K}}$-K | 103.3 | 101.7 |
| PDMS$^{2\text{ K}}$-L | 103.2 | 102.7 |

Table 2 Water contact angles of PDMS[9 K]-derived silicon wafers

| Sample | $\theta_A$ (°) | $\theta_R$ (°) |
| --- | --- | --- |
| PDMS[9 K]-A | 105.0 | 105.0 |
| PDMS[9 K]-B | 106.0 | 105.7 |
| PDMS[9 K]-C | 106.0 | 104.0 |
| PDMS[9 K]-D | 106.0 | 106.0 |
| PDMS[9 K]-E | 106.5 | 106.3 |
| PDMS[9 K]-F | 104.3 | 103.7 |
| PDMS[9 K]-G | 105.7 | 104.2 |
| PDMS[9 K]-H | 105.0 | 104.3 |
| PDMS[9 K]-I | 105.5 | 103.8 |
| PDMS[9 K]-J | 105.8 | 104.5 |

prepared in our laboratory using discrete sterically hindered monochlorosilane compounds, chloro-terminated (difunctional) silicone oligomers and certain mixtures of methylchlorosilanes. We have reported values of $\theta_A/\theta_R = 105°/104°$, 106°/104° and 104°/103° for some of these surfaces[11,13,19] and some of these data have been reproduced many times by multiple researchers including both authors of this paper. We have also reported[18,20,32] values of $\theta_A/\theta_R = 101°/99°$, 104–105°/101–103° and 107°/102° for one of these surfaces (prepared by vapor phase reaction with dimethyldichlorosilane) because these were the data obtained during a particular study by a particular researcher. In all of these studies, a fraction of surfaces prepared with reactive silanes were discarded because of low receding contact angles caused by defects or contaminants. We report the multiple experiments and redundant data in Tables 1 and 2 to emphasize the reproducibility and ease of this modification method.

We were initially surprised by the results that we report here, but in retrospect should not have been. We had ignorantly regarded linear unfunctionalized methylsilicones as unreactive polymers and did not expect that they would react with silica surface silanols. Silicones, however, have long been known to equilibrate[33–36] to most probable molecular weight distributions and equilibrium cyclic oligomer concentrations by both acid and base catalysis. It is reasonable and expected that surface silanols participate in these equilibrations at 100 °C. We have not carried out experiments to distinguish whether the mechanism involves silanols protonating silicone oxygen atoms or silanolates reacting as nucleophiles with silicone chains. Both protonated silicones and silanolates may be involved. Silica has been reported to be acidic[37,38] with an isoelectric point of ~3, but whether this is meaningful at PDMS interfaces at 100 °C is questionable. The hydrolytic cleavage and equilibration of silicones is reported to be Lewis acid-catalyzed by inorganic solids.[39,40]

We plan to report data based on work currently in progress, but make 7 additional observations that are currently only qualitative, involving control experiments that substantiate the results reported here: (1) Ellipsometry and XPS data indicate that PDMS[9 K]-derived grafted layers are about twice as thick as those prepared with PDMS[2 K]. (2) Initial AFM studies suggest that the PDMS[2 K]-derived surfaces are about as smooth as the initial silicon wafer. (3) These surfaces show low contact angle hysteresis with hexadecane and diiodomethane probe fluids, thus are "shear lyophobic." (4) Other linear dimethylsilicones of both higher and lower molecular weight react to hydrophobize silicon wafers. (5) Silicon wafers cleaned using chromic acid[19] react with PDMS[2 K] to form surfaces that exhibit contact angles indistinguishable from those cleaned with oxygen plasma. (6) As-received silicon wafers (not cleaned) react with PDMS[2 K], but receding contact angles reveal that many defects are present in the resulting surfaces. (7) The reaction occurs in vessels other than the disposable vials[28] including Teflon bottles, however the cleanliness of the vessel is important and the disposable vials that are purchased with caps attached are cleaner than vessels that have been exposed to our laboratory and we expect cleaner also than vessels cleaned and dried in any preparative chemistry lab.

## Summary

Contact angle hysteresis on superhydrophobic surfaces, modeled by surfaces containing posts, is due to receding contact line pinning. This pinning is based on the force required for cohesive failure of capillary bridges, thus a tensile analysis is appropriate. The sign of the curvature of the tops of posts is critical to whether hysteresis will be negligible or significant. Low hysteresis smooth surfaces can be conveniently prepared using disposable vials, commercial silicone oil and plasma-cleaned silicon wafers.

## Acknowledgements

We thank Shocking Technologies and the Centers for Materials Research Science and Engineering (DMR-0213695) and Hierarchical Manufacturing (CMMI-0531171) at the University of Massachusetts for support.

## Notes and References

1 T. Young, *Philos. Trans. R. Soc. London*, 1805, **95**, 65. This essay has been digitized and is available at www.google.com/books.
2 F. E. Bartell and C. E. Whitney, *J. Phys. Chem.*, 1932, **36**, 3115.
3 R. N. Wenzel, *Ind. Eng. Chem.*, 1936, **28**, 988.
4 A. B. D. Cassie and S. Baxter, *Trans. Faraday Soc.*, 1944, **40**, 546.
5 A. B. D. Cassie, *Discuss. Faraday Soc.*, 1948, **3**, 11.
6 L. Gao and T. J. McCarthy, *Langmuir*, 2007, **23**, 3762.
7 L. Gao and T. J. McCarthy, *Langmuir*, 2009, **25**, 7249.
8 A. Marmur and E. Bittoun, *Langmuir*, 2009, **25**, 1277.
9 L. Gao and T. J. McCarthy, *Langmuir*, 2008, **24**, 9183.
10 L. Gao and T. J. McCarthy, *Langmuir*, 2006, **22**, 6234.
11 L. Gao and T. J. McCarthy, *Langmuir*, 2006, **22**, 2966.
12 J. P. Youngblood and T. J. McCarthy, *Macromolecules*, 1999, **32**, 6800.
13 W. Chen, A. Y. Fadeev, M. C. Hsieh, D. Öner, J. Youngblood and T. J. McCarthy, *Langmuir*, 1999, **15**, 3395.
14 We use the words, *redefines itself*, because a 3-phase contact line does not have to move. See ref. 10 and 11 and discussion later in this paper.
15 A. Y. Fadeev and T. J. McCarthy, *Langmuir*, 1999, **15**, 3759.
16 A. Y. Fadeev and T. J. McCarthy, *Langmuir*, 1999, **15**, 7238.
17 L. Gao, T. J. McCarthy and X. Xiang, *Langmuir*, 2009, **25**, 14100.
18 K. A. Wier, L. Gao and T. J. McCarthy, *Langmuir*, 2006, **22**, 4914.
19 A. Y. Fadeev and T. J. McCarthy, *Langmuir*, 2000, **16**, 7268.
20 D. Öner and T. J. McCarthy, *Langmuir*, 2000, **16**, 7777.
21 F. E. Bartell and J. W. Shepard, *J. Phys. Chem.*, 1953, **57**, 211.
22 J. B. Brzoska, I. B. Azouz and F. Rondelez, *Langmuir*, 1994, **10**, 4367. This paper gives a "step-by-step foolproof procedure which reproducibly leads to high quality layers", however the silicon wafer cleaning procedure takes 6 steps and the 4th, subsequent to UV treatment at 2 different wavelengths on both sides of the wafer, involves thorough rinsing (3 times) with running tap water! A "thermoregulated bath used for silanization" uses chloroform (b.p. 61 °C) as the heat transfer liquid although a maximum temperature of 60 °C was chosen to be well below the boiling point of $CCl_4$ (76 °C) that was a solvent component. Contact angles were measured after "downward motion was detected by monitoring the displacement of dust particles deposited from air at the liquid/solid interface." There are other issues of concern about this paper including that there are no data presented that are consistent with the phrase, "less than 1°".
23 P. Gupta, A. Ulman, S. Fanfan, A. Korniakov and K. Loos, *J. Am. Chem. Soc.*, 2005, **127**, 4, This paper claims surfaces with zero hysteresis, but reports contact angle data determined "at a tilt angle of ~40°." These data should not be considered factual; sessile droplets on surfaces with zero hysteresis slide at slight tilt angles.

24 D. L. Schmidt, C. E. Cobern, B. M. DeKoven, G. E. Potter, G. F. Meyers and D. A. Fischer, *Nature*, 1994, **368**, 39.
25 D. L. Schmidt, B. M. DeKoven, C. E. Cobern, G. E. Potter, G. F. Meyers and D. A. Fischer, *Langmuir*, 1996, **12**, 518.
26 Silicon wafers were obtained from International Wafer Service (100 orientation, P/B doped, resisitivity from 20 to 40 Ω-cm). Disks (100 mm) were cut into ~1.0 × ~1.0 cm pieces.
27 Samples were exposed to oxygen plasma in a Harrick Expanded Plasma Cleaner at 18 W and 300 mTorr (flowing oxygen) for 30 min.
28 20 mL borosilicate glass scintillation vials capped with pulp-backed metal liners were purchased from Fisher and supplied in boxes of polyethylene film - wrapped trays containing 100 closed vials. These vials are apparently closed in a much cleaner and much more dust-free environment than our laboratories. We regard them as "a clean room in a bottle".
29 $PDMS^{2K}$ and $PDMS^{9K}$ were purchased from Gelest. Product codes are DMS-T12 and DMS-T22.
30 The data reported here were obtained using samples that were rinsed sequentially with copious toluene, acetone and house-purified water from polyethylene squirt bottles. Samples rinsed with only heptane exhibited contact angle values that are indistinguishable from data obtained with toluene/acetone/water rinsing.
31 Contact angle measurements were made with a Rame-Hart telescopic goniometer and a Gilmont syringe with a 24-gauge flat-tipped needle. Dynamic advancing ($\theta_A$) and receding angles ($\theta_R$) were recorded while the probe fluid was added to and withdrawn from the drop, respectively.
32 K. A. Wier and T. J. McCarthy, *Langmuir*, 2006, **22**, 2433.
33 W. Patnode and D. F. Wilcock, *J. Am. Chem. Soc.*, 1946, **68**, 358.
34 D. W. Scott, *J. Am. Chem. Soc.*, 1946, **68**, 2294.
35 J. F. Hyde, U. S. Patent 2,490,357 (December 6, 1949).
36 J. F. Hyde, U. S. Patent 2,567,110 (September 4, 1951).
37 A. Mendez, E. Bosch and M. Roses, *J. Chromatogr., A*, 2003, **986**, 33.
38 G. Busca, *Phys. Chem. Chem. Phys.*, 1999, **1**, 723.
39 J. B. Carmichael and J. Heffel, *J. Phys. Chem.*, 1965, **69**, 2213.
40 I. Rashkov, I. Gitsov and I. Panayotov, *Polym. Bull.*, 1983, **10**, 487.

PAPER

# Amplification of electro-osmotic flows by wall slippage: direct measurements on OTS-surfaces

Marie-Charlotte Audry, Agnès Piednoir, Pierre Joseph† and Elisabeth Charlaix*

Received 23rd December 2009, Accepted 28th January 2010
DOI: 10.1039/b927158a

The control of water flow in Electrostatic Double Layers (EDL) close to charged surfaces in solution is an important issue with the emergence of nanofluidic devices. We compare here the zeta potential governing the electrokinetic transport properties of surfaces, to the electrostatic potential directly measured from their interaction forces. We show that on smooth hydrophilic silica these quantities are similar, whereas on OTS-silanized hydrophobic surfaces the zeta potential is significantly higher, leading to an enhanced electro-osmotic velocity. The enhancement obtained is consistent with an interfacial water slippage on the silanized surface, characterized by a constant slip length of $\sim 8$ nm independent of the salt concentration in the range $10^{-4}$–$10^{-3}$M.

## 1 Introduction

The flow of water close to charged surfaces in solution is of prime importance for electrokinetic transport, widely used in colloidal science, biological analysis, micro- and nanofluidics, and more generally for bringing into motion fluid or solid phases at a small scale.[1-3] Electrokinetic transport takes its origin in the advection by a flow of the ions contained in the non-neutral fluid layer close to a charged surface, the so-called Electric Double Layer (EDL). The zeta potential which determines the amplitude of electrokinetic properties, e.g. electro-osmotic and electrophoretic velocities, streaming currents and potentials, depends indeed not only on the electrostatic properties of the surface which determine the amount of net mobile charges in solution, but also on the hydrodynamics at the solid interface, which governs the efficiency with which these mobile charges are brought into motion.

Typical size of EDLs range between 1 nm to 100 nm in most usual conditions. The emergence of applications involving the manipulation of fluids at small scales, such as micro- and nanofluidic devices for biological analysis, energy storage or conversion, has renewed the interest for a better understanding of the origin of the zeta-potential and its relation with interfacial hydrodynamics.[4,5] Traditional modelling in terms of a no-slip boundary condition (b.c.) at the liquid/solid interface is not sufficient any more in view of the recent findings showing that liquids can undergo substantial slippage onto solid surfaces.[6-8] More specifically, it is now established that water flow onto a range of materials obeys a partial boundary condition, characterized by a slip length which increases with the contact angle, and reaches tens of nanometres on highly hydrophobic surfaces.[9-11] It has been recognized that such

*Laboratoire PMCN Université Lyon 1, CNRS UMR5586, 43 bd du 11 novembre 1918, F-69622 Villeurbanne. E-mail: elisabeth.charlaix@univ-lyon1.fr*

† Present address: LAAS; CNRS; Université de Toulouse; UPS, INSA, INP, ISAE, 7 avenue du Colonel Roche, F-31077 Toulouse, France

interfacial slippage should result in a substantial reduction of the friction in EDLs, and accordingly an enhancement of electrokinetic transport.[12–16] Such enhancement could be of major importance in nanofluidic devices, which take explicit benefit of EDLs overlap to reach new and specific transport functions.[17,18] For instance it has been shown that friction reduction induced by slip lengths of 50 nm could increase the efficiency of mechanical to electrical energy conversion by nanofluidic devices from 3% to 70%.[19,20]

Few experimental works have addressed quantitatively the issue of the respective contribution of surface electrical properties and interfacial hydrodynamics to electrokinetic transport. This requires an independent measurement of the corresponding quantities, however most methods for characterizing surface potential/charges in solution actually rely on measurements of the zeta-potential. Churaev et al.[13] have studied the zeta-potential of hydrophobic methylated quartz capillaries. They have shown that adding a surfactant which changes the wetting into hydrophilic, results in a decrease of the zeta-potential, which they attribute to the suppression of water slippage. From these measurements they deduce a slip length of ∼5–8 nm of water on the methylated quartz, but they have to assume that the surface potential is unchanged by the surfactant.

More recently Bouzigues et al. performed the first independent characterization of surface potential and electro-osmotic flow, and demonstrated an enhancement of 100% of the Schmolukovski velocity in a hydrophobic capillary.[21] They use a nano-PIV technique to study the velocity field close to the wall, and derive the surface potential at the same location from the concentration profile of the colloidal tracors. Their data clearly show that the flow enhancement is due to wall slippage. However due to the finite resolution of the optical method, their study is restricted to a single value of the EDL thickness, ($\simeq 50$ nm) and thus of the salt concentration.

In this work we present and discuss a comparative study of the surface potential and the zeta potential of two type of surfaces in NaCl solutions: hydrophilic silica and OTS-silanized hydrophobic glass. We use the colloidal probe Atomic Force Microscope (AFM) to measure directly the electrical potential of the surfaces from the electrostatic force at equilibrium, and we compare this surface potential to the zeta potential measured by current monitoring. After a brief reminder of the relevant theoretical background in part 2, we describe our experimental procedure in part 3 and discuss our results in part 4. We show an enhancement of the zeta potential with respect to the surface potential on hydrophobic surfaces, which is consistent with a constant slip length of 8 nm independent of the salt concentration in the range $10^{-4}$ to $10^{-3}$M, and in good agreement with the expected slippage of water as a function of the surface wettability.

## 2 Theoretical background

We recall here briefly the principles of electrokinetic transport at a charged liquid–solid interface and the calculation of the zeta-potential in the case of a partial slip flow boundary condition (b.c.). Without loss of generality we take the case of an electro-osmotic flow induced by a stationary solid surface located at $z = 0$ and wearing an electric charge $\sigma$ per unit area (see Fig. 1).

We begin with the Navier expression for the flow partial boundary condition at the solid surface:

$$v_s = b \frac{\partial v}{\partial z}\Big|_{z=0} \qquad (1)$$

where $v_s$ is the slippage velocity of the solution onto the surface, $z$ the distance to the solid surface, $v(z)$ the velocity profile and $b$ the slip length (or Navier length). In the EDL, the solution is not electrically neutral and the charge density $\rho_e(z)$ is related to the electrostatic potential $V(z)$ by Poisson's law:

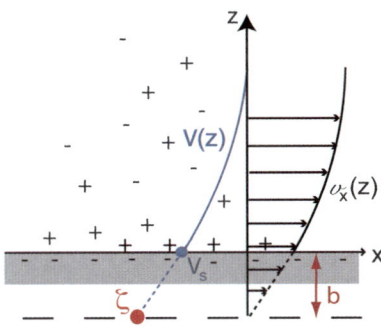

**Fig. 1** Schematic view of the velocity profile $v_x(z)$ near a solid wall where a partial slip occurs, characterized by a slip length $b$. The electrostatic potential of the surface is $V_S$. The zeta potential $\zeta$ can be understood as the linearly extrapolated potential at the depth $z = -b$.

$$\rho_e(z) = -\varepsilon \frac{\partial^2 V}{\partial z^2}$$

with $\varepsilon$ is the dielectric constant of the solution.

If an external electric field $\vec{E} = E\,\vec{e}_x$ is applied parallel to the surface, a body force $\vec{f} = \rho_e \vec{E}$ acts on the non-neutral fluid and induces a fluid flow obeying the Stokes equation

$$\eta \frac{\partial^2 v(z)}{\partial z^2} = -f = \varepsilon E \frac{\partial^2 V}{\partial z^2} \qquad (2)$$

Far from the solid surface the derivative $\partial V/\partial z$ vanishes, and the flow reaches a constant electro-osmotic velocity $v_{os}$. Eqn (2) integrates as:

$$v(z) = v_{os} + \frac{\varepsilon E}{\eta}\left(V(x,z) - V(x,\infty)\right) \qquad (3)$$

The $\zeta$-potential is defined from the electro-osmotic velocity by:

$$v_{os} = -\frac{\varepsilon \zeta}{\eta} E \qquad (4)$$

*No-slip boundary condition*: the flow velocity vanishes at the solid surface so that

$$v_{os} = -\frac{\varepsilon E}{\eta}\left(V_s(x) - V_\infty(x)\right) = -\frac{\varepsilon E}{\eta} V_s \qquad (5)$$

In this case the $\zeta$-potential reduces to the surface potential. It is sometimes considered that the no-slip b.c. should apply not exactly at the solid surface, but at the top of the so-called Stern layer which contains the ions adsorbed at the interface. In this case one has also to consider that the surface potential measured in an AFM colloidal probe experiment should be the potential at the top of this immobile liquid layer, which is not removed when the surfaces come into close contact. Therefore the equality of the zeta and surface potential is expected on a smooth surface in the case of a no-slip boundary condition.

*Partial slip boundary condition*: the flow velocity at the surface is derived from eqn (1) and (3):

$$v_s = b \frac{\varepsilon E}{\eta} \left(\frac{\partial V}{\partial z}\right)_s$$

eqn (3) then gives the expression of the electro-osmotic velocity:

$$v_{os} = -\frac{\varepsilon E}{\eta} V_s \left[1 - \frac{b}{V_s}\left(\frac{\partial V}{\partial z}\right)_s\right] \qquad (6)$$

The $\zeta$ potential is the linear extrapolation of the potential in solution at a depth $b$ under the interface. This enhancement is due to the reduced friction at the solid–liquid interface. The normal electric field at the solid surface is related to the surface charge density by $\sigma = -\varepsilon(\partial V/\partial z)_s$. Thus the relative amplification of $\zeta$-potential due to slippage, $(\zeta - V_s)/V_s$, is the ratio of the slip length $b$ to the EDL characteristic thickness $\kappa_{\mathit{eff}}^{-1}$:

$$\frac{\zeta - V_s}{V_s} = \frac{b}{\kappa_{\mathit{eff}}^{-1}} \qquad (7)$$

$$\kappa_{\mathit{eff}} = -\frac{1}{V_s}\left(\frac{\partial V}{\partial z}\right)_s = \frac{\sigma}{\varepsilon V_s} \qquad (8)$$

In the case of a small electrostatic potential ($V_s < 25$ mV), the thickness $\kappa_{\mathit{eff}}^{-1}$ reduces to the Debye length $\kappa^{-1}$:[22]

$$\kappa_{\mathit{eff}}^{-1} = \kappa^{-1} = \sqrt{\frac{\varepsilon k_B T}{\sum_i n_{io} Z_i^2 e^2}}$$

with $n_{io}$ the number density of ions of type $i$ and valence $Z_i$ in the reservoir solution. In the case of a monovalent electrolyte, the thickness $\kappa_{\mathit{eff}}^{-1}$ is obtained from the non-linear Poisson–Boltzmann equation for all values of $V_s$:[23]

$$\kappa_{\mathit{eff}}^{-1} = \kappa^{-1}\frac{2eV_s/k_B T}{\sinh(2eV_s/k_B T)} \qquad (9)$$

Therefore, if the b.c. does not change with the electrolyte concentration, the relative amplification of the zeta-potential is expected to scale linearly with the inverse of the effective Debye's length.

## 3 Experimental

We perform independent measurements of the surface potential $V_s$ and of the $\zeta$-potential of two types of materials: hydrophilic silica, on which water flow obeys the no-slip b.c.,[9,24] and hydrophobic OTS-silanized glass surfaces, on which substantial slippage is expected.[9–11] For each of these materials the $\zeta$-potential is measured in capillaries by a current monitoring method, and the surface potential is measured by colloidal probe AFM on flat surfaces.

### 3.1 Materials

The following materials are used: silica planes (Suprasil 311, Heraeus, Germany), Pyrex planes (from Pignat, Lyon, France), silica capillaries (Polymicro Technologies) in 75 µm internal diameter.

The flat samples are washed with a detergent solution, rinsed with ultrapure water (Millipore, MilliQ, 18.2 MΩ.cm), and dried with a nitrogen flux. The capillaries are prepared in a similar way by pumping through them successively the detergent solution, the ultrapure water, and clean air under a laminar flow hood. *Silanized materials* are prepared from silica capillaries and Pyrex surfaces in a single silanization operation in order to ensure a similar surface coating for the different samples. An octadecyltrichlorosilane (OTS) solution is prepared from fresh OTS (Aldrich) used as received without further purification, and anhydrous toluene (Roth). The solution is prepared under a dry atmosphere (RH < 3%) by mixing

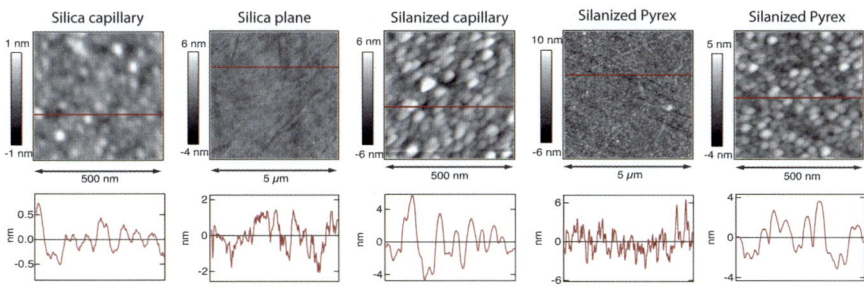

**Fig. 2** Tapping mode topographies of the materials used in this work, from left to right: silica capillary, Pyrex, silica, silanized silica and silanized Pyrex. The topographies of the inside of the capillaries have been made thanks to transversal cuts done after the experiments.

100 µl of OTS with 60 ml of toluene. It is stored in a bottle closed by a silicone cap and cooled under 10 °C. One side of the capillaries is then introduced through the silicone cap into the solution, while the other side is connected, also through a silicone sealing, to a vacuum bottle containing the Pyrex planes to be silanized in the same operation. The OTS solution is then pumped through the capillaries by connecting the bottle to a vacuum pump, until a significant amount of solution covers the flat samples. Thus the internal surface of the capillaries and the flat samples are silanized with the same OTS solution, at the same temperature and hygrometry. The flat samples and the capillaries are then rinsed with chloroform in the same way.

The surface topography of the different samples is recorded by contact mode AFM (Asylum Research MFP-3D) and displayed in Fig. 2. The internal surface of the capillaries are imaged after sawing them at a 45° angle, which generates some debris, therefore the topographies in Fig. 2 and the roughness in Table 1 are restricted to small areas of 500 × 500 nm lying between the debris. The r.m.s. roughness of the plain materials, plane and capillary, are similar. The roughness of the silanized materials is slightly higher and their topographic image is less homogenous, however there is no significant difference between the OTS-plane and the OTS-capillary.

The advancing and receding contact angle of water on the silianized plane are measured with the sessile drop method on several locations of the sample (see Fig. 3). The average contact angle is 98°. The hysteresis is large, about 25°, due to the nanometre scale roughness and defects revealed on the topographic data. However this hysteresis does not vary significantly with the position tested, which shows the overall homogeneity of the OTS-surfaces.

After elaboration and cleaning, all samples are stored under a laminar flux hood. They are immersed for a minimum time of 12 h in a NaCl solution prior to any $\zeta$ or surface potential measurements. Both surface and $\zeta$-potential values are measured in 0.1 mM to 1 mM NaCl solutions without buffer, at pH = 5.8 resulting from the spontaneous dissolution of carbon dioxide of the atmosphere in the solutions.

**Table 1** Surface roughness of the materials: plain capillary, silanized capillary, plain silica and silanized Pyrex

|  | Silica capillary | Silica plane | Silanized capillary | Silanized pyrex |
| --- | --- | --- | --- | --- |
| Area/µm$^2$ | 0.5 × 0.5 | 5 × 5 | 0.5 × 0.5 | 0.5 × 0.5 |
| rms/nm | 0.3 | 0.5 | 2.2 | 1.4 |
| pk–pk | 1.2 | 3.8 | 11.6 | 5.0 |

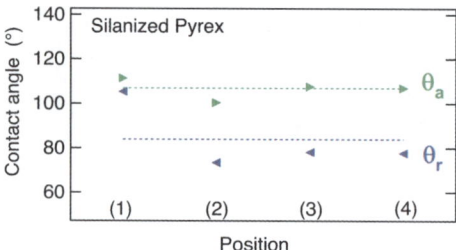

**Fig. 3** Advancing ($\theta_a$) and receding ($\theta_r$) contact angle of water on the OTS-Pyrex measured in different points of the surface. Dashed lines are the mean values.

### 3.2 Zeta potential measurements

The current monitoring set-up[25] used for zeta potential measurements is shown schematically on Fig. 4. Two reservoirs of slightly different NaCl concentrations (about 10%) are connected by the studied capillary, previously filled with the solution of reservoir $n°1$. At time $t = 0$ a high voltage difference $\Delta V = 1500$ V is applied between the reservoirs through platinum electrodes so as to create an electro-osmotic flow from reservoir $n°2$ to reservoir $n°1$. The electrical current is recorded as a function of time and increases linearly as the solution $n°1$ is replaced by solution $n°2$

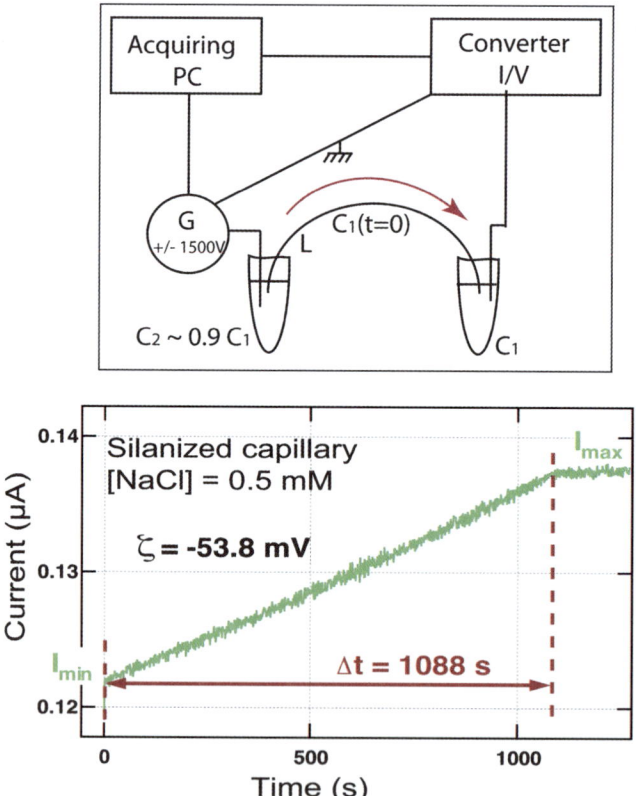

**Fig. 4** Top: schematic description of the current monitoring set-up for the zeta potential measurements. Bottom: Typical current intensity as a function of time.

(Fig. 4). The solutions do not mix due to the very flat velocity profile of the electro-osmotic flow. At time $\Delta t$ the capillary is completely filled by solution $n°2$ and the current stabilizes to a constant value. The process is reversed by changing the voltage sign, and cycles are performed. The $\zeta$-potential is obtained from the electro-osmotic velocity $v_{os} = L/\Delta t$, with $L$ the length of the capillary, and the amplitude $E = \Delta V/L$ of the applied electric field, using eqn (4).

### 3.3 Surface potential measurements

The electrostatic forces between the flat materials and a sphere in solution are measured directly with colloidal probe AFM.[26] The surface potential can in principle be determined from such force curves, but the reliability of the measurements depends strongly on the characterization of the colloidal probe in terms of its roughness, physico-chemistry, cantilever stiffness, and surface potential. For this reason we have built a robust protocol[27] which allows accurate measurements of the surface potential of flat substrates in a dissymmetric configuration. The colloidal probe used is a borosilicate sphere from Duke Scientific, cleaned and glued to a micro-cantilever with a procedure which removes efficiently contamination, and presenting a very low roughness of less than 1 nm r.m.s. on a 5 μm × 5 μm area of the flattened sphere topography. The cantilever stiffness is measured with 3% accuracy with two independent methods.

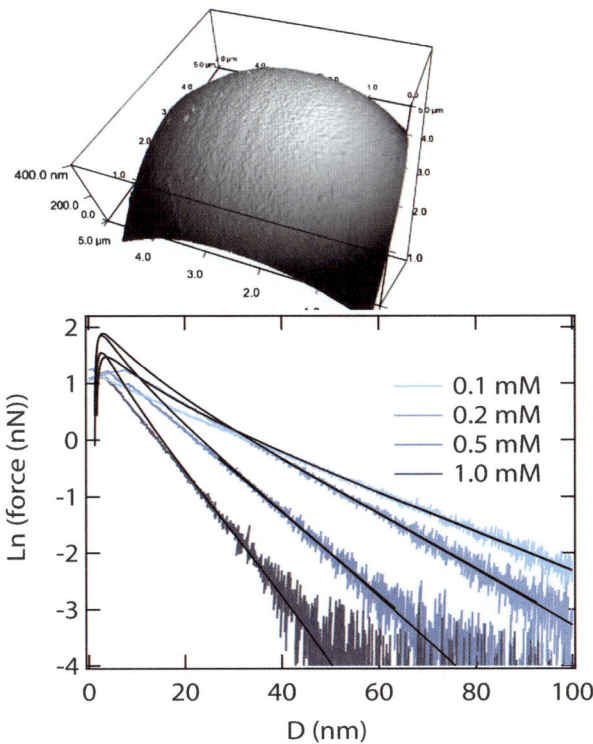

**Fig. 5** Top: AFM topography of the borosilicate colloidal probe used in this study. The particle radius is 8.7 μm. Bottom: force curves measured between the borosilicate probe and the Pyrex substrate, in logarithmic scale, for four different NaCl concentrations. Black curves are fits with the DLVO theory where the non-linear Poisson–Boltzmann equation is solved with a constant charge density condition and with a Hamaker constant fixed to the value of $1. \times 10^{-20}$ J.

**Table 2** Charge density $\sigma$ and surface potential $V_s$ of the colloidal probe issued from AFM force measurements between the borosilicate probe and a flat Pyrex substrate in NaCl solutions of various concentration. The targeted concentrations $c_0$ vary from 0.1 mM to 1.0 mM. The actual salt concentrations measured from the force curves are noted $c_0{}^r$. The values of $pC = -log(c_0{}^r)$ and of the Debye length $\kappa^{-1}$ are mentioned as well

| $c_0$ (mM) | 0.10 | 0.20 | 0.30 | 0.50 | 1.00 |
|---|---|---|---|---|---|
| $|\sigma|$ (mC/m$^2$) | 0.79 | 1.06 | 1.12 | 1.27 | 1.72 |
| $|V_s|$ (mV) | 27.3 | 31.0 | 26.9 | 23.4 | 22.0 |
| $\kappa^{-1}$ (nm) | 24.9 | 21.5 | 17.3 | 13.2 | 9.1 |
| $c_0{}^r$ (mM) | 0.14 | 0.20 | 0.31 | 0.52 | 1.12 |
| $pC$ | 3.83 | 3.70 | 3.511 | 3.28 | 2.95 |

The surface potential of the colloidal probe is calibrated by performing AFM force measurements on a flat Pyrex sample of the same composition as the probe. It is determined by adjusting the curve forces to the Deryaguin–Landau–Verwey–Overbeek (DLVO) theory as explained in the Appendix. Two quantities are adjusted: the Debye length $\kappa^{-1}$ and the common value $V_p$ of the probe/Pyrex surface potential, so that the actual ion content of the solution is checked. The adjusted force curves obtained for different NaCl concentration are shown in Fig. 5 and the calibrated probe potentials are summarized in Table 2.

Once the probe is calibrated, the surface potential of the samples studied is determined by adjusting the force curves measured in each solution with two adjustable parameters, the sample surface potential and the Debye length. The probe potential is interpolated from the calibration curve to its expected value at the electrolyte concentration corresponding to the Debye length found, and kept fixed.

## 4 Results and discussion

The values found for the surface potential and the $\zeta$-potential of plain silica are reported on Fig. 6 as a function of $pC_o = -\log[NaCl]$. They are compared to other $\zeta$-potentials and AFM-measured surface potentials of silica at the same pH, from the literature. A general agreement is found with the previously reported values of the $\zeta$-potential[28–30] as well as the surface potential of silica measured by colloidal probe AFM.[26] We find that the $\zeta$-potential of silica is similar to its surface potential, within the resolution of our determination of those quantities, which is also in good

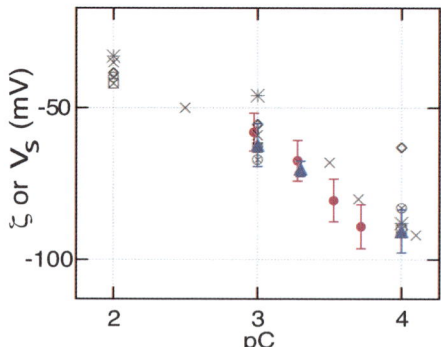

**Fig. 6** Comparison of our surface potential (pink circles ●) and zeta potential (blue triangles ▲) of plain silica, as a function of $pC$. These values are compared as well with others from the literature (grey symbols): ⊗ Weiss, pH 6.0, NaCl; ◇ Ducker, pH 5.7, NaCl; ⊠ Scales pH 5.8, KCl; × Gaudin, pH 7, NaCl; * Giesbers, pH 6.0, NaCl.

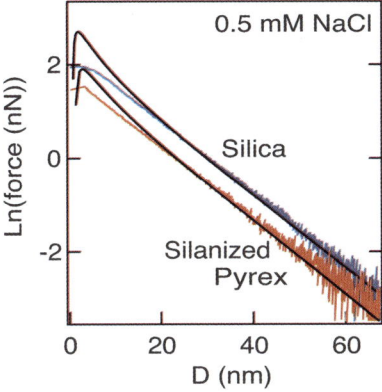

**Fig. 7** Force curves obtained on plain silica and on OTS-silanized Pyrex at 0.5 mM NaCl concentration. The repulsive force shows the negative charge of the OTS surface.

agreement with the previous results. This behaviour is coherent with a no-slip b.c. applied on, or very close to the solid surface, as expected on hydrophilic silica. This result supports the validity of classical hydrodynamics with a b.c. applied on the solid surface, to account for water flow and electrokinetic transport inside the EDL.

In contrast the results on silanized surfaces show a negative $\zeta$-potential whose amplitude is substantially higher than the negative surface potential (see Fig. 7).

Early experiments have indeed shown that interfaces between water and hydrophobic materials, such as liquid alkanes/water interfaces,[31,32] or self-assembled-monolayers made of alkane-chains in water,[33,34] are negatively charged. The negative

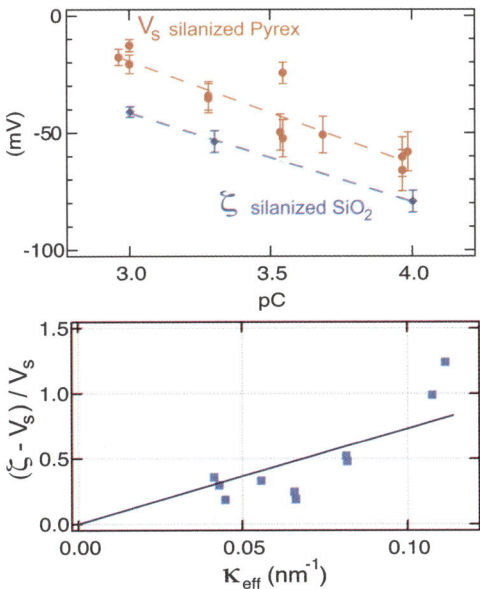

**Fig. 8** (a) Surface potential of OTS-silanized Pyrex (orange circles) and zeta potential of silanized silica (blue diamonds) as a function of $pC = -\log([\text{NaCl}])$. (b) Relative difference between the zeta potential and the surface potential of the OTS-coated surfaces as a function of the inverse of the effective Debye length. The linear fit corresponds to a slip length $b = 7.3 \pm 0.8$ nm.

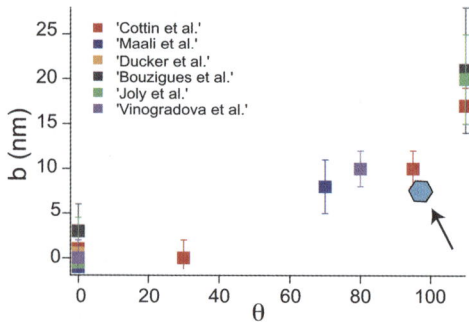

**Fig. 9** Slip length of water on various surfaces as a function of their wettability, from [18]. The arrow shows the value derived from the comparison between the $\zeta$-potential and the surface potential of our OTS surfaces.

charge is attributed to the preferential adsorption of hydroxide ions (OH⁻) relative to hydronium ions (H$_3$O⁺). Theoretically, the preferential adsorption of OH⁻ ions on n-alkane SAMs has been supported by density functional (DFT) simulations[35] with surface charges as large as $-10^{-2}$ C m$^{-2}$, and molecular dynamics simulations have shown that Cl⁻ ions are also likely to adsorb at hydrocarbon/water interfaces whereas Na⁺ are repelled from it.[36] Recent experiments of Tian et al.[37] using sum-frequency-vibrational-spectroscopy on OTS/water interface, evidence clearly the preferential adsorption of OH⁻ ions as well as other negatively charged ions such as Cl⁻. In view of these recent works we conclude that the origin of negative charge in our silanized samples lies at the OTS/water interface, and that the difference in the grafting substrate, Pyrex or silica, does not play a significant role on the charge density.

We can then compare the $\zeta$-potential to the enhancement provided by a partial slip b.c. Fig. 8 plots the relative difference $(\zeta - V_s)/V_s$ as a function of the effective Debye length $\kappa_{eff}^{-1}$ evaluated from eqn (9). The relative enhancement increases with $\kappa_{eff}$, that is when the thickness of the Debye layer decreases, although the absolute value of the $\zeta$- and the surface potential also decrease. This supports the idea that the enhancement is due to a friction reduction effect. The best linear fit corresponds to a slip length $b = 7.3 \pm 0.8$ nm. The magnitude of this slip length is in very good agreement with the slippage expected for these OTS-surfaces wettability, taking into account their nanometric roughness, as illustrated in Fig. 9.

## 5 Conclusion

This work extends the previous results of Churaev et al.[13] and Bouzigues et al.[21] showing the enhancement of electrokinetic transport due to the finite slippage of water on a hydrophobic solid surface. We show by independent measurements of the $\zeta$-potential and the electrostatic surface potential of OTS-coated surfaces that this enhancement is well described by a constant slip length of nanometric amplitude, for a range of electrolyte concentration. The enhancement reaches values of 100% and more at a mM concentration in NaCl, when the slip length becomes larger than the thickness of the Debye layer.

Our approach opens the way to a systematic investigation of the electrostatic and electrokinetic properties of hydrophobic surfaces in solution, and the search of highly slippery and charged surfaces to increase electro-osmotic effects and the efficiency of micro- and nanodevices using them. Work is under progress to study the effect of ion specificity on the electrokinetic properties of hydrophobic surfaces.

## Appendix

The surface potentials are determined from fitting the experimental force curves to DLVO forces calculated using the Derjaguin approximation and the disjunction pressure acting between two parallel surfaces separated by a distance $D$ [23]:

$$F(D) = 2\pi R \varepsilon \frac{(k_B T \kappa)^2}{e^2} (\cosh \psi_m - 1) - \frac{AR}{6D} \qquad (10)$$

where $A$ is the silica–water–silica Hamaker constant (kept fixed to the value $1.10^{-20}$J), $\kappa = \sqrt{2n_o e^2/\varepsilon k_B T}$ is the inverse of the Debye length for the number density $n_o$ of the monovalent electrolyte, and $\psi_m$ is related to the normalized surface potentials $\psi_1 = eV_{s1}/k_B T$ and $\psi_2 = eV_{s2}/k_B T$ by the elliptic integral:

$$\kappa D = \int_{\psi_m}^{\psi_1} \frac{d\psi}{\sqrt{2\cosh\psi - 2\cosh\psi_m}} + \int_{\psi_m}^{\psi_2} \frac{d\psi}{\sqrt{2\cosh\psi - 2\cosh\psi_m}}$$

The calculation of the electrostatic force is directly implemented on the Igor Pro software of the MFP-3D AFM. The DLVO forces are calculated for each value of the distance $D$ at constant surface charge on each surface

$$\sigma_i = \frac{\varepsilon k_B T}{e \kappa^{-1}} \sqrt{2\cosh\psi_{s,i} - 2\cosh\psi_m}$$

The fitting is done by adjusting two parameters: the Debye length $\kappa^{-1}$ and the value of the unknown surface charge (in dissymmetric configurations). The surface potential of the isolated surface in solution is then obtained from the Grahame equation:[22]

$$V_s = \frac{2k_B T}{e} \text{Arg sh}\left[\frac{e\sigma\kappa^{-1}}{2\varepsilon k_B T}\right]$$

## Acknowledgements

This work has been supported by the Region Rhone-Alpes and by the program Synodos of the ANR-PNano.

## References

1. R. J. Hunter, *Foundations of Colloidal Science*. Clarendon Press Oxford, 1989.
2. T. Squires and S. R. Quake, *Rev. Mod. Phys.*, 2005, **77**(3), 977–1026.
3. H. A. Stone, A. D. Strook and A. Ajdari, *Annu. Rev. Fluid Mech.*, 2004, **36**, 381.
4. B. J. Kirby and E. F. Hasselbrink, *Electrophoresis*, 2004, **25**, 187–202.
5. V. Tandon, S. K. Bhagavatula, W. C. Nelson and B. J. Kirby, *Electrophoresis*, 2008, **29**, 1092–1101.
6. N. V. Churaev, V. D. Sobolev and A. N. Somov, *J. Colloid Interface Sci.*, 1984, **97**, 574–581.
7. E. Lauga, M. P. Brenner, and H. A. Stone, *Microfluidics: The No-Slip Boundary Condition*, volume Handbook of Experimental Fluid Dynamics, (chapter 15). Springer, New-York, 2005.
8. J. L. Barrat and L. Bocquet, *Soft Matter*, 2007, **3**, 685–693.
9. C. Cottin-Bizonne, B. Cross, A. Steinberger and E. Charlaix, *Phys. Rev. Lett.*, 2005, **94**, 056102.
10. C. Cottin-Bizonne, A. Steinberger, B. Cross, O. Raccurt and E. Charlaix, *Langmuir*, 2008, **24**, 1165–1172.
11. D. Huang, C. Sendner, D. Horinek, R. R. Netz and L. Bocquet, *Phys. Rev. Lett.*, 2008, **101**, 226101.
12. V. M. Muller, I. P. Sergeeva, V. D. Sobolev and N. D. Churaev, *Kolloid Zh.*, 1986, **48**(4), 606–614.

13 N. V. Churaev, J. Ralston, I. P. Sergeeva and V. D. Sobolev, *Adv. Colloid Interface Sci.*, 2002, **96**, 265.
14 L. Joly, C. Ybert, E. Trizac and L. Bocquet, *Phys. Rev. Lett.*, 2004, **93**, 257805.
15 A. Ajdari and L. Bocquet, *Phys. Rev. Lett.*, 2006, **96**(18), 186102.
16 Todd M. Squires, *Phys. Fluids*, 2008, **20**(9), 092105.
17 W. Sparreboom, A. van den Berg and J. C. T. Eijkel, *Nat. Nanotechnol.*, 2009, **4**(11), 713–720.
18 L. Bocquet and E. Charlaix, *Chem. Soc. Rev.*, 2010, **38**, 1–23, DOI: 10.1039/b909366b.
19 S. Pennathur, J. C. T. Eijkel and A. van den Berg, *Lab Chip*, 2007, **7**(10), 1234–1237.
20 Y. Ren and D. Stein, *Nanotechnology*, 2008, **19**, 195707.
21 C. I. Bouzigues, P. Tabeling and L. Bocquet, *Phys. Rev. Lett.*, 2008, **101**, 114503.
22 J. Israelachvili. *Molecular and Surface Forces*. Academic Press, New York, 1985.
23 D. Andelman, *Membranes: their Structure and Conformations*, chapter Electrostatic properties of membranes: the Poisson–Boltzmann theory. Elsevier, 1995.
24 O. I. Vinogradova and G. E. Yabukov, *Langmuir*, 2003, **19**, 1227–1234.
25 X. Huang, M. J. Gordon and R. N. Zare, *Anal. Chem.*, 1988, **60**, 1837–1838.
26 W. A. Ducker, T. J. Senden and R. M. Pashley, *Langmuir*, 1992, **8**, 1831–1836.
27 M. C. Audry, S. Ramos, A. Piednoir, and E. Charlaix, submitted to Langmuir, 2010.
28 P. J. Scales, F. Grieser and T. W. Healy, *Langmuir*, 1992, **8**, 965–974.
29 M. Giesbers, J. M. Kleijn and M. A. Cohen Stuart, *J. Colloid Interface Sci.*, 2002, **248**, 88–95.
30 A. M. Gaudin and D. W. Furstenau, *Trans. ASME*, 1955, **202**, 66–72.
31 W. Dickinson, *Trans. Faraday Soc.*, 1941, **37**, 140–147.
32 K. G. Marinova, R. G. Alargova, N. D. Denkov, O. D. Velev, D. N. Petsev, I. B. Ivanov and R. P. Borwankar, *Langmuir*, 1996, **12**, 2045–2051.
33 A. Hozumi, H. Sugimura, Y. Yokogawa, T. Kameyama and O. Takai, *Colloids Surf., A*, 2001, **182**, 257–261.
34 R. Schweiss, P. B. Welzel and W. Knoll, *Langmuir*, 2001, **17**, 4304–4311.
35 H. J. Kreuzer, R. L. C. Wang and M. Grunze, *J. Am. Chem. Soc.*, 2003, **125**, 8384–8389.
36 D. Horinek and R. R. Netz, *Phys. Rev. Lett.*, 2007, **99**, 226104.
37 C. S. Tian and Y. R. Shen, *Proc. Natl. Acad. Sci. U. S. A.*, 2009, **106**(36), 15148–15153.

PAPER

# Electrowetting and droplet impalement experiments on superhydrophobic multiscale structures

F. Lapierre,[a] P. Brunet,[a] Y. Coffinier,[ba] V. Thomy,*[a] R. Blossey[b] and R. Boukherroub[ba]

*Received 4th December 2009, Accepted 13th January 2010*
DOI: 10.1039/b925544c

The reversible actuation of droplets on superhydrophobic surfaces under ambient conditions is currently an important field of research due to its potential applicability in microfluidic lab-on-a-chip devices. We have recently shown that Si-nanowire (NW) surfaces allow for reversible actuation provided that the surface structures show certain characteristics. In particular it appears that, for such surfaces, the presence of structures with multiple specific length scales is indeed needed to have a robust reversibility of contact angle changes. Here we report on electrowetting (EW) and impalement experiments on double-scale structured surfaces prepared by a combination of silicon micropillars prepared by an association of optical lithography and silicon etching, and nanowire growth on top of these surfaces. We show that while micropillar surfaces have a low impalement threshold and irreversible EW behaviour, a surface with double-scale texture can show both a very high resistance to impalement and a limited reversibility under EW, provided that the roughness of the micro-scale is large enough - *i.e.* that the pillars are tall enough. The optimal performance is obtained for a space between pillars that is comparable to the height of the nanostructure.

## I. Introduction

The superhydrophobic character of a surface generally arises from an interplay between the surface roughness and its chemical composition. Seminal contributions to an understanding of such surfaces already date to long ago by Adam,[1] and later on formalised by Cassie and Baxter[2] for the situation of a composite interface. Alternatively, Wenzel's approach addresses the state of homogeneous wetting.[3] The recent upsurge of interest in the subject is driven by the development of more and more sophisticated techniques for the production of surface micro- and nanostructures, which make it possible to mimic water-repellent biosurfaces such as Lotus leaves. Such surfaces have already found numerous applications such as textiles, roof and window coatings, coatings in household materials, microelectromechnical systems (MEMS), or lab-on-a-chip (LOC) devices.[4–7]

In order to control the contact angle, electrowetting[8] has emerged as a versatile technique with a number of microsystems being at the stage of commercialization, *e.g.* adjustable lenses[9] or electronic displays.[10] The state of the field from both theoretical and applied aspects has recently been reviewed by Mugele and Baret.[11] For lab-on-a-chip systems and their requirement of low contamination or material loss,

[a]*Institut d'Electronique, de Microélectronique et de Nanotechnologies (IEMN), UMR CNRS 8520, F-59652, Villeneuve d'Ascq, France*
[b]*Interdisciplinary Research Institute (IRI), USR CNRS 3078, 50 Avenue Halley, F-59658 Villeneuve d'Ascq, France*

electrowetting on superhydrophobic substrates has attracted recent interest.[7,12–15] While electrowetting on planar hydrophobic surfaces is usually reversible under ambient conditions, *i.e.*, the contact angle changes back to its original value once the applied voltage is turned off, most published results on superhydrophobic surfaces showed an irreversibility of the contact angle change. There is therefore ongoing activity in order to find surface structures which favor reversibility of electrowetting under superhydrophobic conditions. Recently, Vrancken *et al.* designed a corrugated surface structure for which the transition from a Cassie–Baxter state with quasi-null hysteresis to a Wenzel state with liquid impalement stays reversible for maximum contact angles of 130°.[16]

Apart from this very recently reported case, up to now surfaces with a double texturation (micron-sized posts and nanotexturation) were proposed as the most promising solutions in order to obtain a robust Cassie state for a high resistance to impalement.[17–19] The work reported in this communication is a continuation of our previous work[20–22] on forests of nanowires, here extended on the preparation and wetting properties - including robustness upon external pressure - of double-scale micro-pillar nanowire surfaces. While silicon micropillars are prepared by the combination of optical lithography and silicon etching, the nanotexturation is obtained by silicon nanowire growth leading to unique wetting properties,[20] shown in Fig. 1. In order to compare the hierarchical integration of our specific nanoscale textured surfaces consisting of microscale pillar-like structures with nanowires on top, two types of superhydrophobic surfaces have been realized: micropillars (single-scale roughness) and micropillars covered by nanowires (micro/nano double-scale roughness) (Fig. 2 (a)–(d)). Different parameters such as pillar width, spacing and height, and height and density of nanowires layers have been investigated.

The comparison of the wetting properties of these surfaces is performed using electrowetting (EW) and drop impact experiments. In particular, we determine the robustness of the so-called 'Fakir' (Cassie–Baxter[2]) state against partial or total impalement.[23] Concerning the first method, after an EW cycle the variation of the contact angle (CA) is determined. We assumed that the Cassie state is maintained if a total CA reversibility is observed. Hence, the observation of irreversible behaviour during EW suggests that the liquid has partially or totally impaled the texture (Wenzel state[3]). For the second method, we determined the impalement threshold,

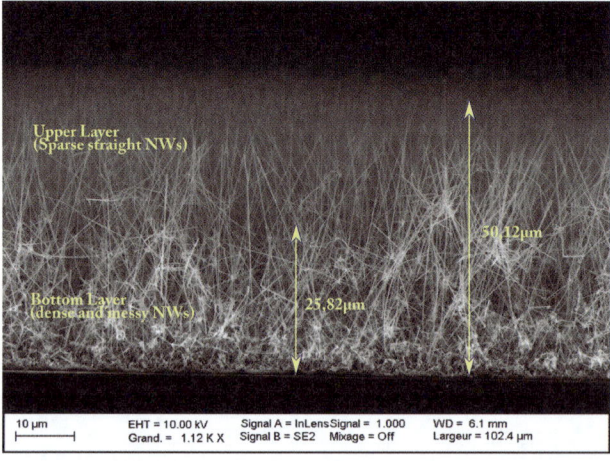

**Fig. 1** SEM image of VLS Nanowire carpet coated with $C_4F_8$. The total height is about 50 μm. A double nanotexturation is observed: at the bottom, a dense and messy NW layer, on the top, sparse straight NW.

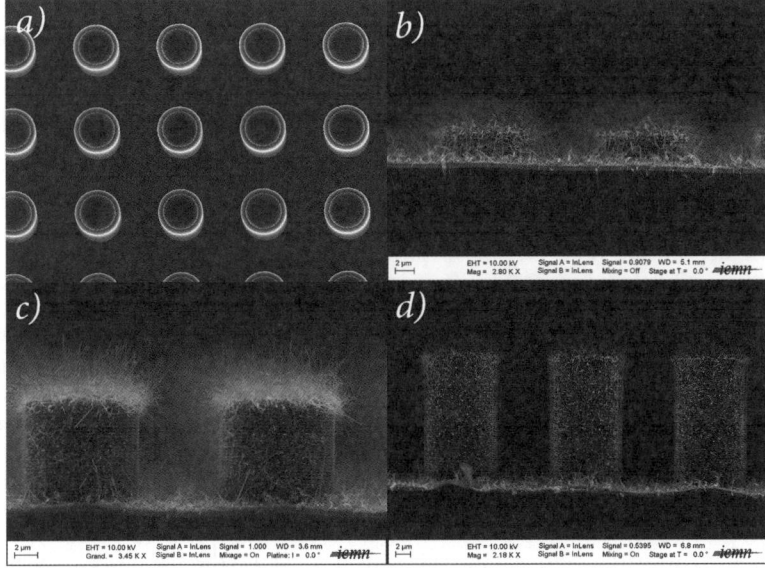

**Fig. 2** SEM images of (a) pillar $P_5^{10}(20)$ (Top view), (b) $P_7^{10}(4)$ + NW-VLS2, (c) $P_7^{10}(10)$ + NW-VLS2, (d) $P_7^{10}(19)$ + NW-VLS2, all coated with $C_4F_8$.

*i.e.*, the maximum drop velocity at impact, beyond which some liquid remains pinned on the surface.

Our experimental results show a correlation between the resistance to impalement for both EW and drop impact, and allows us to draw the following trends:

• Micropillar surfaces have a small impalement threshold and an irreversible EW behavior. Their weak performances make them inappropriate for droplet actuation by EW.

• As previously shown,[20-22] thick double-layered nanotextured surfaces both have a very high resistance impalement to drop impact, and a total reversibility during EW. These offer up to now one of the best performances available for robustness against liquid impalement. Otherwise surfaces made of a thinner layer of NWs are much less robust against impalement.[20,21]

• The double-scale surfaces resulting from a combination of short NWs and micropillars, and specifically investigated in this paper, provide a good performance under the condition that the geometrical parameters fulfil some requirements. More precisely the height of the micropillars has to be large enough and the space between the pillars has to be of the same order at the length of the NWs. A correlation between EW reversibility and drop impact impalement threshold is observed as well.

## II. Surface preparation and characterization

### 1 Fabrication of silicon micropillars

Single-side polished silicon (100) oriented p-type wafers (Siltronix) (phosphorus-doped, 0.009–0.01 Ohm-cm resistivity) were used as substrate. The surface was first degreased in acetone and isopropanol, rinsed with Milli-Q water and then cleaned in a piranha solution (3 : 1 concentrated $H_2SO_4$/30% $H_2O_2$) for 15 min at 80 °C followed by extensive rinsing with Milli-Q water. A negative resist $AZ_n$LOF 2035 (Clariant, France) is spin-coated at 3000 rpm to create a 3.5 µm thick resin layer. A softbake for 1 min on a hot plate at 90 °C is required in order to partially evaporate the solvent and hence to prevent adhesion of resin onto the mask. Then, exposure, post-exposure bake and development are conducted in order to transfer the mask patterns to the resist (latent image). Then, the silicon wafer was etched using Deep Reactive Ion Etching (DRIE), (Silicon Technology System) leading to high aspect ratio micro-pillars. We varied the etching time to obtain different heights of μ-pillars, i.e. 4, 10 and 19 μm.

## 2 Silicon nanowire (NWs) synthesis

The NWs were grown on Si/SiO$_2$ substrates using the vapour–liquid–solid (VLS) growth mechanism. Both the silicon surfaces and silicon surfaces featuring silicon μ-pillars are immersed in 50% hydrofluoric acid to remove the native oxide. Next, a SiO$_2$ layer (300 nm thick) is grown by Chemical Vapour Deposition (CVD). Finally, a gold layer (4 nm thick) is deposited by sputtering onto the Si/SiO$_2$ surface. The metallized interfaces are introduced in a CVD furnace chamber and heated at 500 °C under an SiH$_4$ flow. At this temperature, gold film dewetting leads to the formation of 10–100 nm diameter gold nanodroplets serving as catalysts for preferential silane decomposition. This decomposition leads to the formation of a liquid Au–Si eutectic. Oversaturation of these eutectic droplets by silicon atoms leads to the precipitation of Si and then to the formation of nuclei at the solid–liquid interface. After a certain time wires are obtained (elongation step) with the gold nanoparticle on top. Thus, the nanowire diameter is defined by the catalyst diameter, and the length is a function of time and pressure. All the samples have been prepared using 40 sccm of SiH$_4$ at 500 °C and a total pressure of 0.532 mbar during 60 min and 10 min for the VLS8 and VLS2 interfaces, respectively.[20–22] The protocols correspond to processes P1 (VLS2) and P2 (VLS8) in ref. 21. No doping reagent was used during the synthesis of nanowires.

## 3 Polymer plasma-based coating

A fluoropolymer (C$_4$F$_8$) was then deposited on the silicon nanowires and the silicon nanowires on the μ-pillar interfaces. The interfaces were placed onto a holder and then directly placed in the plasma chamber (Silicon Technology System) leading to a coverage of 30 nm thick of C$_4$F$_8$ [passivation gas, C$_4$F$_8$ (50 sccm); passivation time, 10 s] which is considered an insulator.

The polymer layer (C$_4$F$_8$) presents a contact angle of 105° and a hysteresis of 22° on a planar surface. We chose this polymer for a reason of uniformity between pillars and NWs: because the layer can be plasma-deposited, similarly to that used on NWs. The coating technique used by Bahadur et al.,[15] using a spin-coated Teflon polymer, offers a better hydrophobicity and a better resistance to impalement during EW, but cannot be applied on NWs as the spin-coating would damage them.

## 4 Surface characterization

Scanning electron microscopy (SEM) images were obtained using an electron microscope ULTRA 55 (Zeiss, France) equipped with a thermal field emission emitter and three different detectors (EsB detector with filter grid, high-efficiency In-lens SE detector, Everhart-Thornley secondary electron detector).

Table 1 summarizes the different surfaces we have realized. We characterize them with the notation P$^x_y$(h) where P stands for pillar, x is its diameter, y its spacing and h its height. Fig. 2 shows a few representative cases. The basic setup consists of cylindrical posts, as shown in Fig. 2 (a). These pillars of variable size (see Table 1) are then covered with short NWs according to protocol VLS2 (Fig. 2 b–d). We have also produced surfaces with long NWs, grown on top of the pillars according to protocol VLS8 (not shown). On these surfaces the pillars, whose height is shorter than the height of long NWs, are completely shadowed by a forest of NWs so

**Table 1** Contact angle, hysteresis and impalement degree after EW at 225 $V_{TRMS}$

| $P_x^y/h + C_4F_8$ | $\phi_s$ | $r$ | $\theta_0^{theo}$ (°) | $\theta_0^{Exp}$ (°) | Hysteresis (°) | Impalement EW |
|---|---|---|---|---|---|---|
| $P_2^4(4)$   | 0.35 | 2.74 | 141.3 | —   | 25 | $160 - 98 = 62$ |
| $P_2^{10}(4)$ | 0.54 | 2.09 | 128.9 | —   | 35 | $160 - 99 = 61$ |
| $P_5^{10}(4)$ | 0.35 | 1.69 | 141.3 | —   | 23 | $158 - 97 = 61$ |
| $P_7^{10}(4)$ | 0.27 | 1.54 | 150.9 | —   | 19 | $162 - 84 = 78$ |
| $P_2^4(10)$  | 0.35 | 4.49 | 141.3 | 145 | 23 | $160 - 122 = 38$ |
| $P_5^{10}(10)$ | 0.54 | 3.18 | 128.9 | 134 | 24 | $161 - 108 = 53$ |
| $P_7^{10}(10)$ | 0.35 | 2.39 | 141.3 | 137 | 26 | $160 - 102 = 58$ |
| $P_2^4(10)$  | 0.27 | 2.08 | 150.9 | 148 | 19 | $160 - 106 = 54$ |
| $P_2^4(19)$  | 0.35 | 7.98 | 141.3 | 145 | 23 | $153 - 97 = 56$ |
| $P_2^{10}(19)$ | 0.54 | 5.36 | 128.9 | 144 | 24 | $158 - 96 = 62$ |
| $P_5^{10}(19)$ | 0.35 | 3.79 | 141.3 | 146 | 26 | $161 - 94 = 67$ |
| $P_7^{10}(19)$ | 0.27 | 3.17 | 150.9 | 151 | 19 | $161 - 85 = 76$ |
| NWs | | | | 160 | 0 | see text |

that a drop deposited on top of the surface merely sees the micro-scale modulations induced by pillars, see the discussion below.

## III. Wetting and electrowetting experiments

### 1 Contact angle and hysteresis measurements

We performed contact angle and hysteresis measurements on each surface. For the contact angle measurement, a 3 μl deionised water droplet is deposited on each surface (pillar, pillar + NW) and a goniometer (GBX, France) determines each contact angle. The hysteresis was measured by spreading a 4 μl droplet on the surface (giving $\theta_a$) and then sucking it (giving $\theta_r$). The hysteresis is obtained from $\Delta\theta = \theta_a - \theta_r$, see Table 1. The impalement under EWOD is determined by the difference of the contact angle before and after applying the voltage. We observed that the contact angle evolved as a function of the surface fraction. The contact angle is observed to increase with the pillar spacing. Therefore, the hysteresis measurement showed the same trend; whatever the height of the pillar, $P_7^{10}(h)$ surfaces presented the lowest hysteresis (CAH = 19°). After the NW growth and the hydrophobic coating on each surface, the contact angle went up to 160° and the hysteresis to 0° ± 1°. Hence, this NW texturation on the micro-pillars leads to superhydrophobic surfaces with very low friction.

The contact angles measured on micro pillars surfaces are comparable to those obtained theoretically, see Table 1. To evaluate theoretically the CA, we assume that a drop deposited on the surface before EW, is in the Cassie state,[2]

$$\cos\theta_C^{theo} = \phi_s(1 + \cos\theta) - 1 \qquad (1)$$

with $\phi_s$ as the surface fraction obtained for our cylindrical pillars,

$$\phi_s = \frac{\pi x^2}{4(x+y)^2} \qquad (2)$$

We also calculated the roughness $r$ of the pillar surfaces

$$r = 1 + \frac{\pi x h}{(x+y)^2} = 1 + \frac{4\phi_s h}{x} \qquad (3)$$

## 2 EW experiments

In parallel, we carried out electrowetting experiments (EW) on each surface. The measurements are obtained in the following way: the voltage is applied between the drop and the substrate with a signal generator (CENTRAD GF 265, ELC, France), 0.5 to 21 V output at 1 kHz, coupled to a 50 dBm high-voltage amplifier (TEGAM, USA). This leads to a voltage range between 12.5 V to 250 V at amplifier output. A goniometer (GBX, France) was used to record and measure the CA during EW. To determine the reproducibility of the EW phenomenon, the following protocol is applied:

(1) Droplet formation on the surface (3 ± 0.1 μL) through a metallic, hydrophobic needle;

(2) A single voltage is applied (during 0.5 s) to the droplet through the same conductive needle at the beginning of the first EW cycle;

(3) Return to 0 V during 0.5 s (end of the first EW cycle);

(4) Steps 2 and 3 are repeated at least four times (EWOD cycle) at the same voltage in order to reach a stable value of CA.

These experiments constitute tests for the reversibility during EW, and reflect whether or not the liquid has changed from Cassie–Baxter to Wenzel states. EW allows for applying an external pressure to the liquid depending on some parameters as will be developed in *Section V - Discussion*.

## 3 EW on micropillars

Independently of the surface and the applied voltage (starting from 25 $V_{TRMS}$, the same phenomenon under EW is observed: from a contact angle around 160°, the applied voltage decreases the contact angle, which stays stuck at a constant value when the voltage is turned off (Fig. 3).

The variation of cosines of CA induced by EW *versus* squared voltage is plotted in Fig. 4 (a) for the four surfaces of pillar height equal to 19 μm. The trends for the two other surfaces (*i.e.*, 4 μm and 10 μm height) are similar: on the same surface, the higher the voltage, the larger the decrease of the CA, leading to a deeper impalement with a maximum CA variation of about 80° at 225 $V_{TRMS}$. The EW saturation

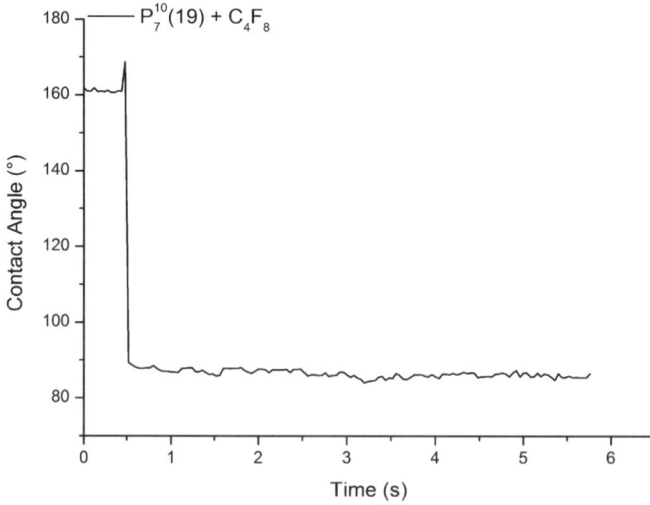

**Fig. 3** Contact angle measurement during the EW cycle (at 225 $V_{TRMS}$) on $P_7^{10}(20) + C_4F_8$. It shows a complete irreversibility with EW, a transition from superhydrophobic to hydrophilic behaviour.

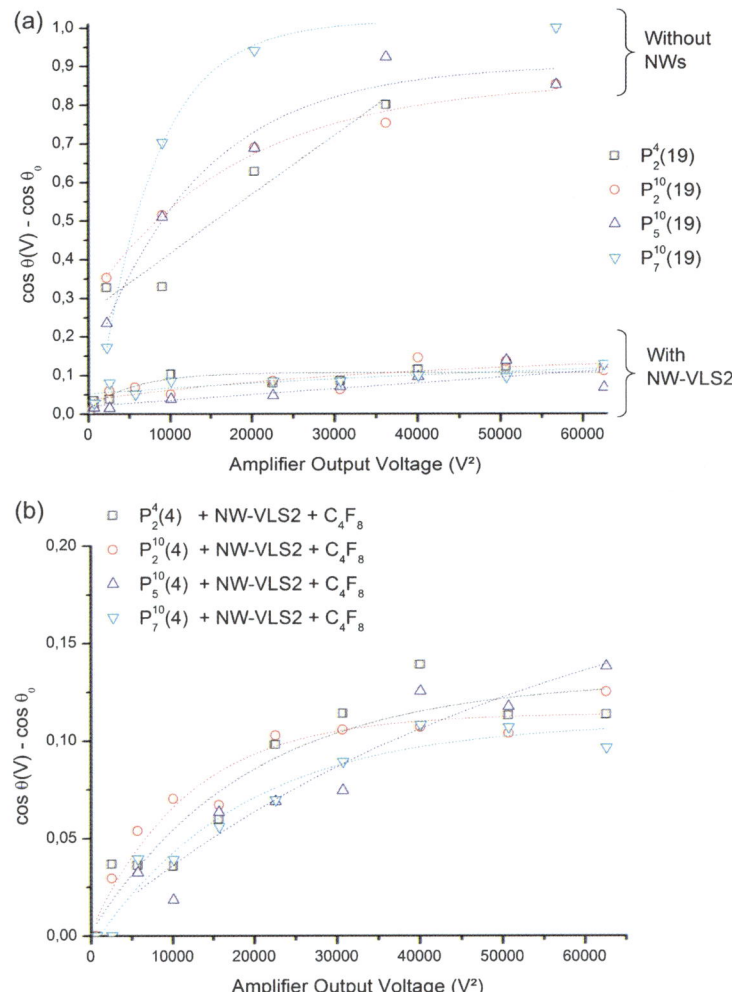

**Fig. 4** Variation of the contact angle *versus* the squared voltage, after the first EW cycle, for different micro/nano double-scale roughness described by $P_x^y(h)$ + NW + $C_4F_8$ where P stands for pillar, $x$ is its diameter, $y$ its spacing and $h$ its height.

appears at a voltage of about 140 $V_{TRMS}$ with a maximum variation of the cosine between 0.7 and 1.0.

When comparing the different surfaces, no clear difference is seen at the same voltage. The level of impalement seems not to depend directly on the characteristics of the microstructures. EW on the microstructured surface leads to a transition from a superhydrophobic to a hydrophobic state (or hydrophilic in a few cases; below 90°). The droplet starts from a Cassie state, then moves to a Wenzel state once the voltage is on, and it never returns to the Cassie state when the voltage is turned off (*i.e.*, after the end of the first cycle).

## 4 EW on micropillars covered by nanowires

Different behaviours have been observed under EW according to the geometric properties of the corresponding surface.

In Fig. 4 (b) we compare the CA variation for the same height of micropillars (*i.e.* 19 μm) with and without nanowire structures, after the first application of the voltage. Here again, as for the surfaces with micropillars only, the trend is the same for the three heights of micropillars. While the cosine variation is between 0.7 and 1.0 for the pillars-only surfaces at 250 $V_{TRMS}$, this variation is inferior to 0.15 for the double-scale surfaces (micropillars covered by short NWs) at the same voltage. The saturation of EW appears at about 200 $V_{TRMS}$ and no clear difference can be ascribed to the different double-scale surfaces after the first EW application. It is to be noted that, according to the measurement accuracy and to the strong heterogeneity of the surface, the error range of CA is about 6°.

To discriminate these surfaces through their robustness, we carried out CA measurements during more than four EW cycles. We performed 4 EW cycles on each surface, and compared their degree of reversibility. To quantify this degree, we defined $CA_1$, the CA during the first EW pulse and $CA_2$, the CA at the end of the EW cycle. $CA_2$–$CA_1$ represents the variation of CA after 4 EW cycles (Fig. 6) at 225 $V_{TRMS}$. In this particular case the difference of the reached values $CA_1$–$CA_2$ during an EW cycle depends on pillar height and on the surface fraction (Fig. 5). This CA difference quantifies the performance of the surface for droplet actuation, as it is related to the capillary force that can be induced by EW under reversible conditions.

- For short pillars $P_7^{10}(4)$ + NW-VLS2 + $C_4F_8$, the CA starts at 160° at 0 V, then decreases to $CA_1$ = 146° when the voltage is prescribed (at 225 $V_{TRMS}$), and remained at this value whenever we turn the voltage off or on ($CA_1 = CA_2$). The CA variation is completely irreversible under EW.
- For intermediate pillars $P_7^{10}(10)$ + NW-VLS2 + $C_4F_8$, the CA starts at 158° at 0 V, then decreases to $CA_1$ = 141° when the voltage is applied (at 225 $V_{TRMS}$) but goes back up to $CA_2$ = 152° when the voltage is off. For each EW following cycle, the CA under EW is constant (CA = 141°) but after each relaxation (voltage turned off), the CA decreases until it reaches the CA after the first EW. After this first cycle, the drop seems to be in a partially impaled state. Depending on the roughness, the Wenzel state is reached after a variable number of EW cycles.

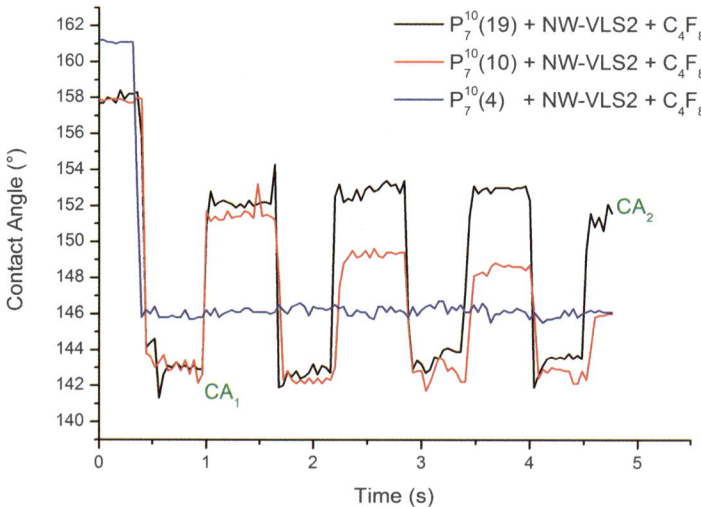

**Fig. 5** Contact angle measurement during the EW cycle (at 225 $V_{TRMS}$) on the different micro/nano double-scale roughness. Notation characterizing the pillar geometry is as before. $CA_1$ corresponds to the contact angle during the first EW cycle, and $CA_2$ is the contact angle when the EW cycle is stopped.

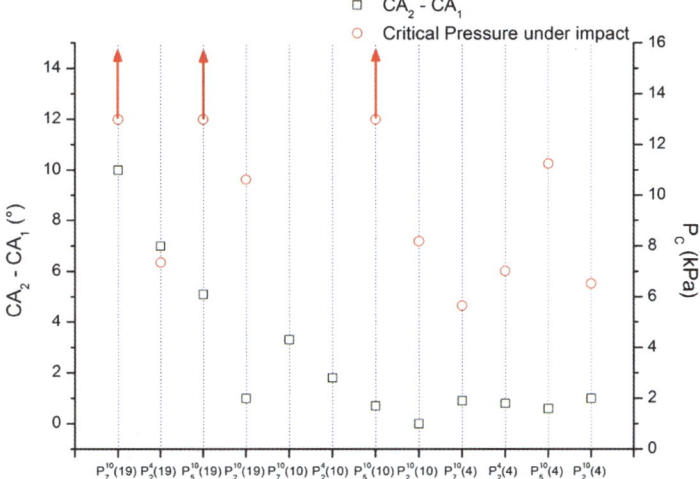

**Fig. 6** $CA_2$–$CA_1$, the degree of reversibility for each surface (square symbols). It increases with the height and the space between the pillars. Comparison with the pressure threshold $P_c$ for impalement obtained by drop impact (circles). The arrows denote that $P_c$ is larger than 13 kPa for these specific surfaces.

• For tall pillars $P_7^{10}(19)$ + NW-VLS2 + $C_4F_8$, the CA starts at 158° at 0 voltage, then decreases to $CA_1 = 141°$ when the voltage is applied (at 225 $V_{TRMS}$) and rises up to $CA_2 = 152°$ when the voltage is off. Meanwhile, for each EW step, the CA relaxes to a permanent value. The surface stays in a semi-reversible state after the EW cycles.

While the NWs introduce a more robust superhydrophobic state on pillars without EW, the height of the pillars acts as a barrier to the total impalement of the droplet under EW: for the 4 μm short pillars + NWs, the Wenzel state is immediately obtained under EW; for the intermediate 10 μm pillars + NWs the CA shows a variation smaller than 5°, indicating that the liquid drop is deeply but partially impaled. Finally, the 19 μm pillars + NWs measurements seem to indicate that the drop is shallowly impaled, with about 10° of difference between $CA_1$ and $CA_2$. A comparison between the 10 μm and the 19 μm height pillars indicates that the $P_7^{10}$ surfaces present the most important degree of reversibility. Furthermore, for these two height values, the quantity $CA_1$–$CA_2$ seems directly linked to the surface fraction $\phi_s$: the $P_7^{10}$ presents the smallest surface fraction, the CA variation under EW is the most important while the $P_2^{10}$ has the largest surface fraction with a quasi-null CA variation under EW. This is consistent with the measurements obtained with the $P_5^{10}$ and $P_2^{10}$ surfaces which offer the same surface fraction: the degree of reversibility is approximately the same (*i.e.*, about 2° and 6° for the pillars with 10 μm and 19 μm height, respectively). Fig. 6 summarises all these results, together with the results of drop impact experiments described hereafter.

## IV. Droplet impalement experiments

### Description of the set-up

To carry out drop impact experiments, we used the same set-up as previously described,[21] which is shown in Fig. 7 (a). A sub-millimetric nozzle releases a drop of liquid (water/glycerin 50/50 mixture) from a height $H$ that prescribes the impact velocity $U = (2gH)^{1/2}$ - which can be up to 4.7 m s$^{-1}$. The diameter of the drop is determined by the capillary length, and is well reproducible at $d = 2.6 \pm 0.1$ mm.

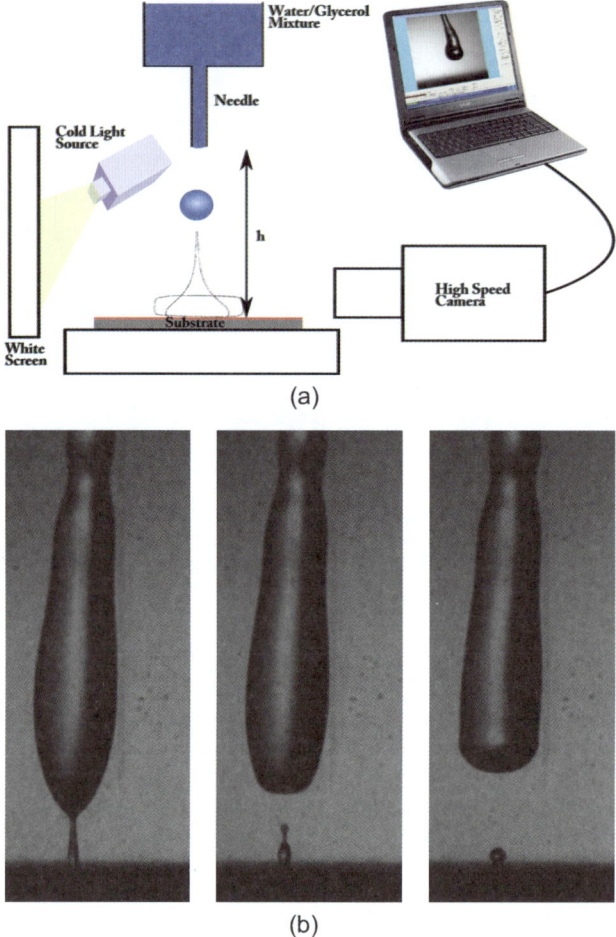

**Fig. 7** (a) Experimental set-up, see text for details. (b) Typical snapshots from the high-speed camera, here showing a bouncing drop after impact and leaving an impaled tiny droplet at the impact location.

The physical parameters of the liquid are: kinematic viscosity $\nu = 6.2$ cSt, surface tension $\sigma = 0.066$ N m$^{-1}$ and density $\rho = 1126$ kg m$^{-3}$. The mixture has a viscosity larger than water, which prevents both splashing - *i.e.*, the splitting of the main drop into tiny ones - and corrugations at the border of the drop during the spreading phase. Hence during the spreading and the retraction, the drop mostly remains axisymmetric. Using backlighting together with a high-speed camera (at a maximal rate of 8000 frames/s, with a resolution of 256 × 512), the shape of the interface during the spreading and bouncing processes can be determined. The magnification allows for an accuracy of about 8 μm per pixel.

To determine the pressure threshold for impalement, we determine the minimal height (or impact velocity) for which some liquid remains at the impact location. The height of fall $H$ is increased until a sequence like in Fig. 7 (b) is observed, *i.e.*, some liquid in the form of a tiny droplet stays impaled at the centre of the impact location. This minimal height $H_{min}$ gives the pressure threshold $P_c = \rho g H_{min}$. Therefore, the key point to prevent the transition to Wenzel state as much as possible is to have the largest possible $P_c$. Our drop impact set-up allows for prescribing a dynamical pressure of $P_{dyn} = \frac{1}{2}\rho U^2$ up to about 13 kPa.

It is noticeable that all pillar+long NWs surfaces (growth with VLS8 protocol, see section II.2 and Fig. 1) had threshold values much larger than the limit of 13 kPa. Actually, the length of NWs (about 40 µm) is much larger than the height of the micro-pillars, and it turns out that these surfaces have impalement properties that are very similar to those of the long NWs alone, without pillars. In this specific case, the micro-structure is shadowed by the NWs and has no influence.

Hence, in the aim for understanding the combined contribution of pillars and NWs on the robustness, it is more suitable to use the shorter NWs (Protocol VLS2) which height is comparable to or smaller than the height of the pillars. The short NWs of average height about 4 µm presented before, and represented in Fig. 2, are adapted to such a study. Therefore in what follows, we report only the quantitative results obtained for the surfaces of micro-pillars covered by the shorter NWs.

**Previous existing results**

Previous experiments have evidenced that droplet impact experiments are suitable for the determination of the impalement threshold on textured SH surfaces,[21,27–29] where the entry pressure is relatively large. By testing various surfaces textured by periodic arrays of micro-pillars, Reyssat *et al.* showed a relationship between geometrical properties of the texture and the impalement threshold [29] that can be written as:

$$P_c = \alpha \frac{\sigma h}{l^2} \tag{4}$$

where $\alpha \simeq 0.048$. The geometrical parameters $h$ and $l$ respectively stand for the height of the posts and the space between them. Therefore, the taller the pillars and the narrower the space between them, the better the robustness to impalement. They found the best robustness for $h = 26$ µm and $l = 3$ µm, corresponding to a dynamical pressure of $P_{dyn} = 4.5$ kPa.[29]

At almost the same time, Bartolo *et al.* carried out the same type of experiments.[28] Although less exhaustive than ref. 29 and with less robust surfaces, their results evidenced two important points: (1) the pillar height $h$ ceases to have a noticeable influence on robustness beyond a certain limit (in their experiments, the robustness was not improved by an increase of $h$ above about 25 µm); (2) by comparing the impalement pressure in both impact and (static) evaporation experiments, the authors concluded that hydrodynamic forces do not play any significant role in the impalement transition. Therefore, the pressure threshold is the same whether it is prescribed dynamically by drop impact, or statically by other means like evaporation, drop squeezing or electrowetting.

Based on these quantitative results on micro-pillars, it is straightforward that a much better robustness would be obtained by using a much smaller, rougher and denser texture. One of our recent studies[21] showed that a dense array of silicon nanowires, coated with an ad-hoc chemical, was robust enough to prevent liquid impalement. Our best surfaces were covered with long NWs that form a lower dense and entangled structure and an upper looser but more regular structure (see Fig. 1), had an impalement threshold $P_c$ larger than 17 kPa.

**Our results on multi-scale textured surfaces**

We tested various surfaces for their robustness on drop impact impalement, and compared them with potential reversibility during EW. The results are presented in Table 2, and plotted in Fig. 6 in comparison with EW experiments. Due to the limited height of fall in our laboratory it was not possible to measure the exact threshold for some of the surfaces as they were too robust. In this case we indicate

**Table 2** Impalement thresholds in drop impact for various double-scale substrates. $P_c^*$ is the pressure that would be found for substrates made only of micro-pillars, according to the results of ref. 29, see eqn (4)

| Pillar height (μm) | Pillar width (μm) | Interspace (μm) | $P_c$ (kPa) | $P_c^*$ (kPa) |
|---|---|---|---|---|
| 4 | 4 | 2 | 7.025 | 1.41 |
| 4 | 10 | 2 | 6.53 | 1.41 |
| 4 | 10 | 5 | 11.264 | 0.225 |
| 4 | 10 | 7 | 5.66 | 0.115 |
| 10 | 10 | 2 | 8.19 | 3.52 |
| 10 | 10 | 5 | ≥13 | 0.562 |
| 19 | 4 | 2 | 7.36 | 6.68 |
| 19 | 10 | 2 | 10.62 | 6.68 |
| 19 | 10 | 5 | ≥13 | 1.07 |
| 19 | 10 | 7 | ≥13 | 0.545 |
| No pillar, only NWs | — | — | 7.14 | 0 |

that the threshold is larger than the maximal accessible dynamical pressure of 13 kPa.

By comparing the different results, it turns out that:
• in comparison to micro-pillars surfaces tested in ref. 29, the adjunction of NWs increases significantly the robustness against impalement as quantified by $P_c$;
• in comparison to surfaces with short NWs only, the combination of pillars and NWs can both significantly increase or decrease $P_c$. The double-scale seems to get its best efficiency when micro-pillars are much taller than the height of the NWs and when the pillar interspace is comparable to NWs height: here, a spacing of 5 μm with NWs of height 4 μm gave the best results.

Therefore, a double-scale structure is not necessarily a guarantee for robustness. One has to take into account the length ratio between micro- and nano-scale elementary units. To some extend, the threshold $P_c$ can be related to the degree of reversibility $CA_2-CA_1$ (see Fig. 6), except for $P_2^4(19)$ and $P_5^{10}(4)$ surfaces.

## V. Discussion

The performance of the droplet actuation by EW on a double-textured (micro/nano) surface depends on two parameters: the roughness $r$ (mostly dependent on the height of the micro-structuration) and the surface fraction $\phi_s$. The total impalement of the liquid inside the pillars is prevented only by the surfaces with the tallest micropillars. As far as it is concerned, the surface fraction can be ascribed to the reversibility of the CA (we have called it degree of reversibility under EW cycles). In other words the transition to the Wenzel state is prevented by the height of the pillars. Once this vertical impalement is limited, a shallower impalement ('quasi-Cassie state') is promoted by a decrease of the surface fraction, and by a space between pillars of the same order as the length of the nano-textutation.

To sum up the previous results, and put them in a broader and more fundamental perspective, we comparatively evaluate the liquid impalement in the texture *via* a prescribed pressure within two configurations:
• The impact of a droplet. In this case, we have a *dynamical* pressure due to the initial kinetic energy of the drop, which equals $P_{dyn} = \frac{1}{2}\rho U^2$, where $U$ is the impact velocity;
• The deformation of the droplet in the vicinity of the contact-line induced by EW.

In this latter case, the pressure is due to the strong local curvature that builds up close to the contact-line. The Laplace pressure $P_L$ balances the Maxwell stress due to the accumulation of charges near the contact-line, as shown by Mugele and Bührle.[24,25] More precisely, under the action of EW the drop *apparent* contact-angle $\theta_{app}$ is modified down to a certain scale $\lambda$. The apparent contact angle is given by the Young–Lippman eqn (8):

$$\cos(\theta_{app}) = \cos(\theta_0) + \frac{1}{2\sigma}\frac{\varepsilon_0\varepsilon V^2}{\lambda} \quad (5)$$

where $V$ stands for the applied voltage, $\varepsilon_0$ stands for the permittivity of the vacuum and $\varepsilon$ stands for the relative effective permittivity of the surface dielectric.

However, accurate visualizations evidenced that the contact angle *always* converges towards the Young's contact-angle - *i.e.*, the angle $\theta_0$ in the absence of voltage - at a distance from the contact-line smaller than $\lambda$.[24] The length $\lambda$ corresponds to an effective width of the dielectric, as shown in Figures 10 and 11 of ref. 24. Another direct evidence of this phenomenon was reported in a still unpublished paper,[26] where it was shown by reflection microscopy that the impalement first starts at the contact-line.

Therefore, the local slope of the drop evolves from the apparent (macroscopic) contact-angle given by eqn (5) to the microscopic contact-angle within a distance of $\lambda$. For textured surfaces, the microscopic angle is the Cassie–Baxter angle if the liquid has not impaled the texture yet.

To check Mugele and Bührle's observations on our surfaces, we acquired magnified images of a drop during electrowetting at the vicinity of the contact line. Fig. 8

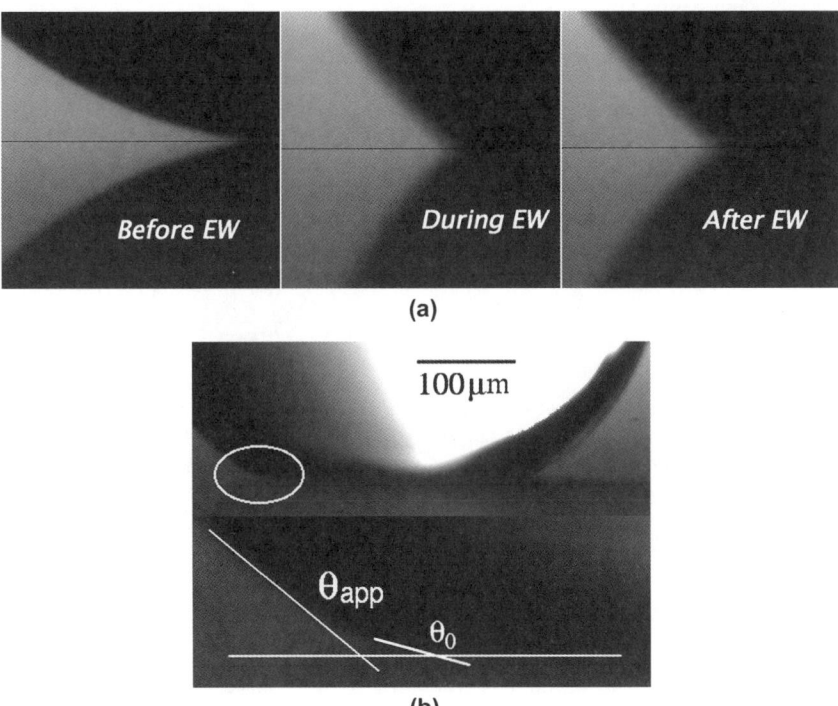

**Fig. 8** Magnified view of the vicinity of the contact-line (a) For a substrate showing irreversibility (impalement) before, during and after electrowetting. (b) For a totally reversible substrate with higher NWs, showing a transition from apparent $\theta_{app}$ to Cassie–Baxter $\theta_0$ angle close to the contact-line.

(a) represents three images before, during and after EW for a surface where liquid is impaled during EW. It turns out that the 'quality' of the surface, *i.e.* its resistance under EW or drop impact impalement, is visible close to the contact-line. When the liquid is impaled during EW (Fig. 8 (a)), the drop profile shows a unique slope. This is understandable considering that, as the liquid has impaled the surface, it leads to a drastic decrease of the effective thickness of dielectric $\lambda$. On the contrary, if the liquid stays shallow and does not enter significantly into the texture, we clearly see a transition from the macroscopic Lippmann (apparent) angle to the Cassie–Baxter angle $\theta_0$ close to the contact-line (Fig. 8 (b) and magnified view therein). This is a direct evidence that the effective thickness of dielectric $\lambda$ is large enough to be observed. We estimated it to about 35 µm from this picture. Again, this is consistent with Mugele's observations of a water drop immersed in oil on flat, non-textured surfaces. Here this is the first time that it was directly observed on a textured surface in ambient air.

The local Gaussian curvature is given by the secondary derivative of the drop profile $r(z)$, with $z$ standing for the vertical coordinate. If $\lambda$ is small enough, one can approximate the second derivative by the finite difference of first derivatives. Therefore, the Laplace pressure close to the contact-line equals

$$P_\text{L} = \sigma\kappa \simeq \sigma\left(\frac{\text{d}^2r}{\text{d}z^2}\right)_{z=\lambda} \simeq \frac{\left(\frac{\text{d}r}{\text{d}z}\right)_{z=\lambda} - \left(\frac{\text{d}r}{\text{d}z}\right)_{z=0}}{\lambda} \quad (6)$$

The derivatives $\frac{\text{d}r}{\text{d}z}$ are related to the apparent and Cassie–Baxter angles, leading to the following expression

$$\kappa = \frac{1}{\lambda}\left(\cot(\theta_\text{app}) - \cot(\theta_0)\right) \quad (7)$$

From eqn (7) and after a bit of algebra, we obtain the expression

$$P_\text{L} = \sigma\kappa = \frac{\sigma}{\lambda}\left(\frac{\frac{\varepsilon_0\varepsilon}{2\sigma\lambda}V^2 + \cos(\theta_0)}{\sqrt{1 - \left(\frac{\varepsilon_0\varepsilon}{2\sigma\lambda}V^2 + \cos(\theta_0)\right)^2}} - \cot(\theta_0)\right) \quad (8)$$

The Laplace pressure *versus* voltage $V$ obtained from eqn (8) is plotted in Fig. 9 for three different values of dielectric thickness $\lambda$. From these plots, it is clear why impalement occurs at low voltage on surfaces with short NWs, as the pressure induced by EW reaches rapidly a few tens of kPa (which is far beyond the measured impalement threshold, see Table 2). Not only are short NWs intrinsically less resistant to impalement (as evidenced by drop impact experiments (21)), but also their thin layer of dielectric produces a larger pressure for the same CA difference. Hence, for the same actuation force under EW, which is quantified by the difference $\cos(\theta_\text{app}) - \cos(\theta_0)$, taller NWs like those produced by the VLS8 protocol have to support a smaller pressure due to the larger effective thickness of dielectric (see eqn (7)). Once the liquid has not impaled, this large effective thickness is preserved.

To conclude, we explained why the height of the pillars and that of NWs are essential to prevent irreversible behaviour under EW. For double-scale surfaces, tall pillars associated with shorter NWs are optimal if the space between pillars $l$ is comparable to the height of the NWs.

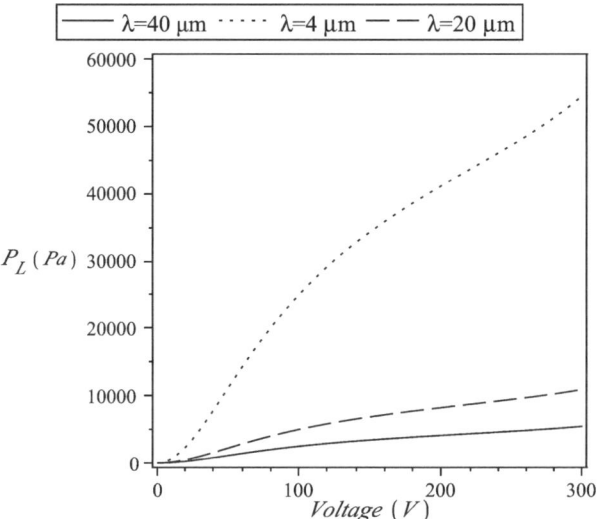

**Fig. 9** Pressure induced by electrowetting, resulting from the equilibrium between the Maxwell stress and capillary forces.

## References

1. N. K. Adam, *The Physics and Chemistry of Surfaces*, 1st Ed. (Oxford Clarendon), 180–181 (1930).
2. A. Cassie and S. Baxter, *Trans. Faraday Soc.*, 1944, **40**, 546.
3. R. Wenzel, *Ind. Eng. Chem.*, 1936, **28**, 988.
4. R. Blossey, *Nat. Mater.*, 2003, **2**, 301.
5. Z. Burton and B. Bushan, *Nano Lett.*, 2005, **5**, 1607.
6. R. Fair, *Microfluid. Nanofluid.*, 2007, **3**, 245.
7. J. Heikenfeld and M. Dhindsa, *J. Adhes. Sci. Technol.*, 2008, **22**, 319.
8. G. Lippmann, *Ann. Chim. Phys.*, 1875, **5**, 494.
9. B. Berge and J. Peseux, *Eur. Phys. J. E*, 2000, **3**, 159.
10. R. A. Hayes and B. J. Feenstra, *Nature*, 2003, **425**, 383.
11. F. Mugele and J.-C. Baret, *J. Phys.: Condens. Matter*, 2005, **17**, R705.
12. A. Torkkeli, 2003.
13. D. L. Herbertson, C. R. Evans, N. J. Shirtcliffe, G. McHale and M. I. Newton, *Sens. Actuators, A*, 2006, **130–131**, 189.
14. N. Verplanck, Y. Coffinier, V. Thomy and R. Boukherroub, *Nanoscale Res. Lett.*, 2007, **2**, 577.
15. V. Bahadur and S. V. Garimella, *Langmuir*, 2008, **24**, 8338.
16. R. J. Vrancken, H. Kusumaatmaja, K. Hermans, A. M. Prenen, O. Pierre-Louis, C. W. M. Bastiaansen and D. J. Broer, *Langmuir*, 2010, **26**, 3335.
17. N. A. Patankar, *Langmuir*, 2004, **20**, 8209.
18. Y. Kwon, N. A. Patankar, J. Choi and J. Lee, *Langmuir*, 2009, **25**, 6129.
19. B. Bhushan, Y. C. Jung and K. Koch, *Langmuir*, 2009, **25**(5), 3240.
20. N. Verplanck, E. Galopin, J.-C. Camart, V. Thomy, Y. Coffinier and R. Boukherroub, *Nano Lett.*, 2007, **7**, 813.
21. P. Brunet, F. Lapierre, V. Thomy, Y. Coffinier and R. Boukherroub, *Langmuir*, 2008, **24**, 11203.
22. F. Lapierre, V. Thomy, Y. Coffinier., R. Blossey and R. Boukherroub, *Langmuir*, 2009, **25**, 6551.
23. S. Moulinet and D. Bartolo, *Eur. Phys. J. E*, 2007, **24**, 251.
24. F. Mugele and J. Buehrle, *J. Phys.: Condens. Matter*, 2007, **19**, 375112.
25. F. Mugele, *Soft Matter*, 2009, **5**, 3377–3384.
26. A. Staicu, G. Manukyan and F. Mugele, arXiv:0801.2683v1, 2008.
27. K. K. S. Lau, J. Bico, K. B. K. Teo, M. Chhowalla, G. A. J. Amaratunga, W. I. Milne, G. H. McKinley and K. K. Gleason, *Nano Lett.*, 2003, **3**, 1701.
28. D. Bartolo, F. Bouamrirene, E. Verneuil, U. Buguin, P. Silberzan and S. Moulinet, *Europhys. Lett.*, 2006, **74**, 299.
29. M. Reyssat, A. Pepin, F. Marty, Y. Chen and D. Quéré, *Europhys. Lett.*, 2006, **74**, 306.

PAPER

# Macroscopically flat and smooth superhydrophobic surfaces: Heating induced wetting transitions up to the Leidenfrost temperature

Guangming Liu and Vincent S. J. Craig*

*Received 26th November 2009, Accepted 28th January 2010*
DOI: 10.1039/b924965f

We present an investigation of the change in wettability of water droplets on 3 different flat, smooth substrates with an elevation in temperature. Two methods were employed. In the first method the droplet was placed on the substrate before it was heated and in the second method the droplets were induced to fall onto a preheated substrate. We find that the intrinsic wettability of the surface is important and that fundamentally different behavior is observed on a hydrophobic surface relative to hydrophilic surfaces. For the hydrophobic surface and employing the first method, we have observed three different regimes over the temperature range of 65 °C to 270 °C. In regime I (65 °C to 110 °C), the contact angle of water droplets exhibit a slight decrease from 108° to 105° and an accompanying significant decrease in droplet lifetime ($\tau$) from ~111 s to ~30 s is observed. In regime II (120 °C to 190 °C), $\tau$ remains constant at ~20 s however the contact angle significantly increases from 127° to 158° - that is we enter a superhydrophobic regime on a flat surface. In this regime the droplet remains stationary on the surface. Regime III (210 °C to 270 °C), is the Leidenfrost regime in which the water droplet exhibits a rapid motion on the solid surface with a contact angle higher than 160°. In comparison, the wetting behavior of a water droplet on two relatively hydrophilic surfaces (Au and GaAs) have also been investigated as a function of temperature. Here no wetting transition is observed from 65 °C up to 365 °C. In the second method, the wetting behavior on the hydrophobic surface is similar to that observed in the first method for temperatures below the Leidenfrost temperature and the water droplet rebounds from the solid surface at higher temperatures. Additionally, the Leidenfrost phenomenon can be observed above 280 °C for the hydrophilic surfaces.

## I. Introduction

Superhydrophobicity is widely observed in nature, the classic example being the surface of the lotus leaf. The contact angle on a flat wax surface is typically ~110°, but the highly textured surface of the leaf increases the apparent contact angle to 150° or more. Based on the model proposed by Cassie and Baxter 65 years ago,[1] air is trapped between the asperities and the apparent contact angle of the water droplet on the rough surface is determined by the average of the contact angles on air and on the solid surface with respect to their respective area fraction.[1] When the water droplet is impaled by the surface structures and the trapped air is replaced

*Department of Applied Mathematics, Research School of Physics and Engineering, The Australian National University, Canberra, ACT 0200, Australia. E-mail: vince.craig@anu.edu.au*

by water, the droplet enters another regime that is defined as the "Wenzel state", in this regime, the apparent contact angle is governed by the contact angle and roughness of the solid surface.[2] Recent discussions indicate that the interactions at the three-phase contact line between droplet and solid are much more important than that within the contact perimeter.[3–8] The two states are distinct but droplets in the Cassie state can be transformed to the Wenzel state by applying pressure,[9] applying an electrical voltage[10] or through vibrations.[11]

Another readily observed example where an extremely high contact angle can be observed is the Leidenfrost effect,[12] whereby a droplet placed on a sufficiently hot surface undergoes a drying transition; as it is suspended on a cushion of evaporating vapour.[13] At the Leidenfrost temperature the droplet lifetime is a local maximum because the evaporation rate of the droplet is decreased due to the reduction in heat flux associated with the formation of the insulating vapour layer.[14] Although a Leidenfrost droplet is not in contact with the solid surface, the apparent contact angle is less than 180° as the vapour cushion necessarily has a finite size. For a nearly spherical water droplet of radius 1 mm the circular flattened area at the bottom has a radius of ∼0.40 mm[13] and hence, from simple geometry, the apparent contact angle is ∼156° which is comparable to most superhydrophobic surfaces. Here also, the inherent wettability of the surface is important as the Leidenfrost temperature decreases as the contact angle increases,[15,16] and is also influenced by surface roughness,[17–19] and the manner by which the droplet impacts the surface.[20,21]

We wish to investigate the contact angle change with temperature as the Leidenfrost temperature is approached from below on both hydrophobic and hydrophilic surfaces. On the hydrophobic surface this will enable us to observe the transition of a droplet from a hydrophobic to the Leidenfrost state which has a contact angle similar to a superhydrophobic surface. An analogy for conventional superhydrophobic surfaces is a transition from the Wenzel to Cassie state which has only recently been observed.[10] On rough hydrophobic surfaces the state of the droplet is binary—it is either in the Cassie or Wenzel state, there is no intermediate condition. We wish to determine if this is also true for flat surfaces undergoing a Leidenfrost transition.

The temperature dependence of the contact angle of water droplets on a solid surface is not well understood and is difficult to predict, although a few theoretical and experimental studies have been conducted.[22–29] Theoretical predictions differ. The statistical theory proposed by Navascués and Berry predicts only an increase in contact angle with increasing temperature,[29] whilst Sullivan (using a van der Waals model) concluded that the contact angle decreases with temperature for high-energy surfaces and increases with temperature for sufficiently low-energy surfaces.[27] Most experimental investigations have demonstrated that the contact angle decreases with increasing temperature for both hydrophilic and hydrophobic surfaces,[22–26] that is, these surfaces have a negative temperature coefficient ($d\theta/dT$). However, Neumann and Renzow have observed that the contact angle of water on silicone treated glass increased with temperature, while hydrocarbons on the same surface gave negative values of $d\theta/dT$.[28] Additionally, some polymer modified surfaces exhibited an increase of contact angle with temperature which was attributed to the change of molecular interactions between water molecules and polymer chains.[30,31] Thus, there is no broad agreement in the literature between experiments and no current theory is able to predict the observed changes in contact angle with temperature.

In the present study, we have fabricated three different flat substrates, which have a similar, very low level of roughness but very different inherent wettability, as demonstrated by the contact angle of water on the surface (shown in Table 1). The fluorinated surface is very hydrophobic, the polished gallium arsenide (GaAs) surface is hydrophilic and the wettability of the gold (Au) surface is intermediate between the other two. We have investigated the temperature dependent wetting behavior of water droplets on these surfaces from well below the Leidenfrost

**Table 1** The surface roughness[a] and contact angles of water on GaAs, Au and fluorinated surfaces at 22 °C. The error bars[b] for the contact angles are shown in brackets. The advancing and receding contact angles were measured using the contact angle goniometer method with a motorized syringe. Specifically, the advancing and receding contact angles were obtained by measurement of the angle immediately before the droplet advanced or receded respectively.

| Surface | Gallium arsenide | Gold | Fluorinated |
|---|---|---|---|
| RMS roughness/nm | 3.8 | 5.3 | 4.7 |
| Advancing Contact Angle ($\theta_A$)/° | 40 (±0.5) | 77 (±0.3) | 121 (±0.3) |
| Receding Contact Angle ($\theta_R$)/° | 13 (±2.7) | 49 (±2.1) | 90 (±4.0) |

[a] Roughness was measured using AFM over an area of 10 μm by 10 μm. [b] The error bars have been set at 3 standard deviations for the measurements.

temperature to above it, with special emphasis on the hydrophobic surface. Interestingly, we find that the flat hydrophobic fluorinated surface at temperatures below but approaching the Leidenfrost temperature is superhydrophobic. As far as we can determine this is the first report of superhydrophobicity on a surface with an insignificant level of roughness.

## II. Experimental section

### Materials

The polished GaAs (001) wafer with a thickness of approximately 380 μm was supplied by American Xtal Technology (AXT) and cleaned before use. The other substrates were obtained by modifying the surfaces of these wafers. The Au surface was obtained by thermal evaporation of a film of gold approximately 180 nm in thickness onto a polished GaAs substrate. In order to produce a stable hydrophobic surface, a $SiO_2$ film approximately 180 nm in thickness was deposited onto a polished GaAs substrate by plasma enhanced chemical vapour deposition at 300 °C under $SiH_4$ (5% in $N_2/N_2O$) flow as preparation for fluorination as described below. Hexadecane (99%) and chloroform (99%) were purchased form Sigma-Aldrich and dried by molecular sieves before use. 1$H$,1$H$,2$H$,2$H$-perfluorooctyltrichlorosilane (97%) was purchased from Alfa Aesar and used as received. The ethanol was purified by distillation prior to use. All water used was filtered through a coarse wool filter, charcoal filter, and reverse osmosis membrane before a final filtration through a Millipore Gradient system (Memtec) giving a resistivity of 18.2 MΩ cm$^{-1}$.

### Fluorinated surface preparation

The hydrophobic modification of the $SiO_2$ coated GaAs wafer was achieved by fluorination. The substrate was washed with ethanol and dried under a stream of pure $N_2$ before RF plasma treatment at a power level of 30 W for 60 s in the presence of water vapour. 50 μL of 1$H$,1$H$,2$H$,2$H$-perfluorooctyltrichlorosilane was added into a solvent mixture of hexadecane (5 mL) and chloroform (1 mL). The clean substrate was placed into the solution for ~10 mins. After surface reaction the substrate was removed from the solution, rinsed with ethanol, dried by $N_2$, and baked at 120 °C in an oven for approximately 30 mins to complete the formation of Si–O bonds.[32] Water droplets on the fluorinated surface exhibited advancing and receding contact angles of 121° and 90°, respectively. This surface treatment was chosen as the fluorinated surface obtained has very good thermal stability. A previous study indicated that the fluorinated layer is essentially inert in air below the temperature of 500 °C.[33] As the substrates used were similar for all surfaces, differences in the heat conductivity due to differences in bulk properties of the substrates were minimized.

**Contact angle measurements**

The contact angles formed by water droplets on the solid substrates were measured using a KSV (Helsinki, Finland) CAM 200 contact angle goniometer at different temperatures. The substrates were heated using a small, locally built, temperature controlled heating stage. The temperature of the heating stage was measured using a thermocouple. Droplets with a volume of ~6 μL were deposited on the surface prior to heating the substrate (sessile droplets) or released from a height of 1 cm onto the preheated substrate. For the former method, the cold substrate with a sessile water droplet was placed on the pre-heated heating stage gently using tweezers. For the latter method, the substrate was placed on the pre-heated heating stage for at least 10 mins before the release of water droplets. The contact angle goniometer is equipped with a Jai CV-M50 monochrome video camera, which captures at 30 fps with a resolution of 512 × 512 pixels. This video camera was used to image and record the dynamic changes in contact angle of the droplets on the hot surface and the lifetime ($\tau$) of water droplets on the hot surface. These images were also analysed using the supplied software to determine the contact angles. All the substrates were rinsed with high purity ethanol and dried by $N_2$ before each use.

**Atomic force microscopy (AFM) measurements**

The surface roughness of the substrates was obtained using a Digital Instruments Multimode Nanoscope III AFM in air using the tapping mode. The root-mean-square (RMS) roughness was evaluated from the AFM images over an area of 10 μm ×10 μm using the supplied software.

## III. Results and discussion

The most convenient means of determining the Leidenfrost temperature is to follow the lifetime of a droplet on the surface.[12,13] If the substrate is well below the boiling point of the liquid the droplets are long lived as the evaporation rate is slow. As the temperature is increased the evaporation rate increases and the droplet lifetime decreases. As the boiling point is reached small bubbles form, are released from the surface and rise inside the droplet. This increases the heat conduction and further shortens the droplet lifetime. As the substrate temperature is further increased larger vapor bubbles are formed and the heat flux decreases due to their thermal insulating effect. Upon further heating a drying transition takes place and the droplet becomes completely separated from the surface by an intervening vapor layer. This occurs at the Leidenfrost temperature and is characterized by a local minimum in the heat flux and a local maximum in the droplet lifetime. Further increases in temperature lead to increases in heat flux due to a higher thermal gradient between the substrate and the droplet and a consequent reduction in droplet lifetime.[12,13]

**Droplets deposited on cool surfaces that are then heated**

In the first water droplet deposition method we employed, water droplets were gently deposited onto the solid surfaces at room temperature and then they were placed on the heater which was preheated to a certain temperature. Because the evaporation of a water droplet on the solid surface is a dynamic process, the contact angle determined in this process is not always equal to the static contact angle. Hence, the contact angle measured here is defined as a quasi-static contact angle ($\theta_{QS}$) and is merely the contact angle that the droplet makes with the horizontal surface during the evaporation process.

In Fig. 1 we present measurements of the lifetime of a sessile drop which has been deposited on the surface before the substrate is heated, on our three chosen substrates. We found that for the hydrophilic surfaces the droplet lifetimes continued to decrease up to 360 °C but for the fluorinated hydrophobic surface the droplet lifetime remained constant from 120 °C up to 190 °C and increased sharply at 210 °C, indicating that the Leidenfrost temperature for the hydrophobic surface at 210 °C is considerably lower than on the hydrophilic substrates. Clearly, wetting plays a significant role in the heat transfer process. Note that it was not possible to accurately determine the lifetime of the water droplets in the Leidenfrost regime because the droplets become mobile and rapidly move out of the field of view.

Due to the rapid rate of evaporation at high temperatures and the non-equilibrium process of droplet shape deformation, it was very difficult to determine the contact angle of a water droplet on the hydrophilic surfaces (qualitatively, the contact angle exhibits a slow decrease with increasing temperature). On the fluorinated surface, the droplet lifetime decreases dramatically at first and then remains constant at around 20 s before entering into the Leidenfrost regime. However, for both Au and GaAs surfaces, the droplet lifetime exhibits a fast decrease in the low-temperature regime followed by a gradual decrease to ∼0 s in the high-temperature regime. There is no wetting to non-wetting transition or Leidenfrost phenomena observed even at the highest temperatures. The decrease in droplet lifetime is also favoured by the lower contact angle which results in a larger contact area between the droplet and the solid surface, thereby leading to a greater heat flux. This result agrees with previous observations.[15,16]

In Fig. 2 we examine the changing profile of sessile droplets on hydrophilic and hydrophobic substrates that are placed on a preheated surface below the Leidenfrost temperature, as they evaporate. It is clear that the water droplets rapidly evaporate without obvious changes in contact angle at 170 °C on the hydrophilic surfaces. Note that only a part of the water droplet can be observed on the GaAs surface in some images as the droplet extends out of the field of view of the video camera due to the low contact angle. However, on the hydrophobic surface at 170 °C, we have very different behavior. Firstly the droplet has a much longer lifetime. The contact angle is initially ∼110° but this is seen to increase to a much greater value—around 150° and the surface becomes superhydrophobic. *That is, we have a surface that is flat, smooth and superhydrophobic.* The transition is rapid and then the droplet gradually shrinks and eventually disappears due to evaporation. As the contact area between the droplet and the surface is much reduced, the heat flux due to convection is reduced and the lifetime of the droplet is increased. The droplet remains stationary on the surface at this temperature, which is very different

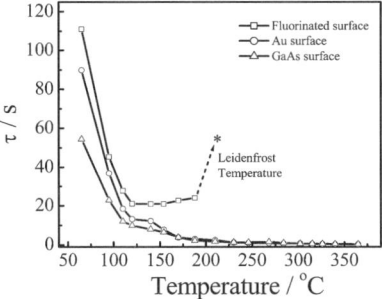

**Fig. 1** The lifetime ($\tau$) of 6 μL water droplets as a function of temperature on GaAs, Au and fluorinated surfaces for a sessile drop which has been deposited on the surface before the substrate is heated. The droplet lifetime at the Leidenfrost temperature could not be accurately determined due to the mobility of the droplets. We have indicated this condition with an asterisk (*).

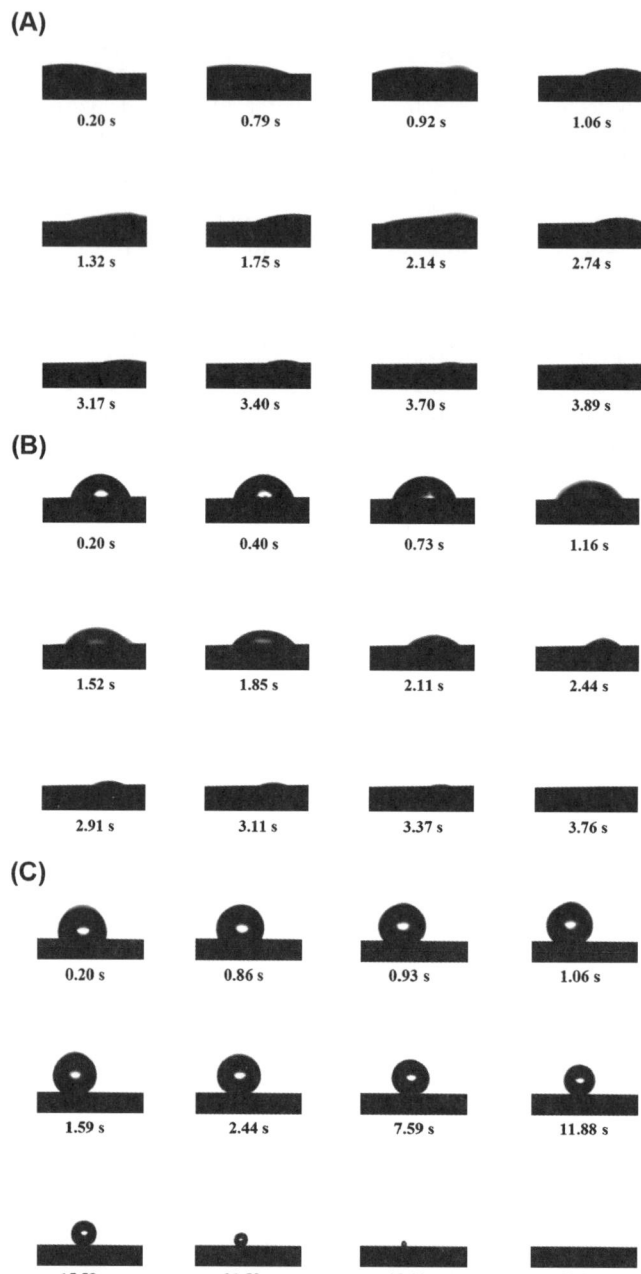

**Fig. 2** Photographic sequences of a droplet on A) GaAs surface, B) Au surface and C) a fluorinated surface at 170 °C. In these cases the droplet was deposited onto the surface before the substrate was heated. Note that the lifetime of the droplet on the fluorinated surface is much greater and a large increase in contact angle is observed.

from the rapid motion that is exhibited by Leidenfrost droplets. This is because the droplet is still in contact with the surface and is pinned at the three phase line. We note that the droplets here are very different to the Cassie droplets as there is no gas or vapour phase between the liquid and solid surfaces.

Fig. 3 shows the heating induced wetting transition of a sessile water droplet on the fluorinated surface, where the substrate is heated following the deposition of a water droplet on the solid surface. In this figure, we observe three different regimes in the temperature range of 65 °C to 270 °C. In regime I (65 °C to 110 °C), $\theta_{QS}$ of the water droplet exhibits a nearly constant contact angle as indicated by a slight decrease from 108° to 105°. In regime II (120 °C to 190 °C), $\theta_{QS}$ significantly increases from 127° to 158°. In this regime, the droplet remains stationary on the surface due to pinning at the three phase line and is in a superhydrophobic state. Regime III (210 °C to 270 °C), is the Leidenfrost regime in which the water droplet exhibits a rapid motion on the solid surface with a $\theta_{QS}$ higher than 160°.

On the hydrophobic surface the lifetime significantly decreases from 111 s to 20 s and then remains constant before increasing and eventually entering into the Leidenfrost regime. The droplet lifetime on the hot surface is determined by the evaporation rate which is determined by the heat flux from the solid surface to droplet. The droplet temperature is known to stay at 100 °C,[13] therefore as the substrate temperature is increased from 120 °C to 190 °C, the temperature gradient is increased and the heat flux per unit area will increase. However, at the same time, the contact angle increases from 127° to 158°, this reduces the contact area between the droplet and the solid surface. Consequently, the energy transfer rate is approximately constant in this regime as indicated by the relatively constant 20 s lifetime.

The changes in contact angle observed in Fig. 3 agree with Sullivans' theory for surfaces with an initial contact angle that is very high. In Sullivans' work, a van der Waals fluid with an initial contact angle of less than 120° will have a decreasing contact angle with increasing temperature, whereas a fluid with an initial contact angle of greater than 120° will have an increasing contact angle with increasing temperature. In our case the initial contact angle is 110° but the system is behaving like the high contact angle state in the Sullivan theory. We note that water is not well approximated by a van der Waals fluid and this may be the reason for the discrepancy in the contact angle at which the system changes behavior. According to Sullivan eventually the contact angle should increase to the non-wetting value of ~180°.[27] Sullivans' theory, also predicts that the contact angle should decrease with increasing temperature (negative $d\theta/dT$) for high-energy surfaces. This is what we find and is also supported by several different research groups.[22–25,28,34,35] Other theories predict the contact angle either increases[29] or decreases[24,36] with temperature—regardless of surface free energy. The agreement with the work of Sullivan suggests that the heating induced wetting transition of a water droplet on the solid surface

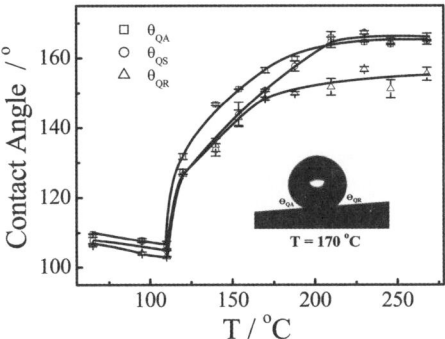

**Fig. 3** The quasi-advancing contact angle ($\theta_{QA}$), quasi-static contact angle ($\theta_{QS}$), and quasi-receding contact angle ($\theta_{QR}$) for a 6 μL water droplet on the fluorinated surface as a function of temperature, where the droplet is deposited on the surface before the substrate is heated. The solid lines are provided to guide the eye. Where error bars are not visible it indicates that the error bars are less than the size of the symbol. Inset: image of a water droplet on a 6 degree slope at 170 °C.

that we observe below the Leidenfrost temperature is controlled by the balance between solid–liquid adhesive forces and liquid–liquid cohesive forces as predicted by Sullivan,[27] and is not a Leidenfrost type effect. More specifically, heating affects both the adhesive forces and the cohesive forces generally to an unequal extent on a hydrophobic surface (low-energy surface). Increasing temperature diminishes the cohesive forces relatively less than the adhesive forces, resulting in a higher contact angle.[27]

When a droplet is deposited on a slope of angle $\alpha$ (critical angle) just before sliding, the maximum liquid advancing contact angle ($\theta_A$) will be shown at the lower side of the droplet and the minimum liquid receding contact angle ($\theta_R$) will be shown at the upper side of the droplet.[37] In the present work, we have measured a series of contact angles of water droplet on a slight slope of approximately 6° at different temperatures. Since the slope used here is not the actual critical angle (which is dependent on temperature), we define the angles at the lower side and the upper side of the droplet as the quasi-advancing contact angle ($\theta_{QA}$) and the quasi-receding contact angle ($\theta_{QR}$), respectively. The inset in Fig. 3 shows the typical state of a water droplet at 170 °C on a slope. The droplet can roll down the slope at this temperature or at higher temperatures. In Fig. 3, it can be seen that the quasi-static contact angles are closer to the quasi-receding contact angles at $T \leq 170$ °C. This is understandable because the quasi-static contact angle obtained in this regime has a retracting contact line during evaporation, which is similar to the situation of a receding contact angle where the three phase line is withdrawn over a pre-wetted surface. On the other hand, the actual critical angle in the low-temperature regime must be larger than the slope we employed, that means we overestimate the receding contact angle. This will also make $\theta_{QR}$ approach $\theta_{QS}$. Note that ideal control of the slope would require that the slope is varied dynamically as the contact angles of the droplet changed with temperature. This was not possible in our experiment. However, the quasi-static contact angles approach more closely to the quasi-advancing contact angles when the temperature is above 210 °C. In this high-temperature regime, the water droplet will rapidly (less than 1 s) transit from a hydrophobic state to a superhydrophobic state with a $\theta_{QS}$ higher than 160° and then roll around on the hot surface. It is worth noting that the system is in the Leidenfrost regime when the temperature is above 210 °C. Therefore, the contact angles measured above 210 °C are not actually contact angles as the water droplets float on a cushion of vapour. Due to the flattening of the droplet in the region of the surface an apparent contact angle is created as described in the introduction. This might be the reason why $\theta_{QS}$ approaches more closely to $\theta_{QA}$ in the high-temperature regime. It is possible that the difference between $\theta_{QA}$ and $\theta_{QR}$ in this regime is partly due to the velocity gradient of fluid in the droplet during sliding of the droplet down the slope, for example, the portion of the droplet away from the solid surface will have a higher speed whereas the portion of the droplet adjacent to the solid surface will have a lower speed. This will make the apparent contact angles differ between the lower and the upper side of the droplet.

### Droplets deposited on surfaces that are hot

Alternatively, water droplets can be deposited on surfaces that are preheated. We released droplets from a height of 1 cm causing them to impact with the surface with a velocity of $\sim$0.44 m s$^{-1}$. Fig. 4 shows the droplet lifetime as a function of temperature for droplets impinging on the Au and GaAs surfaces. Here, we can observe the Leidenfrost regime which leads to the bouncing of water droplets on both surfaces when the temperature is above $\sim$280 °C. Below this temperature, the droplet lifetimes were short, as was found for sessile droplets initially placed on unheated substrates as shown in Fig. 1.

The contact angle and lifetime as a function of temperature for droplets impinging on the fluorinated surface are shown in Fig. 5. Below 190 °C, the contact angle and

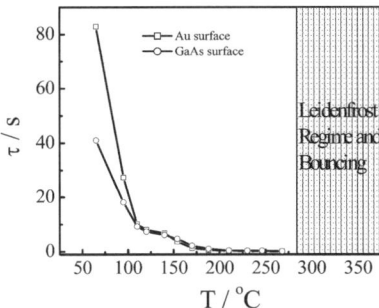

**Fig. 4** Lifetime ($\tau$) of a 6 µL water droplet as a function of temperature on GaAs and Au surfaces, where the droplet is released from a height of 1 cm onto a preheated substrate. It was not possible to measure droplet lifetimes in the Leidenfrost regime as their mobility caused them to rapidly leave the field of view of the video camera.

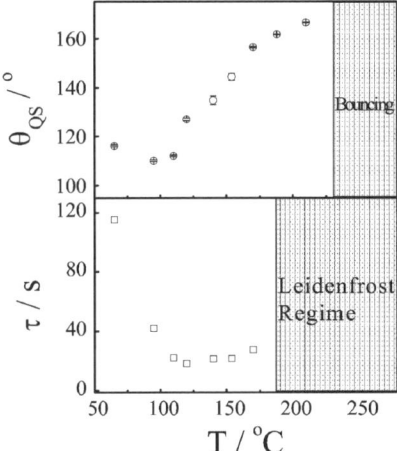

**Fig. 5** The quasi-static contact angle ($\theta_{QS}$) and lifetime ($\tau$) of 6 µL water droplets on the fluorinated surface as a function of temperature, where the droplet is released from a height of 1 cm onto the preheated substrate. Where error bars are not visible it indicates that the error is less than the size of the symbol. It was not possible to determine droplet lifetimes in the Leidenfrost regime due to their high mobility.

lifetime exhibit similar results to droplets placed on the same surface as shown in Figs. 1 and 3. When the temperature is above 190 °C, the system enters into the Leidenfrost regime. The Leidenfrost temperature (190 °C) obtained here is a little lower than that obtained for the sessile drops shown in Fig. 1 (210 °C). Furthermore, the droplet upon impacting with the surface will rebound from the solid surface when the temperature is above 230 °C, which is known as the "dynamic Leidenfrost phenomenon".[38] Between 190 °C and 230 °C, the droplet impacts on the hot surface and transits from a hydrophobic state to the Leidenfrost state immediately and then rolls around on the hot surface whilst evaporating.

Applying this technology to drive the motion of liquid droplets on a solid surface by the heating induced transition from hydrophobic to superhydrophobic is an interesting challenge. This approach has been regarded as an alternative paradigm to the current lab-on-a-chip techniques.[39] Imagine that two or more water droplets are deposited on a tilted hydrophobic surface, one droplet would transit from the hydrophobic state to the superhydrophobic state upon application of heat locally whilst the others will remain in the original state. The transitioned droplet will roll down the slope if the surface has a small contact angle hysteresis. We tested this idea with the flat fluorinated surface on a slope of $\sim 12°$ (data not shown). The water droplet only moved a short distance and then stopped on the slope because of the large contact angle hysteresis of the fluorinated surface (Table 1). If we locally heat the droplet again, the droplet will move another short distance and stop again. Hence, the droplet position can be controlled using a series of "jumps". Alternatively, the movement of the water droplet could be induced on a horizontal surface using the Marangoni effect induced by temperature variations.[40]

## IV. Conclusion

The wetting of a water droplet on solid surfaces has been investigated at elevated temperatures up to the Leidenfrost temperature. Fundamentally different behavior is observed on hydrophilic and hydrophobic surfaces. On hydrophilic surfaces the droplets rapidly evaporate below the Leidenfrost temperature. On the hydrophobic surface, the contact angle increases and the surface becomes superhydrophobic and the lifetime of the droplet remains constant below the Leidenfrost temperature. To our knowledge this is the first report of superhydrophobicity on a flat surface.

### Acknowledgements

V.S.J.C gratefully acknowledges support from the Australian Research Council. We thank Dr Lan Fu of the Electronic Materials Engineering Department for help with the deposition of gold and $SiO_2$ films.

### References

1 A. B. D. Cassie and S. Baxter, *Trans. Faraday Soc.*, 1944, **40**, 546.
2 R. N. Wenzel, *Ind. Eng. Chem.*, 1936, **28**, 988.
3 L. Gao and T. J. McCarthy, *Langmuir*, 2007, **23**, 3762.
4 M. Nosonovsky, *Langmuir*, 2007, **23**, 9919.
5 M. V. Panchagnula and S. Vedantam, *Langmuir*, 2007, **23**, 13242.
6 G. McHale, *Langmuir*, 2007, **23**, 8200.
7 A. Marmur and E. Bittoun, *Langmuir*, 2009, **25**, 1277.
8 L. Gao and T. J. McCarthy, *Langmuir*, 2009, **25**, 7249.
9 A. Lafuma and D. Quéré, *Nat. Mater.*, 2003, **2**, 457.
10 T. N. Krupenkin, J. A. Taylor, E. N. Wang, P. Kolodner, M. Hodes and T. R. Salamon, *Langmuir*, 2007, **23**, 9128.
11 E. Bormashenko, R. Pogreb, G. Whyman and M. Erlich, *Langmuir*, 2007, **23**, 6501.
12 J. G. Leidenfrost, *De Aquae Communis Nonnullis Qualitatibus Tractatus*, Duisburg, 1756. Translation: *Int. J. Heat Mass Transfer*, 1966, 9, 1153.
13 A. L. Biance, C. Clanet and D. Quéré, *Phys. Fluids*, 2003, **15**, 1632.
14 M. Rein, Interactions between Drops and Hot Surfaces, in *Drop-Surface Interactions*, ed. M. Rein, Wien, New York: Springer, 2002, vol. 456, pp. 187.
15 Y. Takata, S. Hidaka, J. M. Cao, T. Nakamura, H. Yamamoto, M. Masuda and T. Ito, *Energy*, 2005, **30**, 209.
16 Y. Takata, S. Hidaka, A. Yamashita and H. Yamamoto, *Int. J. Heat Fluid Flow*, 2004, **25**, 320.
17 M. Prat, P. Schmitz and D. Poulikakos, *J. Fluids Eng.*, 1995, **117**, 519.
18 N. Nagai and S. Nishio, *Exp. Therm. Fluid Sci.*, 1996, **12**, 373.
19 J. D. Bernardin, C. J. Stebbins and I. Mudawar, *Int. J. Heat Mass Transfer*, 1996, **40**, 73.
20 G. P. Celata, M. Cumo, A. Mariani and G. Zummo, *Heat Mass Transfer*, 2006, **42**, 885.
21 G. Castanet, T. Liénart and F. Lemoine, *Int. J. Heat Mass Transfer*, 2009, **52**, 670.
22 F. D. Petke and B. R. Ray, *J. Colloid Interface Sci.*, 1969, **31**, 216.
23 J. H. Lay and V. K. Dhir, *J. Heat Transfer–Trans. ASME*, 1995, **117**, 394.
24 J. D. Bernardin, I. Mudawar, C. B. Walsh and E. I. Franses, *Int. J. Heat Mass Transfer*, 1997, **40**, 1017.
25 H. Schonhorn, *J. Phys. Chem.*, 1966, **70**, 4086.

26 S. Sinha, Molecular Dynamics Simulation of Interfacial Tension and Contact Angle of Lennard-Jones Fluid, Ph.D. Thesis, University of California, Los Angeles, 2004, (Chapter 4), pp. 55–77.
27 D. E. Sullivan, *J. Chem. Phys.*, 1981, **74**, 2604.
28 A. W. Neumann and D. Renzow, *Z. Phys. Chem. (Frankfurt)*, 1969, **68**, 11.
29 G. Navascués and M. V. Berry, *Mol. Phys.*, 1977, **34**, 649.
30 Y. G. Takei, T. Aoki, K. Sanui, N. Ogata, Y. Sakurai and T. Okano, *Macromolecules*, 1994, **27**, 6163.
31 J. Zhang, R. Pelton and Y. Deng, *Langmuir*, 1995, **11**, 2301.
32 G. M. Liu and G. Z. Zhang, *J. Phys. Chem. B*, 2005, **109**, 743.
33 U. Srinivasan, M. R. Houston, R. T. Howe and R. Maboudian, *J. Microelectromech. Syst.*, 1998, **7**, 252.
34 A. W. Neumann, G. Haage and D. Renzow, *J. Colloid Interface Sci.*, 1971, **35**, 379.
35 M. de Ruijter, P. Kölsch, M. Voué, J. De Coninck and J. P. Rabe, *Colloids Surf., A*, 1998, **144**, 235.
36 A. W. Adamson, *J. Colloid Interface Sci.*, 1973, **44**, 273.
37 B. K. Banerji, *Colloid Polym. Sci.*, 1981, **259**, 391.
38 B. S. Gottfried, C. J. Lee and K. J. Bell, *Int. J. Heat Mass Transfer*, 1966, **9**, 1167.
39 K. T. Kotz, K. A. Noble and G. W. Faris, *Appl. Phys. Lett.*, 2004, **85**, 2658.
40 R. Savino, R. Monti and G. Alterio, *Phys. Fluids*, 2001, **13**, 1513.

# Drop dynamics on hydrophobic and superhydrophobic surfaces

B. M. Mognetti,[*a] H. Kusumaatmaja[b] and J. M. Yeomans[a]

Received 14th December 2009, Accepted 21st January 2010
DOI: 10.1039/b926373j

We investigate the dynamics of micron-scale drops pushed across a hydrophobic or superhydrophobic surface. The velocity profile across the drop varies from quadratic to linear with increasing height, indicating a crossover from a sliding to a rolling motion. We identify a mesoscopic slip capillary number which depends only on the motion of the contact line and the shape of the drop, and show that the angular velocity of the rolling increases with increasing viscosity. For drops on superhydrophobic surfaces we argue that a tank treading advance from post to post replaces the diffusive relaxation that allows the contact line to move on smooth surfaces. Hence drops move on superhydrophobic surfaces more quickly than on smooth geometries.

## I. Introduction

The question of how liquid drops move across a solid surface has long caught the interest of academic and industrial communities alike, with applications ranging from microfluidic devices to fuel cells and inkjet printing. In many cases, efficient and effective control of the drop dynamics is highly desirable, and this relies upon understanding the internal fluid motion of the drop.

It has been reported in the literature[1–5] that a liquid drop may move in a variety of ways including sliding, rolling, tank treading and slipping. In some cases, a pearling instability may also be observed at the trailing edge of the drop.[6] Highly viscous drops roll rather than slide,[5] and on superhydrophobic surfaces drops appear to move very easily.[7] Using small particles as tracers, it has recently become possible to access the velocity profile within a drop.[2–5] However it is not yet always clear how the internal fluid motion is related to physical parameters such as fluid viscosity, the strength of the forcing, the equilibrium contact angle and hysteretic properties of the surface. Controlled experiments are typically possible only over a restricted range, and can be difficult for small drops of size below the capillary length. Moreover, certain parameters, such as surface roughness, can be difficult to control and yet may have an important role. Analytical calculations are possible, but they are typically limited to small[8,9] and high[10,11] contact angles. Computer simulations are therefore highly desirable to bridge the knowledge from experiments and analytical theories.

Several simulation methods have been developed to shed light on the dynamics of moving drops.[12–18] The computational tool we use here is a mesoscopic diffuse interface model[19,20] solved using a lattice Boltzmann algorithm.[21,22] This proves a useful approach due to its ability to handle interfacial dynamics and complex geometries

[a]The Rudolf Peierls Centre for Theoretical Physics, 1 Keble Road, Oxford, OX1 3NP, United Kingdom. E-mail: b.mognetti1@physics.ox.ac.uk
[b]Department of Theory and Biosystems, Max Planck Institute of Colloids and Interfaces, 14424 Potsdam, Germany

easily. For example, it has recently been used to study capillary filling,[23] hysteresis of drops on patterned surfaces,[24,25] instabilities and detaching phenomena[15] and driven drops in external flows.[16,17] Diffuse interface models are mesoscopic approaches, which do not resolve the microscopic details of the contact line dynamics. As such, they provide a useful complement to molecular dynamics simulations.

We concentrate on modelling micron-scale drops moving on hydrophobic and superhydrophobic surfaces. We find that, for both hydrophobic and superhydrophobic substrates, the velocity profile across the height of the drop falls into two regimes. Near the substrate the velocity varies quadratically with height corresponding to a sliding motion. This can be characterized in terms of a mesoscopic slip capillary number which depends on the equilibrium contact angle and the mobility of the contact line. Further from the surface the velocity varies linearly with height indicating rolling. There is no dependence of the angular velocity on microscopic details of the contact line motion, such as the mobility, but the rolling becomes faster with increasing viscosity for a given body force $f$.

We identify several major differences between the motion on hydrophobic surfaces, which are smooth, and similar surfaces which are patterned with posts such that they become superhydrophobic. On the superhydrophobic substrates pinning of the contact line at the edges of the posts means that there is a threshold forcing below which the drop will not move. However, once they do start to move, they do so much more easily and drops can reach much larger velocities before being detached from the surface in agreement with the experimental observations. We impose no-slip boundary conditions in the simulations, and on the smooth surfaces there is indeed no slip at the surface. However, the fluid velocity at the surface is non-zero and independent of the mobility for drops moving on superhydrophobic surfaces[26–28] suggesting that the interface moves by tank treading between successive posts, rather than by diffusion-mediated motion of the interface.

The plan of the paper is as follows: In Section II the numerical model is introduced, and in Section III the simulation set-up and the computational parameters are given. Considering first drops on smooth surfaces, in Section IV we describe the crossover from a Poiseuille-like flow to a pure rolling regime with increasing height and point out the relevance of the equilibrium contact angle. In Sections V and VI we investigate the dependence of the Poiseuille and rolling regimes on the viscosity and surface tension of the fluid and on the mobility of the contact line. In Section VII we compare the dynamics of drops on superhydrophobic surfaces. Finally, in Section VIII, we summarise and discuss our conclusions.

## II. The model

We use a binary fluid model, with components $A$ and $B$, described by a Landau free energy[19,20]

$$\Psi = \int_V \left( \psi_b + \frac{\kappa}{2}(\partial_\alpha \phi)^2 \right) dV + \int_S \psi_s \, dS \qquad (1)$$

where the bulk free energy density $\psi_b$ is taken to have the form

$$\psi_b = \frac{c^2}{3} n \ln n + A\left( -\frac{1}{2}\phi^2 + \frac{1}{4}\phi^4 \right) \qquad (2)$$

The functional (1) is discretised on a cubic lattice with lattice spacing $\Delta x$, and $\Delta t$ is the simulation time step. $n$ is the total fluid density $n = n_A + n_B$, $\phi$ is the order parameter $\phi = n_A - n_B$, and $c = \Delta x/\Delta t$. This choice of $\psi_b$ gives binary phase separation into two phases with $\phi = \pm 1$.

The gradient term in eqn (1) represents an energy contribution from variations in $\phi$ and is related to the surface tension between the two phases by $\gamma = \sqrt{8\kappa A/9}$ and to the interface width through $\xi = \sqrt{\kappa/A}$.[20]

The second integral in eqn (1) is taken over the system's solid surface and is responsible for the wetting properties. The surface energy density is chosen to be $\psi_s = -h\phi_s$,[29] where $\phi_s$ is the value of the order parameter at the surface. Minimisation of the free energy shows that in equilibrium the gradient in $\phi$ at the solid boundary is[20]

$$\kappa \partial_\perp \phi = -\frac{\mathrm{d}\psi_s}{\mathrm{d}\phi_s} = -h \quad (3)$$

The value of the phenomenological parameter $h$ is related to the equilibrium contact angle $\theta_{eq}$ ($\theta_{eq}$ is taken with respect to the $\phi = 1$ component) via:[20]

$$h = \sqrt{2\kappa A}\ \mathrm{sign}\left(\frac{\pi}{2} - \theta_{eq}\right) \sqrt{\cos\left(\frac{\alpha}{3}\right)\left\{1 - \cos\left(\frac{\alpha}{3}\right)\right\}} \quad (4)$$

where $\alpha = \cos^{-1}(\sin^2\theta_{eq})$ and the function sign returns the sign of its argument.

The hydrodynamic equations for the binary fluid are

$$\partial_t n + \partial_\alpha(nv_\alpha) = 0 \quad (5)$$

$$\partial_t(nv_\alpha) + \partial_\beta(nv_\alpha v_\beta) = -\partial_\beta P_{\alpha\beta} + \partial_\beta[\eta(\partial_\beta v_\alpha + \partial_\alpha v_\beta + \delta_{\alpha\beta}\partial_\gamma v_\gamma)] + f_\alpha \quad (6)$$

$$\partial_t \phi + \partial_\alpha(\phi v_\alpha) = M\nabla^2 \mu. \quad (7)$$

Eqn (5) is the continuity equation, eqn (6) the Navier–Stokes equation and eqn (7) the convection diffusion equation. $M$ is a mobility coefficient. In eqn (6) we have introduced the dynamic viscosity $\eta$, and a bulk force for unit volume $f_\alpha$. The equilibrium properties of the fluid appear in the equations of motion through the pressure tensor and the chemical potential. Both can be obtained in the usual way from the free energy and are given by[20]

$$\mu = A(-\phi + \phi^3) - \kappa\nabla^2\phi \quad (8)$$

$$P_{\alpha\beta} = \left(p_b - \frac{\kappa}{2}(\partial_\gamma\phi)^2 - \kappa\phi\partial_{\gamma\gamma}\phi\right)\delta_{\alpha\beta} + \kappa(\partial_\alpha\phi)(\partial_\beta\phi) \quad (9)$$

$$p_b = \frac{c^2}{3}n + A\left(-\frac{1}{2}\phi^2 + \frac{3}{4}\phi^4\right)$$

To solve eqn (5), (6) and (7), we use the free energy lattice Boltzmann scheme recently presented by Pooley et al.[30] The main advantage of this method is that the spurious velocities at the contact line are strongly reduced. This is particularly important when the two fluid components have different viscosities.

A word of explanation is in order as to why we are using a binary fluid approach to describe a liquid-gas system. Recently several authors[23,31,32] have investigated capillary filling using liquid-gas and binary lattice Boltzmann algorithms. They found that the binary model is able to reproduce the experimental results, but that the liquid-gas model produces filling that is faster than expected. This is due to excessive condensation of the gas phase at the interface. The effect is suppressed only when the liquid density is much larger than the gas density, a regime difficult to access numerically. Therefore we use a binary model. However, our results are equally applicable to a physical system where a liquid displaces a gas if the important physical parameters are the viscosities, rather than the densities of the fluid components. For convenience we will use liquid/gas terminology in the rest of the paper.

The way in which the contact line moves does depend on the choice of model. Here it is determined by the mobility $M$ which determines how the diffusion across the interface is driven by variations of the chemical potential $\mu$. Given that the model does not include microscopic details we can say nothing about the mechanisms operating at the contact line in a physical system, and we regard $M$ as a phenomenological parameter which controls the rate of slip.

## III. Geometry and simulation parameters

We consider drops of size smaller than the capillary length, so that we can ignore gravity, and sufficiently large that thermal fluctuations can be neglected. For a water drop this corresponds to a range of length scales between $\sim 10^{-6}$m and $10^{-3}$m. On these length scales the motion of the drop depends on its viscosity $\eta$ and surface tension $\gamma$ and a useful dimensionless measure of drop velocity $v$, allowing direct comparison to experiments, is the capillary number $Ca = v\eta/\gamma$. In the simulations we use three different values for the liquid viscosities ($\eta = 0.5, 0.833, 1.1667$) and two for the surface tension ($\gamma = 0.0267, 0.0533$). The gas viscosity $\eta_{gas}$ is always chosen to keep the ratio $\eta/\eta_{gas} = 25$ and a linear interpolation between $\eta$ and $\eta_{gas}$ is used to define the viscosity variation through the interface.

A body force $f$ is applied to the liquid phase. (For example, for a drop moving down an inclined plane with tilt angle $\alpha$, $f = n \cdot g \cdot \sin \alpha$ where $g$ is the acceleration due to gravity.) $f$ is taken to vary with $(1 + \phi(r))/2$ which provides a linear interpolation from $f$ to 0 when moving from the liquid to the gas. Other interpolating functions (e.g. tanh) were tested, without significant differences. Driving by the body force is normally reported in terms of a dimensionless Bond number $Bo = f \cdot V/(\gamma \cdot L_y)$, where $V$ is the volume of the drop, $L_y$ is the length of the cylinder which is equal to the length of the simulation box in the $y$-direction, and $f$ is the body force defined in eqn (6). We choose values of $f$ between $0.05 \times 10^{-6}$ and $1.6 \times 10^{-6}$ and $V = 11 \times 10^3 \cdot L_y$.

Other parameters important in determining the drop motion are the mobility $M$, where we use values from $M = 0.125$ to $M = 2$, and the equilibrium contact angle $\theta_{eq}$, considered below.

We model cylindrical drops placed on a smooth surface or on a surface decorated by regular posts as shown in Fig. 1. For the patterned substrates the use of cylindrical drops is a compromise between two dimensional simulations, which do not well describe the pinning of the contact line on the posts (as they essentially replace the posts by ridges), and full three dimensional simulations which are prohibitively time consuming.

For the smooth surfaces we mainly use a hydrophobic equilibrium contact angle $\theta_{eq} = 145°$, but results at other contact angles are presented for comparison. For the patterned surface the posts were taken to have a square cross section of side $S = 8 \cdot \Delta x$, height $D_h = 10 \cdot \Delta x$, and to be equispaced by $D_x = D_y = 9$ lattice units in both plane directions (see Fig. 1). A hydrophobic surface patterned by posts becomes superhydrophobic. There is a large increase in the contact angle $\theta_C$ of a drop resting on the posts given by the Cassie formula $\cos \theta_C = \phi \cos \theta_{eq} + \phi - 1^{33}$ where $\phi$ is the area fraction of the flat surface covered by posts. For the geometry we consider $\phi = S^2/(S + D_x)^2 \approx 0.22$. Hence using $\theta_{eq} = 100°$ gives a Cassie angle $\theta_C \approx 145°$ equal to the $\theta_{eq}$ used for the flat surface geometry.

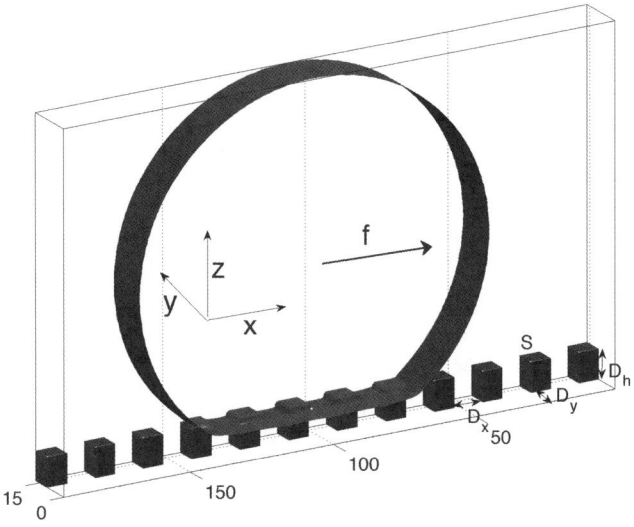

As shown in Fig. 1, the substrate is placed in the $xy$ plane and the sliding motion is along the $x$-direction. Typically we used a simulation box of size $L_x \times L_y \times L_z = 180 \times 1 \times 180$ (lattice units) for the smooth geometry, and $L_x \times L_y \times L_z = 204 \times 17 \times 180$ (lattice units) when posts were present. At small contact angles, for the flat plane, $L_x$ was sometimes increased to $L_x = 360$ in order to minimize the interaction between the drop and its images created by the periodic boundaries conditions. Cylindrical drops were initialized with a radius $R = 60$ centered 50 lattice units above the plane (or the top of the posts in the suspended states) and allowed to relax to equilibrium before the force was applied.

## IV. Velocity profile inside a drop

Understanding the dynamics of a moving drop is more difficult than for rigid objects because of the possibility of the fluid shearing. This means that, in general, the drop

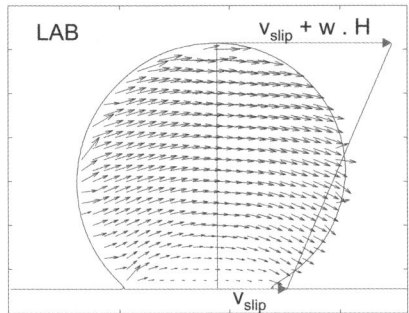
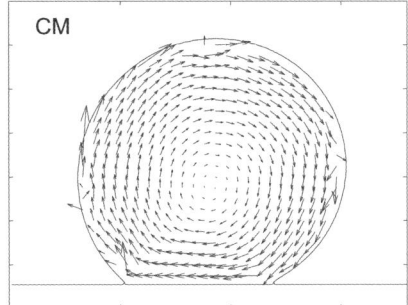

**Fig. 2** Velocity profile for a drop with contact angle $\theta_{eq} = 145°$ (a) in the laboratory reference frame and (b) in the drop's center of mass reference frame. For clarity the velocities have been rescaled and we have not reported the flow of the surrounding gas. In panel (a) we illustrate the phenomenological description of the motion of a drop as superposition of a slip velocity $v_{slip}$ plus a rotational motion of angular velocity $w$.

motion will include both sliding and rolling components. Fig. 2 reports a typical steady velocity profile found using the diffuse interface model eqn (1) for an equilibrium contact angle $\theta_{eq} = 145°$ in the laboratory and drop frames of reference. In the latter (Fig. 2b) the presence of rolling is evident.

In Fig. 3 we report the $x$-component of the drop velocity $v_x$ as a function of height $z$ (computed along the line denoted $H$ in Fig. 2) for different applied forces $f$. There is a clear separation into two regimes. For larger $z$ the velocity scales linearly with height, corresponding to rolling dynamics. For smaller $z$ the behaviour is close to quadratic and is well fit by a Poiseuille-like[34] flow

$$v_x(z) = v_0 + \frac{f\tilde{H}}{\eta}z - \frac{1}{2}\frac{f}{\eta}z^2 \qquad (10)$$

where $\tilde{H}$ would correspond to the centre of the channel in the usual Poiseuille geometry and $v_0$ allows for local slip at the liquid–solid interface. (For forces $f < 0.5 \times 10^{-6}$, where the drops are not too deformed, the quadratic term of the fit is in agreement with the $z^2$ term of eqn (10) to within 10%.) In our simulations we impose non-slip boundary conditions and therefore $v_0 = 0$ for smooth substrates by construction. We will see later that this is no longer the case for superhydrophobic surfaces.

Experimental results[2–4] are often fitted by simplifying the drop dynamics as a superposition of a rotation (with angular velocity $w$) and a constant sliding velocity $v_{slip}$. In this scheme the velocity profile in the $x$ direction is a linear interpolation between $v_{slip}$ and the velocity at the top of the drop (see the arrows in Fig. 2a).

$$v_x = v_{slip} + w \cdot z \qquad (11)$$

Guided by the experimental approach, we identify the angular velocity $w$ as the slope of the linear part of the velocity profiles. Extrapolating the linear fit to $z = 0$ (see the dotted curves in Fig. 3) we define a slip velocity $v_{slip}$ which characterizes the motion of the layers of fluid near the solid wall. (It is important to distinguish $v_{slip}$, and the microscopic slip velocity $v_0$ which is the velocity at $z = 0$.) We will describe how $v_{slip}$ and how the rolling region of the velocity profile, parameterized by $w$, depend on parameters such as viscosity, surface tension, mobility and the shape of the drop in Sections V and VI respectively.

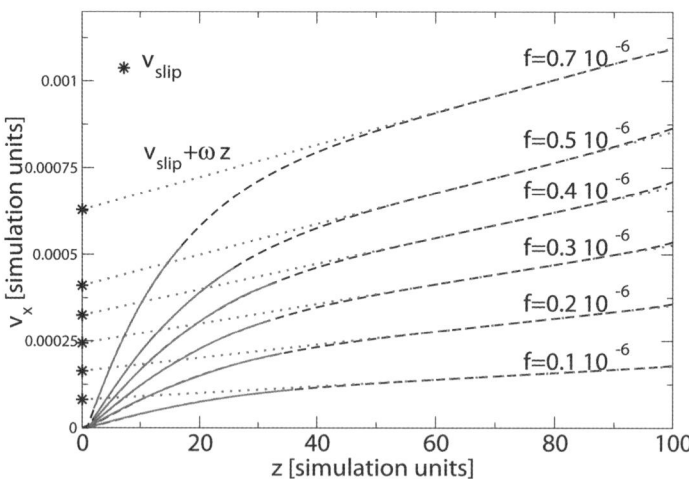

**Fig. 3** Steady velocity in the $x$ direction ($v_x$) as a function of the height $z$. At large $z$ a linear, rolling, regime is present (dashed) which crosses over a quadratic profile for small $z$ (full lines). $v_{slip}$ is defined as the extrapolation of the linear regime to $z = 0$. Data were obtained using $\eta = 0.833$, $\gamma = 0.02667$, $\theta_{eq} = 145°$ and $M = 0.25$.

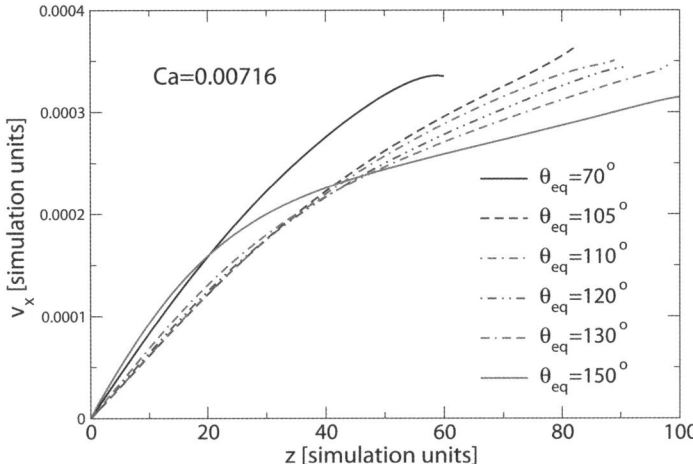

**Fig. 4** Variation of the velocity $v_x$ with height $z$ at different equilibrium contact angles but the same center of mass capillary number $Ca = v_{cm}\eta/\gamma$ (where $v_{cm}$ is the centre of mass velocity). Only for high equilibrium contact angles ($\theta_{eq} > 120°$) is a linear region clearly present. The simulation parameters are the same as in Fig. 3.

However first we present results showing that the equilibrium contact angle $\theta_{eq}$ strongly affects the crossover from the quadratic to the linear regime. In Fig. 4 we report $v_x$ for different equilibrium contact angles $70° \leq \theta_{eq} \leq 150°$ at the same centre of mass capillary number $Ca = v_{cm}\eta/\gamma$, where $v_{cm}$ is the centre of mass velocity. For $\theta_{eq} \leq 120°$, a linear part of the plot cannot be distinguished; only for $\theta_{eq} \geq 130°$ are the two regions clearly visible. The results are consistent with lubrication theory, valid for small contact angles, which predicts a quadratic profile,[35] and scaling calculations that predict rolling at high contact angle.[10,11] The applied force needed to produce the given capillary number in Fig. 4 increases with decreasing contact angle: for example the ratios are $1 : 1.7 : 2.7$ for $\theta_{eq} = 150°$, $105°$ and $70°$. The dissipation is larger for drops of smaller contact angle because the region in which the velocity varies quadratically with height extends further from the substrate.

## V. Slip velocity, $v_{slip}$

We now discuss the dependence of the mesoscopic slip velocity on the parameters of the model. Fig. 5(a) shows a dimensionless measure of $v_{slip}$, $Ca_{slip} = v_{slip} \cdot \eta/\gamma$, plotted against a dimensionless measure of the forcing, the Bond number, $Bo = f \cdot V/(\gamma \cdot L_y)$. There is a data collapse onto curves which depends only on the mobility $M$ and on the equilibrium contact angle $\theta_{eq}$, and $Ca_{slip} = f(\theta_{eq}, M) \cdot Bo$ holds well for $Bo$ sufficiently small that the drop is not significantly deformed. As the mobility $M$ increases $v_{slip}$ increases for a given Bond number as expected, as it becomes easier for the contact line to relax. For smaller equilibrium contact angles $v_{slip}$ decreases for a given Bond number as the dissipation of the drops increases.

Another property of the moving drop that is related to $v_{slip}$ is the degree of deformation caused by the forcing. In Fig. 5(b) we present data that suggest that, for small forcing, $v_{slip}$ is linear in the uncompensated Young stress $\Delta \equiv \gamma (\cos\theta_R - \cos\theta_A)$ where $\theta_R$ and $\theta_A$ are the receding and advancing contact angles. Again the slopes of the curves depend on the mobility and the equilibrium contact angles, but not on the viscosity or surface tension. We caution that the data here are noisy, as the deviations from the equilibrium contact angles are small and the algorithm, for high viscosities, does not precisely reproduce $\theta_{eq}$.[30]

It is plausible and consistent with our results that the mobility coefficient $M$ affects $Ca_{slip}$ only through the deformation of the droplets, in such a way that

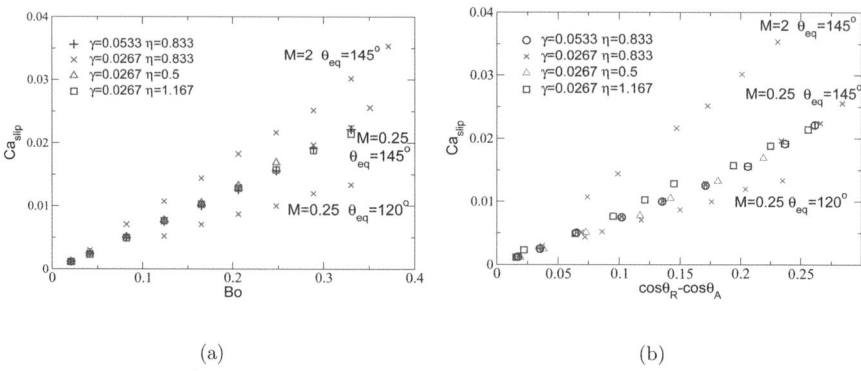

**Fig. 5** Dimensionless slip velocity $Ca_{\text{slip}}$ versus (a) the Bond number $Bo$ and (b) $\cos\theta_R - \cos\theta_A$ where $\theta_R$ and $\theta_A$ are the receding and advancing contact angles. The curves depend on the mobility $M$ and equilibrium contact angle $\theta_{\text{eq}}$ but not, to within the precision of our data, the surface tension $\gamma$ or the viscosity $\eta$.

a more detailed scaling relation for $Ca_{\text{slip}}$ could be written $Ca_{\text{slip}} = \alpha(\theta_{\text{eq}}) \cdot Bo + \beta(\theta_{\text{eq}}, M) \cdot \Delta(Bo)$. Qian et al.[36,37] have also highlighted the importance of $\Delta$ in controlling interface motion. However the relation between our work and theirs is not yet clear as they consider a local microscopic slip near the interface, whereas our parameter $v_{\text{slip}}$ follows from extrapolating from the bulk velocity profile. Work is in progress to investigate this further.

## VI. Rolling

The rolling motion displayed away from the surface by drops of higher equilibrium contact angle has a very different dependence on the model parameters. The rolling contribution to the velocity profile of the drop is $w \cdot z$ in eqn (11), corresponding to the linear part of the velocity profiles in Fig. 3. We estimate a typical rolling velocity as $w \cdot H$, where $H$ is a length comparable to the height of the droplet. A capillary number associated with the rolling motion can then be defined as $Ca_w = w \cdot H \cdot \eta/\gamma$. For the graphs in Fig. 6 we have consistently used $H = 100$ lattice units.

In Fig. 6 we plot $Ca_w$ as a function of the Bond number for different mobilities, viscosities and surface tensions. Fig. 6a shows that different viscosities do not collapse on a single master curve but rather that $Ca_w$ increases with $\eta$. This means that the sliding and the rolling motion of the drops are affected in a different way

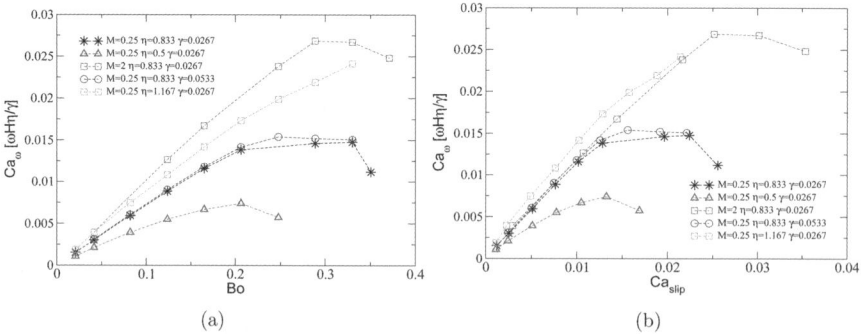

**Fig. 6** Dimensionless angular velocity $Ca_w = w\eta H/\gamma$ against (a) Bond number $Bo$ (b) dimensionless slip velocity $Ca_{\text{slip}} = \eta v_{\text{slip}}/\gamma$. Curves for different surface tensions (but equal viscosity) collapse, while $Ca_w$ increases with $\eta$. In (b) the curves are also independent of mobility.

by the viscosity. In particular the rolling component is favoured for high viscosity. This is not the case for the surface tension $\gamma$: data for drops with different $\gamma$ (but equal $\eta$) collapse onto the same curves to within the precision of the data.

In Fig. 6 $Ca_w$ is also larger for sets of data with high mobility. This is not unexpected because $Ca_{\text{slip}}$ is also $M$ dependent (Fig. 5a). To understand if the mobility affects the rolling part of the motion, we need to compare $Ca_w$ at fixed $Ca_{\text{slip}}$. This is done in Fig. 6(b) where it is apparent that different mobilities collapse onto the same curves showing that $M$ affects the rolling part of the motion only through $v_{\text{slip}}$.

For higher forcing the linear dependence between the capillary number and the Bond number is lost. $w$ decreases, $v_{\text{slip}} \gg w \cdot H$, and sliding dominates the motion. This corresponds to a regime where the drops are highly deformed and a further increase in Bond number causes them to detach from the substrate.[15] Similar behaviour has been observed in experiments[3] for drops on hydrophobic surfaces.

## VII. Drops on patterned surfaces

In this section we investigate the dynamics of a drop on a surface patterned with posts (see Section III and Fig. 1 for the geometric details). In particular we are interested in understanding analogies to, and differences from, motion across a smooth surface. Drops which lie on top of the posts in the suspended or Cassie state exhibit an apparent contact angle $\theta_C$ given by Cassie's relation $\cos \theta_C = \phi \cos \theta_{\text{eq}} + \phi - 1$,[33] where $\theta_{\text{eq}}$ is the equilibrium contact angle for the smooth plane. We consider cylindrical drops on surfaces patterned by obstacles which occupy an area fraction $\phi = 0.22$ of the flat surface, and a value $\theta_{\text{eq}} = 100°$, which gives $\theta_C \approx 145°$, allowing a direct comparison with the results for a smooth surface in Sections V and VI.

In the presence of posts the drop remains stationary for small Bond numbers.[38–40] This occurs because the interface is pinned at the edges of the obstacles.[41] For the geometry we consider here the depinning threshold $Bo_T \sim 0.15$. For higher Bond number we observe oscillations in the velocity which reflect the position of the drop relative to the posts and become smaller further from the pinning threshold $Bo_T$. The steady velocities were therefore computed by averaging over a time interval much longer than the time required for the drop to cross a post.

In Fig. 7(a) we plot the velocity profile $v_x$ as a function of the distance from the top of the posts for Bond numbers $0.29 < Bo < 0.54$. This plot should be compared to Fig. 3 for the smooth surface. Just as for the smooth surface, the velocity profile exhibits a crossover from a quadratic to a linear regime. The main difference is the presence of a microscopic slip velocity at $z = 0$, defined as $v_0$ in eqn (10), for the superhydrophobic substrate.[27,28] These results show that a crossover from

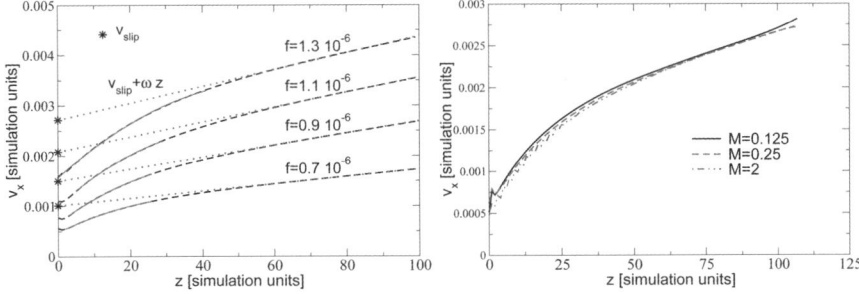

**Fig. 7** (a) The same as Fig. 3 but for superhydrophobic surfaces and for $0.29 < Bo < 0.54$. The plot shows the velocity in the $x$ direction ($v_x$) as a function of the height $z$ (measured from the top of the posts). At large $z$ a linear, rolling, regime is present (dashed) with crossover to a quadratic profile for small $z$ (full lines). (b) $v_x$ for Bond number $Bo = 0.371$ ($f = 0.9 \times 10^{-6}$) and three different mobilities. Data were obtained using $\eta = 0.833$ and $\gamma = 0.02667$.

a quadratic to a linear dependence of velocity on $z$ is not dependent on having zero $v_0$ but is also present if some slip occurs at the solid–liquid interface. We remark that it is difficult to obtain a precise value for $v_0$ from the simulations because of the presence of spurious velocities at the surface.

An important difference between the smooth and the patterned geometry is illustrated in Fig. 7(b) where we plot $v_x$ against $z$ for $Bo = 0.371$ and different values of the mobility $M$. The velocity is unaffected by the mobility, in stark contrast to the smooth surface where $Ca_{\text{slip}}$ depends strongly on $M$ (compare Fig. 5). This provides evidence that the contact line is moving onto successive posts using a tank-treading mechanism, rather than through relaxing the interface distortion by diffusion.

Fig. 8 is a plot comparing the centre of mass capillary number as a function of Bond number for drops on hydrophobic and superhydrophobic surfaces. Full lines refer to the smooth surface, while the broken line is for the patterned geometry. As before, the equilibrium contact angle of the smooth surface is $\theta_{\text{eq}} = 145°$, while for the patterned plane $\theta_{\text{eq}} = 100°$ corresponding to a Cassie angle of $\theta_c \approx 145°$. For $Bo > 0.3$ the drop on the smooth plane is very deformed from its equilibrium shape and a small further increase in $Bo$ leads to the drop detaching from the substrate. For the drop on the posts steady motion is only possible above a threshold Bond number due to pinning, but then the drop moves faster and with much less deformation. Thus it is possible to push the drop much harder and to achieve velocities larger by a factor $\sim 4$ before detachment occurs.

Increasing the bulk force $f$ leads to a drop that detaches from the surface. We have observed that, for superhydrophobic surfaces, there is a long transient regime in which the droplet accelerates before it finally detaches. This was not observed in the case of smooth planes. Similar accelerating drops have been observed in experiments on superhydrophobic surfaces[3] although the acceleration measured in the simulations is much smaller ($0.2\% \times f$) than that in the experiments ($\sim f$).

## VIII. Discussion

We have used a mesoscopic simulation approach to investigate the velocity profile of hydrophobic and superhydrophobic drops subject to a constant body force.

For both hydrophobic and superhydrophobic surfaces we found that the velocity profile perpendicular to the substrate $v_x(z)$ comprises two regions. Close to the surface $v_x(z)$ is quadratic in $z$, as in a Poiseuille flow. This is the velocity profile found using lubrication theory.[35] Further from the substrate $v_x(z)$ is linear in $z$ describing a rolling motion, as predicted for drops of large $\theta_{\text{eq}}$ by scaling arguments.[10,11] The crossover is clearly seen for equilibrium contact angles $\gtrsim 120°$, for smaller contact angles (*e.g.* 70°) the velocity profile is quadratic for all $z$.

Fitting the linear portion of the curve as

$$v_x = v_{\text{slip}} + w \cdot z \qquad (12)$$

defines a mesoscopic slip velocity $v_{\text{slip}}$ which characterizes the quadratic region of the flow field. We find that $Ca_{\text{slip}}$ is independent of the viscosity and the surface tension but depends on the shape of the drop, through the equilibrium contact angle, and on the mobility parameter $M$ which controls the ability of the contact line to move across the surface. This is reminiscent of the 'inner' region, identified by Cox,[42] which encapsulates the microscopic physics.

The rolling part of the flow profile was found to be independent of the mobility. Thus it might be identified as the 'outer' region of the flow which is, as expected, independent of the microscopic details of the contact line motion. The angular velocity is independent of surface tension, as expected, but increases with increasing viscosity.

A similar behaviour is seen for drops on superhydrophobic surfaces, but there is also a microscopic slip velocity, that is, a non-zero velocity of the fluid adjacent to

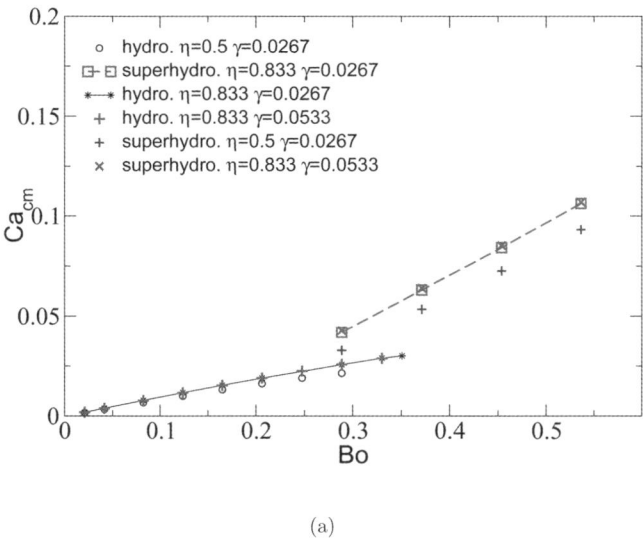

(a)

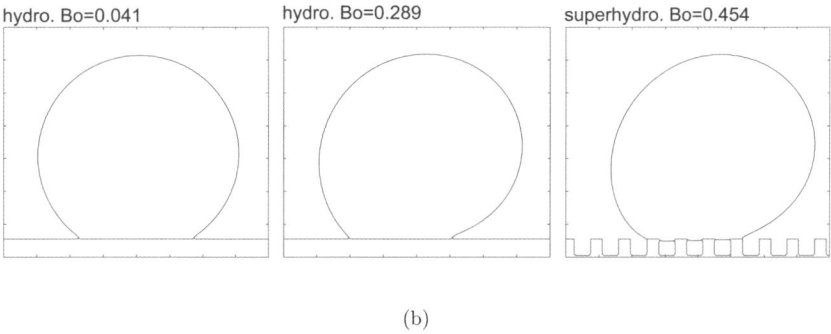

(b)

**Fig. 8** (a) Drop center of mass capillary number $Ca_{cm}$ *versus* Bond number $Bo$ for drops on hydrophobic and superhydrophobic surfaces. The equilibrium contact angle of the smooth surface is $\theta_{eq} = 145°$, while for the patterned plane it is $\theta_{eq} = 100°$ giving a Cassie angle $\theta_c \approx 145°$. (b) Steady state shapes of the drops at different $Bo$.

the surface. We stress that this occurs despite the no-slip boundary conditions imposed in the simulations. It suggests that the contact line moves not by diffusion, but by tank-treading from one post to the next. Further evidence for this is that, for the superhydrophobic surfaces, $v_{slip}$ is independent of the mobility.

Pinning on the posts leads to contact angle hysteresis and it takes a non-zero force to initiate drop motion. Once the drop is moving it can be pushed more easily than on a flat surface and can reach higher velocities before flying off the substrate. The drop slowly accelerates before leaving the surface.

Rolling dynamics have been observed in experiments. In ref. 5 an almost pure rolling motion was measured for glycerol drops with a superhydrophobic coating ($\theta_{eq} = 165°$ with an hysteresis $\Delta\theta \leq 10°$) and diameters between 1.2 and 8 mm. The dominance of rolling was ascribed to the high viscosity of glycerol (950 times bigger than water) and the high equilibrium contact angle. Suzuki *et al.*[2] reported a linear profile for the velocity inside water drops with $\theta_{eq} \approx 110°$ but did not resolve any quadratic regime. However slip at the solid substrate was observed, and identified as $v_{slip}$, and three different cases were reported in which $v_{slip}$ was negligible, comparable and dominant with respect to the rolling part of the motion. In ref. 3 predominantly sliding drops were observed in the superhydrophobic regime ($\theta_{eq} = 150°$), while for lower $\theta_{eq}$ rolling motions were also resolved.

A recent molecular dynamic simulation[14] gave a Poiseuille profile (10) over all the height of the drop. In particular no rolling regime was observed. Possible explanations for the difference between these results and ours are that the equilibrium contact angle in the molecular dynamics simulations was 130°, corresponding to a region in our simulations where rolling was not well resolved, that molecular dynamics typically accesses much smaller length scales than the diffuse interface approach or that the intrinsic slip mechanism at the contact line was different in the two methods.

Clearly more work needs to be done to unify experiments, mesoscopic simulations and microscopic simulations to give a clear picture of how the way in which a drop moves across a surface depends on its geometry, the parameters describing the fluid and the way in which the drop interacts with the substrate. To work in this direction our next aim is to gain a better understanding of how the size of the drop and the degree of intrinsic slip at the surface affect the simulation results.

## Acknowledgements

We thank Rodrigo Ledesma and Matthew Blow for useful discussions.

## References

1 H.-Y. Kim, H. J. Lee and B. Y. Kang, *J. Colloid Interface Sci.*, 2002, **247**, 372.
2 S. Suzuki, A. Nakajima, M. Sakai, Y. Sakurada, N. Yoshida, A. Hasimoto, Y. Kameshima and K. Okada, *Chem. Lett.*, 2008, **37**, 58.
3 M. Sakai, J.-H. Song, N. Yoshida, S. Suzuki, Y. Kameshima and A. Nakajima, *Langmuir*, 2006, **22**, 4906.
4 S. Suzuki, A. Nakajima, M. Sakai, A. Hashimoto, N. Yoshida, Y. Kameshima and K. Okada, *Appl. Surf. Sci.*, 2008, **255**, 3414.
5 D. Richard and D. Quere, *Europhys. Lett.*, 1999, **48**, 286.
6 T. Podgorski, J. M. Flesselles and L. Limat, *Phys. Rev. Lett.*, 2001, **87**, 036102.
7 P. Roach, N. J. Shirtcliffe and M. I. Newton, *Soft Matter*, 2008, **4**, 224.
8 P. A. Durbin, *J. Fluid Mech.*, 1988, **197**, 157.
9 U. Thiele, K. Neuffer, M. Bestehorn, Y. Pomeau and M. G. Velarde, *Colloids Surf., A*, 2002, **206**, 87.
10 L. Mahadevan and Y. Pomeau, *Phys. Fluids*, 1999, **11**, 2449.
11 S. R. Hodges, O. E. Jensen and J. M. Rallinson, *J. Fluid Mech.*, 2004, **512**, 95.
12 A. Dupuis and J. M. Yeomans, *Europhys. Lett.*, 2006, **75**, 105.
13 H. Kusumaatmaja, J. Leopoldes, A. Dupuis and J. M. Yeomans, *Europhys. Lett.*, 2006, **73**, 740; H. Kusumaatmaja, A. Dupuis and J. M. Yeomans, *Math. Comput. Simul.*, 2006, **72**, 160.
14 J. Servantie and M. Mueller, *J. Chem. Phys.*, 2008, **128**, 014709.
15 Q. Kang, D. Zhang and S. Chen, *J. Fluid Mech.*, 2005, **545**, 41.
16 P. Dimitrakopoulos and J. J. L. Higdon, *J. Fluid Mech.*, 2001, **435**, 327.
17 A. D. Schleizer and R. T. Bonnecaze, *J. Fluid Mech.*, 1999, **383**, 29.
18 P. D. M. Spelt, *J. Comput. Phys.*, 2005, **207**, 389.
19 M. R. Swift, E. Orlandini, W. R. Osborn and J. M. Yeomans, *Phys. Rev. E: Stat. Phys., Plasmas, Fluids, Relat. Interdiscip. Top.*, 1996, **54**, 5041.
20 A. Briant and J. M. Yeomans, *Phys. Rev. E: Stat., Nonlinear, Soft Matter Phys.*, 2004, **69**, 031603.
21 J. M. Yeomans, *Phys. A*, 2006, **369**, 159.
22 S. Succi, *The Lattice Boltzmann Equation; for Fluid Dynamics and Beyond*, 2001, Oxford University Press.
23 S. Chibbaro, L. Biferale, F. Diotallevi and S. Succi, *Eur. Phys. J. Spec. Top.*, 2009, **171**, 223.
24 J. Leopoldes, A. Dupuis, D. G. Bucknall and J. M. Yeomans, *Langmuir*, 2003, **19**, 9818.
25 H. Kusumaatmaja, M. L. Blow, A. Dupuis and J. M. Yeomans, *Eur. Phys. Lett.*, 2008, **81**, 36003.
26 C. Cottin-Bizonne, J.-L. Barrat, L. Bocquet and E. Charlaix, *Nat. Mater.*, 2003, **2**, 238.
27 Steinberger, C. Cottin-Bizonne, P. Kleimann and E. Charlaix, *Nat. Mater.*, 2007, **6**, 665.
28 J. Hyväluoma and J. Harting, *Phys. Rev. Lett.*, 2008, **100**, 246001.

29 J. Cahn, *J. Chem. Phys.*, 1977, **66**, 3667.
30 C. M. Pooley, H. Kusumaatmaja and J. M. Yeomans, *Phys. Rev. E: Stat., Nonlinear, Soft Matter Phys.*, 2008, **78**, 056709.
31 C. M. Pooley, H. Kusumaatmaja and J. M. Yeomans, *Eur. Phys. J. Spec. Top.*, 2009, **171**, 63.
32 F. Diotallevi, L. Biferale, S. Chibbaro, G. Pontrelli, F. Toschi and S. Succi, *Eur. Phys. J. Spec. Top.*, 2009, **171**, 237.
33 A. B. D. Cassie and S. Baxter, *Trans. Faraday Soc.*, 1944, **40**, 546.
34 L. D. Landau and E. M. Lifshitz, *Fluid Mechanics*, Butterworth-Heinemann, Oxford, 2nd edition, 1987.
35 A. Oron, S. H. Davis and S. G. Bankoff, *Rev. Mod. Phys.*, 1997, **69**, 931.
36 T. Qian, X.-P. Wang and P. Sheng, *Phys. Rev. E: Stat., Nonlinear, Soft Matter Phys.*, 2003, **68**, 016306; T. Qian, X.-P. Wang and P. Sheng, *Phys. Rev. Lett.*, 2004, **93**, 094501.
37 T. Qian, X.-P. Wang and P. Sheng, *J. Fluid Mech.*, 2006, **564**, 333.
38 E. Raphael and P. G. de Gennes, *J. Chem. Phys.*, 1989, **90**, 7577.
39 H. Kusumaatmaja and J. M. Yeomans, *Langmuir*, 2007, **23**, 6019.
40 M. Reyssat and D. Quere, *J. Phys. Chem. B*, 2009, **113**, 3906.
41 J. W. Gibbs, *Scientific Papers, 1906*, Dover, New York, 1961; P. Concus and R. Finn, *Acta Mathematica*, 1974, **132**, 177.
42 R. G. Cox, *J. Fluid Mech.*, 1986, **168**, 169.

PAPER | www.rsc.org/faraday_d | Faraday Discussions

# Dynamic mean field theory of condensation and evaporation processes for fluids in porous materials: Application to partial drying and drying

J. R. Edison and P. A. Monson*

*Received 8th December 2009, Accepted 22nd January 2010*
DOI: 10.1039/b925672e

We study the dynamics of evaporation for lattice gas models of fluids in porous materials using a recently developed dynamic mean field theory. The theory yields a description of the dynamics that is consistent with the mean field theory of the thermodynamics at equilibrium. The nucleation processes associated with phase changes in the pore are emergent features of the dynamics. Our focus is on situations where there is partial drying or drying in the system, associated with weakly attractive or repulsive interactions between the fluid and the pore walls. We consider two systems in this work: (i) a two-dimensional slit pore geometry relevant to the study of adsorption/desorption or intrusion/extrusion dynamics for fluids in porous materials and (ii) a three dimensional slit pore modeling a pair of square plates in a bath of liquid as used in recent theoretical studies of dewetting processes between hydrophobic surfaces. We assess the theory by comparison with a higher order approximation to the dynamics that yields the Bethe–Peierls or quasi-chemical approximation at equilibrium.

## I. Introduction

The effect of confinement in pores upon fluid properties is a problem with many applications ranging from adsorption separations to catalysis to dewetting processes in protein folding.[1,2] It also provides a rich variety of challenges to theory.[1,3] Confinement may modify or eliminate phase transitions seen in the bulk and metastability can play an important role in the behavior when there is no means of nucleating phase transitions other than through rare event fluctuations. The strength of the surface field (relative to the strength of the fluid–fluid interactions) determines the relative stability of the solid–vapor and solid–liquid interfaces and hence the wetting characteristics and this in turn determines the location of transitions relative to those in the bulk. The occurrence of metastability as well as the potential for mass transfer limitations to equilibrium make studies of the dynamics of substantial importance.

Recently we have developed a dynamic mean field theory (DMFT) that describes the relaxation processes of fluids confined in porous materials.[4–6] The theory is based on a lattice gas model of the system and can be viewed as an approximation to the dynamics averaged over an ensemble of Kawasaki dynamics trajectories of the system. It provides a theory of the dynamics of the system consistent with the thermodynamics in mean field theory.[6] The primary result is the density distribution in the system as a function of time and this can be used to visualize the state of the

*Department of Chemical Engineering, University of Massachusetts, Amherst, MA, 01003, U. S. A*

system and identify the mechanism of phase changes. In particular the nucleation mechanisms associated with confined fluid phase transitions are emergent properties in the dynamics. Our motivation in the development of this method was to study the dynamics of capillary condensation and evaporation, and its relationship to the structure of the porous material at small length scales. This was in turn motivated by the need to understand the characterization methods for porous materials based on gas adsorption or mercury porosimetry. In gas adsorption we are concerned with the dynamics of pore filling and emptying for a fluid which typically wets the pore surfaces and where filling of the pores with liquid occurs at states where the vapor is the stable bulk phase. In mercury porosimetry we are concerned with a fluid which is non wetting and the filling of the pores (intrusion) occurs at states where the liquid is the stable bulk phase. As we will discuss briefly later, the lattice gas model we use here exhibits a symmetry whereby the process of filling for a wetting fluid is isomorphic with that of emptying for a non wetting fluid and *vice versa*.[7–9]

Our first application was to a slit pore system where the solid–fluid interactions would lead to complete wetting of the free surfaces by liquid[6] (some results similar to ours had been obtained earlier using a time dependent Landau–Ginzburg theory[10]). The calculations revealed information about the mechanism of capillary condensation including the formation of undulates and liquid bridges as had been suggested in the work of Everett and Haynes[11] and seen in recent Monte Carlo simulations of nucleation.[12] This was followed by applications to ink bottle pore networks[5] and to partial wetting systems.[4] The case of partial wetting is of interest in connection with systems like water in carbon pores where an interesting question is whether water will condense in such pores in adsorption from undersaturated water vapor.[13,14] In such systems there is an interplay between wetting, capillary condensation and hysteresis.[15] An important observation was that condensation occurs at metastable bulk vapor states and is initiated with the formation of a droplet on only one of the pore surfaces.[4] This is consistent with density functional calculations of nucleation from metastable vapors in a slit pore.[16]

In this paper we focus on pore/fluid systems when the solid–fluid interaction is weakly attractive or even repulsive, giving rise to partial drying or complete drying. This was of interest to us in the first instance in connection with intrusion/extrusion of nonwetting liquids into pores,[9,17,18] such as in mercury porosimetry[9,17] as noted above. It is also of wider interest in connection with the general problem of dewetting processes between hydrophobic surfaces.[2,19–23] Several studies of lattice models of dewetting have been made.[19–23] The DMFT method presented here provides an approximate description of the dynamics of these systems also. We illustrate this with the example of dewetting of a pair of hydrophobic plates immersed in a liquid as we change the bulk chemical potential towards bulk saturation. We focus on the differences in the nucleation mechanism for dewetting for surface fields yielding partial drying *versus* surface fields yielding complete drying. Our DMFT results show that for partial drying cases nucleation can occur by bubble formation at one of the surfaces, analogous to what is seen in the partial wetting case and consistent with Monte Carlo simulations of nucleation by Barrat and coworkers.[24] We also consider the case of the Janus pore where one wall is hydrophobic and the other hydrophilic[25] as well as the effect of surface patterning on the pore wall.[19,22,26]

We consider two slit pore geometries where in both cases the pores are open to the bulk fluid as shown in Fig. 1. In one case we have a slit pore that is of finite length in the x-direction and in contact with the bulk in that direction while infinite in the y-direction. The density distribution in this system is two-dimensional and this greatly reduces the computer time for the MFT and DMFT. In the other case we have a square slit pore finite in both x- and y-directions so that the density distribution is fully three-dimensional. This latter geometry allows the observation of two-dimensional patterning parallel to the pore walls in condensation and evaporation processes and is a more direct model of the surface forces apparatus where

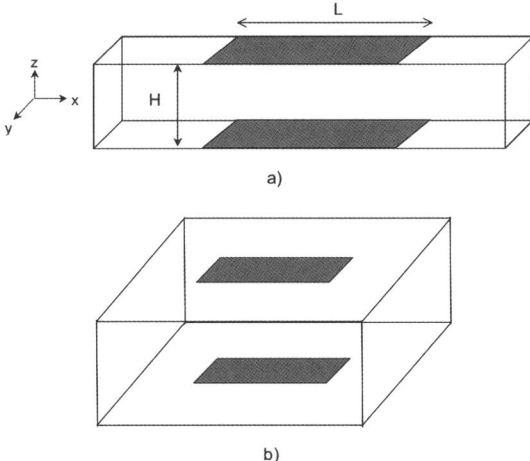

**Fig. 1** The two slit pore geometries considered in this work. In the first model the pore is infinite in the *y*-direction. Periodic boundaries are used in the *x*-direction with the pore in contact with the bulk fluid. In the second model the pore is in contact with the bulk in both the *x*- and *y*-directions.

capillary evaporation of water from hydrophobic pores has been studied experimentally.[27,28] As discussed above for partial wetting capillary condensation may occur for metastable vapor states. Similarly, capillary evaporation for a partial drying system may occur for metastable liquid states. A difficulty with studying the dynamics of capillary condensation/evaporation for metastable bulk states is that the liquid/vapor in the pore can act to nucleate bulk condensation/evaporation. In experiments on water condensation in graphitized carbon black, Easton and Machin[29] were able to suppress condensation of the bulk gas by silanizing the walls of the sample tube before adding the adsorbent material. This made it possible for them to observe condensation in the pores at metastable bulk vapor states. In this work we adjust the surface field at the ends of the pore to suppress this nucleation process. For partial wetting we would make the surface field at the pore edges strongly repulsive and for partial drying we make the surface field at the pore edges strongly attractive. In addition to results from DMFT we will also present some results from a higher order dynamic theory based on the path probability method of Kikuchi[30,31] that yields the Bethe–Peierls or quasi-chemical approximation[32] at equilibrium. The lattice model exhibits a symmetry between pore filling for wetting states and pore emptying for drying states, for both the thermodynamics and dynamics, and we will take advantage of this in the presentation of our results.

## II. Models and methods

### A. Lattice model and static mean field approximation

For a nearest neighbor lattice gas in an external field the Hamiltonian can be written as[6,8,33–35]

$$H = -\frac{\varepsilon}{2}\sum_{\mathbf{i}}\sum_{\mathbf{a}} n_{\mathbf{i}} n_{\mathbf{i+a}} + \sum_{\mathbf{i}} n_{\mathbf{i}} \phi_{\mathbf{i}} \qquad (1)$$

where $n_{\mathbf{i}}$ is the occupancy (0 or 1) at site $\mathbf{i}$, $\mathbf{a}$ is the vector to a nearest neighbor site for any site on the lattice[31] and $\varepsilon$ is the nearest neighbor interaction strength. The second term in the Hamiltonian describes the solid–fluid interactions *via* an external field $\phi_{\mathbf{i}}$.

In this work we use a simple cubic lattice and the interaction with the walls occurs *via* a nearest neighbor interaction with strength $-\alpha\varepsilon$.

The mean field (MFT) Helmholtz energy is written

$$F = kT \sum_i [\rho_i \ln \rho_i + (1 - \rho_i) \ln (1 - \rho_i)] - \frac{\varepsilon}{2} \sum_i \sum_a \rho_i \rho_{i+a} + \sum_i \rho_i \phi_i \quad (2)$$

where $\rho_i$ is the mean density at site **i**. Similarly the grand free energy is given by

$$\Omega = kT \sum_i [\rho_i \ln \rho_i + (1 - \rho_i) \ln (1 - \rho_i)] - \frac{\varepsilon}{2} \sum_i \sum_a \rho_i \rho_{i+a} + \sum_i \rho_i (\phi_i - \mu) \quad (3)$$

where $\mu$ is the chemical potential. By minimizing $\Omega$ at fixed chemical potential or $F$ at fixed overall density we can obtain solutions of the mean field equations for the grand canonical and canonical ensembles, respectively. These yield the free energy and density distributions in these ensembles. In both cases the necessary condition for equilibrium leads to the following equations relating the chemical potential to the local density at site **i**

$$\mu_i = kT \ln\left[\frac{\rho_i}{1 - \rho_i}\right] - \varepsilon \sum_a \rho_{i+a} + \phi_i \quad \forall \; \mathbf{i} \quad (4)$$

In the grand ensemble we may solve these equations iteratively for the density distribution, $\{\rho_i\}$, at fixed uniform $\{\mu_i = \mu\}$, $T$ and $\{\phi_i\}$. In the canonical ensemble the equations are solved with $\mu$ as a Lagrange multiplier for the fixed $N$ constraint, $\sum_i \rho_i = N$.

The lattice model exhibits a symmetry[7-9] in its properties that can be seen by defining

$$\alpha = 1/2 + \delta\alpha \quad \mu = \mu_0 + \delta\mu \quad (5)$$

where $\mu_0$ is the chemical potential at bulk saturation ($\mu_0 = -Z\varepsilon/2$ for the lattice gas model with coordination number $Z$). Positive values of $\delta\alpha$ are associated with the situation where the fluid wets or partially wets the solid and negative values are associated with a drying (non wetting) or partially drying situation. Positive values of $\delta\mu$ correspond to the stable liquid state of the bulk fluid and negative values to the stable vapor state of the bulk fluid. The state of a fluid with positive $\delta\alpha$ and negative $\delta\mu$ (corresponding to gas adsorption/desorption in the wetting case) is isomorphic with that of a fluid with negative $\delta\alpha$ and positive $\delta\mu$ of the same magnitudes as in the wetting case and with all the fluid site occupancies reversed (corresponding to intrusion/extrusion in the drying case). This symmetry is only very roughly followed in off-lattice models or in nature. Nevertheless it is a useful line of thinking and has helped researchers understand, for instance, the relationship between pore characterization methods based on gas adsorption to those based on mercury porosimetry.[9]

### B. Dynamic mean field theory

DMFT[6,31,36] gives an approximation to the time evolution of the density distribution averaged over an ensemble of kinetic Monte Carlo simulations of the lattice gas model using Kawasaki dynamics. The review by Gouyet and co-workers gives an excellent introduction to such methods. The evolution of the ensemble average density at site **i** can be expressed exactly in terms of the net fluxes from site **i** to its nearest neighbor sites **i** + **a** *via*

$$\frac{\partial \rho_i}{\partial t} = -\sum_a <w_{i,i+a}(\{n\}) n_i (1 - n_{i+a}) - w_{i+a,i}(\{n\}) n_{i+a} (1 - n_i)>_t \quad (6)$$

where $w_{i, i+a}(\{n\})$ is the transition probability for transitions from site **i** to site **i** + **a** for a configuration $\{n\}$. The occupancy factors $n_i(1 - n_{i+a})$ and $n_{i+a}(1 - n_i)$ impose the requirement that in a hopping move from site **i** to site **j**, site **i** must be occupied and site **j** unoccupied, and *vice versa*.

In the mean field approximation we obtain

$$\frac{\partial \rho_i}{\partial t} = -\sum_a [w_{i,i+a}(\{\rho\})\rho_i(1 - \rho_{i+a}) - w_{i+a,i}(\{\rho\})\rho_{i+a}(1 - \rho_i)] \quad (7)$$

Given expressions for the transition probabilities, eqn (7) can be solved to obtain $\{\rho_i\}$ as a function of $t$, and is equivalent to that given by Matuszak *et al.*[36] The mean field approximation for the Metropolis transition probabilities in Kawasaki dynamics yields[36]

$$w_{ij}(\{\rho\}) = w_o \exp(-E_{ij}/kT) \quad (8)$$

where

$$E_{ij} = \begin{cases} 0 & E_j < E_i \\ E_j - E_i & E_j > E_i \end{cases} \quad (9)$$

and

$$E_i = -\varepsilon \sum_a \rho_{i+a} + \phi_i \quad (10)$$

$w_0$ is the jump or transition rate in the absence of interactions and can be viewed as defining the timescale. We note that there are two ways in which the mean field approximation appears in this dynamic theory. The first is the expression for the energies in the transition probabilities and is similar to the mean field approximation made in the free energy. The second approximation is in the occupancy factors $\rho_i(1 - \rho_{i+a})$ and $\rho_{i+a}(1 - \rho_i)$, which imply that there are no occupancy correlations in the system dynamics. This approximation in the dynamics appears even in the limit of high temperature where the mean field description of the thermodynamics is exact.

As discussed in our earlier work[6] there are three ways to think about the dynamic theory represented by eqn (7)–(10). In the first case it is a theory of Kawasaki dynamics, yielding an approximation to the density evolution averaged over an ensemble of Kawasaki dynamics trajectories. To the extent that Kawasaki dynamics can be viewed as a description of diffusive processes it is a theory of diffusion.[36] Finally the evolution equation can be linearized and then has the form of a dynamic density functional theory.[37,38] The relationship of the theory to any real system is determined by how accurately Kawasaki dynamics can describe such systems. Evidence from previous work seems to indicate that qualitatively this is the case.[6,39] Of course the theory does not include momentum transport and it will be important to assess the limitations created by that.

We implement DMFT as follows. For both system geometries shown in Fig. 1 we begin by solving the MFT equations at an initial value of the chemical potential to give us the initial density distribution in the system. For the pore geometry shown in Fig. 1a we add a layer of sites at one end of the system where the density is fixed at the value associated with the chemical potential of the state to which we want the system to evolve. The fixed density layer acts as a source/sink of fluid during the dynamics. The system is then evolved by numerical solution of eqn (7). In our initial studies of this approach we used Euler's method but more recently we have also explored using Runge–Kutta methods. We have found Euler's method to be of acceptable accuracy for time steps less than about $\omega_0 \Delta t = 0.1$.

## C. Beyond mean field

We can go beyond mean field theory in the thermodynamics by using the Bethe–Peierls (BP), or quasi-chemical, approximation, as implemented recently, for example, by Salazar and Gelb.[32] This approximation includes occupancy correlations at the pair level. In the dynamics the corresponding approximation is the pair probability method (PPM) developed by Kikuchi,[30,31] which yields a theory of dynamics that reduces to the BP approximation at equilibrium. We will present some results from these methods, primarily as some evidence that the DMFT gives a qualitatively correct picture of the physics. A more detailed presentation will be given elsewhere.[40]

## III. Results

The calculations presented here are for slit pores with a wall spacing of 6 sites, not including the pore walls. This is close to the narrowest pore for which the lattice model yields discernible vapor–liquid menisci. For the two-dimensional pores the pore length was 40 sites with a 10 site length of bulk at each end. For the three-dimensional pores the dimensions were 40 × 40 sites with a 10 site length of bulk on each side. For the dynamic calculations for the partial drying walls the pores were extended by 5 sites at each end with a strong attractive surface field. This suppresses vaporization of the bulk metastable liquid by vapor in the pore as discussed in the introduction. Our calculations are for a single temperature $T/T_c = 2/3$.

### A. Static properties

**1. Contact angles *versus* surface field.** Contact angles from MFT and BP as a function of the surface field can be calculated from Young's equation

$$\cos\theta = \frac{\sigma_{SV} - \sigma_{SL}}{\sigma_{VL}} \quad (11)$$

where $\theta$ is the contact angle and $\sigma_{IJ}$ gives the interfacial tension for the *IJ* interface. The latter are determined from the excess grand free energies from the MFT or BP approximations for the *IJ* interface *via*

$$\sigma_{IJ} = \Omega_{IJ} + \phi_P M \quad (12)$$

For lattice models these calculations should strictly include the effect of the orientation of the interfaces with respect to the lattice planes. However, these effects are expected to be small at the temperature at which our calculations were performed.[24] Fig. 2 shows a plot of $\cos\theta$ *versus* $\alpha = \varepsilon_{sf}/\varepsilon_{ff}$ from the MFT and BP approximations. The results indicate first order wetting and drying transitions for this model at this temperature although we have not studied these transitions in detail. Both lines exhibit the symmetry $\cos\theta(\alpha) = -\cos\theta(1 - \alpha)$ which follows from a symmetry inherent in the lattice gas model[7–9] as discussed earlier. The slope of $\cos(\theta)$ is much steeper in the BP approximation than in the DMFT. This suggested to us that in order to compare the theories for partial wetting systems we should compare them at a fixed contact angle rather than surface field and this is what we have done.

**2. Density *versus* chemical potential isotherms.** Using MFT, and for some cases BP, we have made calculations of the filling/emptying isotherms for the confined systems studied in this work. We present the results for the two-dimensional slit pore. We have made the corresponding calculations for the parallel square plate pores and the results are indistinguishable from those for the two-dimensional pores on the scale of the plots.

Fig. 3 shows isotherms of the density *versus* chemical potential for complete wetting ($\alpha = 2$) and complete drying ($\alpha = -1$) systems from MFT and the BP

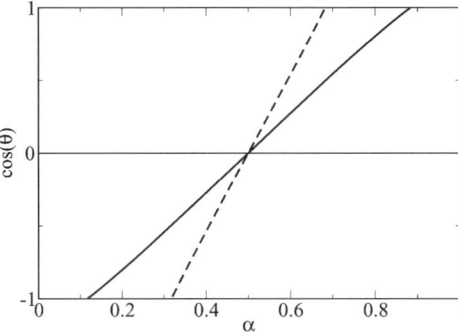

**Fig. 2** Isotherm of $cos(\theta)$ *versus* $\alpha$ for $T/T_c = 0.667$, computed *via* the Mean field (MFT) (full line) and the Bethe–Peierls Approximation (BP) (dashed line).

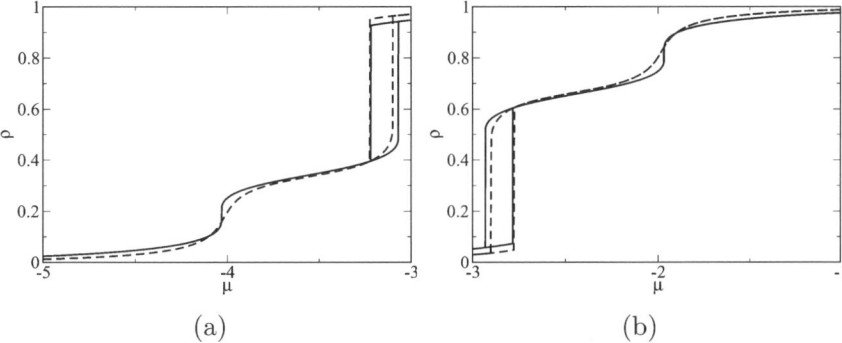

**Fig. 3** Adsorption/desorption isotherms of density *vs.* the chemical potential $\mu$ for (a) a completely wetting [$\alpha = 2$] and (b) a completely drying [$\alpha = -1$] slit pore at $T/T_c = 0.667$ computed with the Mean Field (full line) and the Bethe–Peierls approximations (dashed line).

approximations. In these and all other results, we are presenting the density at the center of the pore. This reduces the pore end effects upon the density and the results are then quite insensitive to pore length. For complete wetting the pore fills at a chemical potential less than the bulk saturation chemical potential, $-3\varepsilon$, while for complete drying the pore fills at a chemical potential above bulk saturation and there is hysteresis in both cases. In the hysteresis region for the complete wetting case the states on the adsorption branch are metastable while those on desorption are essentially at equilibrium since there is contact with the bulk vapor on desorption eliminating any nucleation barrier. This was shown in our earlier work[6] for a complete wetting case using a plot of the grand free energy. Conversely in the hysteresis region for the complete drying case the states on the emptying or evaporation branch are metastable while those on filling are essentially at equilibrium since there is contact with the bulk liquid. The step-like features in the isotherms at very low or very high chemical potentials are associated with the formation of a dense monolayer (wetting case) and dilute monolayer or vapor layer (drying case). The symmetry in the isotherms is evident. The comparison between MFT and BP is done at the same value of $T/T_c$ which reduces the effect of the differences in the critical temperatures ($kT_c^{(MFT)}/\varepsilon = 1.5$ *vs.* $kT_c^{(BP)}/\varepsilon = 1.233$) in comparing the results. The qualitative agreement is very good when the results are plotted in this way. Fig. 4 presents the same results but with the density *versus* the relative activity, $\lambda/\lambda_0$ where $\lambda = e^{\mu/kT}$. In this case the symmetry is no longer evident. The relative activity is of the

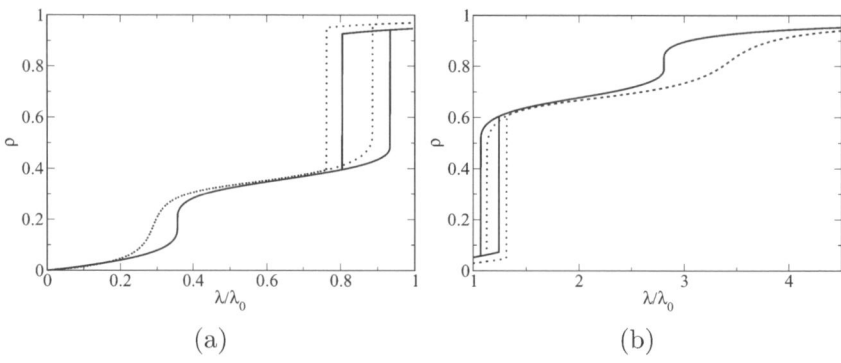

**Fig. 4** Adsorption/desorption isotherms of density *vs.* the relative activity $\lambda/\lambda_0$ for (a) a completely wetting [$\alpha = 2$] and (b) a completely drying [$\alpha = -1$] slit pore at $T/T_c = 0.667$ computed with the Mean Field (full line) and the Bethe–Peierls Approximations (dashed line).

same order of magnitude as the relative bulk pressure and the plot shows the large values of bulk pressure required to bring about intrusion of the non wetting liquid into the pores for the complete drying case. Fig. 5 shows a series of visualizations of the density distribution from MFT in the complete drying case during filling and emptying. We note that the intrusion of the liquid into the pore begins with the formation of a meniscus near the pore entrance. Fig. 5g and 5h show the states either side of the drying or dewetting transition. Our DMFT calculations will reveal the nature of the states encountered during the dynamics of this transition. Isotherms of density *versus* chemical potential for the partial wetting and partial drying cases are shown in Fig. 6. The partial wetting case corresponds to a contact angle of 60° while the partial drying case corresponds to 120°. Notice that the hysteresis loops associated with capillary condensation extend beyond $\mu/\varepsilon = -3$. Thus for partial wetting condensation occurs for metastable bulk vapor states even though the equilibrium transition between vapor and liquid in the pore occurs for $\mu/\varepsilon < -3$ as discussed recently.[6,13,14] The experiments of Easton and Machin[29] on water condensation in carbon also show this behavior. Similarly for partial drying capillary evaporation occurs for $\mu/\varepsilon < -3$ even though the equilibrium transition between vapor and liquid in the pore occurs for $\mu/\varepsilon > -3$. Fig. 7 shows visualizations of the density distribution in the partial drying case during filling and emptying.

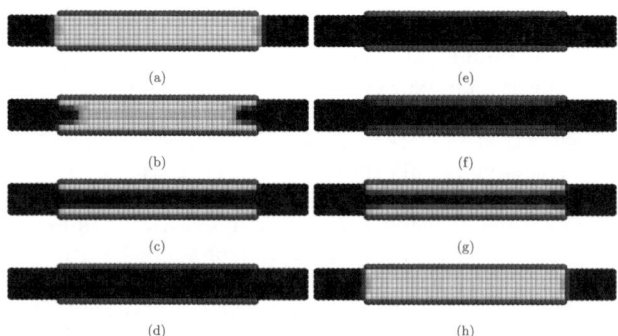

**Fig. 5** Visualizations of the density distribution for intrusion (a–d) and extrusion (e–h) states on the isotherm (completely drying slit pore [$\alpha = -1$]) shown in Fig. 3: (a) $\lambda/\lambda_0 = 1.041$ (b) $\lambda/\lambda_0 = 1.241$ (c) $\lambda/\lambda_0 = 1.291$ (d) $\lambda/\lambda_0 = 4.0$ (e) $\lambda/\lambda_0 = 4.0$ (f) $\lambda/\lambda_0 = 3.0$ (g) $\lambda/\lambda_0 = 1.075$ (h) $\lambda/\lambda_0 = 1.065$.

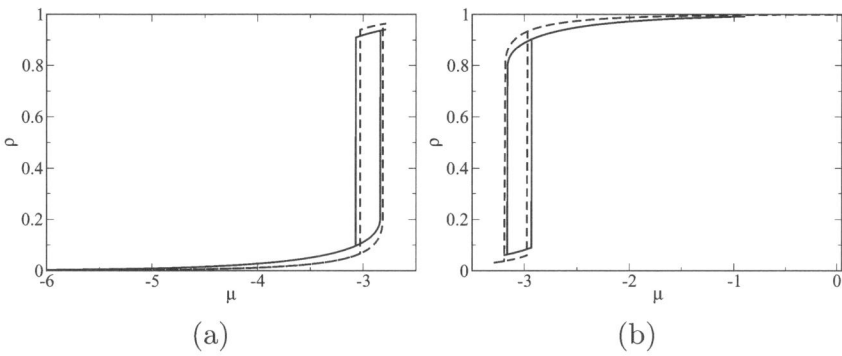

**Fig. 6** Adsorption/desorption isotherms of density *vs.* the chemical potential $\mu$ for (a) a partially wetting [$\theta = 60°$] and (b) a partially drying [$\theta = 120°$] slit pore with $H = 6$ at $T/T_c = 0.667$ computed with the Mean Field (full line) and the Bethe–Peierls Approximations (dashed line).

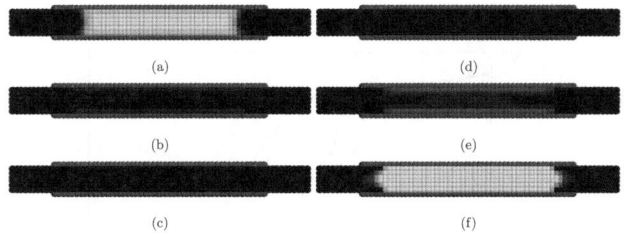

**Fig. 7** Visualizations of the density distribution for intrusion (a–c) and extrusion (d–f) states on the isotherm (partially drying case $\theta = 120°$) shown in Fig. 6: (a) $\mu = -2.941$ (b) $\mu = -2.904$ (c) $\mu = -1.128$ (d) $\mu = -1.908$ (e) $\mu = -3.1278$ (f) $\mu = -3.186$.

### B. Dynamics of evaporation from pores

**1. Two-dimensional slit pore.** We now consider the dynamics associated with changes of state between closely spaced states either side of the filling/emptying transitions shown in Fig. 3. Fig. 8 shows the average density over the pore *versus* time from DMFT for the dynamics of filling of the complete wetting pore together with that for the dynamics of emptying of the complete drying pore. We see that the symmetry of the lattice model is preserved in the dynamics - the uptake dynamics for complete wetting is the same as the emptying dynamics for the complete drying pore. We return to this point later. In the filling dynamics for the complete wetting pore the dominant event is the formation of a liquid bridge between the walls of the pore, which is associated with the cusp in the uptake curve, as is evident in the visualizations shown in our earlier work.[6] Once the nucleation process of bridge formation has occurred the dynamics of pore filling proceeds quite quickly. Conversely, in the emptying dynamics the dominant event is the appearance of a vapor bridge. A comparison of the dynamics of emptying from the DMFT with the higher order PPM are shown in Fig. 9 and the agreement is qualitatively good. For each case we show both the density averaged over the entire pore and the density averaged over $z$ at a value of $x$ equidistant from the pore ends. The density decreases more rapidly at the center of the pore since this is where vaporization is nucleated. Visualizations of the states appearing in the dynamics of emptying from DMFT in the complete drying case are shown in Fig. 10, where the nucleation *via* the formation of a vapor bridge is evident. We note that the dynamics for the PPM is a little different from that in DMFT in that we do not see well defined vapor undulates near the pore walls prior to the bridge

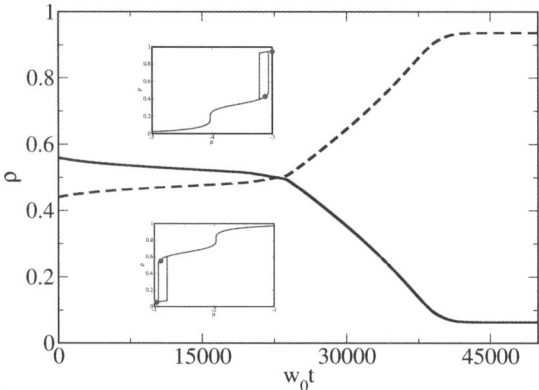

**Fig. 8** Symmetry preserved in dynamics of pore filling/emptying (dashed/full lines) for completely wetting/drying slit pores.

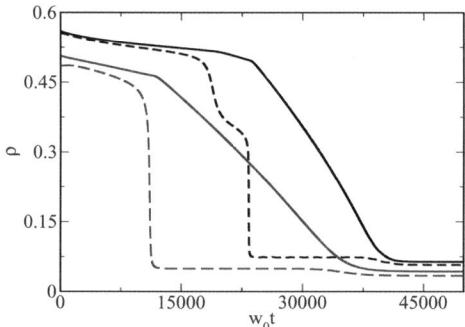

**Fig. 9** Density *vs.* time during a pore emptying process for a quench of $\Delta\mu = 0.03$ across the capillary evaporation transition for a completely drying slit pore. Black curves are from DMFT and blue curves are from the PPM approximation. For each case we show both the density averaged over the entire pore (full line) and the density averaged over $z$ at a value of $x$ equidistant from the pore ends (dashed line).

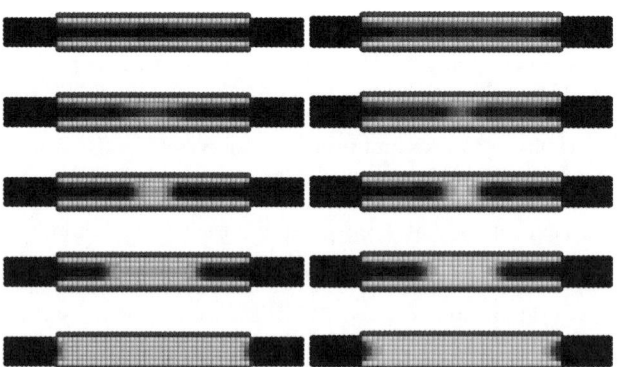

**Fig. 10** Visualizations showing a sequence of states along an evaporation process for a completely drying slit pore computed *via* the DMFT (left side) and PPM (right side).

formation. This may be a consequence of the difference in temperatures used in the two methods in order to use the same value of $T/T_c$.

A brief comment about the influence of the model pore geometry is in order here. We are considering a pore with periodic boundaries in the $y$-direction so that in our system there are no density variations in that direction. This means that condensation and evaporation nuclei will have the form of bridges or cylindrical droplets with axes in the $y$-direction. Later when we consider slit pores that are finite in both the $x$- and $y$-directions we will see droplets with circular symmetry and bridges which are cylinders with axes normal to the pore walls.

At first glance the symmetry in the dynamics for the system may seem unphysical. One would expect perhaps that vapor bridge formation in a liquid would be slower than liquid bridge formation in a vapor, given the greater potential for hydrodynamic effects in the latter. Nevertheless the symmetry will likely occur for Kawasaki dynamics without the DMFT or PPM approximations and we are currently checking this with dynamic Monte Carlo simulations. We also plan to investigate the relative timescales for condensation and evaporation processes using grand canonical molecular dynamics calculations such as those of Sarkisov and Monson.[39,41]

We now turn to the case of partial wetting and partial drying systems. Having shown the symmetry of the lattice model for both thermodynamics and dynamics at this point we simply focus on the partial drying case. Our results for pore emptying for the partial drying case with surface field strength $\alpha\varepsilon$ can be interpreted as also applying to pore filling for the partial wetting case with surface field strength $(1-\alpha)\varepsilon$. In Fig. 11 we show the average density in the pore during evaporation from a filled pore in the partial drying case from DMFT. We see that initially there is only a slow change in the density with time until a point is reached where the density starts to fall rapidly. As shown in the visualizations presented in Fig. 12 this point is associated with the appearance of a vapor bubble that is attached to one of the surfaces. This behavior contrasts with the complete drying case where the vapor bubble bridges the two surfaces and it is consistent with the results obtained in Monte Carlo simulations of nucleation in a lattice model of a fluid in a cylindrical pore by Barrat and coworkers.[24] The breaking of the symmetry of the density distribution during the DMFT dynamics is a quite remarkable result and is observed in the PPM approximation also. We previously observed the analogous behavior for the partial wetting case where a liquid droplet is formed on one of the walls in the condensation process[4] and this has also been seen in density functional studies of nucleation for an off-lattice model of a fluid, a slit pore.[16]

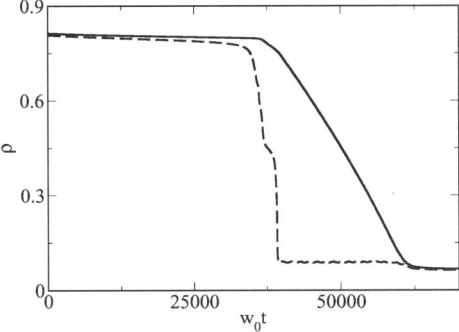

**Fig. 11** Density *vs.* time for a pore emptying process for a quench of $\Delta\mu = 0.005$ across the capillary evaporation transition for a partially drying slit pore. We show both the density averaged over the entire pore (full line) and the density averaged over $z$ at a value of $x$ equidistant from the pore ends (dashed line).

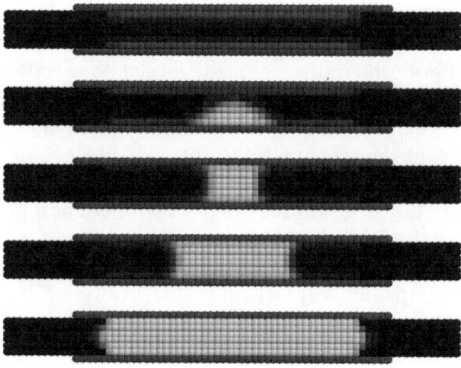

**Fig. 12** Visualizations showing a sequence of states along an evaporation process for a partially drying slit pore from DMFT.

**2. Three-dimensional slit pore.** We now consider the dynamics of emptying from the three-dimensional slit pore for complete drying ($\alpha = -1$) and partial drying ($\alpha = 0.3165$) cases calculated from DMFT. Fig. 13 shows the density versus time for a change in the chemical potential between two neighboring states either side of the emptying transition for the complete drying case. The shape of the curve is similar to that seen in Fig. 9. Visualizations of the behavior are shown in Fig. 14 for each point in time we show a two-dimensional slice through the density distribution half way across the pore. In addition we show the density in the third layer in the $x$–$y$ plane. Evaporation proceeds via the formation of a vapor cylinder at the center of the pore, analogous to the vapor bridge seen in the two-dimensional case. Fig. 15 shows the corresponding behavior for the partial drying case with visualizations in Fig. 16. In this case evaporation proceeds via the formation of a vapor bubble attached to one of the pore walls in the center of the square. This is analogous to the cylindrical bubble seen in the two-dimensional case.

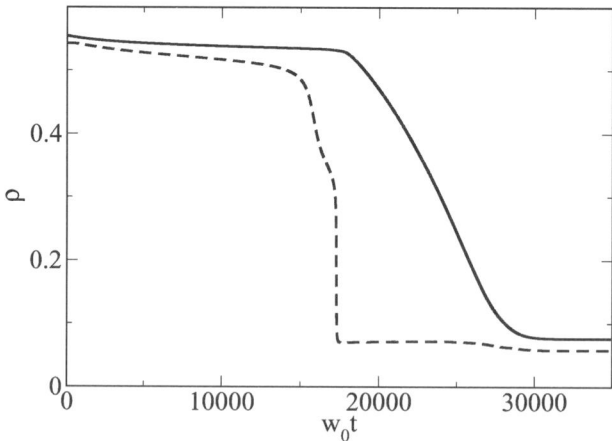

**Fig. 13** Density vs. time for a quench across the capillary evaporation transition $\Delta\mu = 0.011$ for a completely drying three-dimensional slit pore. We show both the density averaged over the entire pore (full line) and the density averaged over $z$ at a point equidistant from the pore sides (dashed line).

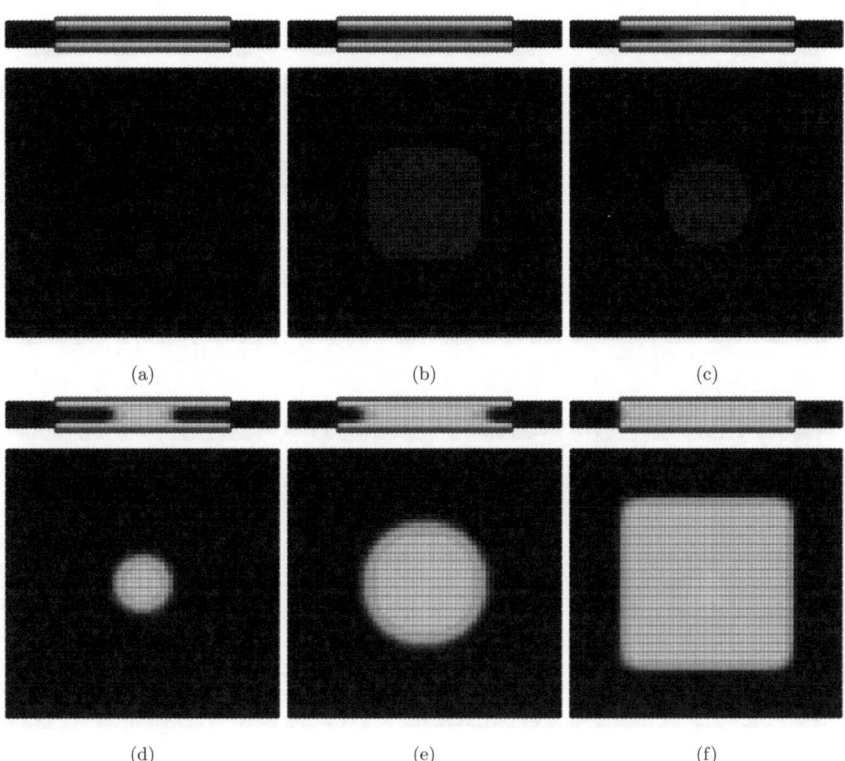

**Fig. 14** Visualizations showing a sequence of states along an evaporation process for a complete drying three-dimensional slit pore. Each panel shows at the top a visualization from a two-dimensional slice into the distribution half way across the pore and parallel to one of the pore sides, and below that the density in the $x$–$y$ plane in the third layer.

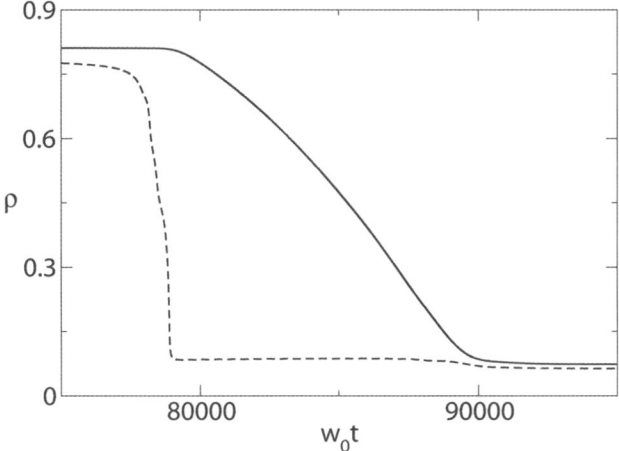

**Fig. 15** Density *vs.* time for a quench across the capillary evaporation transition $\Delta\mu = 0.0001$ for a partially drying three-dimensional slit pore. We show both the density averaged over the entire pore (full line) and the density averaged over $z$ at a point equidistant from the pore sides (dashed line).

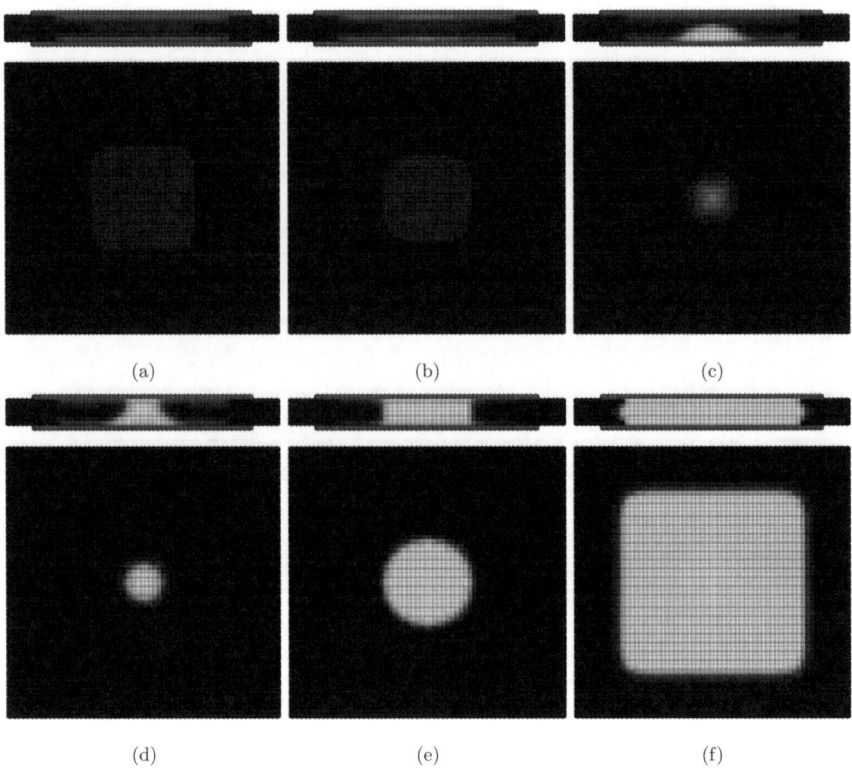

**Fig. 16** Visualizations showing a sequence of states along an evaporation process for a partially drying three-dimensional slit pore. Each panel shows at the top a visualization from a two-dimensional slice into the distribution half way across the pore and parallel to one of the pore sides, and below that the density in the $x$–$y$ plane in the third layer.

**3. Patterned surfaces.** As an illustration of the wider utility of the approach presented here we consider some cases where the surface is patterned with both wetting and drying regions. We consider two cases. The first is a so-called Janus pore[25] where one pore wall has a partial drying (hydrophobic) surface field and the other a complete wetting (hydrophilic) surface field. We implement this by having a surface field with $\alpha = 0.3165$ on one wall and a surface field $\alpha = 3$ on the other. The second case is a patterned surface in which the pore walls have both hydrophilic and hydrophobic regions. In the present case we consider a checkerboard pattern on each wall similar to one of the patterns considered by Berne and coworkers[26] although we use a somewhat larger length scale on the pattern.

Fig. 17 shows visualizations of evaporation from the Janus pore. The process is nucleated by the appearance of a vapor bubble at the partial drying surface. The growth of this bubble in a direction normal to the pore wall is limited by a film of liquid on the complete wetting wall. Once the bubble stops growing in that direction the evaporation process proceeds by the movement of the vapor liquid interface towards the external surfaces of the pore. During this process the liquid film on the complete wetting surface is largely unchanged. The structure of the interfaces is broadly consistent with that seen in the experiments of Granick and coworkers.[25]

The visualizations in Fig. 18 show the evaporation process from the checkerboard. The evolution of the density distribution is significantly more complex with both changes in the density and the symmetry of the density distribution with time. We note especially the appearance of multiple vapor cylinders. The location of these

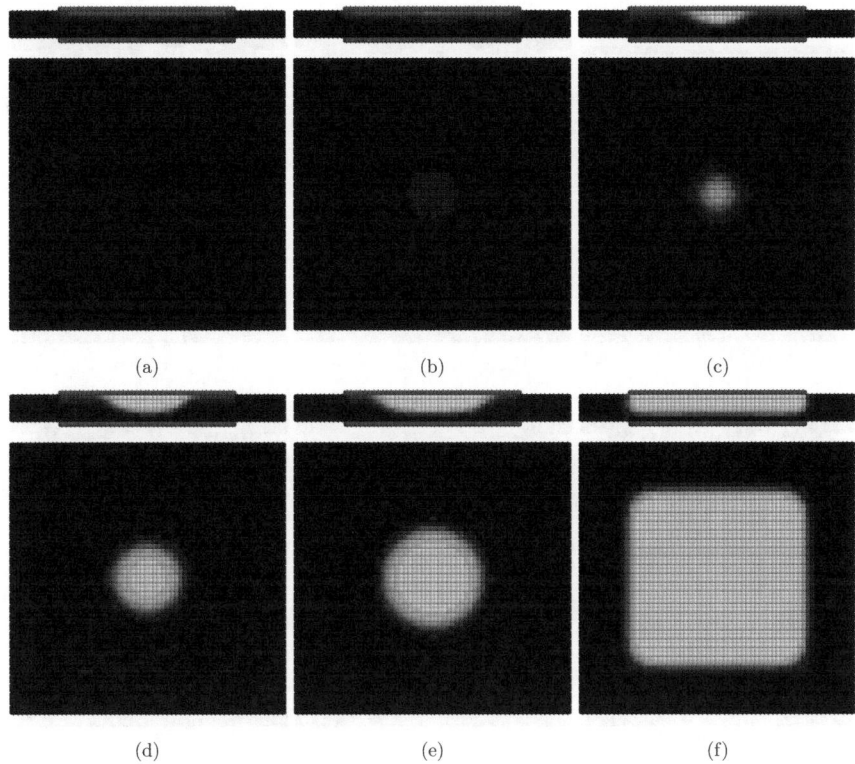

**Fig. 17** Visualizations showing a sequence of states along an evaporation process for the Janus pore. Each panel shows at the top a visualization from a two-dimensional slice into the distribution half way across the pore and parallel to one of the pore sides, and below that the density in the $x$–$y$ plane in the third layer.

cylinders coincides with the location of the drying regions of the pore surfaces but they emerge at different times in the dynamics. The first to emerge are those at the corners of the pore, then those at the sides of the pore with those at the interior of the pore appearing later. Also note that at the center of the pore there is liquid maintained between two drying regions until the very last stage of the emptying process. Clearly we are seeing the role of the mass transfer limitations on the development of the density distribution in the system during the dynamics.

## IV. Summary and conclusions

We have presented results for the dynamics of evaporation for a lattice model of fluids confined in slit pores with a focus on partial drying and complete drying cases. The dynamics is calculated *via* a DMFT which is consistent with the thermodynamics from the mean field for equilibrium or metastable equilibrium states. DMFT is in essence a theory of the average dynamics over an ensemble of Kawasaki dynamics trajectories but it can also be interpreted in other ways, including as a dynamic density functional theory.[37,38] Comparison with results from the PPM approximation[30] reveals that the qualitative results from DMFT are unchanged by going to a higher order theory. The lattice model exhibits a symmetry between pore filling for a wetting (hydrophilic) pore and emptying for a drying (hydrophobic) pore and *vice versa*. This symmetry also carries over to the dynamics in DMFT and PPM and likely to Kawasaki dynamics as well.

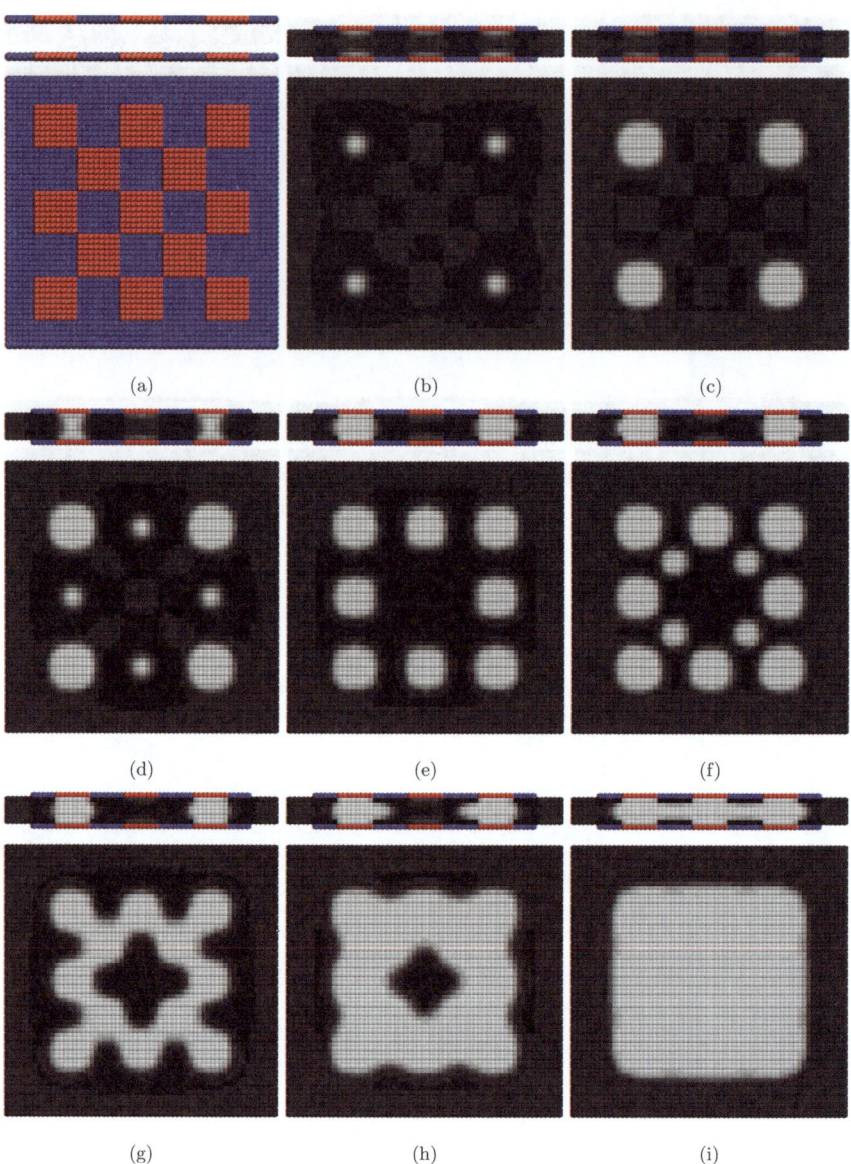

**Fig. 18** Visualizations showing a sequence of states along an evaporation process for the checkerboard pore. The first panel shows the patterning with hydrophobic in red and hydrophilic in blue. Each of the other panels show at the top a visualization from a two-dimensional slice into the distribution half way across the pore and parallel to one of the pore sides, and below that the density in the $x$–$y$ plane in the third layer.

We have presented results for two slit pore geometries (two-dimensional and three-dimensional) where the pore is in contact with the bulk liquid. In each geometry we compared the evaporation processes for partial drying pore walls with those for a complete drying case. In the former the evaporation process studied proceeds by the formation of a vapor bubble attached to one of the pore walls, representing a symmetry breaking in the density distribution during the dynamics. This has also been seen in the complementary case of condensation in pores with partial wetting walls[4] and observed in nucleation calculations for fluids in pores *via* density functional theory[16] and Monte Carlo simulations.[24] For the complete drying case the evaporation proceeds *via* the formation of a vapor bridge (two-dimensional pore) or vapor cylinder (three-dimensional pore). These observations are for processes with small changes in chemical potential. For larger chemical potential quenches we may encounter multiple vapor bridges during the dynamics analogous to the multiple liquid bridges formed during condensation in a wetting pore.[6]

Our results for patterned surfaces (Janus pore and checkerboard pore) illustrate the potential wider utility of the DMFT method. The Janus pore shows an evaporation process that proceeds *via* the motion of the vapor liquid interface with a fairly stable liquid film on the hydrophilic surface. For the checkerboard pattern a quite complex evolution of the density distribution takes place during the evaporation with multiple vapor cylinders developing and growing at different times due to mass transfer limitations upon the dynamics. It will likely be fruitful to relate the phenomena seen here to those encountered in other studies of fluids confined between patterned surfaces *via* density functional theory.[42] Also the effect of patterning in these systems is somewhat analogous to the effects of pore structure in complex porous materials[8,35,43] and in particular it will be interesting to look at the effects of disorder on the dynamics.[22,35,43]

There are some theoretical issues about the DMFT approach that deserve additional study. For instance the calculations do not include momentum transfer and it is important to determine the impact this has on the relevance of the results to real systems. Comparison with molecular dynamics simulations of comparable off-lattice models, which are in progress, should be of assistance here. The symmetry of the lattice model dynamics is also worth exploring further and this is also underway. DMFT and PPM methods do not describe the fluctuation events associated with individual Kawasaki dynamics (or molecular dynamics) trajectories since the theories are built by averaging over an ensemble of trajectories. For this reason the nucleation processes studied here are limited to those where the final state lies beyond the stability limit of the confined liquid (confined vapor for wetting and partial wetting systems). One can add fluctuations to the evolution equation (eqn (7)) in an ad hoc manner, and this might be useful for studying nucleation from metastable states closer to the equilibrium transition.[10] On the other hand the theoretical status of the resulting calculations is questionable.[44]

The lattice model approach as used by us and others is primarily a qualitative modeling tool.[8,21,22,35] Its great advantage is the scope of the calculations that are accessible with only modest computing resources. The development of DMFT greatly extends the scope of the dynamics calculations that can be made for these models. We believe it will prove a powerful tool for studying the phase transition dynamics of confined fluids, especially for systems with more complex confinement geometries and surface patterns.

## Acknowledgements

This work was supported by the National Science Foundation (Grant No.'s CBET-0649552 and CBET-0849552).

## References

1 L. D. Gelb, K. E. Gubbins, R. Radhakrishnan and M. Sliwinska-Bartkowiak, *Rep. Prog. Phys.*, 1999, **62**, 1573–1659.
2 B. J. Berne, J. D. Weeks and R. Zhou, *Annu. Rev. Phys. Chem.*, 2009, **60**, 85–103.
3 R. Evans, *J. Phys.: Condens. Matter*, 1990, **2**, 8989–9007.
4 J. R. Edison and P. A. Monson, *J. Low Temp. Phys.*, 2009, **157**, 395–409.
5 P. A. Monson, *Characterization of Porous Solids VIII*, 2009, pp. 103–110.
6 P. A. Monson, *J. Chem. Phys.*, 2008, **128**, 084701.

7 R. Pandit, M. Schick and M. Wortis, *Phys. Rev. B: Condens. Matter*, 1982, **26**, 5112–5140.
8 E. Kierlik, P. A. Monson, M. L. Rosinberg, L. Sarkisov and G. Tarjus, *Phys. Rev. Lett.*, 2001, **87**(5), 055701.
9 F. Porcheron, P. A. Monson and M. Thommes, *Langmuir*, 2004, **20**, 6482–6489.
10 F. Restagno, L. Bocquet and T. Biben, *Phys. Rev. Lett.*, 2000, **84**, 2433–2436.
11 D. H. Everett and J. M. Haynes, *J. Colloid Interface Sci.*, 1972, **38**, 125–137.
12 A. Vishnyakov and A. Neimark, *J. Chem. Phys.*, 2003, **119**, 9755–9764.
13 J. C. Liu and P. A. Monson, *Langmuir*, 2005, **21**, 10219–10225.
14 J. Liu, P. Monson and F. van Swol, *J. Phys. Chem. C*, 2007, **111**, 15976–15981.
15 P. A. Monson, *Langmuir*, 2008, **24**, 12295–12302.
16 V. Talanquer and D. Oxtoby, *J. Chem. Phys.*, 2001, **114**, 2793–2801.
17 F. Porcheron and P. A. Monson, *Langmuir*, 2005, **21**, 3179–3186.
18 F. Cailliez, M. Trzpit, M. Soulard, I. Demachy, A. Boutin, J. Patarin and A. H. Fuchs, *Phys. Chem. Chem. Phys.*, 2008, **10**, 4817–4826.
19 K. Leung and A. Luzar, *J. Chem. Phys.*, 2000, **113**, 5845–5852.
20 K. Leung, A. Luzar and D. Bratko, *Phys. Rev. Lett.*, 2003, **90**, 065502.
21 K. Lum and D. Chandler, *Int. J. Thermophys.*, 1998, **19**, 845–855.
22 A. Luzar and K. Leung, *J. Chem. Phys.*, 2000, **113**, 5836–5844.
23 L. Maibaum and D. Chandler, *J. Phys. Chem. B*, 2003, **107**, 1189–1193.
24 A. Saugey, L. Bocquet and J. Barrat, *J. Phys. Chem. B*, 2005, **109**, 6520–6526.
25 X. Zhang, Y. Zhu and S. Granick, *Science*, 2002, **295**, 663–666.
26 L. Hua, R. Zangi and B. J. Berne, *J. Phys. Chem. C*, 2009, **113**, 5244–5253.
27 H. Christenson and P. Claesson, *Science*, 1988, **239**, 390–392.
28 P. Claesson and H. Christenson, *J. Phys. Chem.*, 1988, **92**, 1650–1655.
29 E. B. Easton and W. D. Machin, *J. Colloid Interface Sci.*, 2000, **231**, 204–206.
30 R. Kikuchi, *Prog. Theor. Phys., Suppl.*, 1966, **35**, 1.
31 J. F. Gouyet, M. Plapp, W. Dieterich and P. Maass, *Adv. Phys.*, 2003, **52**, 523–638.
32 R. Salazar and L. Gelb, *Phys. Rev. E: Stat., Nonlinear, Soft Matter Phys.*, 2005, **71**, 041502.
33 M. J. DeOliveira and R. B. Griffiths, *Surf. Sci.*, 1978, **71**, 687–694.
34 C. Ebner, *Phys. Rev. A: At., Mol., Opt. Phys.*, 1981, **23**, 1925–1930.
35 H. J. Woo and P. A. Monson, *Phys. Rev. E: Stat., Nonlinear, Soft Matter Phys.*, 2003, **67**, 041207.
36 D. Matuszak, G. L. Aranovich and M. D. Donohue, *J. Chem. Phys.*, 2004, **121**, 426–435.
37 U. M. B. Marconi and P. Tarazona, *J. Chem. Phys.*, 1999, **110**, 8032–8044.
38 A. J. Archer and R. Evans, *J. Chem. Phys.*, 2004, **121**, 4246–4254.
39 L. Sarkisov and P. A. Monson, *Langmuir*, 2001, **17**, 7600–7604.
40 J. Edison and P. A. Monson, *In preparation*, 2009.
41 L. Sarkisov and P. A. Monson, *Langmuir*, 2000, **16**, 9857–9860.
42 H. Bock, D. Diestler and M. Schoen, *Phys. Rev. E: Stat., Nonlinear, Soft Matter Phys.*, 2001, **64**, 046124.
43 R. Valiullin, S. Naumov, P. Galvosas, J. Karger, H. J. Woo, F. Porcheron and P. A. Monson, *Nature*, 2006, **443**, 965–968.
44 A. J. Archer and M. Rauscher, *J. Phys. A: Math. Gen.*, 2004, **37**, 9325–9333.

# Molecular dynamics simulations of urea–water binary droplets on flat and pillared hydrophobic surfaces

Takahiro Koishi,[*a] Kenji Yasuoka,[b] Xiao Cheng Zeng[c] and Shigenori Fujikawa[d]

*Received 22nd December 2009, Accepted 10th February 2010*
DOI: 10.1039/b926919c

We performed molecular dynamics (MD) simulations to investigate equilibrium behavior of urea–water binary droplets on flat (graphitic) and pillared surfaces. The contact angles as a function of urea concentration on the flat surface are computed. It is found that the contact angle decreases as the urea concentration increases. At the equilibrium state, the urea molecules in the droplet tend to be located near the hydrophobic graphite surface. This behavior is consistent with the denaturing effects of urea in protein solutions. We also performed MD simulations of collision between a urea–water droplet and the pillared surface to examine the tendency for the droplet being in the Cassie state (droplet staying on top of the pillared surface) or in the Wenzel state (droplet staying at the bottom of the pillared surface), at various urea concentrations.

## 1 Introduction

Urea is one of the most commonly used denaturing agents for promoting protein denaturation. Despite the denaturing effects having been studied for almost sixty years,[1] the detailed mechanism by which urea unfolds proteins is still not fully understood. To achieve a better understanding of the denaturing effects of urea on proteins, a careful study of the behavior of a urea–water droplet in contact with a hydrophobic surface can be helpful. We are aware that many researchers have used tools of molecular dynamics (MD) simulation to study molecular behavior of urea in water.[2–11] These studies indicate that one needs to carefully construct a model of urea for MD simulation because many properties of aqueous urea solutions can be strongly dependent on the model selected. Several force fields of urea for MD simulations have been reported in the literature.[2–4,8] The KBFF model was developed by Weerasinghe and Smith[4] who parametrized the charge distribution on urea to reproduce the experimental Kirkwood–Buff integrals. This model can well reproduce solution thermodynamics, *e.g.* the activity of urea as well as other properties such as density, diffusion constant and compressibility. In addition, MD simulations with the KBFF model have been employed to examine the concentration-dependent solvation[9] and hydrogen-bonding properties.[10]

Investigation of the equilibrium state of a liquid droplet on a solid surface offers a way to assess hydrophobicity of the surface. In a recent computational study, we performed MD simulations of water droplets on pillared hydrophobic surfaces.[12]

[a]*Department of Applied Physics, University of Fukui, 3-9-1 Bunkyo, Fukui, 910-8507, Japan*
[b]*Department of Mechanical Engineering, Keio University, Yokohama, 223-8522, Japan*
[c]*Department of Chemistry, University of Nebraska, Lincoln, Nebraska, 68588, USA*
[d]*Nanostructure Research Laboratory, RIKEN, Wako, Saitama, 351-0198, Japan*

We found that when a water nanodroplet is on the pillared surface, it can either fill the gaps between pillars (the Wenzel state[13]) or be in contact with peaks of the pillared surface (the Cassie state[14]). Furthermore, we found that the Wenzel and Cassie states can be coexistent on the same pillared surface at certain pillar heights because of a modest free-energy barrier separating the two states.

In this paper, we extend our previous simulation study to a binary urea–water droplet system on graphitic surfaces. The concentration dependence of the contact angle of the urea–water droplet on the flat graphite surfaces was estimated in order to assess the trend of affinity between the droplet and the surfaces. When the concentration of the urea increases, the contact angle becomes smaller because the urea molecules increase the affinity of the urea–water droplet to the flat graphite surfaces. We also computed the free-energy barrier separating the Wenzel state and Cassie state for the urea–water droplet on a rugged surface.

## 2 Simulation method

The simulation system consists of a solid surface, either flat or rugged, and a nanodroplet of urea–water mixture. The flat surface is the (0001) graphite surface with hexagonally arranged carbon atoms. Carbon atoms of the graphite were fixed during the entire simulation runs. The rugged surface is an artificial pillared surface. Quadrangular pillars with a lateral size of 12.3 Å × 12.8 Å were arranged with spatial intervals of 12.3 Å and 12.8 Å with $x$ and $y$ direction, respectively. This square-lattice-like pillar arrangement has been used previously by Lundgren et al.[15,16] The height of the nanopillars is 13.4 Å which corresponds to the height of 4 graphene layers.

The MD simulations were carried out at constant-volume and constant-temperature (298 K) conditions. The temperature was controlled by using the velocity scaling method. The periodic boundary condition was applied in all three spatial dimensions. A rigid-body model of water, the SPC/E[17] model, was employed. The potential function of the SPC/E model includes two terms, a Coulomb term and a Lennard-Jones (LJ) term. We also used the rigid-body model for urea. The potential functions of each atom of the urea molecule also consists of the Coulomb term and the LJ term. The potential parameters for the urea molecule were defined according to the Kirkwood–Buff theory.[4] The long range charge–charge interaction between water and urea molecules was calculated by using the Ewald method. The carbon atoms of the graphite were simply assumed to be LJ particles whose size and energy parameters are 3.4 Å and 0.2325 kJ mol$^{-1}$, respectively.[18] Table 1 lists all the LJ parameters and the charges. The time integration for the translational and rotational motion was undertaken using the velocity Verlet method and time-reversible algorithm.[19] The MD time step was set at 2.0 fs.

To derive the initial coordination of the urea–water droplet, we performed constant-temperature and constant-pressure ($NPT$) simulations of aqueous urea

**Table 1** LJ parameters and charges

|  | $\sigma$/Å | $\varepsilon$/kJ mol$^{-1}$ | $q(|e|)$ |
|---|---|---|---|
| H (urea) | 0.158 | 0.0880 | 0.285 |
| N (urea) | 3.110 | 0.5000 | −0.693 |
| C (urea) | 3.770 | 0.4170 | 0.921 |
| O (urea) | 3.100 | 0.5600 | −0.675 |
| C (graphite) | 3.400 | 0.2328 | 0.000 |
| O (water) | 3.166 | 0.6500 | −0.8476 |
| H (water) | 0.000 | 0.0000 | 0.4238 |

**Table 2** Number of urea and water molecules for each concentration. $N_u$ is the number of urea, $N_w$ is the number of water and $N = N_u + N_w$ is the total number of molecules. The values of molarity are derived from interpolation and extrapolation of the molarity in the literature.[4]

| Mol (%) | $N_u$ | $N_w$ | $N$ | Molarity/M |
| --- | --- | --- | --- | --- |
| 1 | 58 | 5774 | 5832 | 0.54 |
| 5 | 292 | 5540 | 5832 | 2.56 |
| 10 | 583 | 5249 | 5832 | 4.82 |
| 15 | 875 | 4957 | 5832 | 6.81 |
| 20 | 1166 | 4666 | 5832 | 8.52 |

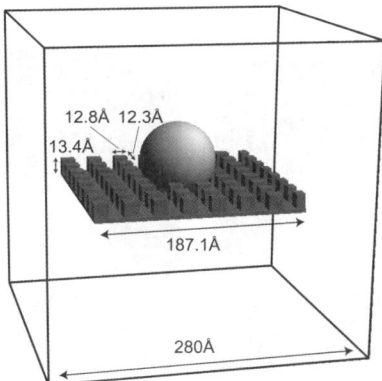

**Fig. 1** A schematic plot of the simulation system.

solution for 100 ps. The total number of urea and water molecules were 5832 (= 18 × 18 × 18). The number of urea molecules were determined by the concentration of urea (Table 2). After the *NPT* simulation, the aqueous urea cube was put on the graphite surface as the initial configuration of the urea–water droplet. Note that the length of the graphite surface (187.1 Å) is almost three times the length of the initial urea–water cube. The length of the simulation cell (280 Å) is about five times the length of the initial urea–water cube. Fig. 1 shows a schematic plot of the simulation system.

Because this simulation entails a large system size, we used a special-purpose computer "MDGRAPE-3"[20–22] to perform the MD simulations. The peak performance of a MDGRAPE-3 board at 250 MHz is 2.16 TFLOPS. We used two special-purpose computers for the MD simulations: one is for the real part of the Ewald calculation and the other is for the reciprocal-space part of the Ewald calculation. A 1.0 ns MD simulation of the urea–water droplet on the pillared surface system took nearly half a day using the MDGRAPE-3.

## 3 Results and discussion

We first estimated the contact angle of the urea–water droplet on the flat surface, and compared with that of a pure water droplet (calculated from our previous study).[12] To obtain the contact angle, we determined the surface locus of a urea–water droplet and their radial projection, fitted by a circle. The contact angle, $\theta_c$, is defined as the angle between a tangential line of the droplet surface at any three-phase contact point and the line in the graphite surface. The detailed method is given elsewhere.[12] Fig. 2 shows snapshots of the urea–water droplet at $t = 5.0$ ns

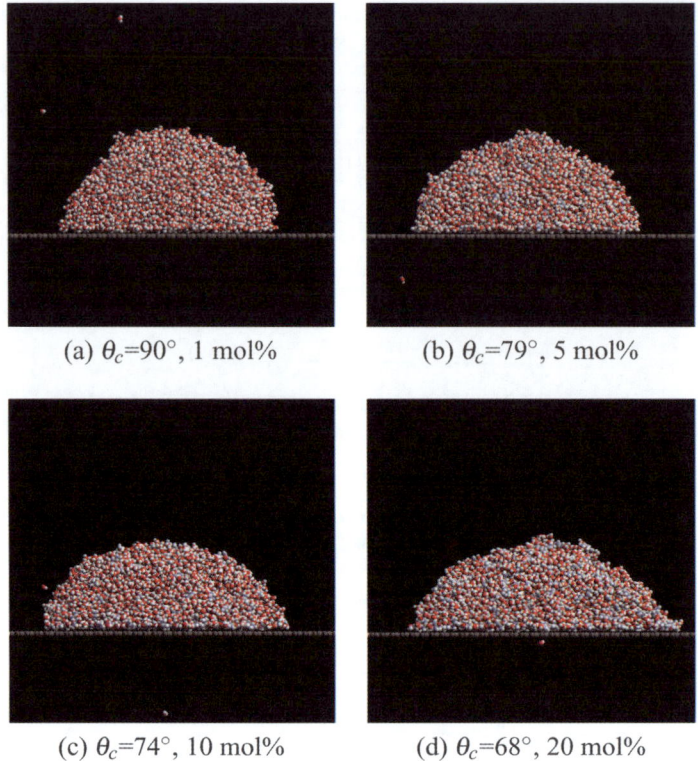

**Fig. 2** Snapshots of the urea–water droplet at $t = 5.0$ ns with the averaged contact angle, $\theta_c$. The concentration of the urea is (a) 1, (b) 5, (c) 10 and (d) 20 mol%, respectively.

with 1, 5, 10 and 20 mol% of urea, respectively. Table 2 lists the number of urea and water molecules. In general, the contact angle of the urea–water droplet is less than that of the pure water droplet (92°). We also calculated the contact angle of the urea–water droplet, with urea concentration ranging from 1 mol% to 20 mol% at 1 mol% interval. Fig. 3 shows that the contact angle of the urea–water droplet gradually decreases with increasing concentration. Hence, the affinity between the urea–water droplet and the graphite surface also increases with increasing concentration.

To describe the local distribution of urea and water molecules in the droplet near the surface, we defined local concentration, $\rho_l(z)$, as

$$\rho_l(z) = \frac{n(z)}{N_z dz}, \qquad (1)$$

where $z$ is the distance from the flat surface, $n(z)$ is the number of urea (or water) molecules from $z$ to $z + dz$ ($dz = 0.2$ Å), $N_z$ is the total number of molecules from $z$ to $z + dz$. Fig. 4 shows the local concentration of urea when the concentration of urea is 1, 5, 10 and 20 mol%, respectively. With the increase of the concentration, the first peak $\rho_l(z)$ of urea increases. These results indicate that the urea molecules are preferably near the graphite surface, compared to the water molecules. Because the graphite surface is weakly hydrophobic, this affinity of urea towards the solid surface is understandable. In a similar fashion, urea molecules tend to be attracted to the hydrophobic core of a protein and denature it.[23,24] In addition, urea molecules can break the long-range order of bulk water.[25] These characteristics of the urea molecules are manifested in the change of the contact angle as well as the local distribution near the surface. Since the urea molecules favor to be near the graphite

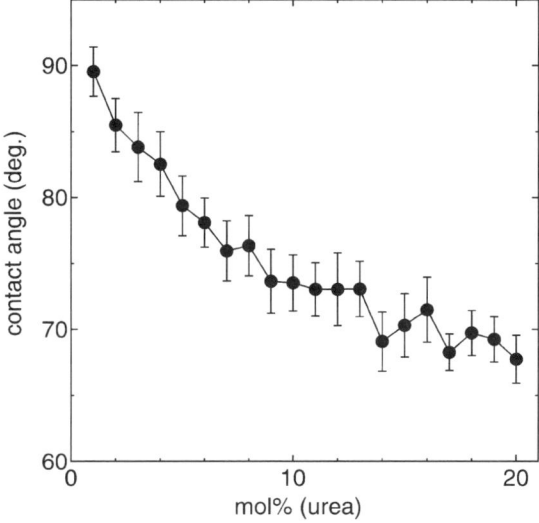

**Fig. 3** The concentration dependence of the contact angle.

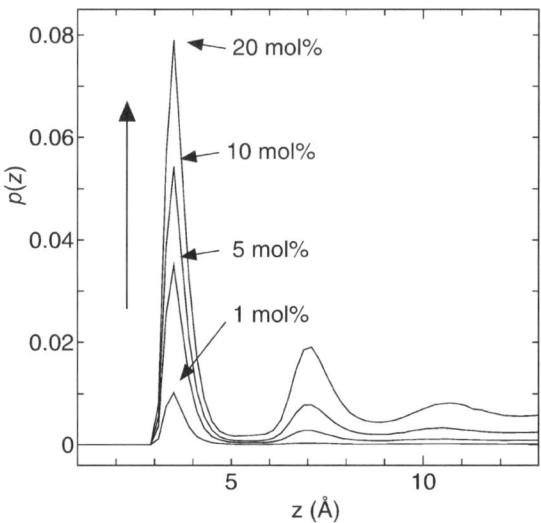

**Fig. 4** Local concentration of $p(z)$ of urea molecules.

surface, they reduce the intermolecular interactions among water molecules near the surface, hence the contact angle of the urea–water droplet.

We also calculated the probability distribution function $p(r_c)$ in order to examine the adsorption of urea at the liquid–vapor surface. This function represents the probability of water (or urea) within $r_c$ and $r_c + dr_c$, where $r_c$ is the distance from the center of the droplet (i.e., the center of the fitting sphere for estimating the contact angle). Fig. 5 illustrates the definition of $p(r_c)$ and Fig. 6 shows the calculated $p(r_c)$. The probabilities are normalized by the probability of water near the surface and $r_c$ is normalized by the radius of the fitting sphere $R$. Since the probabilities of urea are slightly smaller than those of water at the liquid–vapor surface, it is expected that there is no adsorption of urea at the surface. Note that urea molecules preferably distribute inside the droplet. These results are consistent with those of $p(z)$.

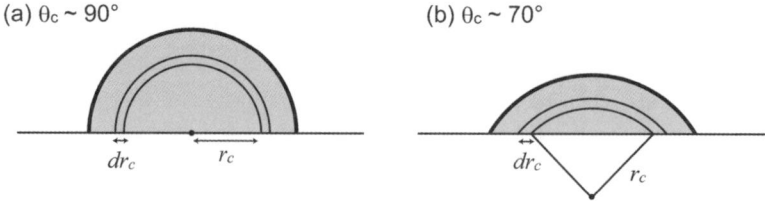

**Fig. 5** The definition of the probability distribution function from the center of the droplet, $p(r_c)$. The contact angle $\theta_c$ is about (a) 90° and (b) 70°.

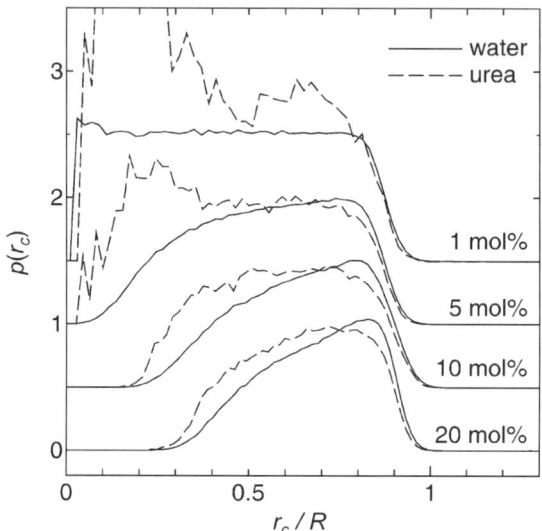

**Fig. 6** Calculated probability distribution function $p(r_c)$ from the center of the droplet with the concentration of urea being 1, 5, 10 and 20 mol%, respectively.

We also performed the "raining" simulations to estimate the free-energy barrier separating the Wenzel and Cassie state for a droplet on a rugged surface, using the same approach as our previous study.[12] Initially, we performed several test simulations of a urea–water droplet locating on the pillared surface to examine the relative stability of the Wenzel versus the Cassie state. Two initial locations for the urea–water cube were considered, one at the top of the pillars and another at the bottom of the grooves. It is found that at a certain height of the pillars, the droplet can reach two equilibrium positions (bistability), depending on the initial position of the urea–water cube. We examined the height range from 2 to 5 graphite-layers and found that the height of 4 graphite-layers is the minimum to exhibit bistability for the urea–water droplet. Hereafter, all "raining" simulations were carried out using the 4 graphite-layer pillared surface.

A urea–water cube was initially located 60 Å away from the top of the pillars and was equilibrated there for 100 ps at 298 K. At time $t = 100$ ps, a downward velocity $v_d$ was imposed instantly to all urea and water molecules. The kinetic temperature (accounting for the downward velocity) was controlled at 298 K during the fall of the urea–water droplet (Fig. 7(a)). After the droplet collided with the pillared surface, it eventually settled down either on top of the pillars (in the Cassie state, Fig. 7(b)) or at the bottom of the grooves (in the Wenzel state, Fig. 7(c)). The total time of the "raining" simulations was 500 ps. The downward velocity $v_d$ was

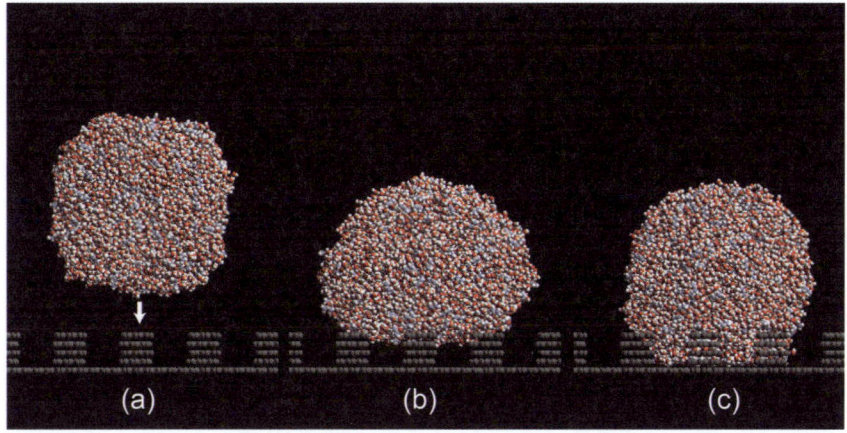

**Fig. 7** A snapshot of a urea–water droplet at 10 mol% as it is (a) falling down to the pillars, or (b) at the Cassie state and (c) the Wenzel state after the droplet colliding with the pillared surface.

carefully selected to attain a sufficient number of statistical events for achieving both the Cassie and Wenzel states. If $v_d$ is too high, the water droplet can easily go over the free-energy barrier and reaches the Wenzel state. If $v_d$ is too low, the water droplet favors the Cassie state. We performed 98 independent MD simulations with a given $v_d$, and recorded the number of events for the urea–water droplet in the Wenzel and Cassie states, respectively. The probability for the urea–water droplet in the Wenzel state $P_w$ was estimated by fitting the ratio of the number of events to an exponential equation

$$P_w = P_0 \exp\left(-\frac{\Delta G_{cw}}{e_k}\right) \qquad (2)$$

where $P_0$ is a preexponential factor, $\Delta G_{cw}$ is defined as the free-energy barrier separating the Cassie and Wenzel state, and $e_k$ is the kinetic energy of the center of mass of the droplet (per molecule), given by

$$e_k = \frac{1}{2} M v_d^2 \qquad (3)$$

and

$$M = \frac{m_w N_w + m_u N_u}{N_w + N_u} \qquad (4)$$

where $m_w$ is the mass of a water molecule, $N_w$ is the number of the water molecules, $m_u$ is the mass of a urea molecule and $N_u$ is the number of the urea molecules.

Fig. 8 shows the fitting probabilities, $P_w$, using inverse- and log-plot for the $x$-axis and $y$-axis respectively. Table 3 lists the calculated $\Delta G_{cw}$'s, namely, the free-energy barrier separating the Cassie and Wenzel state. It can be seen that $\Delta G_{cw}$ increases with increasing urea concentration. These results indicate that the droplet with more urea molecules will favor the Cassie state. The transition from the metastable Cassie state to the stable Wenzel state becomes more difficult for urea-rich droplets due to an increasingly higher free-energy barrier. On the other hand, the results of the contact angle show that the affinity between the urea–water droplet and the flat surface is enhanced with increasing urea concentration. Hence, the contact-angle result and the free-energy barrier result seemingly contradict to each other. However, the results of distribution of urea molecules give us another interpretation.

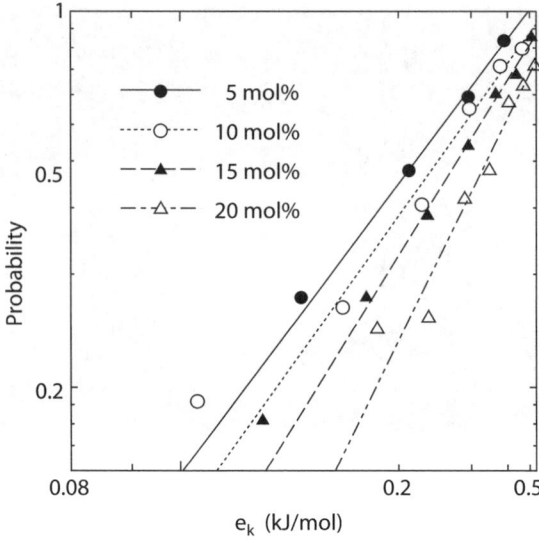

**Fig. 8** The probabilities $P_w$ (at the Wenzel state) and the fitting results (figure was revised).

**Table 3** Concentration dependence of the free-energy barrier[a]

| Concentration (mol%) | $\Delta G_{cw}$/kJ mol$^{-1}$ |
|---|---|
| 0[b] | 0.334 |
| 5 | 0.355 |
| 10 | 0.373 |
| 15 | 0.430 |
| 20 | 0.544 |

[a] Table 3 was revised following a question from Mr Stirnemann, see DOI:10.1039/c005416j
[b] Pure water droplet reported in our previous study.[12]

Since the urea molecules are preferably near the graphite surface, they will favor the top surface of the pillars as well. As the urea concentration increases, the stronger affinity of the urea molecules to the top surface prevents the droplet going down to the bottom of the pillared surface. The urea molecules, of course, also favor the bottom surface of the grooves. The placement priority of the urea molecules at the top surface of the pillars is ahead of that of the bottom surface of the grooves, since the urea–water droplet reaches the top surface first. Once the urea–water droplet reaches the bottom surface, however, we expect that it would be even harder for the droplet to re-approach the top of the pillars with higher urea concentration in the droplet.

Finally, we performed additional raining simulations for which the relaxation time for the droplet (with 15 mol% urea) was extended from 100 ps to 200 ps before the collision. The computed free energy barrier was slightly reduced to 0.353 kJ/mol from 0.430 kJ/mol. A reason for this reduction of the free energy barrier is that the shape of raining droplets become more spherical compared with that shown in Figure 7(a). Nevertheless, the trend for the urea-concentration dependence of the free-energy barriers is expected to be the same as that shown in Table 3.

## Acknowledgements

This work was supported by the Japan Science and Technology Corporation, the Ministry of Education, Culture, Sport and Technology in Japan, and the Grant-in-Aid (KAKENHI) for Young Scientists (B). XCZ is supported by US National Science Foundation and the Nebraska Research Initiative.

## References

1  J. A. Schellman, *Biophys. Chem.*, 2002, **96**, 91–101.
2  R. A. Kuharski and P. J. Rossky, *J. Am. Chem. Soc.*, 1984, **106**, 5786–5793.
3  B. Kallies, *Phys. Chem. Chem. Phys.*, 2002, **4**, 86–95.
4  S. Weerasinghe and P. E. Smith, *J. Phys. Chem. B*, 2003, **107**, 3891–3898.
5  L. J. Smith, H. J. C. Berendsen and W. F. van Gunsteren, *J. Phys. Chem. B*, 2004, **108**, 1065–1071.
6  D. K. Klimov, J. E. Straub and D. Thirumalai, *Proc. Natl. Acad. Sci. U. S. A.*, 2004, **101**, 14760–14765.
7  A. Caballero-Herrera, K. Nordstrand, K. D. Berndt and L. Nilsson, *Biophys. J.*, 2005, **89**, 842–857.
8  A. Caballero-Herrera and L. Nilsson, *THEOCHEM*, 2006, **758**, 139–148.
9  R. D. Mountain and D. Thirumalai, *J. Phys. Chem. B*, 2004, **108**, 6826–6831.
10 H. Kokubo and B. M. Pettitt, *J. Phys. Chem. B*, 2007, **111**, 5233–5242.
11 R. Zangi, R. Zhou and B. Berne, *J. Am. Chem. Soc.*, 2009, **131**, 1535–1541.
12 T. Koishi, K. Yasuoka, S. Fujikawa, T. Ebisuzaki and X. C. Zeng, *Proc. Natl. Acad. Sci. USA*, 2009, **106**, 8435–8440.
13 R. N. Wenzel, *Ind. Eng. Chem.*, 1936, **28**, 988–994.
14 A. B. D. Cassie and S. Baxter, *Trans. Faraday Soc.*, 1944, **40**, 546–551.
15 M. Lundgren, N. L. Allan, T. Cosgrove and N. George, *Langmuir*, 2002, **18**, 10462–10466.
16 M. Lundgren, N. L. Allan, T. Cosgrove and N. George, *Langmuir*, 2003, **19**, 7127–7129.
17 H. J. C. Beredensen, J. R. Grigera and T. P. Straatsma, *J. Phys. Chem.*, 1987, **91**, 6269–6271.
18 V. R. Bhethanabotla and W. A. Steele, *Langmuir*, 1987, **3**, 581–587.
19 N. Matubayasi and M. Nakahara, *J. Chem. Phys.*, 1999, **110**, 3291–3301.
20 M. Taiji, T. Narumi, Y. Ohno, N. Futatsugi, A. Suenaga, N. Takada and A. Konagaya, *Proceedings of the SC2003 (High Performance Networking and Computing)*, Phoenix, USA, 2003.
21 M. Taiji, *HOT CHIPS 16*, Stanford, USA, 2004.
22 T. Narumi, Y. Ohno, N. Okimoto, T. Koishi, A. Suenaga, N. Futatsugi, R. Yanai, R. Himeno, S. Fujikawa, M. Ikei and M. Taiji, *Proceedings of the SC06 (High Performance Computing, Networking, Storage and Analysis)*, Tampa, USA, 2006.
23 D. B. Wetlaufer, S. K. Malik, L. Stoller and R. L. Coffin, *J. Am. Chem. Soc.*, 1964, **86**, 508–514.
24 J. Jan Hermans, *J. Am. Chem. Soc.*, 1966, **88**, 2418–2422.
25 E. G. Finer, F. Franks and M. J. Tait, *J. Am. Chem. Soc.*, 1972, **94**, 4424–4429.

# General Discussion

**Professor Bratko** opened the discussion of the paper by Dr Thomy: In your electrowetting experiments on pillared surfaces, for given voltage ranges, the reversibility of contact angle modulation is shown to depend on surface characteristics. Does the degree of reversibility also vary with the voltage and concomitant amplitude of contact angle?

**Dr Thomy** responded: All our electrowetting experiments on microstructured surfaces lead to the same droplet behavior: for all applied tensions, from 25 to 250 V, no reversibility is observed. The contact angle decreases irreversibly when the applied tension increased as shown in Figure 4(a) of our paper.[1] The introduction of a nanotexture on top of the microstructured surfaces can induce a degree of reversibility in the variation of the contact angle. For a given voltage this degree of reversibility seems linked to the surface fraction of the surfaces: the smaller the surface fraction, the higher the degree of reversibility.

1 F. Lapierre, P. Brunet, Y. Coffinier, V. Thomy, R. Blossey and R. Boukherroub, *Faraday Discuss.*, 2010, **146**, DOI: 10.1039/b925544c

**Professor Butt** opened the discussion of the paper by Professor Craig: Vince, let me ask a question about the increasing contact angle at high temperature on the hydrophobic surface. When you increase temperature, the solubility of gas decreases. Thus, gas may nucleate at the solid–liquid interface, form tiny bubbles and replace the solid–liquid by a gas–liquid interface. This might be one explanation of your superhydrophobic effect. Could you get a similar effect when first exposing the liquid drop to a high gas pressure, say 10 bar. When you release the pressure the dissolved gas would be set free. May be you can create superhydrophobicity in this way.

**Professor Craig** answered: That would be a novel way to produce superhydrophobicity — pressure dependent superhydrophobicity. My experience with nucleation of gas is that it a very random process and somewhat more difficult to achieve than one might think. So the question is would the gas equilibrate by diffusion out of the droplet or by nucleation. I feel that if it was by nucleation the first bubble nucleated would tend to swallow up the supersaturated gas and in doing so would (a) grow and (b) prevent further nucleation. Our observations didn't reveal any bubbles below the Leidenfrost regime, so based on the little evidence we have I feel that our droplet is probably supersaturated with respect to gas. To look at contact angle response associated with outgassing due to a pressure drop would be an interesting experiment. Perhaps the same phenomenon could be investigated by chemically supersaturating a drop with $CO_2$. This might be experimentally more accessible.

**Professor Mugele** asked Dr Thomy: Recent experiments in our laboratory indicate that local field enhancement close to the contact line plays an important role in the electrowetting driven transition from the Cassie to the Wenzel state for micrometre-sized post arrays. Do your experiments display any indication that the local behavior in the vicinity of the contact line is different from the rest of the drop-substrate interface?

**Dr Thomy** replied: The nanotexture of our silicon based surfaces does not allow us to make any optical characterization of the droplet impalement under electrowetting. Nonetheless, one of our recent papers [ref. 22[1] of our Faraday paper] details two different observations that can indicate the local Wenzel state at the contact line:

—During the droplet suction for the receding contact angle measurement, the contact angle decreased considerably until the droplet took off the substrate indicating a partial impalement at the triple line contact of the droplet.

—After electrowetting tests, SEM images have been taken on the triple air–water–C4F8 coated nanowire interface. Nanowires localized at the periphery of the droplet are bent compared to the ones present at the center. This local deformation of the nanostructure can be ascribed to the high applied electrical field at the contact line. Even if we do not have any direct observations of the droplet impalement, these macro and nanoscales characterizations are consistent with the observations made by the Mugele *et al.* [ref. 26[2] of our paper].

F. Lapierre, V. Thomy, Y. Coffinier and R. Boukerroub, *Langmuir*, 2009, **25**, 6551.
A. Staicu, G. Manukyan and F. Mugele, 2008, arXiv:0801.2683v1

**Professor Evans** asked: Professor Craig refers to the important paper of D. E. Sullivan (1981). Sullivan makes interesting general predictions for the variation of the equilibrium contact angle as a function of temperature for both solvophilic and solvophobic (homogeneous) surfaces based upon a very simple density functional theory for a model with Yukawa type fluid–fluid and substrate–fluid potentials. Professor Craig finds, in his experiments where droplets are deposited on the surface before it is heated, some agreement with the predictions of the Sullivan theory. To what extent does he believe that the results he finds are generic? Would other systems exhibit similar temperature variation of contact angle $\theta$ on a solvophobic substrate? How does the contact angle of liquid mercury change with temperature for droplets on clear glass or stainless steel?

**Professor Craig** replied: We are using water in our experiments and the theory of Sullivan is specifically for van der Waals fluids. Further, Sullivan himself made it clear that the theory would need to be extended to account for more complex fluids. So it would be unfair to say that our experiments are verification of his theory. Are our results generic? Well we have to do more work to answer that with confidence. We would have to find a suitable system to study that would give the requisite contact angle at low temperatures and this is not a simple matter. Using mercury is a nice idea but experimentally it would be challenging as the boiling point of mercury is 357 °C—and then there are the inevitable OH and S issues of dealing with mercury (which will be exacerbated at high temperatures). We are looking at using different substrates at the moment, we will add the mercury experiment to the list of planned experiments.

**Professor McCarthy** said: What is the temperature of the drop? Does the heat of vaporization cool it?

**Professor Craig** answered: Yes the drop is cooled by the evaporation process. This is an important point. We didn't measure the temperature of the drop but it has previously been established by the group of David Quéré that the drop is at 100 °C (see reference 13[1] from our paper), even when the substrate is much hotter. So for the drop to remain at 100 °C it must be losing heat and it does this primarily through evaporation.

1 A. L. Biance, C. Clanet and D. Quéré, *Phys. Fluids*, 2003, **15**, 1632.

**Dr Perkin** opened the discussion of the paper by Professor Charlaix: Regarding the measurement of $V_s$ (surface potential) and $\xi$ (zeta-potential) on hydrophilic surfaces, have you observed a situation where $V_s$ and $\xi$ are not the same? If $V_s$ and $\xi$ are identical, does this imply that the ions in the Stern layer next to the hydrophilic surface are not mobile (since the zeta-potential and $V_s$ are then both measured outside the Stern layer)? Could there be a case where the Stern layer is mobile and therefore the $\xi$ is different to $V_s$?

**Professor Charlaix** answered: In principle from eq. (7) in the text $V_s$ is equal to $\xi$ if the surface potential is measured on the plane where a no-slip boundary condition holds for the flow. If the Stern layer is highly mobile with an apparent slip onto the solid surface, due for instance to a low friction coefficient of the hydrated ions on the solid, then one should get $\xi$ larger than $V_s$ (in absolute value). We have not yet found such a situation on an hydrophilic surface, but at the present our current monitoring set-up for measuring zeta potentials limits the type of samples we can investigate. We are presently working on a new set-up to extend the $\xi$ measurement to flat open surfaces.

**Professor Chen** asked: Can you explain the deviation from the linear theory in Figure 8 (bottom graph)? In the theoretical derivation, is the relationship between the surface charge density ($\sigma$) and the surface potential ($V_s$) in equation (8) valid only for small Debye length?

**Professor Charlaix** responded: The equation (8) is valid for all Debye lengths because $\kappa_{\text{eff}}$ is not the Debye length, but the length defined by the ratio of the electrical field at the surface divided by the surface potential. Only the unnumbered equation after equation (8) is restricted to small Debye lengths. We do not use this unnumbered equation in the plot of Figure 8; we use equation (9) which gives the value of $\kappa_{\text{eff}}$ as a function of the surface potential and the Debye length, which we both determine from the DLVO interaction force. We believe that the deviation from the linear theory observed in Figure 8 has two origins: first when the Debye length is small the range of the DLVO force decreases, therefore the uncertainty on the measurement of the surface potential increases and we do not know if there are systematic bias. Second, the two smallest Debye lengths investigated (corresponding to the two experimental points at the right of the figure where the linear relation fails at most) are less than 13 nm, that is very comparable to the peak-to-peak amplitude of the surface roughness. In this limit the theoretical derivation for a smooth surface is less and less adapted to our data (see also the questions of J. Klein and W. Ducker).

**Professor J. Klein** said: In this very interesting paper you deduce, in Figure 8b, from a linear fit to the data, that there is a slip length of 7.3 ± 0.8 nm associated with the flow of water past a silanized silica surface or a silanized Pyrex surface. But in Fig. 2 of your paper you show AFM micrographs of the same surfaces with a (peak-to-peak) roughness of order 7–10 nm in both cases. How do you clearly disentangle the slip length from the (exactly comparable) surface roughness?

**Professor Charlaix** responded: Certainly the roughness of our hydrophobic samples is a reason why in Fig. 8 the data do not show a nice linear behaviour. To disentangle the slip length from the surface roughness, we have to remember that only the samples are rough, the colloidal probe has a low roughness (1 nm peak-to-peak). On Figure 2 one sees that the highest peaks of the sample roughness are about 6 nm above the mid-plane (defined as the middle of the surface profile), and are typically separated by hundreds of nanometers. The prevalent roughness pattern consists of densely packed clusters about 50 nm in lateral size and extending about 3 to 4 nm above the mid-plane. Taking into account the radius of the colloidal probe (10 μm) the most frequent contact configuration is when the probe touches the top of these dense packed clusters, *i.e.* when it is located about 4 nm above the mid-plane. Therefore we can consider (see answer to question of W. Ducker) that we measure as the surface potential $V_s$ the average potential at a distance 4 nm above the mid-plane. From eq (7) in the text, we then deduce from the zeta potential measurements that the no-slip plane lies 7.3 nm underneath the contact plane, that is, about 3 to 4 nm UNDER the mid-plane surface. Now, on a rough non-slippery surface with a rather symmetric roughness (*i.e.* not made of very sparse high peaks) hydrodynamic calculations show that the effective no-slip plane lies between the mid-plane and the top of the roughness (the closest to the top the highest the slope of the roughness) but not under the mid-plane. For our type of roughness made of densely packed clusters, if these clusters did enforce a no-slip boundary condition on their outer surface (as imaged by AFM) one would expect an effective no-slip plane lying quite close to their summit, may be 2 to 3 nm above the mid-plane.

Therefore our data cannot be explained with a purely geometrical effect due to the surface roughness. However it is a real issue to realize smooth hydrophobic surfaces on which we can both measure directly the surface potential and perform trustable zeta potential measurements. We are currently working in this direction.

**Dr Vollmer** asked Professor Craig: The authors placed a droplet on a cold substrate and rapidly heated the substrate to the desired temperature. Can you estimate the time needed until the substrate reaches the desired temperature? The lifetime of a droplet placed on a hydrophilic substrate may last only a few seconds. Do the results for the lifetime of the droplet depend on the heating rate?

**Professor Craig** responded: From some rough calculations I estimate the time taken for the substrate to heat up is less than 0.01 s. The time is short because GaAs is a very good heat conductor and the wafer is only 380 microns thick. As this is far less than the droplet lifetime I don't think the heating rate is having any significant effect.

**Dr Wan** addressed Professor McCarthy: will the elasticity of the pillars effect the pitch off process of the water droplet?

**Professor McCarthy** answered: We have not made surfaces with pillars of different elasticity, so there are no facts that I can use to address the question. Using a simple contact line analysis of the sort made in Figures 2–4 of the paper, I would expect that deformation of pillars at the receding contact line would decrease receding contact angles (and increase hysteresis). If the pillars were elastic enough (and long enough), they would coalesce in bunches as the drop recedes and provide significant pinning sites.

**Mr Wu** asked Professor Craig: During the heating process towards the Leidenfrost temperature, would the local convection and heat gradient within the droplet affect the surface tension enough to explain the observed increase in contact angle?

Also, from a thermodynamics perspective, fluid surface tension is often affected by temperature. However, would it be possible to measure or infer changes in surface energy of the substrate through temperature? Specifically, if one was to deposit a SAM with other molecules of varying degrees of freedom, packing density *etc.*, would you expect to see changes in the observed behaviour?

**Professor Craig** replied: Just below the Leidenfrost regime the droplet will be at 100 °C as it evaporates. As Professor McCarthy has just pointed out to me the evaporation process will be cooling the surface—whilst the base is being heated so there is evidently heat flow throughout the droplet. I don't think this alone can explain what we observe. For example the same process will be occuring on the hydrophilic surfaces and very different behavior is seen there.

The surface energy of the solid surfaces will change with temperature. The difficulty in determining surface energies involving the solid is that the measurables in the Young equation yield the difference in the surface energy of the solid–vapor and solid–liquid interfaces rather than absolute values. Unless one can determine one the other cannot be determined. A Zisman plot approach is sometimes used but I don't think that is possible here.

I think you are saying that the surface energy of different surfaces will change with temperature to varying degrees and we may be able to observe this in these experiments. To an extent I agree, but as our droplet is remaining at 100 °C the interface with the solid must be very near 100 °C so the temperature of this interface is not changing in the region of interest. However the temperature of the solid–vapor interface is changing and differences may be revealed there.

**Professor Evans** asked: Professor McCarthy describes an elegant methods for constructing 'smooth' hydrophobic surfaces that exhibit rather small contact angle hysteresis. My question is a conceptual one. For such surfaces can one assume that the equilibrium contact angle lies between the measured advancing and the receding angle? If this is the case, and given that these angles are quite close, it would follow that one knows the equilibrium angle quite accurately for these systems—see Tables 1 and 2. Professor McCarthy will surely agree that this equilibrium contact angle is that determined by Thomas Young's equation. Would he also agree that for these systems at least, Young's notion of 'an appropriate angle of contact' for a liquid–solid pair is correct and is not a faulty assumption? I allude to the first few sentences of his paper. More generally, does he challenge the whole notion of an equilibrium (thermodynamic) contact angle which many of us believe to be inherent to the statistical mechanics of wetting?

**Professor McCarthy** answered: If one is concerned about an "equilibrium contact angle," that is one that cannot be measured and that is not important to real liquid droplets on real surfaces, then one could make the assumption that it must be between the measurable advancing and receding values. Young's notion of "an appropriate angle of contact" has been disproved for almost every liquid–solid pair ever measured and is thus a very poor assumption. I do not doubt that the statistical mechanics of wetting is important to some, but I deal with real liquids on real solids and make real measurements that I interpret. Perhaps my interpretations are wrong, but the data are correct within error.

**Dr Sedev** communicated: Real solid surfaces, in my experience, always show some contact angle hysteresis. In that sense, the Young equation and the pertaining thermodynamic formalism are an approximation. In fluid systems though, *e.g.* a liquid drop on a mercury surface or a foam film and its meniscus, this formalism applies strictly. Liquid drops on superhydrophobic surfaces can be treated in much the same way as they are often approaching the ideal limit of zero hysteresis. The microscopic events at the contact line are relevant, of course, but not considered explicitly. Therefore, it appears unnecessary to introduce "tensile" and "shear" hydrophobicity.

**Professor McCarthy** communicated in reply: A drop can be applied to the underneath side of a horizontal surface (for instance to a ceiling) and can remain in adhesive contact with sufficient tensile strength. If the surface is tilted slightly, however, and the drop slides there is little shear strength.

**Mr Boreyko** asked Professor Craig: You mention that there "is no gas or vapour phase between the liquid and solid surfaces", but isn't partial film boiling underneath the drop necessary to account for the dramatic increase in apparent contact angle?

**Professor Craig** responded: Just to clarify, above the Leidenfrost temperature we certainly do have a vapour phase between the liquid and solid surfaces—I take it that you are referring to the high contact angles observed just below the Leidenfrost temperature.

The contact angle is governed by what is happening at the three phase line so even if there were bubbles underneath the droplet (but away from the perimeter) these cannot explain the high contact angles observed. In any case we saw no evidence of such bubbles. One could postulate that we are nucleating bubbles at the three phase line and this results in the high contact angles observed. We can't rule out that possibility. However if these bubbles were all around the perimeter then the droplet would not be pinned as we observed and if they are occuring randomly around the perimeter one might expect the contact angle value to be "noisy" reflecting the production and elimination of bubbles at the boundary and the droplet to be partially mobile, but one can imagine situations where bubbles are forming at the perimeter that fit with our observations. More experiments are required to identify the mechanism.

**Ms Mishchenko** remarked: I was wondering how your observations of static droplets could translate to dynamic impact experiments with surfaces near the Leidenfrost temperature. I mistakenly thought that the following paper was by one of your co-authors.[1] It relates surface temperature to changes in droplet dynamics:

1 Xiying Li, Xuehu Ma, and Zhong Lan, *AIChE J.*, 2009, **55**(8), 1983.

**Professor Craig** replied: Droplet dynamics are influenced by surface roughness, contact angle and heat transfer. What our work shows is that the contact angle can be strongly dependent on temperature and it is well known that the surface roughness influences the macroscopic contact angle. The fact that the parameters are not independent will complicate any understanding of the dynamic behavior. For example near the Leidenfrost temperature we show that the contact angle increases on a hydrophobic surface—so changes in temperature near the Leidenfrost temperature will be influenced by both changes in heat flux and changes in contact angle and this should be borne in mind when trying to understand or model droplet dynamics.

**Professor Quéré** addressed Professor McCarthy: You show this nice footprint phenomenon—how does it depend, on surfaces decorated by microposts, on the velocity of the liquid front, and on the pinning strength? Could it be used, in your opinion, to characterize the degree of adhesion (or even contact) between textured solids and liquids, in a superhydrophobic state?

**Professor McCarthy** replied: We have not examined the "footprint phenomenon" (that droplets are left on posts due to capillary bridge failure) in any detail, other than to observe the droplets. We do not know whether the droplet size depends on the liquid velocity, but this is an interesting question. I believe that it should measure the "degree of adhesion (or even contact)." The droplet size will reflect the work of adhesion in terms of the surface tension of the liquid and the cosine of the receding contact angle.

**Mr Sherwood** communicated to Professor Craig: Re: Page 10, paragraph 1, line 9, you mention here that on the 12° slope the water droplet moved a short distance.

Did you attempt to quantify this distance? (Your response at the conference was that the drop stopped, as soon as the drop left the part of surface that had been contacted with the solder iron.) So as the drop left this area, was there a time lag *i.e.* did the surface of the drop require cooling before stopping or was the elimination of the vapour layer required? Were these observations on the stopping of the droplet the same for the flat surface?

**Professor Craig** responded: The aim of this experiment was to show that we could selectively move one droplet whilst leaving the other stationary—as a proof of concept for microfluidic control. This was done by locally applying heat under one droplet using the tip of a soldering iron on the back of the substrate. What is seen in the video I showed was that the drop moved a short distance and then stopped when it reached a portion of the surface that is not as hot. We made no attempt to ascertain the temperature profile of the surface.

We were not interested in the distance the drop moved or how long it took so we have not quantified it, but I estimate that it moved about 4 mm. But of course this distance will depend on how the surface is heated. The drop could be moved a shorter distance by applying the heat to a smaller area or a greater distance by applying it to a greater area. We know that if the drop is not in the Leidenfrost state then it becomes pinned and will not move. So we can surmise that as the droplet moves to a cooler part of the surface the heat flux is insufficient to maintain the vapour layer that suspends the drop and the droplet becomes pinned.

We have no information on the temperature of the droplet during this motion. We did not try the same experiment on a horizontal surface. In the experiments that we did on a horizontal surface the whole surface was heated so the droplet did not come to rest before completely evaporating.

**Dr Honig** asked: Why correlate slip length to contact angle measurements (a 3 phase measurement) when fluid flow over a solid surface is a 2 phase problem?

Would solid–liquid interfacial energy be a more appropriate parameter for the correlation?

**Professor Craig** replied: The degree of slip is thought to be related to the strength of the interaction between the liquid and the solid so the solid–liquid interfacial energy would indeed be the most appropriate parameter to use in looking for a correlation with the slip length. However this value is generally not known, hence the contact angle is used as a *defacto* measure of the solid–liquid interfacial energy.

**Professor Charlaix** responded: In the Barrat *et al.* paper[1] where the correlation between the slip length and the contact angle is suggested, they introduce this notion for a given liquid surface tension and a given solid surface tension, and investigate only the effect of the liquid–solid interaction—thus of the liquid–solid interfacial tension. In this case, correlating the slip length to the liquid–solid surface tension is exactly the same as to the contact angle. I guess it is somewhat abusive to extend the concept to experimental data where changing the solid surface changes both solid–vapor and liquid–solid surface tension and yes, I agree that the solid–liquid interfacial energy should be more appropriate. Up to now the correlation seems to work well with subsequent numerical simulation or Huang *et al.*[2] on various solids, as well with the experimental data available today. More work is certainly needed to draw any conclusions.

1 Jean-Louis Barrat and Lydéric Bocquet, *Faraday Discuss.*, 1999, **112**, 119–127.
2 David M. Huang, Christian Sendner, Dominik Horinek, Roland R. Netz, and Lydéric Bocquet, *Phys. Rev. Lett.*, 2008, **101**, 226101.

**Professor Ducker** asked: In your work, you extract the slip length from zeta potential measurements by using the surface potential that was obtained from a fit to a double-layer force obtained by AFM measurement. The fit of the double-layer force to the measured AFM data is sensitive to the definition of zero separation used in the AFM measurement. It appears that your definition of zero separation is where asperities touch each other. However, the charge on the surface, from which the double-layer force arises, is probably distributed all over the surface, and not just on the asperities. How do you account for the distribution of the charge, and how sensitive is your slip length to the effective position of the charge?

**Professor Charlaix** replied: That is right: by using the Poisson–Boltzmann theory to derive a surface potential from the DLVO force we implicitly assume that we measure the average potential $V_s$ on the plane where the contact occurs, that is at the top of the asperities located about 4 nm above the mid-plane of the hydrophobic sample (see answer to J. Klein question). We indeed fit the part of the DLVO force at distances larger than the Debye length, where the constant charge and constant potential solutions overlay; at this distance the corrugations of the potential induced by surface irregularities vanish due to its exponential decay, so that the double layer charges at this distance actually "see" a uniform charged surface located at the contact plane. The slip length derived in Figure 8 then means that the no-slip plane for the flow occurs 7.3 nm underneath the contact plane.

Now about the validity of this picture, I expect that for values of the Debye length larger than the peak-to-peak amplitude of the roughness, the uneven repartition of the charge is not screened over the (vertical) scale of the roughness: the long range character of the Coulomb interaction should then smear out this uneven repartition, and our implicit assumption that ALL the double layer behaves as if it were associated to a plane of uniform charge should not be too bad. But for values of the Debye length comparable to the roughness, the charges located in the holes of the roughness are screened and less visible in the DLVO force, and still should attract in their vicinity some free ions contributing to the electrokinetic transport. So our model of the flat surface should not work well in this case, and it is why I think the data obtained for these low values of the Debye length do not agree with the linear fit in Figure 8.

**Professor Evans** asked: In the discussion much was spoken of the long slip lengths found in various experiments using hydrophobic surfaces. Professor Charlaix's paper states that it is established that for a range of materials the slip length increases with contact angle and can be tens of nanometres on highly hydrophobic surfaces (see Introduction and references). I believe that no one doubts that if the contact angle is extremely high so that drying or near drying is occurring, the slip length could become very large indeed. However, in real experiments the contact angle is rarely larger than 120–140°, and in these circumstances it is difficult to see why there should be a correlation between slip length and contact angle. Indeed the same point was made earlier in the Discussion. Could it be, and this is a very speculative suggestion, that a better 'static' indicator of a surface's propensity to foster a large slip length would be the local compressibility of the fluid near the substrate? This quantity was mentioned in Professor Rossky's introductory lecture. I am referring to the work of Professor S. Garde and colleagues. We might hear more about this in paper 23.[1] If the local compressibility is high then in some sense the molecular layers near the substrate are 'soft' which might favour slip. For solvophilic surfaces one would not expect this.

1 Hari Acharya, Srivathsan Ranganathan, Sumanth N. Jamadagni and Shekhar Garde, *Faraday Discuss.*, 2010, **146**, DOI: 10.1039/b927019a.

**Professor Charlaix** answered: The correlation between the slip length and the contact angle mentioned here is limited to water and represents only the actually available results obtained from molecular dynamics simulation (Huang *et al.*[1]) and from experiments on a range of solids. This correlation should be understood as a first trial to connect the slip length to a macroscopic characteristic of the solid–liquid (SL) interface, but clearly the contact angle does not capture all the details of the interface which can play a role for the slip length. The paper of Barrat *et al.*[2] for instance mentions some of those parameters, such as the structure factor of the liquid in the direction parallel to the surface, its commensurability with the solid, the molecular size, *etc.*. The compressibility of the interface is certainly a property which depends also on these structural parameters and could be correlated with the slip length. I have however a concern—the same as with the contact angle correlation indeed—in the case of a composite SL interface embedding some (nano-) bubbles or gas pockets trapped onto the solid surface. Gaseous structures will have a high impact on the average interface compressibility and may have a lower impact on the slippage. For instance our work about slippage on a bubble mattress[3] has shown that gas trapped at the SL interface does not always favor slippage and may even favor high friction, whereas it always increases highly the surface compressibility.[4] This restriction is however the same with the contact angle: super-hydrophobic surfaces in the Cassie state have a very high c.a. but are not always more slippery than usual surfaces.

1 David M. Huang, Christian Sendner, Dominik Horinek, Roland R. Netz, and Lydéric Bocquet, *Phys. Rev. Lett.*, 2008, **101**, 226101.
2 Jean-Louis Barrat and Lydéric Bocquet, *Faraday Discuss.*, 1999, **112**, 119–127.
3 Audrey Steinberger, Cécile Cottin-Bizonne, Pascal Kleimann and Elisabeth Charlaix, *Nat. Mater.*, 2007, **6**, 665–668.
4 A. Steinberger, C. Cottin-Bizonne, P. Kleimann and E. Charlaix, *Phys. Rev. Lett.*, 2008, **100**, 134501.

**Professor Panchagnula** asked Professor Craig: It appears that there is evidence of vapor nucleation and nucleate boiling in the data that you present. Here are two pieces of evidence:
1. The drop video clearly shows oscillations indicative of vapor nucleation causing pressure fluctuations. At the same time, the triple line is pinned on the substrate causing the drop to only oscillate.
2. In both Figures 1 and 4, drop lifetime *versus* temperature plots, the slope of the curve is decreasing with increasing temperature indicating a source of thermal resistance that is increasing with temperature. One possible source of this thermal resistance is a vapor film whose resistance (read area fraction) is increasing with temperature.

**Professor Craig** replied: These could both be cited as evidence of vapor nucleation but I think there are also other explanations for these observations. Regarding (1); The disturbances in the stationary drop could result from both the momentum change when the three phase line retracts and Marangoni flows on the surface due to thermal gradients. Regarding (2); I agree the thermal resistance is increasing. Whilst this is occuring the three phase line is receding (the contact angle is increasing) and the droplet footprint on the surface is decreasing. This is another and significant source of increasing thermal resistance.
I think this issue requires further investigation. I'm thinking that we might be able to confirm or rule out vapor nucleation by listening closely to the droplets.

**Mr Peter** commented: What makes the difference between the onset of the Leidenfrost temperature for GaAs and Au surface on preheated substrates (Fig. 4) and cool substrates with subsequent heating (Fig. 1)? Therefore, I have the question of how accurate your temperature measurement is? Which temperature measurement technique was used? —Answer: IR camera. Is it possible that the IR camera measured the temperature behind the GaAs which is transparent for IR wavelength and that there was a temperature gradient from the hot plate to the surface? Did the size of the droplet influence the transition to the Leidenfrost regime because the drops on the cool substrates with subsequent heating show evaporation until surface reaches the Leidenfrost temperature regime. Would therefore be a faster temperature increase help to generate similar Leidenfrost regimes as for the pre-heated samples?

**Professor Craig** responded: In Figure 1 the droplet is placed onto the surface before heating whereas in Figure 4 the droplet is allowed to fall onto a pre-heated substrate. It is well known that the Leidenfrost temperature depends on the conditions under which the droplet is deposited—the latter case is often called the dyanmic Leidenfrost effect.

We used an IR thermometer to measure the temperature of the substrate. Perhaps you are correct in that we may have been measuring the temperature of the substrate as GaAs is transparent to IR. However, the substrate is thin (380 microns) and Gallium Arsenide is a very good thermal conductor (hence the interest of the semiconductor manufacturers) so I don't believe there is any significant thermal gradient across the wafer. Essentially the wafer should be equilibrated with the substrate in far less than a second and remain so.

We didn't investigate different drop sizes. The difference between the dynamic and static Leidenfrost effects is likely to depend on droplet size, but we have not investigated this.

**Dr Christenson** remarked: I think that nucleation of bubbles, *i.e.* boiling, in the water drop on the hydrophobic surface is extremely unlikely. Bubbles do not easily nucleate on a smooth surface, even a hydrophobic one.

Your work is very interesting indeed, but I think that a lot more work is needed in order to clarify what is going on.

**Professor Craig** answered: I agree that more work is required to understand the origin of the change in contact angle as the Leidenfrost temperature is approached. Currently I feel (guess!) that an explanation based on surface energy alone may be sufficient to explain the effect. As the temperature is increased the liquid–vapor interfacial energy and the liquid–solid interfacial energy will remain somewhat constant (as the droplet temperature does not rise above 100 °C). If the solid–vapor interfacial energy drops with increasing temperature this may explain the effect.

**Mr Pacheco Benito** addressed Dr Thomy: Do you think a minimum micro or nanostructured layer thickness is needed to obtain superhydrophobic surfaces or is the distance between the micro-pillars the parameter that determines superhydrophobicity or total bounce of a drop?

**Dr Thomy** responded: The total bounce of the drop depends on the height of the pillars, as already measured by Bartolo *et al.* and Reyssat *et al.* It was found that the pressure threshold for water impalement depends on the distance between pillars ($l$) and the height ($h$) as:

$P_c = \alpha\sigma h/l^2$ which is referred to in eq. (4) of our paper.

The pressure threshold also depends on the nanostructure: covering the pillar surface by about 3 microns-length nanowires leads to a better robustness against impalement, as shown in Table 2. However, this influence could not be quantitatively predicted because of the complex structure of nanowires. From our experiments, it seems that the length of nanowires has to be comparable to the spacing between pillars for the best efficiency.

Our previous experiments on nano-structured surfaces (no micron-scale pillars) evidenced that the best robustness was obtained with a dense sub-layer of entangled wires and a loose upper layer of straight nanowires.

**Mr Wu** answered: While we believe that there exists a set of criteria to induce superhydrophobicity, the intrinsic interdependency of nanoroughness, microroughness, aspect ratios and chemistry would render this set of criteria extremely flexible.

To clarify this further, I wish to cite one of many comprehensive studies, namely, a study conducted by L. Barbieri,[1] which indicated that different sets of geometric parameters were required to induce superhydrophobicity on photolithographically patterned micro-pillars depending on the pitch, shape and diameter of the pillars.

Due to this interdependency, it is difficult to assign a specific parameter or sets of parameters that determines superhydrophobicity and any associated properties.

1 L. Barbieri, E. Wagner and P. Hoffmann, *Langmuir*, 2007, **23**, 1723–1734.

**Ms Mishchenko** asked Professor McCarthy: My question is regarding your silicon/silica surfaces. As far as I understand, this is a single layer of linear, covalently-bound PDMS chains. What is the temperature stability of this substrate? Does it retain its contact angle hysteresis properties even if the surface is scratched?

**Professor McCarthy** replied: These surfaces contain monolayers of PDMS chains, that is all chains present are covalently attached. Physisorbed chains would have been removed by rinsing. These should have thermal stabilities that are similar to those of linear silicones. Scratching (at length scales much larger than the film thickness) would introduce defects that cause contact line pinning and lower receding contact angle values.

**Professor Patankar** communicated: The experiments by McCarthy and co-workers showing liquid droplets left behind by a receding contact line and the explanation based on liquid bridge break up were insightful. Another recent paper claims to have shown the presence of a "liquid film" left behind by a receding contact line.[1] It is claimed that the mechanism by which the residual liquid film emerges on the pillars' tops can be ascribed to the pinch-off of liquid threads. This was verified by comparing the characteristic times of movement of the receding contact line with the time estimated theoretically for pinch-off of liquid threads.

1 Xiying Li, Xuehu Ma and Zhong Lan, *Langmuir*, 2010, **26**, 4831–4838.

**Professor Giovambattista** asked Professor Craig: In Fig. 3 of your paper it is shown the contact angle of water on fluorinated surfaces as function of temperature. In particular, it is shown that the water contact angle increases as temperature decreases from ~100 °C down to ~25 °C. Did you do experiments or is there any experimental result showing water contact angle at lower temperatures.

**Professor Craig** answered: No we didn't look at temperatures below room temperature. We were interested in the influence of elevated temperatures in this study.

**Professor Evans** asked: Professor Craig speaks of a wetting transition. It is not precisely clear what he means by this term in the particular context of his present experiments, *i.e.* the results in Fig. 3? In the phase transition community, wetting transition has a very specific meaning. This term refers to the onset of either complete wetting $\theta = 0$ or complete drying $\theta = \pi$. Only in these circumstances is there singular behaviour in the surface thermodynamic functions.

**Professor Craig** replied: I have evidently used the term incorrectly. Our intended meaning of the term 'wetting transition' was to describe the change in contact angle from a typical value on a hydrophobic surface (*e.g.* 100–120°) to that akin to one that usually corresponds to a superhydrophobic (or Cassie) state (>150°). I gather we should only use the term wetting transition to describe the eventual change to a Leidenfrost condition which is complete drying.

**Professor Evans** opened the discussion of the paper by Professor Yeomans: Professor Yeomans introduces a binary fluid model that is described by a simple Landau free energy functional constructed with a single (concentration) order parameter $\Phi$. The fluid-substrate coupling is treated in Cahn fashion; the surface field $h$ couples only to concentration $\Phi$. However, in the hydrodynamics the total density $n$ appears, as it does in Eq. (2). One could carry through the whole analysis in terms of the local densities of each species, *i.e.* use a full two component treatment. Could you comment on what, if any, new features might arise in such a treatment? Is it easy to see why a one-component treatment is adequate?

**Professor Yeomans** replied: The fluid is, to a good approximation, incompressible so the total density is constant and $\Phi$ is therefore the most natural variable to use in the free energy.

We wrote down the hydrodynamic equations in terms of the densities of the two components.[1] The new physics is that the velocities of the two components can be different. If these velocities relax to the same value on a short time scale we regain the $\Phi$ model.

1 A. Malevanets and J. M. Yeomans, *Comput. Phys. Commun.*, 2000, **127**, 105.

**Dr Willmott** communicated: A microscopic slip length $v_0$ was identified in the simulations on the patterned surface, even though a non-slip boundary condition was used. Did this slip length apply to the fluid adjacent to the posts, or the gaps, or is it averaged over both? Figure 7b suggests that $v_0$ as calculated gives a slip length 5–10% of the drop size; can you suggest how that number would change when the length scales of the simulation (that is, if the ratios between drop size, post width and post spacing) are changed?

**Professor Yeomans** communicated in reply: The slip length is an average over posts and gaps.

Ybert *et al.*,[1] discuss how the slip length depends on the post dimensions for flow in a channel; essentially it is an average over no slip on the posts and the slip between them. I expect a similar behaviour for a drop, as we find that dissipation at the contact line does not play a large role on superhydrophobic surfaces.

1 C. Ybert, C. Barentin, C. Cottin-Bizonne, P. Joseph and L. Bocquet, *Phys. Fluids*, 2007, **19**, 123601.

**Mrs Seyed-Yazdi** opened the discussion of the paper by Professor Monson: Can the lattice model provide information (direct or indirect) about supercooled samples to see where the formation of ice will take place in hydrophobic and hydrophilic pores?

**Professor Monson** replied: The lattice model does not predict a first order transition between fluid and solid. As such it would not be appropriate in that context.

**Professor Debenedetti** commented: The two mechanisms that you described (formation of a gap-spanning vapor cylinder; symmetry breaking and formation of a bubble at one of the surfaces) are nucleation events. Have you ever performed quenches deep enough to see spinodal decomposition?

**Professor Monson** replied: Spinodal decomposition is associated with a quench from a single-phase state into a region of the phase diagram where the equilibrium state is two phases in coexistence. In the pore geometry (where the pore fluid is in contact with the bulk) and processes we have considered in this work, the final state is always a single-phase state so we would not expect to see a spinodal decomposition mechanism. We could, however, consider a different system where we have the pore fluid in the canonical ensemble without contact with the bulk. Then we could study the behavior in processes where the system was taken into the two-phase region from a single-phase state *via* a lowering of temperature at fixed overall density. Under those circumstances we could study spinodal decomposition under confinement. Such studies have been done in the past for binary liquid–liquid systems (see, for example, the review by Binder in ref. 1).

1 K. Binder, *J. Non-Equilibrium Thermodyn.*, 1998, **23**, 1–44.

**Dr Patel** opened the discussion of the paper by Professor Koishi: The simulations of raining urea droplets imply that the barrier to go from the Cassie to the Wenzel state increases on increasing the concentration of urea (Fig. 8). Since urea prefers the surface, one expects the affinity of the droplet for both the top and bottom of the pillars (and consequently the Cassie and Wenzel states) to increase equally on increasing the concentration of urea. However, the increasing barriers imply that the transition state is destabilized relative to the Cassie state. Can you comment on what the transition state for the Cassie to Wenzel transition looks like and why it is destabilized relative to the Cassie state on increasing the concentration of urea?

**Professor Koishi** answered: Indeed, the urea molecules are preferably near the graphene surface, and the affinity of the urea–water droplet for the surface increases with increasing the concentration of urea. The transition state for the droplet to cross the barrier is when a small liquid "finger" extended from the droplet touches the bottom of the pillar. Because the urea–water droplet arrives at the top surface of pillars first in raining simulation, the stronger affinity of the urea molecules to the top surface tends to reduce the chance for the liquid "finger" to reach to the bottom of the pillar. As such, it would be harder for the droplet to cross the barrier from the Cassie to Wenzel state.

**Mr Acharya** asked Professor Monson: Pores have some roughness. Can you include roughness/complex geometries in your modelling system? How does the observed hysteresis relate to the contact angle and would this change with the addition of roughness?

**Professor Monson** answered: The present approach can be applied to any system for which a lattice model can be developed. For example, we have previously studied lattice gas models of fluids in Vycor glass[1] and porous Silicon.[2] The present method would in principle be applicable to those systems. One difficulty with systems that exhibit a high degree of disorder in terms of pore size and rough pore surfaces is that the relaxation dynamics in the hysteresis will be very slow due to the roughness of the free energy landscape.[3] To model the dynamics in this case one must include a fluctuation mechanism that allows the system to cross barriers between local free energy minima. This is not included in the present version of DMFT but it could be done in an *ad hoc* manner, as discussed in a response to a question from Professor Evans elsewhere in this discussion, or *via* Kawasaki dynamics simulations.

For slit-like or cylindrical pores the hysteresis is primarily associated with metastability of the confined vapor (capillary condensation) or confined liquid (capillary evaporation). For capillary condensation this is well established from theory and simulation of model systems (see, for example ref. 4) and from analysis of experiments on gas adsorption in mesoporous silica.[5] For capillary evaporation we have evidence from recent work on mercury porosimetry that reaches the same conclusion.[6] For a discussion of the relationship between contact angles and hysteresis we would refer you to a recent paper covering this topic.[7]

1 H.-J. Woo and P. A. Monson, *Phys. Rev. E*, 2003, **67**, 041207.
2 S. Naumov, A. Khokhlov, R. Valiullin, J. Karger and P. A. Monson, *Phys. Rev. E*, 2008, **78**, 060601.
3 R. Valiullin, S. Naumov, P. Galvosas, J. Kärger, H-J. Woo, F. Porcheron and P. A. Monson, *Nature*, 2006, **443**, 965–968.
4 P. A. Monson, *J. Chem. Phys.*, 2008, **128**, 084701.
5 P. I. Ravikovitch, S. C. O'Domhnaill, A. V. Neimark, F. Schuth and K. K. Unger, *Langmuir*, 1995, **11**, 4765–4772.

6 F. Porcheron, M. Thommes, R. Ahmad, P. A. Monson, *Langmuir*, 2007, **23**, 3372–3380.
7 P. A. Monson, *Langmuir*, 2008, **24**, 12295–12302.

**Professor Law** addressed Professor Koishi: I don't understand how Figure 3 can be compatible with Figure 6. If one extrapolates the contact angle *versus* concentration plot of Figure 3 to 100% urea concentration, this figure implies that urea should have a lower surface tension than water. This would in turn imply that urea, not water, should preferentially adsorb at the liquid–vapor surface, which is opposite to that shown in Figure 6.

**Professor Craig** pointed out: that urea is a solid at room temperature and that urea has attracted interest for it's ability to enhance the solvation of proteins and is said to reduce the structure of water (structure breaker). Relevant here is the property that the surface tension of urea–water mixtures increase with increasing urea concentration as the urea component is depleted from the liquid–vapor surface. These observations explain this apparent mystery.

**Professor Koishi** answered: Since the solubility of urea is 27 mol% at 25 °C, we cannot extrapolate to 100 mol% for Figure 3.

In our interpretation, the surface free energy between solid (graphene) and liquid (urea–water) is important to explain Figure 3. The urea molecules are preferably adsorbed to the graphene surface than water, as shown in Figure 4. This local concentration enhancement of urea near the graphene seems to decrease the contact angle of the droplet. In addition, we found the original reference[1] that reported increasing of the liquid–vapour surface tensions of urea–water mixtures with increasing urea concentration, as pointed out by Professor Craig above.

1 M. Siskova, J. Hejtamankova and L. Bartovska, *Coll. Czech. Chem. Commun.*, 1985, **50**, 1629–1635.

**Professor Luzar** asked Professor Monson: In your studies on patterned surfaces, did you try to play with varying patterns, *i.e.* random *vs.* regular ones? What we have found in our own studies[1,2] of kinetics of capillary evaporation was that the effect of different surface patterning on the evaporation rate could be drastic. Evaporation was up to two orders of magnitude faster in the case of random patterned ones, showing the efficiency of the physics of "Lifshitz traps"[3] aiding the vapor bridge formation.

1 A. Luzar and K. Leung, *J. Chem. Phys.*, 2000, **113**, 5836.
2 K. Leung and A. Luzar, *J. Chem. Phys.*, 2000, **113**, 5845.
3 I. M. Lifshitz, *Sov. Phys. Usp.*, 1965, **7**, 549; R. Friedberg and J. M. Luttinger, *Phys. Rev. B*, 1975, **12**, 4460.

**Professor Monson** responded: Thus far we have only considered one type of pattern as an illustration of what might be possible with the method. A wider study of different kinds of patterns including the effect of disorder would clearly be interesting. For disordered patterns we may need to assess the potential influence of quenched disorder upon the dynamics as discussed in our response to the question from Mr Acharya.

**Mr Wang** asked Professor Koishi: I was wondering how the researchers actually introduced the additional velocity to their water drop during the "raining" experiments in computer simulation. And if they have seen the drops land in different positions on the surfaces for different trajectories?

**Professor Koishi** replied: We first relaxed a urea–water droplet which was located initially 60 Å above the pillared surface. The droplet temperature was kept at 298 K. We then added the constant downward velocity (in the $z$-direction) to every water molecule. To maintain the constant temperature condition, we rescaled the velocity of all molecules every 50 MD step. The $z$-component of temperature, $T_z$, is higher than 298 K due to the downward velocity, while the $x,y$-component of temperature, $T_x$ and $T_y$, are lower than 298 K; the averaged temperature, $T = (T_x + T_y + T_z)/3$, is kept at 298 K during the raining simulation.

The trajectories are dependent on the landing position as pointed out because the difference between the droplet size and the pillar size is not very large. In a previous work, we performed test simulations to estimate the free energy barrier for a pure water droplet. The initial position (in $x,y$ coordination) of the water droplet was $(x,y) = (0,10)$, $(0,-10)$, $(10,0)$, $(-10,0)$ and $(10,10)$, respectively. All droplets reached the bottom of the pillars after the impact except one condition of $(0,-10)$. So we expect the landing position has a minor effect on the estimation of the barrier in this case as well.

**Dr Christenson** asked Professor Monson: Both your paper and Professor Rossky's introductory lecture have mentioned experimental work on capillary evaporation. I would here like to give what I believe is the correct account of such work. Yushchenko et al.[1] gave an account of "cavity formation" between hydrophobic surfaces and calculated that the energy barrier would be prohibitive at any separation in excess of molecular dimensions. The same authors also described an experiment[2] where a glass test tube was pushed against the bottom of a glass beaker containing mercury. Mercury was seen to recede spontaneously from the contact point of the two glass vessels as dewetting occurred. In this case, the capillary evaporation could occur without an energy barrier, but I do not know if any one has further studied this phenomenon in mercury, which is not a popular material for experimental work these days. I am not aware of any work with the surface force apparatus where capillary evaporation at finite separations has been recorded. It was first noted by Pashley et al. in a 1985 Science paper,[3] and Per Claesson and I later made more extensive observations[4,5] described in another Science paper. Several co-workers and I have made further scattered observations over the years, but I do not recall ever seeing capillary evaporation with the surfaces out of contact. This fits with the early predictions of too great an energy barrier.

Unfortunately, the phenomenon has been confused in the minds of many by the coalescence of pre-existing "nanobubbles", and there are many publications describing this.

1 V. S. Yushchenko, V. V. Yaminsky and E. D. Shchukin, *J. Colloid Interface Sci.*, 1983, **96**, 307–314.
2 V. V. Yaminsky, V. S. Yushchenko, E. A. Amelina and E. D. Shchukin, *J. Colloid Interface Sci.*, 1983, **96**, 301–306.
3 R. M. Pashley, P. M. McGuiggan, B. W. Ninham and D. F. Evans, *Science*, 1985, **229**, 1088–1090.
4 H. K. Christenson, P. M. Claesson and R. M. Pashley, *Proc. Indian Acad. Sci. (Chem. Sci.)*, 1987, **98**, 379–389.
5 H. K. Christenson and P. M. Claesson, *Science*, 1988, **239**, 390–392.

**Professor Monson** answered: We welcome your clarification of the experimental work on capillary evaporation for the surface force apparatus and we plan to make a detailed study of these papers and their relationship to the model systems we have studied. For additional clarification purposes, we note that we are studying capillary evaporation for fixed wall separation by varying the chemical potential in the bulk, rather than varying the wall separation at fixed chemical potential, as in the surface force apparatus. The experimental situations most closely related to this would be systems like intrusion/extrusion of mercury into porous materials[1] and of water into hydrophobic porous materials. In these systems capillary evaporation is presumably occurring on extrusion of the liquid.

1 F. Porcheron, M. Thommes, R. Ahmad, P. A. Monson, *Langmuir*, 2007, **23**, 3372–3380.
2 F. Cailliez, M. Trzpit, M. Soulard, I. Demachy, A. Boutin, J. Patarin and A. H. Fuchs, *Phys. Chem. Chem. Phys.*, 2008, **10**, 4817–4826.

**Professor Bratko** asked: In a number of cases you discussed, the pathway of capillary evaporation involved hemispherical bubbles growing toward the opposite wall. By following the energetics of the transition path, a saddle point associated with the critical vapor nucleus could in principle be identified.[1]. For supersaturated gas solutions in water, our open ensemble molecular simulations illustrate the possibility of single-wall vapor nucleation[2] while the transition state vapor pockets spanned the slit approximately symmetrically in the absence of supersaturation. Can you describe the geometry of the critical nucleus in asymmetric vapor-nucleation scenarios you have analyzed?

1 Kevin Leung, Alenka Luzar and Dusan Bratko, *Phys. Rev. Lett.*, 2003, **90**, 065502.
2 D. Bratko and A. Luzar, *Langmuir*, 2008, **24**, 1247.

**Professor Monson** responded: Our calculations do not specifically identify critical nuclei in the sense of locating a maximum in the free energy of vapor bubble formation. Our use of the term nucleation arises from the similarity between states we encounter in the dynamics and the critical nuclei identified in static Monte Carlo simulations and DFT calculations of nucleation (see, for example, reference 24[1] of our paper).

1 A. Saugey, L. Bocquet and J. Barrat, *J. Phys. Chem. B*, 2005, **109**, 6520.

**Professor Weeks** asked: In your dynamical mean field theory you replace local "hard core" (single occupancy) constraints by a mean field treatment based on average occupancies of the sites. I believe this is a much more severe approximation than what is typically done in equilibrium uses of mean field theory where one averages over weaker longer ranged and slowly varying interactions. A similar problem could arise in your description of nucleation, where an unlikely connected series of local "facilitated" events has to occur. These may not be well described by products of local averages. Do you know of any way to deal with these issues and still have a tractable dynamical theory? It also seems to me that higher order pair and cluster approximations, which do improve mean field theory away from critical points, *etc.* in equilibrium, are less effective in correcting errors from hard core constraints. Would you agree?

**Professor Monson** replied: If we analyze the theory from the point of view of comparing it with Cahn-Hilliard or dynamic DFT[1] then we find that the largest impact of the averaging approximation you mention is in determining the mobility. However, our approach is an advance over these methods in that it provides a locally varying mobility and it is not limited to the linear regime. It would be very useful to go beyond this averaging approximation but we do not know how to do that except *via* the PPM which adds as much as a factor of ten to computation time of the calculations. The PPM is a significant quantitative improvement over DMFT but the qualitative results are essentially the same as can be seen in Figures 9 and 10 of our paper. We are also making comparisons of DMFT and PPM with Kawasaki dynamics Monte Carlo simulations and after averaging over several Kawasaki dynamics trajectories and we are finding that the behavior predicted from DMFT and PPM is qualitatively (if not quantitatively) accurate. We would of course welcome theoretical developments that could improve on these approximations while preserving the computational tractability that characterizes them, especially in the case of DMFT.

1 P. A. Monson, *J. Chem. Phys.*, 2008, **128**, 084701.

**Dr Hummer** asked Professor Koishi: Could differences in the shape between droplets of low and high urea content explain part of the observed small differences in the probability $P_w$ of forming a Wenzel state upon impact? I noticed that at impact with the surface, the droplets in your simulation appear to be non-spherical. As I understand from your paper, the initial droplets have a cubic shape, and during their flight path toward the surface they may not have had enough time to relax to a spherical shape. The rate of this relaxation could itself be a function of the urea concentration (because of differences in surface tension, viscosity *etc.*), and the resulting urea-dependent shape differences could possibly account for the small differences in $P_w$.

**Professor Koishi** responded: We first relaxed a urea-water droplet for 100 ps. Note that this droplet was located 60 Å above the pillared surface during the 100 ps relaxation. It appears that this relaxation time (100 ps) is too short, which is why the droplet had a non-spherical shape. We will utilize a relaxation time of 200 ps to see whether the non-spherical shape of droplet prior to impact would have a marked effect on the $P_w$ value. This is a very good point.

**Dr Christenson** asked Professor Monson: I wonder to what extent your simulations include the equivalent of a Laplace pressure across the emerging liquid–vapour interface as the nucleation of a cavity occurs? Depending on the external pressure and precise contact-angle values there may be many configurations were thermodynamically stable cavities cannot exist. Experimentally, we have found that the expected restrictions imposed by the fact that the pressure in the cavity can only be from 0 to 1 atm (at an external pressure of 1 atm) seem to be followed. This means that very small cavities can only exist if their principal radii of curvature are very nearly equal but of opposite sign. Annular or bridge shaped cavities are only permitted above a certain size. Often, capillary evaporation can be seen to occur on separation of the surfaces from contact (perhaps its formation is facilitated by transient tension between the surfaces). If the surfaces are brought back together again the cavity vanishes. Alternatively, many small and hour-glass shaped cavities may form at discrete sites around the contact zone.[1,2]

1 H. K. Christenson, P. M. Claesson and R. M. Pashley, *Proc. Indian Acad. Sci. (Chem. Sci.)*, 1987, **98**, 379–389.
2 H. K. Christenson and P. M. Claesson, *Science*, 1988, **239**, 390–392.

**Professor Monson** responded: The interfacial structure and associated thermodynamics are emergent properties of both the static MFT and DMFT calculations presented here—the only input is the Hamiltonian and pore geometry, as well as the MFT and DMFT approximations. We would expect the interfacial thermodynamics to be qualitatively consistent with macroscopic thermodynamic concepts, except that we are considering pore widths of only a few nm and so the rigorous definition of Laplace pressure may be problematic. Bubbles and bridges found during the DMFT dynamics would be saddle points (conditionally stable) in the free energy in the grand ensemble.[1] They can be rendered stable in studies of systems with constrained average density in the canonical ensemble.[2] We wonder whether your observation of multiple vapor cavities in the surface force apparatus may be related to what is seen in our simulations of capillary evaporation between square plates for larger changes in the external chemical potential. For a small change in the chemical potential we see only a single vapor bridge (see Figure 14 of our paper[3]) but when the change is substantially larger the evaporation proceeds through the formation of multiple bridges. This is shown here in Fig. 1. In individual Kawasaki dynamics trajectories these cavities would be distributed randomly over the surface.

1 D. W. Oxtoby and R. Evans, *J. Phys. Chem.*, 1985, **89**, 7521–7530; V. Talanquer and D. W. Oxtoby, *J. Chem. Phys.*, 2001, **114**, 2793–2801.

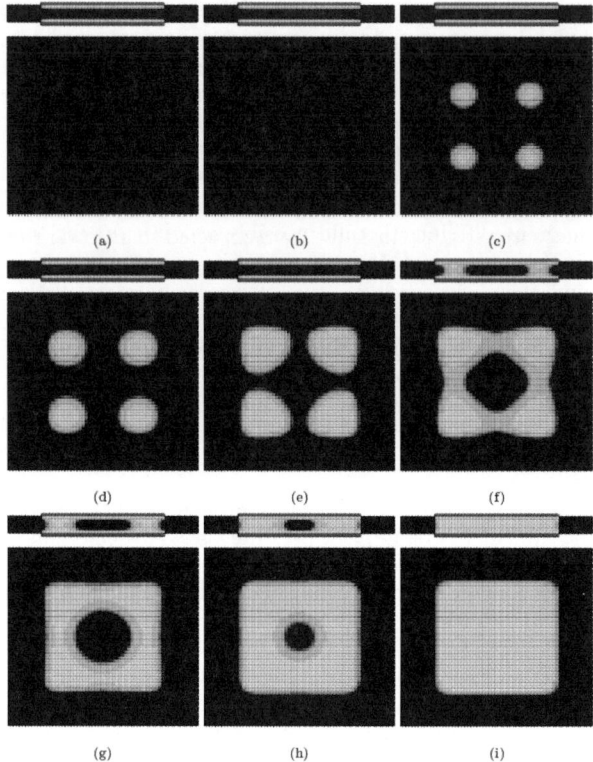

**Fig. 1** Visualizations showing a sequence of states along an evaporation process for a complete drying three-dimensional slit pore for a large quench. Each panel shows at the top a visualization from a two-dimensional slice into the distribution half way across the pore and parallel to one of the pore sides, and below that the density in the $x$–$y$ plane in the third layer.

2 P. A. Monson, *J. Chem. Phys.*, 2008, **128**, 084701.
3 J. R. Edison and P. A. Monson, *Faraday Discuss.*, 2010, **46**, DOI:10.1039/b925672e

**Professor Debenedetti** asked: Are you planning to study the dependence of nucleation times on the separation between plates?

**Professor Monson** answered: Our emphasis to this point has been in investigating how the DMFT reveals qualitative mechanisms of phase transformations rather than in quantifying the dynamics. Your suggestion is certainly one that could be taken up in future work.

**Mr Stirnemann** addressed Professor Koishi: The free energy barriers for the transition between a Cassie and a Wenzel state presented in Table 3 are small compared to the thermal energy $kT$. One would therefore expect spontaneous thermal transitions between the two states. Have such transitions been observed? Since these transitions mostly involve the interfacial waters, normalizing the free energy barrier by the number of molecules in the drop as done in Table 3† may lead to values which are dependent on the drop size and lower than the actual barrier.

---

† The values in Table 3 were amended at proof following this question, see Takahiro Koishi, Kenji Yasuoka, Xiao Cheng Zeng and Shigenori Fujikawa, *Faraday Discuss.*, 2010, **146**, DOI:10.1039/b926919c

**Professor Koishi** responded: In this paper, the free-energy barrier that separates the Cassie state and the Wenzel state is defined through Eq. (2). It was meant to quantify the capability of the falling droplet to reach the Wenzel state in the raining experiments. It was not meant to provide an absolute value of the thermal barrier (defined *via* the Boltzmann equation) for the droplet to cross *via* thermal fluctuation. Because the droplet is in a dynamic "falling" state, in some sense, the droplet may be viewed as being driven to a state that is closer to the transition state, compared to a non-falling droplet with zero downward velocity and being initially in the Cassie state. Hence, the barriers shown in Table 3 are expected to be lower than the thermal barriers.

In our previous work,[1] we computed the free energy barriers (through Eq. (2)) for pure water droplets with 5823 and 1728 water molecules, respectively. Nearly the same value of the free-energy barrier was attained. So, the normalization was meant to compare the capability for the falling droplet to reach the Wenzel state without concerning its actual size.

1 T. Koishi, K. Yasuoka, S. Fujikawa, T. Ebisuzaki and X. C. Zeng, *Proc. Natl. Acad. Sci. USA*, 2009, **106**, 8435–8440.

**Dr Christenson** addressed Professor Monson: The symmetry between vapour-bridge formation in liquid and liquid-bridge formation in vapour is quite contrary to results with the surface force apparatus. I have already noted that capillary evaporation has not been experimentally observed, but a capillary condensation transition at finite separations has been well documented between hydrophilic, *e.g.* mica surfaces in vapours of water[1] and many simple liquids.[2–5] It has also been seen below the bulk melting point of the liquid,[6,7] and in capillary condensation of water from nonpolar liquids.[8,9] The surface separation at which this happens[10,11] is consistent with a mechanism proposed by Derjaguin and Churaev,[12] involving the thickening of adsorbed films due to van der Waals forces. A nucleation event is hence not necessary. Could the lack of experimental symmetry hence be related to the fact that a hydrophilic surface will adsorb a multilayer film in vapour, but if there is a vapour film on a hydrophobic surface it is of submolecular thickness? Could you build this difference into your simulations?

1 M. M. Kohonen and H. K. Christenson, *Langmuir*, 2000, **16**, 7285–7288.
2 E. J. Wanless and H. K. Christenson, *J. Chem. Phys.*, 1994, **101**, 4260–4267.
3 J. E. Curry and H. K. Christenson, *Langmuir*, 1996, **12**, 5729–5735.
4 H. K. Christenson, *Colloids Surf. A*: *Physicochem. Eng. Asp.*, 1997, **123–124**, 355–367.
5 M. M. Kohonen and H. K. Christenson, *J. Phys. Chem. B*, 2002, **106**, 6685–6695.
6 Y. Qiao and H. K. Christenson, *Phys. Rev. Lett.*, 1999, **83**, 1371–1374.
7 P. Barber, T. Asakawa and H. K. Christenson, *J. Phys. Chem. C*, 2007, **111**, 2141–2148.
8 H. K. Christenson, *J. Colloid Interface Sci.*, 1985, **104**, 234–249.
9 H. K. Christenson, J. Fang and J. N. Israelachvili, *Phys. Rev. B*, 1989, **39**, 11750–11754.
10 H. K. Christenson, *Phys. Rev. Lett.*, 1994, **73**, 1821–1824.
11 J. Crassous, E. Charlaix and J.-L. Loubet, *Europhys. Lett.*, 1994, **28**, 37.
12 B. V. Derjaguin and N. V. Churaev, *J. Colloid Interface Sci.*, 1976, **54**, 157.

**Professor Monson** responded: The symmetry seen in our calculations is built into the lattice gas model and we would not expect it to be quantitatively accurate for real systems. However, evidence from studying the relationship between gas adsorption and mercury porosimetry in terms of static behavior suggests that there is a qualitative symmetry.[1] The question of the symmetry of the dynamics remains an open question. Our description of the dynamics of capillary condensation is consistent with other modeling studies of fluids in porous materials. The consensus in that field is that there is a large nucleation barrier to condensation of vapor and this is the origin of the hysteresis seen in experiments.[2]

Longer ranged solid-fluid interactions can certainly be built into the model and this would remove the exact symmetry that we have for nearest neighbor interactions, although we believe that an approximate symmetry will remain. Our calculations are for quite narrow pores corresponding to just a few nm. Investigations of wider pores may also help clarify the picture. An interesting question may be whether the dynamics of condensation or evaporation is different in the surface force apparatus, where there is changing surface separation and fixed chemical potential, is different from that in porous materials, where there is fixed surface separation and changing chemical potential.

1 F. Porcheron, M. Thommes, R. Ahmad, P. A. Monson, *Langmuir*, 2007, **23**, 3372–3380.
2 See, for example, P. I. Ravikovitch, S. C. O'Domhnaill, A. V. Neimark, F. Schuth and K. K. Unger, *Langmuir*, 1995, **11**, 4765–4772.

**Professor Evans** said: Lines of capillary condensation or evaporation terminate at capillary critical points, for sufficiently small wall separations. Such critical points are special as they lie in the two-dimensional Ising universality class. Your mean-field treatment will capture the existence of such points but will not account for the proper critical behaviour. Nevertheless it might be very interesting to investigate the dynamics in the neighbourhood of such critical points where there are long-ranged, albeit mean-field correlations. One might find rich behaviour.

**Professor Monson** replied: As yet we have not studied behavior near the critical points but we agree that it could be very interesting.

**Professor Weeks** asked: You briefly discuss the broken symmetry in the partial drying case as illustrated in Fig. 12. Since the governing equations are symmetric, what do you think gives the asymmetry. Could it be round-off errors? Do you have nucleation on different surfaces from different runs?

**Professor Monson** responded: We do not yet understand the origin of the symmetry breaking in the dynamics. Tests thus far carried out suggest that it is not due to round-off error or the accuracy of the integration scheme. It is important to note that such symmetry breaking in cavity and bubble states have been seen in both static Monte Carlo simulations and DFT studies of nucleation (see, for example, reference 24[1] of our paper). This bolsters our confidence in the DMFT results even though we do not understand the symmetry breaking. We hope to have an improved understanding of this behavior in the near future.

1 A. Saugey, L. Bocquet and J. Barrat, *J. Phys. Chem. B*, 2005, **109**, 6520–6526.

**Professor Evans** asked: I believe Professor Monson's DMFT captures essentially all the key ingredients in the dynamics of adsorption/desorption in simple model pores. It is a most versatile tool for searching for new phenomena in more complex geometries. The issue of symmetry breaking is most intriguing and I do not pretend to understand this. However, I wonder whether Professor Monson has attempted to implement what he mentions in the penultimate paragraph of his paper. Adding fluctuations by hand to the evolution equations is indeed *ad hoc*, and difficult to justify from any formal approach but it might be interesting to see if this altered the nature of symmetry breaking.

**Professor Monson** responded: We welcome this suggestion. We also plan to study the same systems with Kawasaki dynamics Monte Carlo simulations.

**Professor Corrales** communicated: In your gridded pattern simulation the particles evaporate sequentially first from the corner blocks, then the outer edge blocks, and the evaporation process continues inward in a sequential manner. It appears then that the corner blocks have the highest free energy, then the edge blocks have a lower free energy and each subsequent block has a lower free energy as it approaches the center one with the lowest free energy. One would think that there would be a process where particles from the lower free energy blocks would jump up to a higher free energy block before leaving the system, and therefore there would be some transient population always occupying the higher free energy blocks until the lowest free energy block was completely evaporated. Can you explain the mechanism by which the particles depart the system? Are you allowing for adsorption processes to occur, in an adsorption/desorption equilibrium fashion, as the desorption process is taking place?

**Professor Monson** communicated in reply: We interpret the pattern formation during the dynamics in terms of the mass transfer resistances encountered in the process. These mass transfer resistances are created by lack of contact of the inner hydrophobic regions with the bulk and the liquid held in the hydrophilic regions which slows vapor transfer from the system. The corner hydrophobic blocks empty first since they have the closest contact with the bulk (contact on two sides), followed by the edge hydrophobic blocks (contact with the bulk on one side). Following that the four symmetrically-placed interior hydrophobic regions empty followed finally by the one in the center of the system.

The fluxes in the process are quite complex and evolve during the process. We make no assumptions about particular processes in the system. Both adsorption and desorption processes will be seen to occur in different parts of the system at different times.

PAPER

# First- and second-order wetting transitions at liquid–vapor interfaces

K. Koga,[*a] J. O. Indekeu[*b] and B. Widom[*c]

Received 8th December 2009, Accepted 13th January 2010
DOI: 10.1039/b925671g

Wetting transitions, in which one liquid wets, or spreads at, the interface between a second liquid and their common vapor, are defined and first- and second-order transitions are distinguished. The mean-field density-functional models of fluid interfaces are recalled. A criterion is noted for determining when the wetting transitions in those models are required to be of first order or may be of second order. It is seen how two examples of such density-functional models that have been treated in the past, one leading to a first-order and the other to a second-order wetting transition, provide examples of the application of the criterion.

## 1. Introduction

We consider two liquid phases (called phases 2 and 3) in equilibrium with their common vapor (phase 1). The three phases may meet at a line of common contact with three non-vanishing contact angles, or alternatively one of the phases, say 2, may "wet" (spread at) the interface between 1 and 3. Let the three interfacial tensions be denoted $\sigma_{12}$, $\sigma_{23}$, and $\sigma_{13}$, and suppose the liquid–vapor tension $\sigma_{13}$ to be the greatest of the three. Then the condition for the three phases to meet at a line of common contact ("non-wetting") is the triangle inequality

$$\sigma_{13} < \sigma_{12} + \sigma_{23} \text{ (non-wetting).} \tag{1}$$

The alternative ("wetting") is that phase 2 spreads as a macroscopically thick film at the 1,3 interface, with

$$\sigma_{13} = \sigma_{12} + \sigma_{23} \text{ (wetting).} \tag{2}$$

A transition between these two modes of three-phase contact is termed a wetting transition.

Such a transition might occur as the temperature or a chemical potential varies, leading to two continuous ranges of three-phase states (triple points), with (1) holding over one range and (2) over the other. Between those two ranges of triple points there would then be a boundary at which (1) goes over continuously to (2) or the reverse. According to the phase rule, to have a continuum of three-phase states instead of just isolated triple points requires that the system be a mixture of at least two components. In a one-component system there are only isolated triple

[a]Department of Chemistry, Faculty of Science, Okayama University, Okayama, 700-8530, Japan. E-mail: koga@cc.okayama-u.ac.jp
[b]Institute for Theoretical Physics, Katholieke Universiteit Leuven, BE-3001 Leuven, Belgium. E-mail: joseph.indekeu@fys.kuleuven.be
[c]Department of Chemistry, Baker Laboratory, Cornell University, Ithaca, NY, 14853-1301, USA. E-mail: bw24@cornell.edu

points, where either one of the phases wets the interface between the other two, or not. There is no transition between those two circumstances because no field variable can then vary without one or two of the three phases disappearing.

For simplicity we consider here systems with the minimum number, two, of components. By the phase rule, the system of three phases and two components then has one degree of freedom; $i.e.$, there is then a one-dimensional manifold of triple points parameterized by a single field variable, which we henceforth call $b$, and which we may think of as some function of the temperature and the two chemical potentials. The wetting transition between (1) and (2) occurs when $b$ takes some value $b_w$. We shall take $b > b_w$ to be the range of non-wet states, where (1) is then satisfied, and $b < b_w$ to be the range of states where the 1,3 interface is wet by 2, so that (2) holds.

In reality, many such liquid–liquid–vapor three-phase equilibria occur in systems of $c > 2$ components. To model such systems and still have the simplicity of having only one variable field we would imagine holding fixed $c - 2$ of the system's $c + 1$ field variables. Then with the two phase-equilibrium constraints we would still be considering only a one-dimensional manifold of triple points, parameterized by a single varying field.

The distinction between first- and second-order wetting transitions is illustrated in Fig. 1, where $\sigma_{12} + \sigma_{23} - \sigma_{13}$, which we call $\Delta\sigma$, is shown schematically as a function of $b$. In the first-order transition $\Delta\sigma$ vanishes linearly with $b - b_w$, and has a metastable extension, shown by the dashed line in Fig. 1(a), in which the non-wet interface for $b < b_w$ is at a local but not global minimum of the free energy. At the global minimum, $\Delta\sigma = 0$ over the range $b < b_w$. At a classical second-order wetting transition (in the sense of the Ehrenfest classification), $\Delta\sigma$ vanishes proportionally to $(b - b_w)^2$ as $b$ approaches $b_w$ from above [Fig. 1(b)], and does not have a metastable extension to states with $b < b_w$. A wetting transition may be very weakly first order, when the curve of $\Delta\sigma$ in Fig. 1(a) crosses the $b$-axis at $b = b_w$ so shallowly that it may be difficult to locate $b_w$ or to distinguish the slope of the curve at $b_w$ from 0.[1]

## 2. Mean-field density-functional theories recalled

In each of the three interfaces, the densities $\rho_1$ and $\rho_2$ of the two components vary in the direction $z$ perpendicular to that interface. We define an excess free energy of inhomogeneity $\Psi[\rho_1(z), \rho_2(z);b]$, which is a functional of the two spatially varying densities and depends on the single field variable $b$. We take it to be of the classical square-gradient form

$$\Psi = F[\rho_1(z), \rho_2(z); b] + \frac{1}{2}\left[\rho_1'(z)^2 + \rho_2'(z)^2\right] \qquad (3)$$

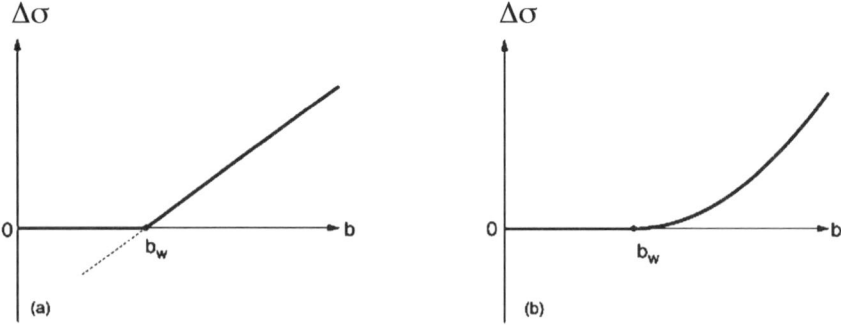

**Fig. 1** (a) First-order transition; (b) second-order transition.

where $F(\rho_1, \rho_2; b)$ is such that it, along with $\partial F/\partial \rho_1$ and $\partial F/\partial \rho_2$, vanish when, for the given $b$, the densities $\rho_1$ and $\rho_2$ are those of any of the three coexisting bulk phases; while $F > 0$ otherwise. The tension $\sigma$ of each of the three interfaces is then

$$\sigma = \min_{\rho_1(z), \rho_2(z)} \int_{-\infty}^{\infty} \Psi \, dz \qquad (4)$$

i.e., the variational minimum of the integrated free-energy density, minimized with respect to the densities $\rho_1(z), \rho_2(z)$ subject to their taking their respective bulk-phase values at $z = \pm \infty$. The minimizing $\rho_1(z)$ and $\rho_2(z)$ satisfy the Euler–Lagrange equations

$$\frac{\partial F}{\partial \rho_1} = \frac{d^2 \rho_1}{dz^2}, \quad \frac{\partial F}{\partial \rho_2} = \frac{d^2 \rho_2}{dz^2} \qquad (5)$$

These are to be solved subject to the boundary conditions at $z = \pm \infty$ noted above.

When $z$ is eliminated between $\rho_1(z)$ and $\rho_2(z)$ for any of the interfaces the result is a trajectory in the $\rho_1, \rho_2$ plane that describes how $\rho_1$ and $\rho_2$ vary with each other through that interface. There are three such trajectories, one for each of the three interfaces, as shown schematically in Fig. 2. The density variables are taken to have the values $\rho_1 = -1, \rho_2 = 0$ in phase 1; $\rho_1 = 0, \rho_2 = b$ in phase 2; and $\rho_1 = 1, \rho_2 = 0$ in phase 3. Here $\rho_1$ and $\rho_2$ are not to be taken literally as densities but rather as two independent functions of the physical densities, chosen to make the notation and the representation in Fig. 2 simple.

The curve in Fig. 2 extending from $-1,0$ to $0, b$ is the 1,2 interfacial trajectory; that from $0, b$ to $1,0$ is the 2,3 interfacial trajectory; that from $-1,0$ to $1,0$ that does not go via $0, b$ is that of a 1,3 interface that is not wet by phase 2; while the composite trajectory that does go from $-1,0$ to $1,0$ via $0, b$ is that of a 1,3 interface that is wet by phase 2.

Each trajectory corresponds to a solution $\rho_1(z), \rho_2(z)$ of the Euler–Lagrange eqns (5) with the appropriate boundary conditions. When these are substituted in the integrand on the right-hand side of (4) the result is the tension $\sigma$ of the corresponding interface. The three tensions are functions of $b$.

The equilibrium 1,3 interface is or is not wet by phase 2 according to whether the tension $\sigma_{13}$ that corresponds to the direct 1,3 trajectory is respectively greater or less than the sum $\sigma_{12} + \sigma_{23}$, which is the tension of the wet 1,3 interface. The wetting transition occurs at that $b(= b_w)$ for which $\sigma_{13}$ of the non-wet interface equals $\sigma_{12} + \sigma_{23}$. If the transition is of second order, there is no solution of the Euler–Lagrange equations corresponding to a non-wet 1,3 interface when $b < b_w$ [cf. Fig. 1(b)], hence no direct 1,3 trajectory in the $\rho_1, \rho_2$ plane, only the indirect trajectory via $0, b$. The latter is always present because it is a composite of the 1,2 and 2,3 interfacial

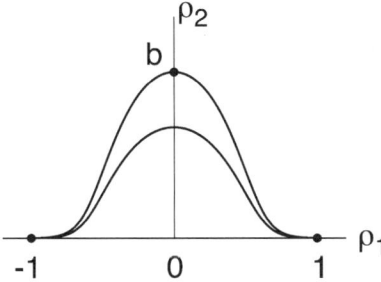

**Fig. 2** Wetting and non-wetting trajectories in the $\rho_1, \rho_2$ plane. The bulk-phase densities are $\rho_1 = -1, \rho_2 = 0$ (phase 1); $\rho_1 = 0, \rho_2 = b$ (phase 2); $\rho_1 = 1, \rho_2 = 0$ (phase 3).

trajectories, and those interfaces are present irrespective of whether phase 2 wets the 1,3 interface or not.

## 3. First- and second-order transitions

Wetting transitions, which here are transitions in the structure of the 1,3 interface, are in many ways analogous to the transitions between bulk phases in three dimensions. For the latter, while the free energy is necessarily continuous at the transition, its first derivatives with respect to the thermodynamic fields, *viz.*, the energy, entropy, and mass densities and the chemical composition, are discontinuous for first-order transitions (except for accidental azeotropies), but are still continuous at transitions of higher order. Discontinuity in the thermodynamic densities reflects discontinuity in the structures of the phases; at higher-order transitions the structures are continuous.

The relevant free energy in the wetting transitions is $\Delta\sigma$. Its derivative with respect to the field variable $b$ is discontinuous at $b = b_w$ for the first-order transition in Fig. 1(a) but continuous for the second-order transition in Fig. 1(b). In our mean-field density-functional models the "structures" of the two alternative 1,3 surface phases are the alternative chemical composition profiles $\rho_1(z)$, $\rho_2(z)$ of the wet and non-wet 1,3 interfaces. These correspond to the two alternative 1,3 trajectories in the $\rho_1$, $\rho_2$ plane of Fig. 2, one going directly from $-1,0$ to $1,0$ (non-wet) and the other indirectly, *via* 0, $b$ (wet). If the transition is of first order the structure is discontinuous; *i.e.*, the two alternative trajectories are still distinct at $b = b_w$. If it is of second order the two trajectories coincide at the transition, where there is then only the one, which goes *via* 0, $b_w$.

As noted earlier, the function $F(\rho_1, \rho_2; b)$ in (3) and its derivatives $\partial F/\partial \rho_1$ and $\partial F/\rho_2$ are 0 at the points $\rho_1$, $\rho_2$ in Fig. 2 representing the compositions of the bulk phases, while $F > 0$ everywhere else. Near those points contours of constant $F$ are ellipses. For simplicity we suppose $F(\rho_1, \rho_2; b)$ to be an even function of $\rho_1$, so the trajectories are symmetric in $\rho_1$, as in the example in Fig. 2. Then in the immediate neighborhood of the phase-2 point at 0, $b$ we will have

$$F(\rho_1, \rho_2; b) \sim \left(\frac{\rho_1}{a_1}\right)^2 + \left(\frac{b-\rho_2}{a_2}\right)^2 \tag{6}$$

where $a_1\sqrt{F}$ and $a_2\sqrt{F}$ are the semi-axes of the elliptical contours of constant $F$ in the directions of the $\rho_1$ and $\rho_2$ axes, respectively, with $a_1$ and $a_2$, in general, functions of $b$. In related work[2] it is shown from the Euler–Lagrange eqns (5) and their first integral, and from (6), that when the wetting transition is of higher than first order, $d^2\rho_2/d\rho_1^2$ evaluated at the maximum on the non-wetting 1,3 trajectory in Fig. 2 must diverge at the wetting transition, while $d^2\rho_2/d\rho_1^2$ on the wetting trajectory near 0, $b$ diverges or remains finite as $b \to b_w$ according to whether, at $b = b_w$, $a_1/a_2 < 2$ or $a_1/a_2 \geq 2$, respectively. Then since the wetting and non-wetting trajectories coincide at a higher-order wetting transition, a necessary condition for the transition to be of higher order is that $a_1/a_2 < 2$ at $b = b_w$, and then a sufficient condition that it be of first order is that $a_1/a_2 \geq 2$.

We shall now recall two wetting-transition models that have been treated in the past,[1,3] in one of which the transition is of first order and in the other of which it is of second order, and we shall see how the stated criterion is satisfied.

In the first,[1]

$$F(\rho_1, \rho_2; b) = 16\rho_2^2(\rho_2 - b)^2 + [(\rho_2 - b\rho_1)^2 - b^2]^2 + [(\rho_2 + b\rho_1)^2 - b^2]^2 \tag{7}$$

The coexisting phases are again of compositions $(-1,0)$, $(0, b)$, and $(1,0)$, as in Fig. 2. In the immediate neighborhood of 0, $b$ this $F$ is again of the form (6), now with $a_1/a_2 = \sqrt{3}/b$. There is a first-order wetting transition at $b = b_w \simeq 0.51$,[1] where then $a_1/a_2 \simeq 3.4$. Thus, the wetting transition in this model had to be of first order, as it is. Indeed, the non-wetting trajectory in this model just runs along the $\rho_1$ axis itself, from $-1,0$ to $1,0$, nowhere near the wetting trajectory, which passes through $0, b$.

In the second of the two earlier models,[3]

$$F(\rho_1, \rho_2; b) = [(\rho_1 + 1)^2 + \rho_2^2][(\rho_1 - 1)^2 + \rho_2^2][\rho_1^2 + (\rho_2 - b)^2] \qquad (8)$$

This, too, is of the form (6) near $0, b$, now with $a_1/a_2 = 1$. It thus satisfies the necessary condition for its wetting transition to be of higher order, and indeed what is found is a classical second-order transition as in Fig. 1(b). (In this case the diverging second derivative $d^2\rho_2/d\rho_1^2$ at $b = b_w$ is manifested as a discontinuous first derivative: $b - \rho_2 \sim |\rho_1|$.)

From (6) it is apparent that the ratio $a_1/a_2$ plays the role of an asymmetry, or anisotropy parameter, $a_1/a_2 = 1$ corresponding to the symmetric (isotropic) model. In general, the thermodynamic singularities at wetting are expected to be non-universal with respect to this ratio, in the sense that the critical exponents associated with higher than first-order wetting transitions vary continuously with $a_1/a_2$. Indeed, in earlier work on models of wall wetting with a two-component order parameter[4] or anisotropic vector order parameter[5] this non-universality was demonstrated.

Particularly interesting cases arise when the ratio $a_1/a_2$ reflects the ratio of characteristic physical length scales of the problem. These two lengths can already appear in a single-component order parameter model, in the form of decay lengths of the wall-fluid and the fluid–fluid potential, respectively.[6] In this model, the non-universality of the critical exponents at wetting, found at the classical (van der Waals theory) level, was found to be robust when thermal fluctuations are taken into account using functional renormalization group calculations.[7]

In the entirely different physical system of a type-I superconductor, in which a "wetting" or interface delocalization transition was uncovered theoretically[8] and verified experimentally,[9] these two lengths are the superconducting coherence length $\xi$ and the magnetic penetration depth $\lambda$. For the case in which the theoretically predicted interface delocalization transition is critical, the critical exponent of the surface excess free energy was shown[10] to vary continuously with the Ginzburg–Landau parameter $\kappa$, which is the ratio of $\lambda$ to $\xi$.

Now, returning to our model defined through (3) and (6), higher than first-order wetting transitions are found in the entire interval $0 < a_1/a_2 < 1$.[2] This is again consistent with the criterion we exposed. Moreover, for these transitions, too, the critical exponents of the surface free energy at wetting are found to be non-universal and vary continuously with the asymmetry or anisotropy parameter $a_1/a_2$ (ref. 2). It is of interest that the order of the wetting transition in mean-field density-functional models with excess free-energy density $\Psi$ of the form (3), with two spatially varying densities $\rho_1(z)$ and $\rho_2(z)$, depends on the structure of the function $F$ only in the immediate neighborhood of the composition $\rho_1, \rho_2$ of the wetting phase.

## Acknowledgements

We acknowledge support by a Grant-in-Aid for Scientific Research and the Next Generation Super Computing Project from MEXT, Japan; by grant G.0115.06, FWO-Vlaanderen; and by the U.S. National Science Foundation.

## References

1 I. Szleifer and B. Widom, *Mol. Phys.*, 1992, **75**, 925.
2 K. Koga, J. O. Indekeu and B. Widom, *Phys. Rev. Lett.*, 2010, **104**, 036101.
3 K. Koga and B. Widom, *J. Chem. Phys.*, 2008, **128**, 114716.
4 E. H. Hauge, *Phys. Rev. B: Condens. Matter*, 1986, **33**, 3322.
5 C. J. Walden, B. L. Gyorffy and A. O. Parry, *Phys. Rev. B: Condens. Matter*, 1990, **42**, 798.
6 T. Aukrust and E. H. Hauge, *Phys. Rev. Lett.*, 1985, **54**, 1814.

7 E. H. Hauge and K. Olaussen, *Phys. Rev. B: Condens. Matter*, 1985, **32**, 4766.
8 J. O. Indekeu and J. M. J. van Leeuwen, *Phys. Rev. Lett.*, 1995, **75**, 1618; J. O. Indekeu and J. M. J. van Leeuwen, *Physica C*, 1995, **251**, 290.
9 V. F. Kozhevnikov, M. J. Van Bael, P. K. Sahoo, K. Temst, C. Van Haesendonck, A. Vantomme and J. O. Indekeu, *New J. Phys.*, 2007, **9**, 75.
10 J. M. J. van Leeuwen and E. H. Hauge, *J. Stat. Phys.*, 1997, **87**, 1335.

# Hierarchical surfaces: an *in situ* investigation into nano and micro scale wettability

Alex H. F. Wu,[a] K. L. Cho,[a] Irving I. Liaw,[a] Grainne Moran,[b] Nigel Kirby[c] and Robert N. Lamb[*a]

*Received 22nd December 2009, Accepted 2nd February 2010*
DOI: 10.1039/b927136h

Two scales of roughness are imparted onto silicon surfaces by isotropically patterning micron sized pillars using photolithography followed by an additional nanoparticle coating. Contact angles of the patterned surfaces were observed to increase with the addition of the nanoparticle coating, several of which, exhibited superhydrophobic characteristics. Freeze fracture atomic force microscopy and *in situ* synchrotron SAXS were used to investigate the micro- and nano-wettability of these surfaces using aqueous liquids of varying surface tension. The results revealed that scaling different roughness morphologies result in unique wetting characteristics. It indicated that surfaces with micro, nano or dual scale roughness induced channels for the wetting liquid as per capillary action. With the reduction of liquid surface tension, nano-wetting behaviour differed between superhydrophobic and non-superhydrophobic dual-scale roughness surfaces. Micro-wetting behaviour, however, remained consistent. This suggests that micro- and nano-wetting are mutually exclusive, and that the order in which they occur is ultimately governed by the energy expenditure of the entire system.

## 1. Introduction

The influence of roughness on the surface wetting behaviour, first reported by Wenzel,[1] highlighted a reduction in contact angle as a result of the increase in surface area. The postulation of a fully dewetted state, as reported by Cassie and Baxter,[2] revealed, beyond a certain level of roughness, the existence of a composite interface (air and solid) in contact with the liquid.

The combination of low energy surface with surface roughness results in superhydrophobicity, where water contact angle exceeds 150° with a sliding angle of less than 10°. This is an extreme case of the Cassie–Baxter dewetted state. Since its discovery,[3] efforts into synthetically replicating this phenomenon boomed with the development of various techniques to induce stable superhydrophobic characteristics on surfaces through irregular roughness,[4–11] hierarchical roughness[12,13] and arrayed nanotubes.[14–19] Recent studies have shown that, regardless of fabrication technique, superhydrophobic characteristics are stabilized by hierarchical roughness.[13,20–22] This suggests that both micro-roughness (RMS roughness 100 nm – several microns) and nano-roughness (RMS roughness <100 nm), and its

[a] *School of Chemistry, The University of Melbourne, Melbourne, Victoria, Australia*
[b] *UNSW Analytical Centre, The University of New South Wales, Sydney, New South Wales, Australia*
[c] *The Australian Synchrotron, Clayton, Victoria, 3168, Australia*

corresponding wetting behaviour, contributes to observable macroscopic wettability.

Despite comprehensive theoretical studies such as thermodynamic[23–27] and mathematical[28,29] modelling, experimental measurements in controlled texturing,[30–32] SANS[33] and AFM force measurements,[34,35] the mechanism of hierarchical wetting behaviour and its contribution to macroscopic wettability is still unclear. In an attempt to elucidate the effects of hierarchical wetting behaviour, our previous studies have revealed a correlation between nanoroughness and antifouling behaviour.[36] We have also demonstrated that two superhydrophobic surfaces (SHS) with identical contact angle and hysteresis underwent different nanowetting phenomena,[37] which also correlated with antifouling characteristics.[38] This is a strong indication that both micro- and nanowetting behaviour govern different macroscopic properties of SHS, which is indistinguishable using macroscopic wettability measurements.

We recently adapted two complimentary, non-invasive techniques; freeze fracture microscopy[39] and synchrotron small-angled X-ray scattering (SAXS)[37] to observe the fluid/surface interface of SHS at both the micro- and nanoscale.

Freeze fracture microscopy (FFM) has long been used in morphological studies in microbiology.[40–42] A recent adaptation of this technique was conducted by Switkes[43] to observe nanobubbles at the interface of a smooth hydrophobic surface. FFM achieves this by rapidly cryofixating the fluid/surface interface followed by replicating this interface using metal sputter coating, thus imprinting the fluid's morphology at the interface onto a thin metal replica. Applying this technique to SHS, we have been able to quantify the morphology of air inclusion at the microscale.[39]

SAXS is a technique where the scattering of X-rays through a material is caused by electron density inhomogeneities due to the presence of particles or voids. The scattered X-rays form a pattern which is commonly represented as scattering intensity ($I$) as a function of the scattering vector ($q = 4\pi\sin(\varphi/2)/\lambda$). Where $2\varphi$ is the angle between the incident and scattered X-ray beam and $\lambda$ is the wavelength of the X-ray.

In a dual media system of differing electron densities ($\rho_1$ and $\rho_2$), the scattering intensity ($I_{12}$) is directly related to the electron density contrast ($\delta_P$):[44]

$$I_{12} \propto (\rho_1 - \rho_2)^2 = \delta_P^2 \tag{1}$$

Wetting involves the replacement of an air/surface interface with a liquid/surface interface. Since liquid has a much higher electron density than air, a reduction in $\delta_P$, and subsequently, the scattering intensity is observed. To quantify this effect, we make use of the invariant $Q$, which is a measure of the overall scattering power of the system, and is approximated by

$$Q = \int I(q)\, q^2\, dq \tag{2}$$

The raw calculated value of invariant $Q$ is affected by the scattering media's thickness, and is subsequently a source of error as the thickness between each SHS may vary. By calculating the percentile changes in invariant $Q$ between dry and wet substrates ($Q_{wet}/Q_{dry}$), we can remove the effects of thickness variations and standardize invariant $Q$, thus allowing the direct comparison of the percentage interface that undergoes nanowetting between SHS.

In this work, photolithographically patterned micron-sized pillars with thin layers of nanoparticles are used systematically to vary micro- and nano-roughness. The effect of varying pillar size, pitch and nanoparticle size on micro- and nano-scaled wetting behaviour by immersing the surfaces in fluids with different surface tension was studied. Both FFM and synchrotron SAXS were performed *in situ* to reveal changes in the wetting behaviour at the micro- and nanoscale, respectively.

## 2. Materials and methods

### 2.1 Preparation of microrough surfaces

Microroughness on silicon wafers (8 in. diameter, P/N doped, <100> oriented, 35 μm thick) was fabricated by standard photolithography techniques.[30] The desired patterns of a master mask were printed (Bandwidth Foundry Pty Ltd.) on a 4½ in. square quartz plate. The masks were used to irradiate a photoresist coated wafer with a maximum precision of about 1 μm. After irradiation and development, the wafers were etched using reactive ion etching (RIE) with $SF_6$ gas for 30–45 min depending on the desired heights of the pillars.

The surface chemistry of photolithographically patterned silicon substrates was altered by immersing in piranha solution (3 : 1 $H_2SO_4$ : $H_2O_2$) for 30 min at 90 °C, followed by rinsing with MilliQ water (0.05 μS cm$^{-1}$, Millipore), drying with a gentle stream of nitrogen, then refluxing in a 1 M solution of trimethylchlorosilane in hexane for 24 h. The functionalized surfaces were cleaned of excess silanes by sonication in n-hexane for 10 min, and stored under argon until use.

### 2.2 Preparation of nano and dual-scaled roughness surfaces

Flat and photolithographically patterned silicon substrates were both spin-coated with a dilute solution of nanoparticles to generate nano-scaled and dual-scaled roughness, respectively. To investigate the effect of nanoroughness on wetting behaviour, three nanoparticle sizes (7 nm, 20 nm, 40 nm) were used. The solutions were prepared by mixing silica nanoparticles (Degussa, 0.25 g), n-hexane (AR, 40 mL), methyltrimethoxysilane (MTMS, Aldrich, 0.36 g) and HCl (10 M, 0.2 mL) in a reaction flask and sonicating the mixture for 30 min at 40 kHz. Prior to coating, the mixture was diluted with additional n-hexane (80 mL). The coated substrates were baked in a furnace at 150 °C for 15 min.

### 2.3 *In situ* synchrotron transmission small angled X-ray scattering measurements

All transmission SAXS experiments were performed at the SAXS/WAXS beamline at the Australian Synchrotron, Clayton, Australia. Measurements were taken with a camera length of 7000 mm, providing a data collection $q$ range of 0.001–0.1 Å$^{-1}$. All samples were mounted on a sealed fluid cell (Specac) with a pathlength of 1 mm, which was bolted tightly onto a sample stage, such that the same position was analyzed for each test fluid used. The wetting fluids used were aqueous solutions of sodium dodecyl sulfate (SDS, Aldrich) with different concentrations (0%, 20%, 60% and 100% critical micelle concentration (CMC)).

### 2.4 Freeze fracture atomic force microscopy measurements

The freeze fracture technique was performed using a Moor-type freeze fracture unit manufactured by Balzers AG. Cryogens used in the process were liquid nitrogen and Freon 22 (chlorodifluoromethane, DuPont). Further details of this technique are described elsewhere.[39] The resulting surface replicas were analyzed using AFM section analysis (Digital Instruments 3000 AFM) to determine the penetration depth of the fluids into the troughs of textured surfaces.

## 3. Results and discussion

### 3.1 Microrough surfaces

The micro-scale roughness was fabricated by photolithographically patterning the surface with pillars (Fig. 1). The roughness of the surface was then geometrically controlled through the variance of pillar size and pitch whilst maintaining a consistent height throughout.

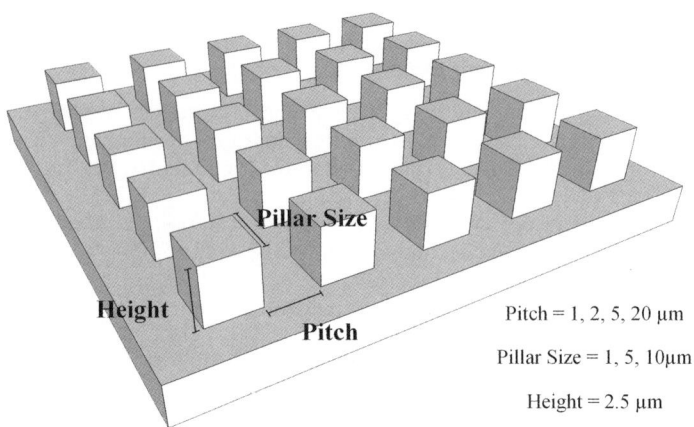

**Fig. 1** Schematic of photolithographically fabricated microrough surfaces.

**Table 1** Contact angle and roughness characteristics of test surfaces

| Micro-roughness | Nano-roughness | Contact angle/° | Hysteresis/° |
| --- | --- | --- | --- |
| 1 μm × 1 μm, 1 μm pitch | — | 110 | 18 |
| 1 μm × 1 μm, 2 μm pitch | — | 108 | 28 |
| 1 μm × 1 μm, 5 μm pitch | — | 94 | 50 |
| 1 μm × 1 μm, 20 μm pitch | — | 96 | >70 (Pinned) |
| 5 μm × 5 μm, 1 μm pitch | — | 92 | 15 |
| 10 μm × 10 μm, 1 μm pitch | — | 102 | 17 |
| — | 12 nm | 126 | 6 |
| — | 20 nm | 94 | 8 |
| — | 40 nm | 85 | 32 |
| 1 μm × 1 μm, 1 μm pitch | 12 nm | 157 | 4 |
| 1 μm × 1 μm, 2 μm pitch | 12 nm | 155 | 6 |
| 1 μm × 1 μm, 5 μm pitch | 12 nm | 121 | 12 |
| 1 μm × 1 μm, 20 μm pitch | 12 nm | 115 | 23 |
| 5 μm × 5 μm, 1 μm pitch | 12 nm | 123 | 10 |
| 10 μm × 10 μm, 1 μm pitch | 12 nm | 119 | 9 |

Photolithograpy affords a control over the process ensuring the consistency across the surface with little or no variance. The penetration depth, as measured by freeze fracture AFM, is defined by the depth of which the wetting fluid penetrates into the troughs between each pillar. Ideally, the walls of the pillar are synonymous to the boundaries of a capillary tube with a diameter equivalent to the pitch.

Subsequent functionalization with trimethylchlorosilane rendered the surface features hydrophobic as shown in Table 1.

**3.1.1 Wetting of microrough surfaces.** The observed microwetting phenomena conforms to behaviour described by capillary action in eqn (3), where $h$ is the height of a liquid in a column, $\gamma$ is the liquid–air surface tension, $\theta$ is the contact angle of the surface chemical group, $\rho$ is the density of the fluid, $g$ is gravity and $r$ is the radius of the tube, or in this case, pitch.

$$h = 2\gamma\cos\theta/\rho g r \quad (3)$$

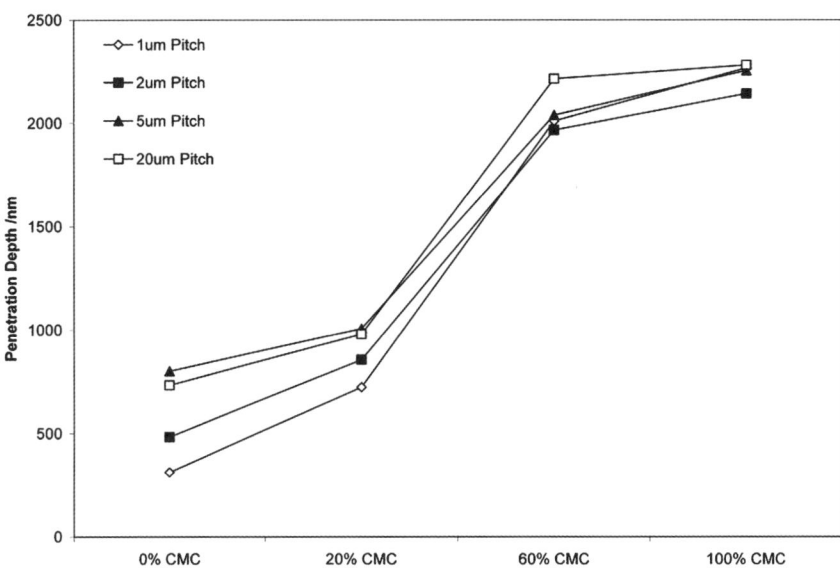

**Fig. 2** Correlation between penetration depth and pitch of microrough surfaces with 1 μm × 1 μm pillar size, under fluid immersion of varying surfactant concentration.

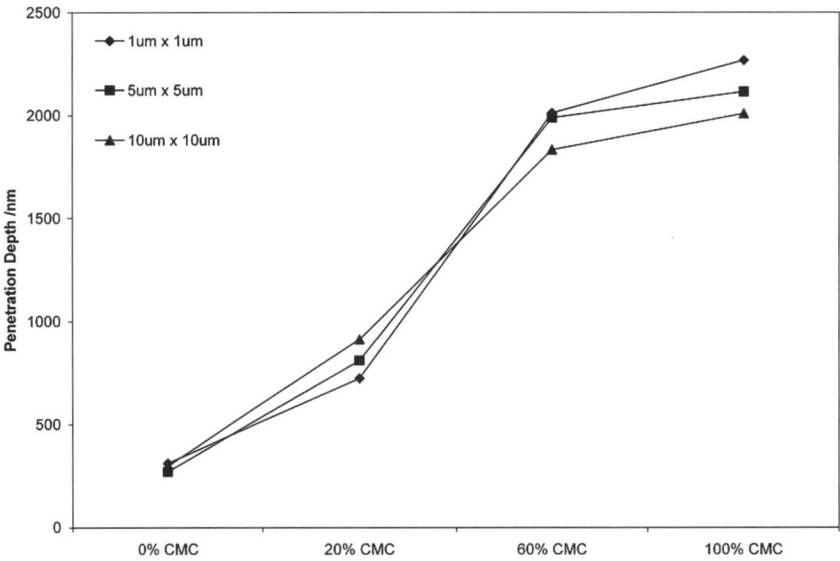

**Fig. 3** Correlation between penetration depth with pillar size of microrough surfaces with 1 μm pitch, under fluid immersion of varying surfactant concentration.

Due to the surface's chemical hydrophobicity, the $\cos\theta$ term is negative. Thus, the reduction in height is equivalent to the increase in penetration depth. This is demonstrated in Fig. 2 where the correlation between fluid penetration depth and surfactant concentration is obtained using freeze fracture AFM where the wetting of the surface is illustrated to be proportional to the pitch. The initial penetration of the wetting solution is increased with larger pitch sizes which sits well with the concept of increased roughness resulting in lower wettability. Standard capillary action wetting proceeds as per normal between 20%–60% before, the penetration depth plateaued at approximately 2.5 μm, which is the pillar height, indicating complete microwetting. As expected, increasing levels of surfactant concentration also show increasing levels of fluid penetration.

This is further confirmed by trends observed in Fig. 3, which showed there was no observable increase in the penetration depth with increasing pillar size as the pitch was kept constant at 1 μm.

### 3.2 Nanorough surfaces

Thin layers of nanoparticles were spin-coated onto flat silicon surfaces to generate a rough hydrophobic surface (Table 1). Roughness measurements, as summarized in Table 2, indicate that these coated substrates are rough only at the nanoscale.

#### 3.2.1 Wetting of nanorough surfaces.

Fig. 4 compares the SAXS profile of a 20 nm-coated silicon substrate immersed in fluids of increasing surfactant concentration. The percentile changes in $Q$ reflect the nanowetting behaviour of the fluids. It is therefore possible to elucidate the nanowetting behaviour of rough surfaces by observing percentile changes in $Q$ as a function of surface tension.

The effect of varying surface tension on nano-wetting can be seen in Fig. 5. The most notable trend is the plateauing of $Q$ despite variations in surfactant concentration for 12 nm and 20 nm nanoparticle coatings. This observation can only be

**Table 2** Macroscopic wettability and roughness characteristics of nanorough surfaces

| Coating | Roughness factor | Roughness RMS/nm |
|---------|------------------|------------------|
| 12 nm   | 1.28             | 97               |
| 20 nm   | 1.22             | 88               |
| 40 nm   | 1.14             | 64               |

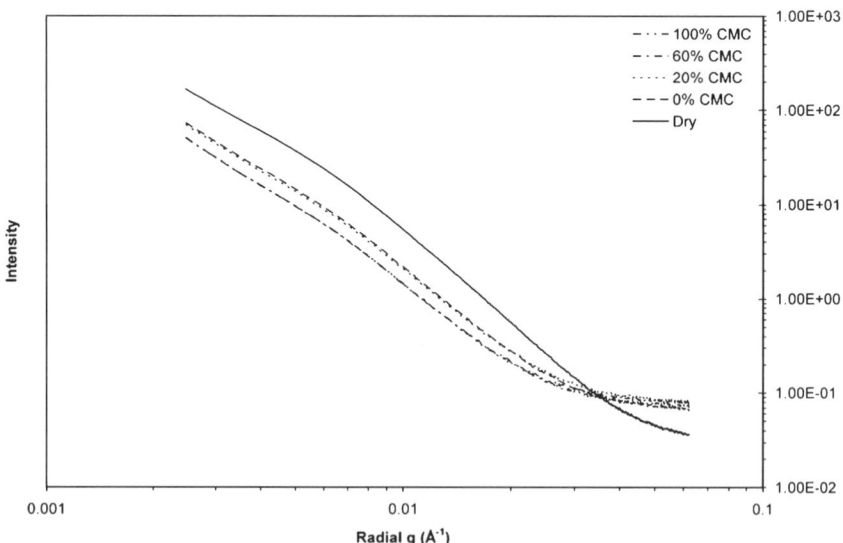

**Fig. 4** Synchrotron SAXS profile of 20 nm-coated silicon substrate under fluid immersion of varying surfactant concentration.

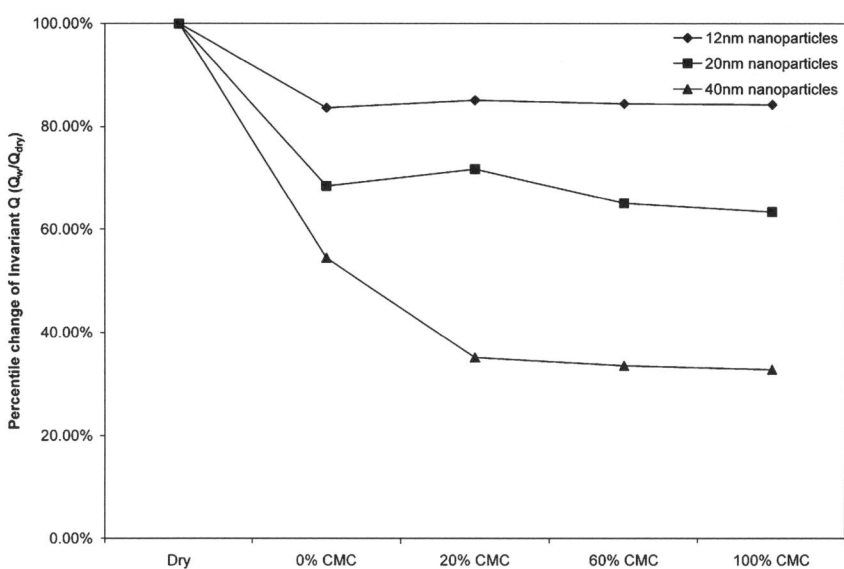

**Fig. 5** Percentile change in $Q$ as a function of surfactant concentration of nanorough surfaces.

interpreted as wetting remained constant throughout reduction in fluid surface tension after initial contact with the test fluid. Although complete wetting of the nanoroughness (Wenzel regime) was expected, hysteresis measurements of less than 8° for both coatings is indicative of a Cassie regime. This suggests that a composite interface exists for both 12 nm and 20 nm coatings. Furthermore, the percentile reduction in $Q$ for a 40 nm nanoparticle coating is substantially greater, which corresponds a large exchange between air/surface and fluid/surface interfaces. This is in agreement with the hysteresis measurements (Table 1) as a 40 nm coating was observed to possess lower roughness (Table 2), thus a higher degree of pinning.

### 3.3 Dual-scale roughness surfaces

Dual scale roughness surfaces were fabricated through a combination of the two previous scaled surfaces. A further layer of nanoparticles is spun onto a lithographically micro-patterned surface. An electron microscope image of this dual-scaled roughness surface, as shown in Fig. 6, reveals that the nanoparticle coating successfully imparted nanoroughness in the troughs of the substrate, and most importantly, atop pillars.

The contact angle of several photolithographically patterned surfaces increased dramatically into the superhydrophobic regime after applying a 12 nm nanoparticle coating as seen in Table 1. Of the tested substrates, only two exhibited superhydrophobic characteristics after the addition of nano-roughness. The aspect ratio (height : pitch) of the two superhydrophobic surfaces were 1 : 1 and 1 : 2, respectively. Typically, microrough surfaces will only exhibit superhydrophobic characteristics at aspect ratios above 4 : 1,[30] therefore the addition of nanoroughness can be seen to increase resilience against wetting.

If we consider that the behaviour of micro- and nanowetting are mutually exclusive and remains unchanged between single and dual-scale roughness surfaces, the progression of wetting in dual-scale roughness surfaces can be predicted as follows:

At 0% CMC, the wetting fluid would first partially wet the nanoroughness atop the microscale pillars and reduce invariant $Q$ from SAXS measurements. Penetration depth would be identical to microrough surface measurements of ∼250 nm as the pitch of the microroughness is unchanged.

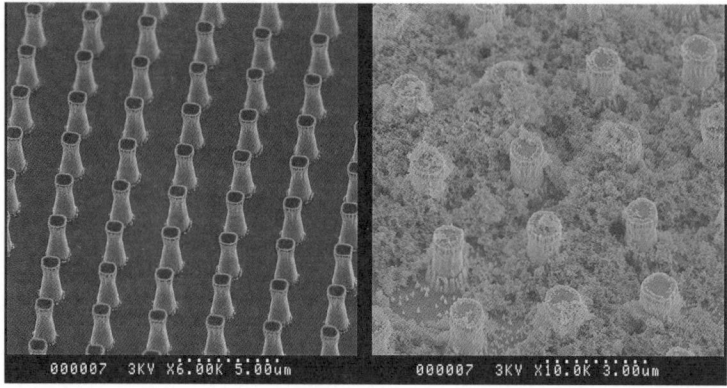

**Fig. 6** Electron microscope image of (left) microrough surface and (right) dual-scaled roughness surface.

Increasing surfactant concentration to 20% CMC, according to Fig. 5, the invariant $Q$ would remain unchanged as no further nanowetting proceeds, due to the resilience against wetting observed earlier in nano-rough SAXS measurements. However, since the sidewalls of the pillars are uncoated and remains nanoscopically smooth, microwetting would continue as described by capillary action.

According to Fig. 2, at 60% CMC, a large jump in penetration depth is observed as a result of increases in the surfactant concentration, where the wetting fluid reached the bottom of the microroughness. Therefore, the same is expected for the dual-scaled roughness surface as the sidewalls are smooth. Considering that the nanocoating is approximately 1 μm thick, the maximum penetration depth attainable for the wetting fluid would be approximately 1.5 μm, at which point, nanowetting would begin at the base of the microroughness, causing a further decline of invariant $Q$. Since the majority of the surface area is in the nanoroughness at the base of the substrate, the corresponding drop in invariant $Q$ would be comparably larger than the initial wetting observed with 0% CMC atop pillars.

Finally at 100% CMC, considering the resilience of nanoroughness against wetting, the invariant $Q$ would be expected to remain unchanged.

From Fig. 7a and 7b, the respective micro- and nanowetting behaviour is observed to correspond with the predicted model of mutually exclusive wetting behaviour of dual-scale roughness surfaces. There is, however, one discrepancy. In Fig. 7b, the onset of the 2nd decline of invariant $Q$, which is indicative of nanowetting at the base of the substrate, for 1 μm and 2 μm pitch is earlier than that of 5 μm and 20

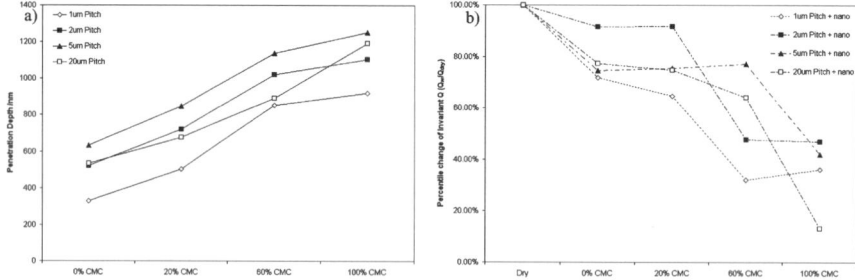

**Fig. 7** a) Comparison in microwetting behaviour of microrough and dual-scaled rough surfaces and b) corresponding nanowetting behaviour of varying pitch as a function of surfactant concentration.

μm pitch. Since both 1 μm and 2 μm pitch dual-scale roughness surfaces are superhydrophobic, a correlation exists where nanowetting at the base of the substrate occurs earlier for superhydrophobic surfaces. If we take into consideration that the only variable is the pitch between the two sets of nanowetting behaviour, this transition can be attributed to a competition between micro- and nanowetting to achieve the lowest energy state.

Wetting, as mentioned earlier, is the interaction between interfaces to achieve the lowest energy state. Thus, when wetting occurs on a dual-scale roughness surface, the fluid can either wet microscopically or nanoscopically, or even both simultaneously. From the observed transition in Fig. 7b, it is believed that the fluid, when in contact with 1 μm and 2 μm pitch SHS, achieves a lower energy state by nanowetting, however, for 5 μm and 20 μm pitch dual-scale roughness surface, a lower energy state is first achieved by microwetting. In other words, the energy expenditure of microwetting is fine-tuned by the pitch of the microroughness and ultimately governs the preferential wetting order of the entire system.

## 4. Conclusion

Microroughness was imparted onto silicon substrates using photolithographic techniques, allowing precise control of microroughness by adjusting pillar size and pitch. By spin-coating a thin layer of nanoparticles, a systematic method of generating dual-scaled roughness was possible, providing an ideal platform for wetting studies. Micro and nanowetting behaviour was successfully measured using freeze fracture AFM and *in situ* Synchrotron SAXS, respectively, with fluids of varying surface tension. The progression of wetting was tracked by correlating the penetration depth and invariant $Q$ analysis of substrates with dual-scale roughness.

By combining the wetting behaviour at both micro- and nanoscale, a wetting model was proposed to explain the transition between preferential micro- and nanowetting behaviour for dual-scaled roughness surfaces, which was observed to be tunable according to microroughness. The demonstration provided herein suggests while the wetting behaviour at both micro- and nanoscale appears to be mutually exclusive, their respective roughness ultimately governs the progression of wetting of a hierarchical roughness surface.

## Acknowledgements

The authors would like to thank Dr. Simon Lam and Dr. Jia Du (CSIRO Sydney) for their assistance provided. This research was undertaken on the SAXS/WAXS beamline at the Australian Synchrotron, Victoria, Australia. This research was supported under Australian Research Council's Discovery Projects funding scheme.

## References

1 R. N. Wenzel, *Ind. Eng. Chem.*, 1936, **28**, 988–994.
2 A. B. D. Cassie and S. Baxter, *Trans. Faraday Soc.*, 1944, **40**, 546–551.
3 W. Barthlott and C. Neinhuis, *Planta*, 1997, **202**, 1.
4 R. N. Lamb, H. Zhang and C. L. Raston, Hydrophobic Coatings; Au. Pat., 97-5789, 1997. PCT Int. Appl. WO 9842452, 1998.
5 A. W. Jones, R. N. Lamb and H. Zhang, Hydrophobic Coating Material containing Modified Gels; Au Pat., 99-2345, 1999. PCT Int. Appl. WO 0114497, 2001.
6 A. W. Jones, R. N. Lamb and H. Zhang, Durable Superhydrophobic Coating; Au. Pat., 2003901735, 2003. PCT Int. Appl. WO 090065, 2004.
7 A. Nakajima, A. Fujishima, K. Hashimoto and T. Watanabe, *Adv. Mater.*, 1999, **11**, 1365–1368.
8 A. Hozumi and O. Takai, *Thin Solid Films*, 1997, **303**, 222–225.
9 H. Y. Erbil, A. L. Demirel, A. Yonca and O. Mert, *Science*, 2003, **299**, 1377–1380.
10 K. Tadanaga, N. Katata and T. Minami, *J. Am. Ceram. Soc.*, 1997, **80**, 3213.

11 M. Miwa, A. Nakajima, A. Fujishima, K. Hashimoto and T. Watanabe, *Langmuir*, 2000, **16**, 5754–5760.
12 T. Onda, S. Shibuichi, N. Satoh and K. Tsujii, *Langmuir*, 1996, **12**, 2125–2127.
13 H. E. Jeong, S. H. Lee, J. K. Kim and K. Y. Suh, *Langmuir*, 2006, **22**, 1640–1645.
14 H. Li, X. Wang, Y. Song, Y. Liu, Q. Li, L. Jiang and D. B. Zhu, *Angew. Chem.*, 2001, **113**, 1793.
15 L. Feng, S. H. Li, Y. S. Li, H. J. Li, L. J. Zhang, J. Zhai, Y. L. Song, B. Q. Liu, L. Jiang and D. B. Zhu, *Adv. Mater.*, 2002, **14**, 1857.
16 K. K. S. Lau, J. Bico, K. B. K. Teo, M. Chhowalla, G. A. J. Amaratunga, W. I. Milne, G. H. McKinley and K. K. Gleason, *Nano Lett.*, 2003, **3**, 1701.
17 J. G. Fan, X. J. Tang and Y. P. Zhao, *Nanotechnology*, 2004, **15**, 501.
18 S. Tsoi, E. Fok, J. C. Sit and J. G. C. Veinot, *Langmuir*, 2004, **20**, 10771.
19 T. N. Krupenkin, J. A. Taylor, T. M. Schneider and S. Yang, *Langmuir*, 2004, **20**, 3824.
20 N. Michael and B. Bhushan, *Microelectron. Eng.*, 2007, **84**, 382–386.
21 B. Cortese, S. D'Amone, M. Manca, I. Viola, R. Cingolani and G. Gigli, *Langmuir*, 2008, **24**, 2712–2718.
22 L. Feng, Y. Zhang, J. Xi, Y. Zhu, N. Wang, F. Xia and L. Jiang, *Langmuir*, 2008, **24**, 4114–4119.
23 R. E. Johnson and R. H. Dettre, *J. Phys. Chem.*, 1964, **68**, 1744–50.
24 F. E. Bartell and J. W. J. Shepard, *J. Phys. Chem.*, 1953, **57**, 211.
25 S. Herminghaus, *Europhys. Lett.*, 2000, **52**, 165.
26 N. A. Patankar, *Langmuir*, 2003, **19**, 1249.
27 C. W. Extrand, *J. Colloid Interface Sci.*, 1998, **207**, 11–19.
28 E. Cheng, M. W. Cole and P. Pfeifer, *Phys. Rev. B: Condens. Matter*, 1989, **39**, 12962.
29 C. W. Extrand, *Langmuir*, 2002, **18**, 7991.
30 L. Barbieri, E. Wagner and P. Hoffmann, *Langmuir*, 2007, **23**, 1723–1734.
31 D. Öner and T. J. McCarthy, *Langmuir*, 2000, **16**, 7777.
32 Z. Yoshimitsu, A. Nakajima, T. Watanabe and K. Hashimoto, *Langmuir*, 2002, **18**, 5818–5822.
33 D. Broseta, L. Barre, O. Vizika, N. Shahidzadeh, J. P. Guilbaud and S. Lyonnard, *Phys. Rev. Lett.*, 2001, **86**, 5313.
34 J. W. G. Tyrrell and P. Attard, *Phys. Rev. Lett.*, 2001, **87**, 176104.
35 S. Singh, J. Houston, F. van Swol and C. J. Brinker, *Nature*, 2006, **442**, 526.
36 H. Zhang, R. Lamb and J. Lewis, *Sci. Technol. Adv. Mater.*, 2005, **6**, 236–239.
37 H. Zhang, R. N. Lamb and D. J. Cookson, *Appl. Phys. Lett.*, 2007, **91**, 254106.
38 A. J. Scardino, H. Zhang, D. J. Cookson, R. N. Lamb and R. de Nys, *Biofouling: The Journal of Bioadhesion and Biofilm Research*, 2009, **25**, 757–767.
39 A. H. F. Wu, G. Moran, N. Roberts and R. N. Lamb, *Mater. Res. Soc. Symp. Proc.*, 2009, **1146E**, 1146-NN06-21.
40 J. H. M. Wilison, A. J. Rowe, Replica, Shadowing and Freeze-Etching Techniques, Elsevier North-Holland Biomedical Press, The Netherlands.
41 D. Lang, *Philos. Trans. R. Soc. London, Ser. B*, 1971, **261**, 151.
42 L. Bachmann, W. W. Schmitt-Fumian, Spray-freezing-etching of dissolved macromolecules, emulsions and subceullar components, in: Freeze-etching, techniques and applications, E. L. Benedetti and P. Favard, ed. (Soc. Fr. Microsc. Elect., Paris), p.63.
43 M. Switkes, *Appl. Phys. Lett.*, 2004, **84**, 4759.
44 G. Porod, *Small Angle X-ray Scattering*, edited by O. Glatter and O. Kratky, New York, 1982, pp. 17.

PAPER

# An experimental study of interactions between droplets and a nonwetting microfluidic capillary

Geoff R. Willmott,[*a] Chiara Neto[b] and Shaun C. Hendy[ac]

*Received 4th December 2009, Accepted 28th January 2010*
DOI: 10.1039/b925588e

We present a detailed experimental study of water drops coming into contact with the end of vertical polytetrafluoroethane (PTFE) capillary tubes. The drops, supported on a superhydrophobic substrate, were between 0.06 and 1.97 mm in radius, and the inner radius of the vertical tube was 0.15 mm. These experiments expand on our recent work, which demonstrated that small water droplets can spontaneously penetrate non-wetting capillaries, driven by the action of Laplace pressure within the droplet, and that the dynamics of microfluidic capillary uptake are strongly dependent on the size of the incident drop. Here we quantitatively bound the critical drop radius at which droplets can penetrate a pre-filled capillary to the narrow range between 0.43 and 0.50 mm. This value is consistent with a water–PTFE contact angle between 107.8° and 110.6°. Capillary uptake dynamics were not significantly affected by the initial filling height, but other experimental factors have been identified as important to the dynamics of this process. In particular, interactions between the droplet, the substrate and the tubing are unavoidable prior to and during droplet uptake in a real microfluidic system. Such interactions are classified and discussed for the experimental set-up used, and the difficulties and requirements for droplet penetration of a dry capillary are outlined. These results are relevant to research into microfluidic devices, inkjet printing, and the penetration of fluids in porous materials.

## 1 Introduction

The penetration of liquid into a capillary tube, driven by surface forces, has been a topic of active research over the last century, and is important in a wide range of fields. Examples of processes utilizing capillarity include transport of liquids in plants and imbibition of fluids by porous media such as powders, soils and granular materials. Capillary tubes in contact with fluid reservoirs have been studied in depth, and the dynamics of uptake are well understood. When the reservoir is replaced by a droplet of comparable size to the tube, the physical behaviour is surprising and relatively little work has been done in this case. In a recent study,[1] we showed that small water droplets can penetrate and fill dry, hydrophobic capillaries. Further, we showed that the drop size can affect both the direction of meniscus motion in a pre-filled capillary, and the speed of capillary uptake. This work was carried out using tubing and droplets on the scale of hundreds of microns, reflecting

[a]*The MacDiarmid Institute for Advanced Materials and Nanotechnology, Industrial Research Limited, 69 Gracefield Rd, PO Box 31-310, Lower Hutt, 5040, New Zealand. E-mail: g.willmott@irl.cri.nz*
[b]*School of Chemistry, The University of Sydney, NSW, 2006, Australia*
[c]*School of Chemical and Physical Sciences, Victoria University of Wellington, Wellington, 6140, New Zealand*

recent interest and possible applications in the growing field of microfluidics. The observed effects can be explained by considering the Laplace pressure within the liquid droplet. The same general phenomenon has been studied in the context of metallic liquid droplets and carbon nanotubes using theory and molecular dynamics simulations.[2,3] This interaction is entirely general, and is relevant to many aspects of capillary research.

At a curved liquid interface, surface tension ($\gamma$) gives rise to a pressure difference across that interface, known as the Laplace pressure. This well-understood phenomenon underpins our understanding of conventional capillary uptake.[4] When the tip of a cylindrical capillary (radius $r_t$) comes into contact with the liquid–air interface of a fluid reservoir, a liquid meniscus forms inside the capillary. If the fluid wets the capillary, the contact angle between the fluid and the inner wall of the tube is $\theta_c <$ 90°, and the meniscus takes on a concave curvature (Fig. 1(a)). The pressure on the liquid side of the meniscus is lower than the ambient air pressure. This pressure difference across the curved interface ($\Delta P$) is given by the Laplace equation:

$$\Delta P = \frac{2\gamma \cos\theta_c}{r_t} \qquad (1)$$

This pressure difference drives the liquid into the capillary tube, and the meniscus stops rising when an equal and opposing pressure is applied, for example due to gravity. In this conventional description of capillary uptake, liquids with contact angles greater than 90° will form a meniscus with convex curvature at the base of the tube (Fig. 1(b)). The sign of cos $\theta_c$ is reversed, and therefore $\Delta P$ favours drainage of fluid from the capillary. The liquid cannot rise in the tube by capillary action in this case.

When the capillary is in contact with a liquid droplet instead of a reservoir, the situation is different. The Laplace pressure within a spherical droplet of radius $r$ is $\Delta P = 2\gamma/r$. Including this factor, the dynamics of capillary uptake are described by the equation:[3]

$$\rho \frac{d}{dt}\left(h\frac{dh}{dt}\right) + \frac{8\eta}{r_t^2} h \frac{dh}{dt} = \frac{2\gamma \cos\theta_c}{r_t} - \rho g h + \frac{2\gamma}{r} \qquad (2)$$

Here $\rho$ is fluid density, $\eta$ is the dynamic viscosity of the fluid, $t$ is time and $g$ is gravitational acceleration. Eqn (2) contains terms representing (from left to right, respectively) fluid inertia and viscous drag; the capillary force, gravity and the Laplace pressure within the droplet. The potential effects of the Laplace pressure term on capillary uptake of droplets were first recognised by Marmur,[5] then Schebarchov and Hendy.[2,3] There have been numerous other theoretical studies of eqn (2) (or equivalent expressions) for fluid reservoirs, taking the limit as $r \rightarrow \infty$. In this limit, the well-known Lucas–Washburn law[6,7] applies when the viscous term dominates; the Bosanquet result[8] applies when inertia dominates. The topics exciting most interest at present include dimensional analysis, recently summarised and extended

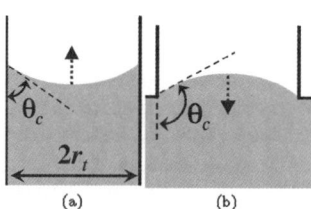

Fig. 1 Schematic cross-sectional diagrams showing meniscus configurations for a fluid within a cylindrical wetting capillary (Fig. 1(a)) and a fluid reservoir in contact with the end of a cylindrical non-wetting capillary (Fig. 1(b)). Dashed arrows indicate the direction that Laplace pressure across the curved interface would push the meniscus.

by Fries and Dryer.[9] There has also been considerable interest in identifying conditions under which different terms in eqn (2) dominate,[10,11] and analytic solutions have been explicitly derived for various particular experimental configurations.[12–14]

Considering eqn (2) with finite $r$, it is clear that the meniscus will move up the tube ($dh/dt$ will be positive) as long as the terms on the right hand side are positive. Experiments in the present work take place in a regime in which the gravity term in eqn (2) is not important, although gravity significantly affects the shape and dynamics of the drop outside the tube (discussed below). For a particular capillary tube and fluid, the direction of meniscus motion is therefore controlled by one variable: the drop size $r$. We define a critical drop radius $r_{max}$, below which theory predicts that the meniscus will move up into the tube, and above which fluid should drain from the tube:

$$r_{max} = -\frac{r_t}{\cos\theta_c} \qquad (3)$$

This result indicates that it is possible to obtain capillary rise for particular droplet/capillary combinations even in a non-wetting case. While our previous study[1] showed that smaller droplets are taken up by hydrophobic capillaries, and larger droplets are not, experiments were not able to effectively determine $r_{max}$. In fact, the data suggested that $r_{max}$ changed depending on the initial filled height of the capillary $h_0$. Experiments were carried out using vertical hydrophobic capillaries coming into contact with water droplets supported by superhydrophobic surfaces. To the best of our knowledge, the most relevant of other previous experimental work was carried out by Bazilevsky et al.[15] While developing a high-speed imaging system, these researchers studied the volume of water protruding from a glass capillary ($r_t = 0.3$ mm), and presented data showing a weak drop-size dependence of the capillary wetting rate. Other experimental work has included studies of water draining from a horizontal, non-wetting perfluoroalkoxy tube[16] and wetting capillaries which were chemically treated to increase $\theta_c$.[17,18] More conventional capillary studies have employed high-speed imaging as an experimental technique.[19–22]

In the present work, the phenomena associated with droplet uptake are studied in more detail. Experiments are focussed on PTFE tubing, with the aim of collecting enough data to quantitatively determine the value of $r_{max}$. The dependence of $r_{max}$ on $h_0$ is investigated in depth. Close attention is paid to the experimental conditions, and the effect they have on the data. The various interactions of droplets with the superhydrophobic and PTFE surfaces prior to and during uptake are classified and discussed. Finite capillary wall thickness, droplet deformation and dynamics, and the relative difficulty of penetrating a dry capillary are also discussed. Such processes are important in real world systems, in contrast to the configuration used in the development of theoretical approaches such as eqn (2). This ideal system consists of a stationary droplet, suspended in space, and in perfectly-aligned contact with the end of a tube which has walls of infinitesimal thickness.

## 2 Experimental

The PTFE capillary tube (Cole-Parmer) used in experiments was cylindrical, with inner and outer diameters of 0.31 and 0.76 mm respectively (Fig. 2(a)). A clean stainless steel razor blade was used to cut as-supplied tubing, and to score the outer surface of the tubing with five fiducial marks, approximately evenly spaced at 1 mm distance from the end of the tube and from each other (Fig. 2(b)). These marks, labelled 1 to 5, were used to guide the initial filled height of capillaries $h_0$ in experiments. Fig. 2(c) shows a magnification of the meniscus inside a tube containing a stationary plug of water. The meniscus is approximately flat, indicating a contact angle close to 90°, which is consistent with the observation that the tubing is not wet by a reservoir of water. The interior and the end of the tubing were cleaned

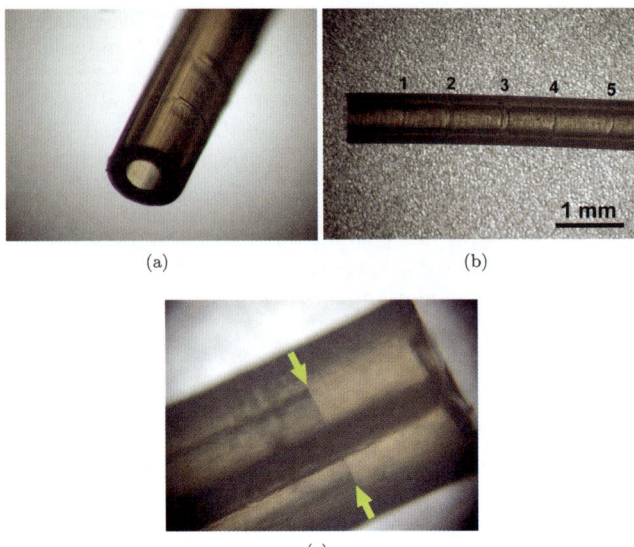

**Fig. 2** Optical microscopy images of the cylindrical PTFE tubing used in experiments. Fig. 2(a), taken with the tube at an oblique angle, demonstrates the finite thickness of the tube walls; the inner and outer diameters are 305 and 762 μm respectively. Five fiducial marks were superficially scored on the tube exterior using a clean razor blade, as shown in Fig. 2(b). The meniscus of water within the horizontal tube is visible in Fig. 2(c). The edges of the meniscus are indicated by arrows and the contact angle is indistinguishable from 90°.

in a laminar flow bench using a syringe-pumped stream of water, followed by drying in a stream of nitrogen gas. The water used for this process and in all other aspects of the experiments was deionised to MILLI-Q standard (18.2 MΩ cm). Cleaning was carried out immediately prior to the commencement of experiments each day, and at regular intervals during the experimental work.

The capillary tube was brought into contact with water droplets supported by a freshly prepared superhydrophobic substrate, fabricated using a slightly modified version of a published electroless deposition methodology.[23] Briefly, the superhydrophobic surface is a rough electrodeposited silver surface coated with a self-assembled monolayer of 1-dodecanethiol. The measured advancing and receding contact angles of water on these substrates were 173° and 161° respectively.[1] Literature values[24,25] for the contact angle of water on PTFE usually range between 105° and 115°. Droplets were periodically generated on the substrate by dispensing water from a syringe. The flow rate and angle of dispensing were controlled so that the stream of water striking the surface was broken into numerous droplets, covering the range of suitable sizes. The droplets dispersed over the superhydrophobic surface, and became preferentially (but not always) immobilised at small imperfections. The typical appearance of a droplet is shown schematically in Fig. 3(a). The affinity between any particular droplet and the substrate is best demonstrated by the contact angle on the substrate $\theta_s$ prior to uptake, which is significantly greater than $\theta_c$ in all experiments. Therefore, the droplets have greater affinity for the PTFE tubing than for the superhydrophobic surface. As described below, the drop is considered to be a partial sphere ('spherical cap') for the purposes of measurement and comparison with the theoretical analyses. The Laplace pressure within the droplet does not depend on $\theta_s$.

Fig. 3(b) is a typical photographic image taken from an experiment, and demonstrates the experimental setup. A particular droplet on the substrate is aligned so that the centre of the drop is directly below the centre of the vertical capillary tube. Alignment of the tube is aided by attaching it to a straight-edged support.

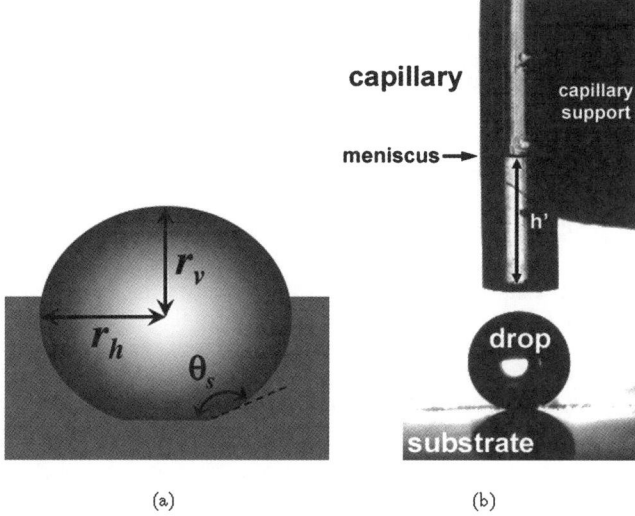

**Fig. 3** The experimental configuration. Fig. 3(a) is a schematic diagram of a water droplet on a superhydrophobic surface, as observed in experiments. The horizontal radius ($r_h$) and the vertical radius ($r_v$) of the upper half of the droplet and the contact angle on the substrate are indicated. Fig. 3(b) is an example of a photographic frame such as those captured in experiments. The capillary is prefilled to an initial height $h'_0$. The substrate can move upwards to bring the drop into contact with the capillary.

The meniscus of water in the pre-filled capillary is clearly visible, and has height $h'$. The experiment is initiated by slowly raising the stage until the droplet interacts with the capillary tube. Note that the approach velocity is insignificant in comparison with the dynamics of drop motion and deformation observed in experiments. Uptake was observed using a contact angle measurement apparatus (CAM 200, KSV Instruments), with a minimum interframe time of 16 ms. Photographs were often overexposed to ensure clear observation of the meniscus.

The setup was enclosed in a confined aluminium box (6 × 6 × 5 cm) containing a water reservoir added some time before experiments commenced to create a saturated vapour pressure in the box and to slow the evaporation of droplets. The box, which had an open top to provide access for the capillary and perspex windows to allow photography, was placed on the stage of the contact angle measurement system. Otherwise, experiments were carried out under ambient laboratory conditions.

Experiments were analysed by studying the photographic frames captured before, during and after the capillary came into contact with the droplet. Photos were spatially calibrated using the known diameter of the tubing. The key quantitative measurements were the following:

1. The drop radius ($r_{avge}$), taken to be the average of the horizontal radius and the vertical radius of the upper half of the droplet ($r_h$ and $r_v$ respectively, Fig. 3(a)). This quantity was measured using any frame prior to contact in which the drop shape was not affected by the presence of the tubing, but may have been aspherical due to gravity.

2. The initial filling height of the water in the PTFE tubing ($h'_0$). This was the measured value of $h'$ (Fig. 3(b)) in any frame prior to contact. As in previous work,[1] the position at which the liquid meniscus meets the internal edge of the tube is not visible in photographs. The measurement from the base of the tube to the visible central height of the meniscus $h'$ is used as an estimate for the height of the meniscus at the inner tube wall $h$. The difference between $h'$ and $h$ is small because the contact angle is close to 90° (Fig. 2(c)).

3. The change in the meniscus height $\Delta h'$, measured by comparing any frame prior to contact with a frame some time after contact, when meniscus motion was effectively arrested. This comparative technique resulted in a spatial resolution for $\Delta h'$ limited only by photographic resolution.

## 3 Results and discussion

### 3.1 Critical drop size

Data showing the dependence of $\Delta h'$ on the average radius of the droplets is plotted alongside previous data in Fig. 4(a). This figure shows that, independently of the initial meniscus height $h'_0$, $\Delta h'$ is positive (*i.e.* the droplet is taken up by the hydrophobic capillary) as long as the droplet is small enough. At larger droplet size, $\Delta h'$ becomes negative, in which case the droplet is not taken up, but instead empties the capillary. The maximum droplet radius at which uptake occurs is $r_{max}$, as defined in eqn (3). Data for values of $r_{avge}$ between 0.3 and 0.7 mm, which certainly bound

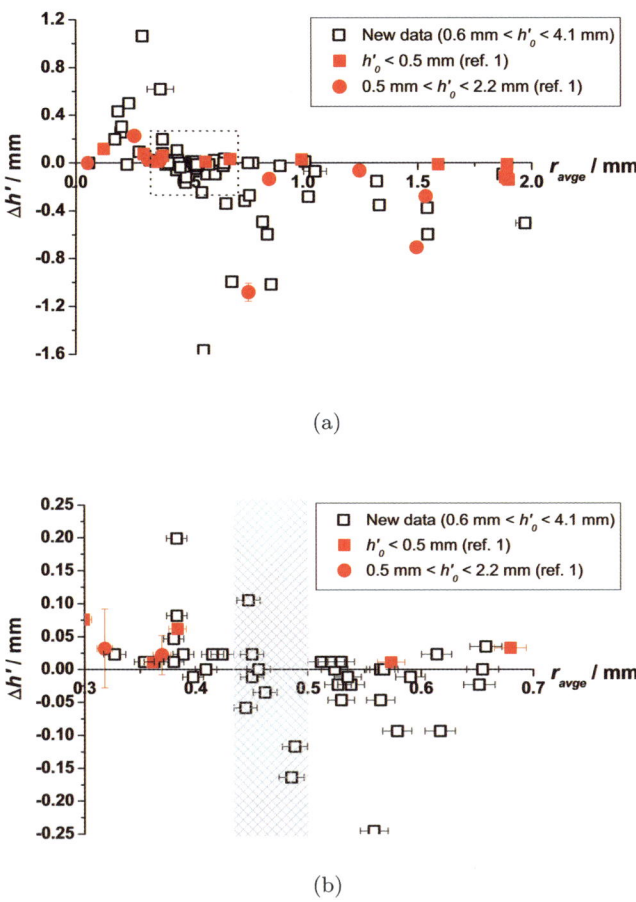

(a)

(b)

**Fig. 4** The change in meniscus height in prefilled capillaries is plotted as a function of drop size. Fig. 4(a) shows all the new data, with the most interesting, data-rich region of Fig. 4(a) (bounded by a dashed-line box) magnified in Fig. 4(b). The shaded grey area indicates the bounding of $r_{max}$ using those data for which $|\Delta h'| > 0.05$ mm, as discussed in the text. The error bars, plotted only when they clearly exceed the size of the data point, are largely due to scaling and resolution of photographs.

$r_{max}$, are magnified in Fig. 4(b). Outside this region, there are some cases in which little or no meniscus movement occurs. Note that $\Delta h'$ can be limited by the finite droplet size or initial filling height.

In Fig. 4(b), little or no meniscus movement is often observed, and there is some inconsistency in the trend between $r_{avge}$ and the sign of $\Delta h'$ in this magnified region. Nonetheless, the transition from filling to drainage with increasing $r_{avge}$ is evident. By considering only data for which $|\Delta h'| > 0.05$ mm, $r_{max}$ can be restricted to between 0.43 and 0.50 mm, as indicated by the shaded grey area in Fig. 4(b). Using eqn (3), the equivalent static contact angle between water and the PTFE tubing is between 107.8° and 110.6°, in agreement with expected values. The extent of this bounded range is comparable with the usual experimental uncertainty associated with contact angle experiments, and is much smaller than the variation in literature values for the PTFE–water contact angle.

In order to investigate the possible influence of initial filling height on $r_{max}$, experiments were divided into five groups for which $h'_0$ was approximately equivalent. Each group corresponded roughly to a fiducial mark on the outside of the tube (Fig. 2). Fig. 5 compares data for the various groups, which are divided by horizontal dashed lines. The data is further categorized based on whether the level of the meniscus was observed to increase, decrease, or not discernibly change in the drop uptake experiment. Within each group, the range of $r_{max}$ has been identified, indicated by vertical dotted lines. For each of the groups 1 through to 5, this range is bounded by 0.23 mm and 1.01 mm and is consistent with the range for $r_{max}$ derived from all of the data in Fig. 4. Perhaps the most striking feature of Fig. 5 is that the extent of the identified range of $r_{max}$ is much larger in groups 4 and 5, which have the greatest initial filling height.

Previously,[1] it was observed that drops with radii between 0.5 and 1.0 mm appeared to fill capillaries near $h'_0 = 0.2$ mm, but drain them for higher values of $h'_0$. This dependence of $h'_0$ is not explained by the force balance analysis described in the Introduction, in which $r_{max}$ is independent of $h'$. The data for groups 1 to 3 in

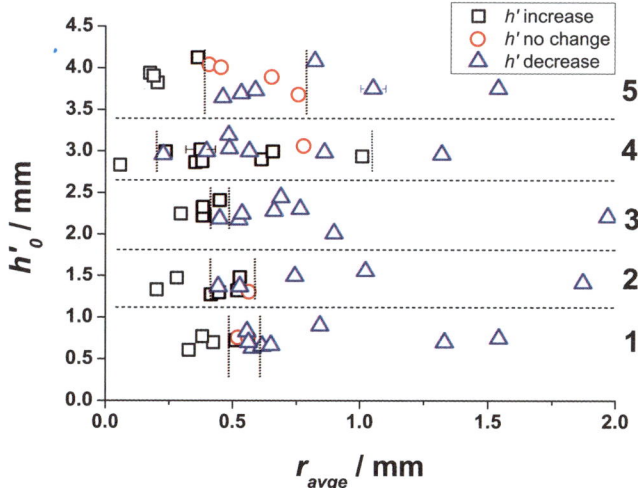

**Fig. 5** Data showing the influence of the initial filling height $h'_0$ on the change in meniscus height. Horizontal dashed lines separate five groups, roughly defined according to meniscus alignment with the fiducial markings (Fig. 2(b)), and numbered on the right hand side of the plot. Categories in the legend refer to the direction of meniscus motion after the drop comes into contact with the capillary. Vertical dotted lines span the region (within each group) where data from the three categories coexist. Error bars are plotted when they clearly exceed the size of the data point.

Fig. 5 suggest that $r_{max}$ increases slightly with decreasing $h'_0$, but not to the extent observed in the previous study, where $r_{max}$ was observed to lie between 0.99 and 1.59 mm for $h'_0 < 0.5$ mm. Moreover, the previous data is relatively sparse (Fig. 4), and the inconsistency in $r_{max}$ can credibly be attributed to scatter in those results rather than any particular trend.

The transition between filling and drainage has been effectively, and quantitatively, addressed using the extensive data set presented here. However, some of the most interesting aspects to be drawn from these experiments relate to factors underlying the scatter in the data. One major type of experimental variation is discussed in the next Section; a further important experimental factor to recognise is the detail of the interaction between the meniscus and the capillary inner walls. This interaction is likely to be responsible for the wide variation in the magnitude of $\Delta h'$, arising from apparent 'arrest' of the meniscus motion in many cases. As noted above, this occurs even for drops of radius significantly greater or smaller than $r_{max}$. Arrest of fluid drainage could arise from pinning of the meniscus to microscopic asperities on the inner wall of the capillary. There is a parallel with the hysteresis observed in advancing and receding contact angle measurements, a phenomenon usually attributed to surface roughness. Prewetting of a surface, resulting in a liquid film precursor, is also known to affect wetting properties.

### 3.2 Modes of interaction

The process of bringing a droplet into contact with the end of a capillary necessarily involves interactions between the droplet, the tubing and any other supporting material, such as the substrate in the present experiments. Such interactions do not form part of the modelled, theoretical system, but are unavoidable in any practical nano- or microfluidic system. Our experiments have closely replicated the theoretical system by supporting the drop on a superhydrophobic substrate, so that (i) the approach velocity could be constrained to negligible levels, (ii) the drop could be effectively aligned with the end of the tube, and (iii) the drop could retain high curvature and be almost spherical, with higher affinity for the PTFE than for the substrate. In this Section we identify five distinct modes of interaction which occur in our experiments, and comment on how they affect the data. These different modes of interaction account for the scatter in the data in Fig. 4 and 5. Generally, the drop's affinity for the substrate depends on the roughness and imperfections of the superhydrophobic surface. The effect of gravity is also variable, depending on the drop size. Gravity can distort the drop geometry from a spherical cap, and can also work against the drop leaving the substrate.

The modes, demonstrated pictorially in Fig. 6, are classified as follows:

Mode A (Fig. 6(a)): in this case, the droplet 'jumps into contact' with the capillary tube, resulting in complete uptake within one or two frames. This mode is most commonly observed for small droplets with very low affinity for the substrate, so that attraction to the capillary (and the water within it) easily overcomes gravity. The velocity of the drop as it comes into contact with the capillary may not be negligible, and the drop shape may also be slightly distorted.

Mode B (Fig. 6(b)): the droplet jumps into contact with the capillary tube, but is not completely absorbed into the tube even though its radius may be smaller than $r_{max}$.

Mode C (Fig. 6(c)): at contact, the droplet does not jump, but remains in contact with the substrate, although the drop shape may be distorted due to affinity for the PTFE. The stage is lowered slightly after contact so that the drop is pulled off the surface. The drop is then attached solely to the tubing, as in the ideal case, but undergoes some elongation during the pull off process, as in the frame at 432 ms. This deformation reduces the curvature (and thus internal pressure) in the droplet, encouraging fluid drainage from the tube.

Mode D (Fig. 6(d)): as for mode C, but in this case the attempt to remove the drop from the surface by lowering the stage fails. This occurs when the droplet is located

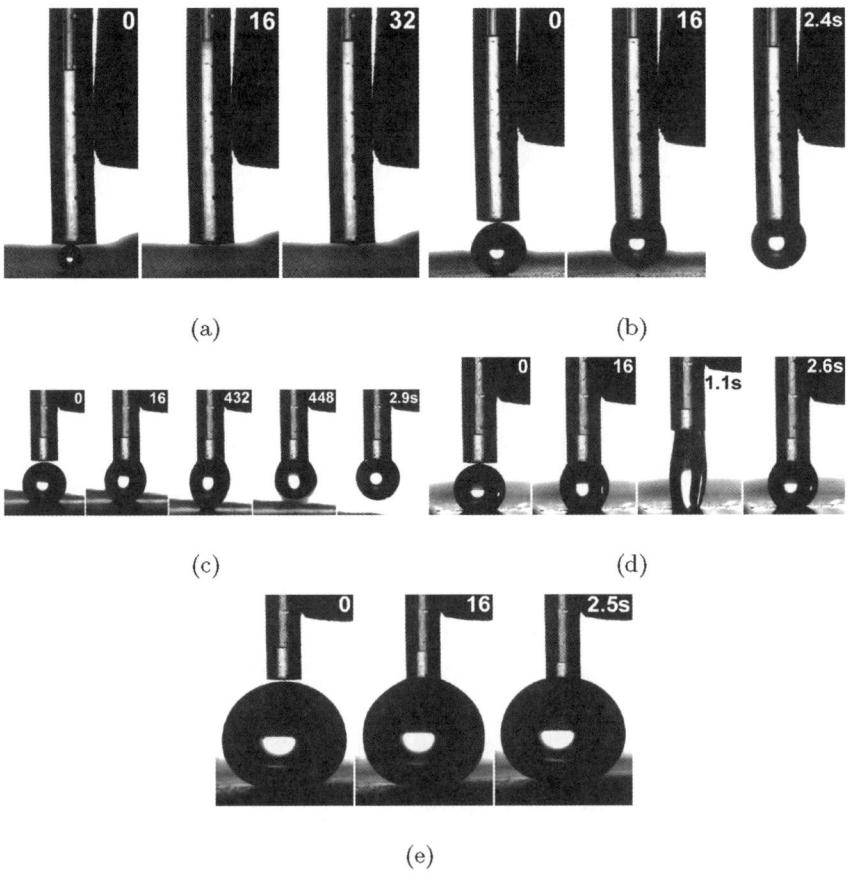

**Fig. 6** Photographic sequences demonstrating the five modes of interaction between droplet, substrate and prefilled capillary described in the main text. Each frame is labelled with a relative time in milliseconds, unless explicitly labelled as seconds. For each sequence, the first two or three frames capture the moment of contact. In Fig. 6(a) (mode A), complete uptake of the drop is almost immediate. In Fig. 6(b) (mode B), the drop jumps into contact and persists on the end of the tubing. In mode C (Fig. 6(c)), the droplet is pulled off from the surface, as in the frames at 432 and 448 ms, eventually coming to rest attached to the end of the tube. The frame at 1.1 s in Fig. 6(d) demonstrates extreme elongation of the droplet during an attempt to remove it from the surface; the drop is later returned to a near-spherical shape in contact with both substrate and tubing. In Fig. 6(e) (mode E), little droplet deformation occurs either at contact or afterwards, and the drop remains in contact with both tube and substrate.

on a particularly large or hydrophilic patch on the substrate, and/or the droplet is relatively large. The stage is lowered after contact, then returned to a height so that the drop is approximately spherical. This is when the final measurement of $h'$ is made. In some Mode D experiments, compression of the droplet also occurs.

Mode E (Fig. 6(e)): there is no jump into contact; the drop is simultaneously in contact with the substrate when it comes into contact with the tubing, but no attempt is made to remove the drop from the substrate. Therefore, there is little or no distortion of the drop shape. This mode was typically observed for large drops.

The data from Fig. 4 is replotted according to these modal groups in Fig. 7. There is strong correlation between the different modes and the drop size, and the different modes and the observed $\Delta h'$. This is unsurprising, given that gravity plays a role in determining how easily the droplet detaches from the substrate. The division of data into modes does not significantly affect the main trend of $\Delta h'$ versus $r_{avge}$ or the value

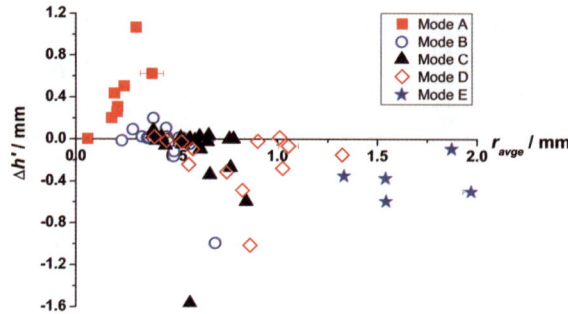

**Fig. 7** Data from the present work, replotted according to the five modes of interaction described in the text. Error bars are plotted when they clearly exceed the size of the data point.

of $r_{max}$. It is interesting that some relatively large mode B drops jump into contact with the capillary, but immediately extract fluid from the tubing (*e.g.* Fig. 6(b)), suggesting that the drop velocity caused by the initial jump into contact does not significantly affect drainage dynamics.

The relative sizes of $\Delta h'$ for the various modes are of particular interest. The division between modes A and B does seem significant, because the greatest volumetric uptake in all experiments is observed for mode A droplets. The boundary between droplets exhibiting mode A and B behaviours is approximately equal to the tube inner diameter (0.31 mm). It is likely that mode A droplets do not significantly interact with the PTFE prior to absorption into the fluid at the end of the tube. In contrast, mode B drops initially protrude from the end of the capillary, so that there is a triple boundary between water, air and PTFE around the capillary opening. If the triple boundary remains pinned and the drop is small enough to fill the tubing, the curvature of the droplet surface will gradually decrease, so that filling is eventually arrested. The drop size ranges for modes A and B overlap, suggesting that the same droplet can be entirely or partially absorbed, with probable dependence on alignment of the tubing with the droplet or possibly the relative velocity of tubing and droplet at contact.

The differences between the modes are less clear when it comes to fluid drainage. It might have been expected that the drained volume would be greatest for modes C and D, in which drops can undergo significant elongation and obtain low surface curvature. In fact, as Fig. 6(d) shows, fluid is drained from the capillary during the elongation stage, but the tube is again filled when the drop is returned to a spherical geometry and the final measurement of $h'$ is made. The overall trend for fluid drainage is supported by data for each of the modes from B through to E.

### 3.3 Uptake from dry capillaries

Previously,[1] it was shown that small droplets can penetrate a dry, nonwetting capillary. As part of the experimental programme presented above, we attempted to study the critical drop radius for penetration when $h_0 = 0$. It became clear that penetration of a dry capillary is more difficult to achieve than adding fluid to a pre-filled capillary. A typical attempt to fill a dry capillary is shown in Fig. 8: the droplet adheres to the rim of the capillary in preference to wetting the bore. In contrast, a droplet coming into contact with a pre-filled tube will preferentially coalesce with the fluid already in the tube. A jump into contact is prevalent in these experiments, because droplets pinned to the substrate by gravity or surface forces are likely to be too large to wet the capillary.

Filling of dry, nonwetting capillaries is likely to be important in some applications. It is worth considering the likely requirements to achieve such penetration in a reproducible fashion. Firstly, the droplet should be precisely aligned with the

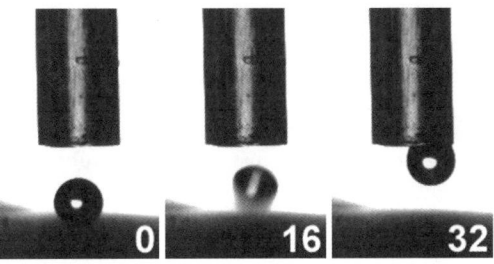

**Fig. 8** An experiment in which a small droplet jumps into contact with a dry capillary. The droplet adheres to the outer rim of the tube base, and does not wet the bore. Frames are labelled with the relative time in milliseconds.

end of the tubing, so that it is not preferentially attracted to any particular part of the rim. The general standards for alignment in the present work are apparent in Fig. 8 — the tubing is vertical to within 2°, and centred over the top of the drop to within one photographic pixel (∼12 μm). Future attempts to achieve drop penetration of a dry capillary should exceed this alignment precision. Secondly, the surface area of the tube rim should be minimised. This surface area lies between the outer and inner tube diameters, clearly shown for the PTFE tubing in Fig. 2(a). Most of the theoretical analyses describing critical uptake, including all of those which have considered droplets, assume that this rim area is negligible.

## 4 Further experimental considerations

The nature of capillary uptake dynamics changes significantly depending on the particular experimental parameters involved. In particular, dynamics are determined by the force terms in eqn (2). Parameters used to assess the dominance of the various forces, and to justify the assumptions underlying key theoretical approaches, were introduced and discussed previously.[1] Here we present and discuss calculated values of these parameters.

Throughout this work, it is assumed that gravity has no significant effect on the fluid within the capillary. The ratio of gravitational to interfacial forces, known as the Bond number (Bo = $(\rho g r_t^2)/\gamma$) is low (3.12 × 10$^{-3}$). The hydrostatic limit would be 98 mm for a fully wetting capillary, far in excess of the filling heights used.

When the fluid first enters a dry tube, inertia and capillary forces dominate, because the viscous term is zero at $h = 0$. In this regime, entrance effects such as meniscus reorientation and pre-linear inertial acceleration are important; these terms are not included in the derivation of eqn (2) and (3). The meniscus height below which entrance effects are significant is $h_1 = r_t/2 = 76$ μm. With the exception of those experiments in which dry capillaries were specifically studied, $h'_0 > 0.5$ mm in the present work.

It is also of interest to determine whether the viscous force dominates inertia, in which case the dynamic regime is equivalent to the Lucas–Washburn description (for filling a wetting capillary from a reservoir) and the theoretical approach employed by Schebarchov and Hendy[2,3] applies. The time scale at which viscous effects are not important is $t_2 = \rho r_t^2/\eta = 23.1$ ms. In all the experiments described above, the final capillary height is measured at least two frames (32 ms) after contact, although often the observed meniscus motion predominantly (or entirely) takes place during the first two photographic frames. Note that the theoretical value of $r_{max}$ is independent of whether the dynamics are dominated by viscosity or inertia. We conclude that eqn (2), (3) and Schebarchov and Hendy's analysis[2,3] all apply to the present experiments.

The experiments most significantly differ from the idealised theoretical case in the following ways:

1. The contact angle at the interior wall $\theta_c$ is likely to depend on contact angle hysteresis, precursor wetting and possibly meniscus dynamics.

2. The droplet deforms rapidly at the time of contact, or jumps into contact, due to the interaction between the drop, the substrate and the tube. This potentially introduces dynamic or curvature effects.

3. Curvature of the droplet can change due to the effect of gravity on drop shape and post-contact deformation caused by relative motion of the tube and substrate.

Scatter in the experimental data is likely to be attributable to a combination of these factors. By way of example, consider the experiments in which the meniscus is observed to stop moving (arrest) following little or no initial motion. This observation is not consistent with theory, which predicts that filling or drainage, once started, should continue until the drop shape changes significantly. Several likely causes of meniscus arrest have been identified. As discussed above, the meniscus may be pinned to the capillary internal wall. In mode B interactions, it is possible for droplet curvature to significantly change shortly after contact. High jump velocities or short-lived drop deformation could instigate motion at the point of contact, before quickly disappearing. Perhaps most curiously, arresting menisci (and associated scatter) appear more frequently for greater $h_0$, suggesting some resistance to fluid motion that depends on initial filling height. In general, meniscus arrest is responsible for much of the observed scatter, and several explanations for this behaviour are possible.

The droplet Laplace pressure has been inferred using $r_{avge}$, a measurement made when the droplet is at rest, not interacting with the capillary tube, and assumed to be a spherical cap. It would be more accurate to calculate this pressure based on the precise deformed droplet shape at the point of contact with the tube; and it would be still more rigorous to dynamically follow the droplet morphology as it deforms with time. The nature of droplet deformation varies from experiment to experiment, so such a detailed analysis would preclude a consistent comparison of many experiments, which has been enabled here by using $r_{avge}$. A likely direction for future work would be comparison of droplet uptake for single experiments with a detailed, dynamic computational prediction.

Further, more specific, aspects of these experiments lend themselves to discussion. For example, the fluid penetration dynamics have not been closely followed as a function of time, and comparison of uptake speeds could be compared with the relevant theory in future work. Also, the no-slip boundary condition is assumed to apply throughout the present work. Non-zero slip at the water–PTFE interface would affect the uptake dynamics,[3] and studies of slip using capillary uptake are possible in future work.

## 5 Conclusions

This paper has detailed experiments that were robust enough to measure (with some precision) $r_{max}$, the maximum droplet radius for which a droplet can spontaneously penetrate a non-wetting capillary tube, due to the Laplace pressure within the droplet. In the process, drop size-dependent phenomena hypothesized in theory, and briefly demonstrated previously, have been confirmed and explored. There is no evidence for dependence of $r_{max}$ on initial filling height, but scatter in the data slightly increases with filling height. The promise of this work is perhaps best demonstrated by the process used in order to pre-fill the tubing near to a predetermined height, defined by a fiducial mark. This was achieved using droplet–capillary interactions. The tube could be filled using small droplets, for which $r < r_{max}$. Slightly larger droplets could also be used to fill the tube, by raising the stage so that the water surface curvature was increased, effectively 'squashing' the droplet until water was pressure-driven into the capillary. If overfilled, water could be extracted from the tube by bringing it into contact with a larger droplet. Simple fluid manipulations such as these are likely to be useful in microfluidics and a wide range of other application areas.

## 6 Acknowledgements

The Royal Society of New Zealand's International Science and Technology Linkages Fund and Marsden Fund have supported this work. The authors would like to thank members of the IRL nano- and microfluidics team for assistance and discussions, especially Rod Stanley for preparing the substrates.

## References

1 G. R. Willmott, C. Neto, and S. C. Hendy, 2009, submitted.
2 D. Schebarchov and S. C. Hendy, *Nano Lett.*, 2008, **8**, 2253.
3 D. Schebarchov and S. C. Hendy, *Phys. Rev. E: Stat., Nonlinear, Soft Matter Phys.*, 2008, **78**, 046309.
4 P. G. Gennes, F. Brochard-Wyart, and D. Quéré, *Capillarity and Wetting Phenomena. Drops, Bubbles, Pearls, Waves*, Springer, New York, 2004.
5 A. Marmur, *J. Colloid Interface Sci.*, 1988, **122**, 209.
6 R. Lucas, *Kolloid-Z.*, 1918, **23**, 15.
7 E. W. Washburn, *Phys. Rev.*, 1921, **17**, 273.
8 C. H. Bosanquet, *Philos. Mag., Ser. 5*, 1923, **45**, 525.
9 N. Fries and M. Dreyer, *J. Colloid Interface Sci.*, 2009, **338**, 514–518.
10 M. Stange, M. E. Dreyer and H. J. Rath, *Phys. Fluids*, 2003, **15**, 2587.
11 N. Fries and M. Dreyer, *J. Colloid Interface Sci.*, 2008, **327**, 125–128.
12 N. Fries and M. Dreyer, *J. Colloid Interface Sci.*, 2008, **320**, 259–263.
13 R. Chebbi, *J. Colloid Interface Sci.*, 2007, **315**, 255–260.
14 A. Hamraoui and T. Nylander, *J. Colloid Interface Sci.*, 2002, **250**, 415–421.
15 A. V. Bazilevsky, K. G. Kornev, A. N. Rozhkov and A. V. Neimark, *J. Colloid Interface Sci.*, 2003, **262**, 16–24.
16 N. Ichikawa and Y. Satoda, *J. Colloid Interface Sci.*, 1994, **162**, 350.
17 B. V. Zhmud, F. Tiberg and K. Hallstensson, *J. Colloid Interface Sci.*, 2000, **228**, 263–269.
18 A. Siebold, M. Nardin, J. Schultz, A. Walliser and M. Oppliger, *Colloids Surf., A*, 2000, **161**, 81.
19 D. Quéré, *Europhys. Lett.*, 1997, **39**, 533.
20 A. Hamraoui, K. Thuresson, T. Nylander and V. Yaminsky, *J. Colloid Interface Sci.*, 2000, **226**, 199–204.
21 H. T. Xue, Z. N. Fang, Y. Yang, J. P. Huang and L. W. Zhou, *Chem. Phys. Lett.*, 2006, **432**, 326.
22 A. A. Jeje, *J. Colloid Interface Sci.*, 1979, **69**, 420.
23 I. A. Larmour, S. E. J. Bell and G. C. Saunders, *Angew. Chem., Int. Ed.*, 2007, **46**, 1710.
24 A. B. Ponter and A. P. Boyes, *Nature*, 1971, **231**, 354.
25 W. Loh, J. R. Lopes, and A. C. S. Ramos, in *Polymer Interfaces and Emulsions*, Marcel Dekker, New York, 1999, p. 409.

PAPER

# Hydrophobic interactions in model enclosures from small to large length scales: non-additivity in explicit and implicit solvent models

Lingle Wang, Richard A. Friesner and B. J. Berne

*Received 3rd December 2009, Accepted 22nd January 2010*
DOI: 10.1039/b925521b

The binding affinities between a united-atom methane and various model hydrophobic enclosures were studied through high accuracy free energy perturbation methods (FEP). We investigated the non-additivity of the hydrophobic interaction in these systems, measured by the deviation of its binding affinity from that predicted by the pairwise additivity approximation. While only small non-additivity effects were previously reported in the interactions in methane trimers, we found large cooperative effects (as large as $-1.14$ kcal mol$^{-1}$ or approximately a 25% increase in the binding affinity) and anti-cooperative effects (as large as 0.45 kcal mol$^{-1}$) for these model enclosed systems. Decomposition of the total potential of mean force (PMF) into increasing orders of multi-body interactions indicates that the contributions of the higher order multi-body interactions can be either positive or negative in different systems, and increasing the order of multi-body interactions considered did not necessarily improve the accuracy. A general correlation between the sign of the non-additivity effect and the curvature of the solute molecular surface was observed. We found that implicit solvent models based on the molecular surface area (MSA) performed much better, not only in predicting binding affinities, but also in predicting the non-additivity effects, compared with models based on the solvent accessible surface area (SASA), suggesting that MSA is a better descriptor of the curvature of the solutes. We also show how the non-additivity contribution changes as the hydrophobicity of the plate is decreased from the dewetting regime to the wetting regime.

## 1 Introduction

Hydrophobic interactions (HI) play a very important role in the formation and stability of many self-assembled aggregates and biological structures,[1] and are considered to be the driving force for protein folding.[2,3] A fundamental understanding of HI is crucial to the study of many important biological phenomena, such as protein folding, micelle formation and protein–ligand binding.

An important question concerning the protein folding problem is whether hydrophobic associations are pairwise additive, cooperative, or anti-cooperative.[4,5] In other words, is the potential of the mean force (PMF) holding a cluster of $n$ hydrophobic particles together equal to the sum of pairwise interaction free energies (or pair pmfs), or is it more negative (cooperative) or more positive (anti-cooperative) than what is predicted by the pairwise additivity approximation.

Nemethy and Scheraga[6] found that the hydrophobic interactions between more than two solute particles can't be expressed as a sum of pairwise solute–solute

*Department of Chemistry, Columbia University, 3000 Broadway, New York, NY, 10027, USA*

interactions. Palma also observed the non-additivity of solvent induced potential of mean force for hydrophobic particle solutions by molecular dynamics simulation.[7] In 1997, Rank and Baker[8] studied three methane molecules in an isosceles triangle geometry with two methane molecules at contact distance forming a fixed base, and they found that the three-body PMF was anti-cooperative for distances up to 6.5 Å. The conclusions of this work were not reliable, however, because the error associated with the baseline of the PMF was of the same order as the non-additive term itself. After that, Chan[4,9] and Scheraga[10–12] performed studies of methane trimers using the weighted histogram analysis method (WHAM) and the test particle insertion method respectively, and found contradictory results. After exchange of comments between these two groups,[13–16] they found that the disagreement came from the assignment of the baseline for the PMFs. Comparing these two methods Scherage[17] stated that the methane dimer seemed to be the largest system that can be treated by the test particle insertion method. The effect of size, pressure, temperature, and salts on the non-additivity of the three particle system was also investigated.[18–21] Recently, cooperative effects on the association of four methane molecules were also investigated by Scheraga's group.[22]

In this paper, we used the FEP method to study the binding affinities between a united-atom methane and various model hydrophobic enclosures. The binding affinities were compared with the predictions of the pairwise additivity approximation to assess the non-additive contributions to the hydrophobic interactions in these systems. Two different empirical models were also used to predict the binding affinities and non-additivity effects, and comparisons were made with the FEP reference data. We found that the molecular surface area (MSA, or Connolly surface area) model performed much better than the solvent accessible surface area (SASA) model for the predictions of both the binding affinities and the non-additivity effects, which is consistent with previous findings.[4,8,10,11] Detailed analysis of these two models indicates that the SASA model can't predict the cooperative effects, but the MSA model can. Decomposition of the total PMF into increasing orders of multi-body interactions indicated that the higher order multi-body interactions can be either positive or negative, and increasing the order of the multi-body interactions considered did not necessarily improve the accuracy.

## 2 Definition of cooperative and anti-cooperative effects

In general, for a solution with $n$ solute particles forming a cluster, the potential of mean force (PMF), $W(1, 2,..., n)$, is related to the $n$-particle correlation function $g^{(n)}(1, 2,..., n)$ by the following definition:[23]

$$g^{(n)}(1, 2,..., n) = e^{-\beta W(1, 2,..., n)} \quad (1)$$

where $\beta^{-1} = k_B T$, $k_B$ is Boltzmann's constant, and $T$ is the temperature. $W(1, 2,..., n)$ is a short-hand notation for $W(r_1, r_2,..., r_n)$, where $r_i$ denotes the position of the $i$-th particle, and corresponds to the free energy required to bring $n$ particles from infinitely far apart to the current configuration.

The $n$-particle PMF can be decomposed into single-body, pairwise and multi-body contributions:

$$W(1,2,...,n) = F(1,2,...,n) - \sum_{i=1}^{n} F(i) \\
= \sum_{i<j} \delta F(i,j) + \sum_{i<j<k} \delta F(i,j,k) + ... + \delta F(1,2,...,n) \quad (2)$$

$$W(1, 2,..., n) = W_2 + W_3 + ... + W_n \quad (3)$$

Where $F(1, 2,..., n)$ is the hydration free energy for the specified configuration of the solute particles, $F(i)$ is the hydration free energy of solute article $i$ in an infinitely dilute solution, $\delta_F(i, j)$ is the same as the normalized two-body PMF $W(i, j)$, and $\delta_F(i,j,...)$ corresponds to subsequent higher order multi-body interactions. $W_2$ is the sum of pairwise contributions, and $W_m$ is the sum of $m$-body interactions. Truncating the series to an $n$-body term leads to the Generalized Kirkwood Superposition Approximation (GKSA) to the $n$-th order.[24,25] Specifically, the hypothesized pairwise additivity of PMF is obtained by truncating the series at the pairwise term. In clusters this approximation neglects shielding effects where a third particle could shield a pair from other particles and from the solvent.

For the three-particle case, pairwise additivity is equivalent to the Kirkwood superposition approximation,[26–28] and $\delta_F(i, j, k)$ measures the non-additive part of the three body interactions. Cooperativity is defined when $\delta_F(i, j, k)$ is negative, meaning the free energy between the third particle and the remaining two particles is more negative and the configuration is more favorable than pairwise additivity predicts, and similarly, anti-cooperativity is defined when $\delta_F(i, j, k)$ is positive, meaning the free energy between the third particle and the remaining two particles is more positive and the configuration is less favorable than pairwise additivity predicts. For the $n$-particle case, if the interaction between a specific particle and the remaining $n - 1$ particles is equal to the sum of the pairwise interactions between that specific particle with each of the other $n - 1$ particles, and if this condition holds for each of the $n$ particles, the total $n$-body PMF will be equal to the sum of the pairwise interactions. So we further generalize the cooperative and anti-cooperative concepts to the $n$-particle case: when the interaction energy between one specific particle and the remaining $n - 1$ particles is more negative than the sum of pairwise free energies between the specific particle and each particle in the remaining $n - 1$ cluster, we term it "cooperative"; "anti-cooperative" is similarly defined. In other words, if we label the specific particle as $n$, and introduce the notation $\delta W(1, 2,..., n - 1;n)$, which is the sum of all higher than two body interactions involving particle $n$,

$$\delta W(1, 2, ..., n-1; n) = F(1, 2, ..., n) - F(1, 2, ..., n-1) - F(n) - \sum_{i=1}^{n-1} \delta F(i, n) \quad (4)$$

$$\delta W(1, 2, ..., n-1; n) = \sum_{i<j \leq n-1} \delta F(i,j,n) + \sum_{i<j<k \leq n-1} \delta F(i,j,k,n) + ...$$
$$+ \delta F(1, 2, ..., n-1, n) \quad (5)$$

then, cooperativity or anti-cooperativity is defined when $\delta W(1, 2,..., n - 1;n)$ is negative or positive, respectively. In eqn (4), the term, $F(1, 2,..., n) - F(1, 2,..., n - 1) - F(n)$, gives the interaction energy between particle $n$ and the remaining $n - 1$ particles, and the term, $\sum_{i=1}^{n-1} \delta F(i, n)$, gives the sum of pairwise interactions between particle $n$ and each of the other $n - 1$ particles. The difference between these two terms provides a measure of the non-additive contribution to the interaction.

## 3 Simulation details

In this paper, molecular dynamics simulations were performed using the DESMOND program[29] to study the binding affinities between a united-atom methane and 13 model hydrophobic enclosures depicted in Fig. 1. The geometry of the model hydrophobic plate in these systems is displayed in the bottom right of Fig. 1. It consists of 19 single-layer atoms arranged in a triangular lattice with a bond length of 3.2 Å. For systems consisting of two plates, the two plates were parallel and in-registry with a separation distance of $D = 7.46$ Å. The LJ atoms forming the enclosures were uniformly represented with Lennard Jones parameters $\sigma = 3.73$ Å

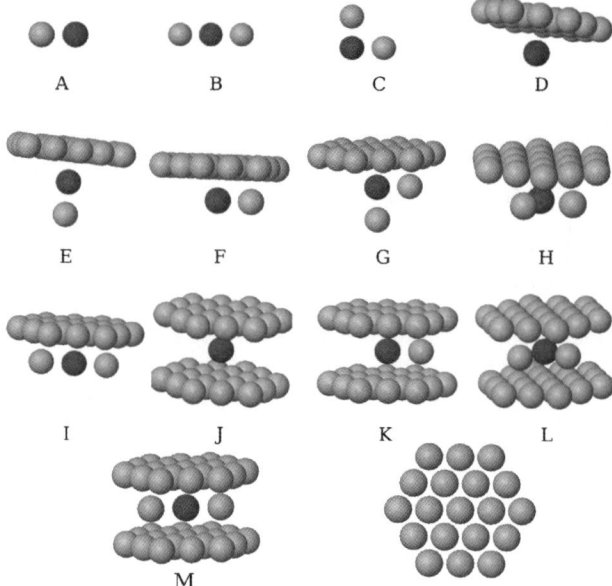

**Fig. 1** The 13 model systems studied. The hydrophobic enclosures are depicted in gray. The location of the methane molecule when bound to the respective hydrophobic enclosure is depicted in black. The geometry of the hydrophobic plate is depicted in the right bottom of the figure.

and $\varepsilon = 0.294$ kcal mol$^{-1}$, which are the same as the united-atom methane parameters used in these simulations.[30] The inserted methane particle displayed in black (online version green) in Fig. 1 was placed at the contact distance ($d = 3.73$ Å) with the other atoms and plate(s) forming the enclosures. In order to study the non-additivity of hydrophobic interactions between the insertion methane and the enclosures, another 10 systems corresponding to the subsystems of the 13 model enclosures (depicted in Fig. 2, and named after the corresponding system in Fig. 1 by adding a "prime") were also studied. The binding affinities for the other four systems in Fig. 2 [G″, H″, I″, K″] were calculated by combining the binding affinities for related systems in a thermodynamic cycle. For example, the binding affinity for system G″ was calculated by combination of binding affinities calculated for systems G′, G and E.

The free energy perturbation (FEP) method was used to determine the binding affinities between the inserted methane and each of the enclosures. The Maestro System Builder utility[31] was used to insert each enclosure into a cubic water box with a 10 Å buffer. The SPC water model[32] was used to describe the solvent. The atoms of the enclosures were constrained to their initial positions throughout the dynamics, and only the solvent degrees of freedom were sampled. The united-atom methane was "turned on" inside the model enclosures over 9 lambda windows with $\lambda = [0, 0.125, 0.25, 0.375, 0.50, 0.625, 0.75, 0.875, 1]$, where $\lambda$ is the coupling parameter to turn on/off the LJ interaction between the methane and the rest of the system with initial and final states corresponding to $\lambda = 0$ and $\lambda = 1$ respectively. In these simulations, the core of the LJ potential is made softer[33] as $\lambda \to 0$ to avoid singularities and numerical instabilities. For each of the $\lambda$ windows, molecular dynamics simulations were performed. The energy of the system was minimized, and then equilibrated to 298 K and 1 atm with Nose–Hoover[34,35] temperature and Martyna–Tobias–Klein[36] pressure controls over 100 ps of molecular dynamics. A cut-off distance of 9 Å was used to model the Lennard-Jones interactions, and the particle-mesh Ewald method[37] was used to model the electrostatic interactions.

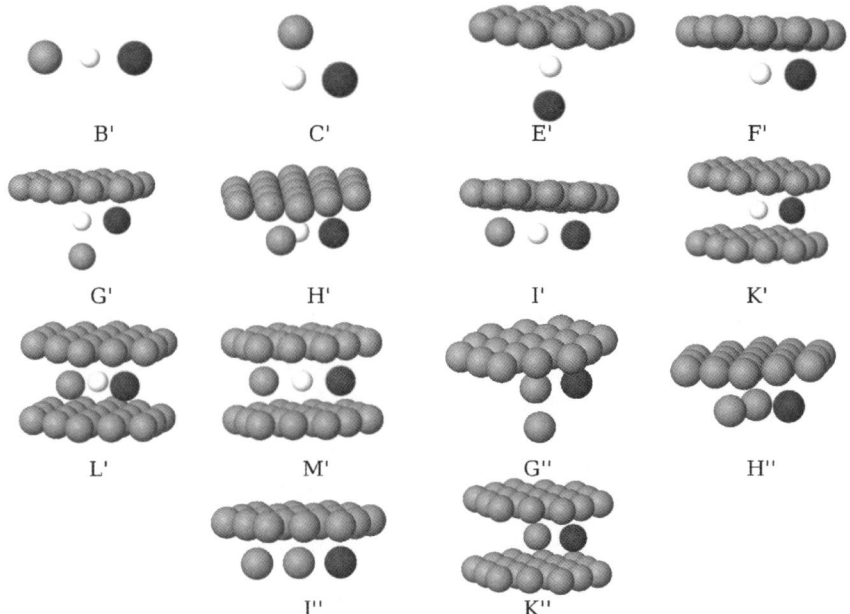

**Fig. 2** The 14 model systems corresponding to subsystems depicted in Fig. 1. The black particle in each system denotes the methane that will bind to the enclosure which is depicted in gray, and the small white particle denotes a pseudo-particle that specified the position of the binding methane for the corresponding system in Fig. 1. (The binding affinity for system G″, H″, I″ and K″ were calculated through thermodynamic cycles by combination of the binding affinities calculated for related systems).

Following the equilibration, a 20 ns production molecular dynamics simulation was performed and configurations of the system were collected every 1.002 ps. The energy difference between neighboring $\lambda$ windows for each configuration saved was calculated and the Bennett acceptance ratio method[38] was used to calculate the free energy difference between neighboring states. The sum of the free energy difference between neighboring states gave the solvation free energy of methane in the enclosures. The same procedure was followed to calculate the solvation free energy of methane in bulk water. The difference between the two solvation free energies gave the binding free energy to bring a methane from infinitely far to inside the hydrophobic enclosure, which is the potential of mean force (PMF) between the methane and the enclosure. The error associated with these binding affinities was of the order of $\pm 0.02$ kcal mol$^{-1}$.

### Implicit solvent model calculations

Due to the large computational cost of running explicit solvent model simulations, in protein folding or protein–ligand binding problems, one is often forced to use implicit solvent models to reduce the computational cost. Many of the implicit solvent models attempt to account for hydrophobicity in terms of nonpolar surface exposure to water,[39] among which solvent accessible surface area (SASA)[40] and molecular-surface-area (or Connolly surface area, MSA)[41] models are the most popular. The solvent accessible surface (SAS) is traced out by the probe sphere center as it rolls over the solute, and the molecular surface (MS) is the surface traced by the inward-facing surface of the probe sphere.

In this paper, the SASA and MSA of each enclosure, both with and without the bound methane, were computed with the Connolly molecular surface package,[42] as was the SASA and MSA of the methane particle by itself. From this data the buried surface area upon methane-enclosure complexation was determined. The direct Lennard-Jones interaction energy upon the binding of methane to each enclosure was similarly computed. The buried surface area multiplied by a surface tension like quality is often used to approximate to the solvent induced potential of the mean force. Together with the direct Lennard-Jones interaction energy, the total binding affinity between the methane and each enclosure can then be calculated, as has been done in many empirical methods for calculating binding affinities.

In addition, to investigate whether these two implicit solvent models can predict the non-additivity of the hydrophobic effect, the predicted pairwise additive buried surface area upon methane binding to each enclosure was also calculated. The deviation of the actual buried surface area from the pairwise additivity predicted allowed us to estimate the non-additive part of hydrophobic interactions based on these models.

## 4 Results and discussion

The binding free energies between methane and the model hydrophobic enclosures, as measured by FEP, are reported in Table 1. It is found that the range of binding free energies of the methane for the model enclosures is nearly 5 kcal mol$^{-1}$. Also reported in Table 1 are the pairwise additivity predicted binding affinity upon complexation, the buried surface area upon complexation, (both SASA and MSA), the pairwise additivity predicted buried surface area upon complexation, and the deviation between corresponding terms which gives the non-additive contributions.

**Comparison for different implicit solvent models in predicting the binding affinity**

From the data presented in Table 1, we can determine how well the buried surface area/molecular mechanics model predicts the binding affinity. Tuning the surface tension coefficient to minimize the mean-average-error (MAE) of fit with FEP reference data, we obtained an optimal surface tension coefficient of $\gamma = 0.00763$ kcal mol$^{-1}$ Å$^{-2}$ for the SASA and $\gamma = 0.03767$ kcal mol$^{-1}$ Å$^{-2}$ for the MSA models of these enclosures. For comparison, the surface tension for SASA determined by fitting to experimental solvation free energy for linear or branched alkanes by Honig[43] was 0.005 kcal mol$^{-1}$ Å$^{-2}$, whereas the macroscopic water–alkanes surface tension was 0.070 kcal mol$^{-1}$ Å$^{-2}$. Since there are no overlaps between the inserted methane and the enclosures, the buried van der Waals area is zero for all these systems. So the van der Waals surface model would just predict the binding affinity to be the direct LJ interaction energy. The predicted binding affinities *versus* the FEP reference data are reported in Fig. 3. From this figure we see that, for the most part, the MSA model performed better than the SASA model. This is indicated by a higher $R^2$ value (0.89 vs. 0.76) and smaller MAE (0.40 vs. 0.57), which is consistent with previous findings.[4,8,10,11] Both of these models performed much better than the van der Waals surface area model, which has $R^2 = 0.70$ and MAE = 0.94. The SASA based model cannot differentiate the hydrophobicity between systems [J, K, K′, K″, L, L′, M, and M′] (predicting similar binding affinities of about −4.1 kcal mol$^{-1}$) nor between systems [D, E, F, F′, G, G′, G″, H, H′, H″, I, I′, and I″] (predicting similar binding affinities of about −2.2 kcal mol$^{-1}$), while the MSA model performed much better for these systems and predicted the right order of hydrophobicity among these systems to some extent.

The SASA model found enclosures J and K′ in which a methane molecule is bound between two hydrophobic plates to be the most hydrophobic of all the systems [J, K, K′, K″, L, L′, M, M′]. The buried SASAs for these two systems, upon methane complexation, were the largest because large swaths of formerly accessible surface area on the faces of the plates are buried by the presence of the binding methane. For the other enclosures, several methane molecules already lie

**Table 1** The binding thermodynamics of methane for the various model hydrophobic enclosures[a]

|    | $\Delta G_{FEP}$ | $\Delta G_2$ | $\Delta\Delta G$ | $\Delta SASA$ | $\Delta SASA_2$ | $\Delta\Delta SASA$ | $\Delta MSA$ | $\Delta MSA_2$ | $\Delta\Delta MSA$ | Sign |
|----|------|------|------|---------|---------|--------|--------|--------|--------|------|
| A  | −0.60 | —    | —    | −57.44  | —       | —      | −4.15  | —      | —      | —    |
| B′ | −0.06 | —    | —    | 0.00    | —       | —      | 0.00   | —      | —      | —    |
| C′ | 0.18  | —    | —    | −25.75  | —       | —      | 3.54   | —      | —      | —    |
| D  | −1.66 | —    | —    | −89.09  | —       | —      | −14.44 | —      | —      | —    |
| E′ | −0.06 | —    | —    | 0.00    | —       | —      | 0.00   | —      | —      | —    |
| F′ | −1.64 | —    | —    | −88.54  | —       | —      | −14.16 | —      | —      | —    |
| B  | −1.15 | −1.21 | 0.06 | −114.88 | −114.88 | 0.00   | −8.29  | −8.29  | 0.00   | anti |
| E  | −2.17 | −2.26 | 0.09 | −146.53 | −146.53 | 0.00   | −18.58 | −18.58 | 0.00   | anti |
| G′ | −1.44 | −1.46 | 0.02 | −114.29 | −114.29 | 0.00   | −10.61 | −10.61 | 0.00   | anti |
| J  | −2.86 | −3.31 | 0.45 | −178.17 | −178.17 | 0.00   | −28.87 | −28.87 | 0.00   | anti |
| K′ | −2.83 | −3.27 | 0.44 | −177.08 | −177.08 | 0.00   | −28.31 | −28.31 | 0.00   | anti |
| C  | −1.41 | −1.21 | −0.20 | −95.61 | −114.88 | 19.27  | −11.46 | −8.29  | −3.16  | coop |
| F  | −2.63 | −2.26 | −0.37 | −115.03 | −146.53 | 31.50 | −25.14 | −18.58 | −6.56  | coop |
| G  | −3.41 | −2.86 | −0.54 | −153.20 | −203.97 | 50.77 | −24.48 | −14.76 | −9.72  | coop |
| G″ | −2.68 | −2.06 | −0.62 | −120.96 | −171.73 | 50.77 | −24.48 | −14.76 | −9.72  | coop |
| H  | −3.44 | −2.86 | −0.57 | −129.81 | −203.97 | 74.15 | −38.00 | −22.73 | −15.27 | coop |
| H′ | −1.97 | −1.46 | −0.51 | −101.36 | −114.29 | 14.16 | −13.01 | −10.61 | −2.40  | coop |
| H″ | −2.77 | −2.06 | −0.71 | −116.14 | −171.73 | 55.59 | −25.88 | −14.76 | −11.12 | coop |
| I  | −3.47 | −2.86 | −0.60 | −140.97 | −203.97 | 63.00 | −35.84 | −22.73 | −13.11 | coop |
| I′ | −1.74 | −1.69 | −0.05 | −88.54 | −88.54  | 0.00   | −14.16 | −14.16 | 0.00   | coop |
| I″ | −2.57 | −2.29 | −0.28 | −114.49 | −145.98 | 31.49 | −24.86 | −18.31 | −6.55  | coop |
| K  | −4.59 | −3.92 | −0.67 | −172.62 | −235.61 | 62.99 | −46.13 | −33.46 | −13.11 | coop |
| K″ | −4.56 | −3.77 | −0.68 | −171.53 | −234.52 | 62.99 | −45.57 | −32.46 | −13.11 | coop |
| L  | −5.24 | −4.52 | −0.72 | −164.01 | −293.05 | 129.04 | −64.10 | −37.17 | −26.93 | coop |
| L′ | −4.23 | −3.09 | −1.14 | −176.97 | −202.83 | 25.86 | −29.57 | −24.77 | −4.80  | coop |
| M  | −5.45 | −4.52 | −0.94 | −167.06 | −293.05 | 125.99 | −63.39 | −37.17 | −26.22 | coop |
| M′ | −3.80 | −3.33 | −0.48 | −177.09 | −177.09 | 0.00   | −28.32 | −28.32 | 0.00   | coop |

[a] $\Delta G_{FEP}$ denotes the binding free energies from FEP, (free energy perturbation) $\Delta G_2$ denotes the predicted binding affinities by assuming pairwise additivity, and $\Delta\Delta G$ denotes the deviation of the FEP binding affinity results from the pairwise additivity predicted binding affinities (which corresponds to non-additivity of hydrophobic interactions defined by formula 5). $\Delta SASA$ denotes the change of SASA (solvent accessible surface area) upon methane binding, the $\Delta SASA_2$ denotes the pairwise-additive contribution to the change of SASA. $\Delta\Delta SASA$ denotes the deviation of $\Delta SASA_2$ from $\Delta SASA$. The notations for MSA (molecular surface area) are similarly defined. "anti" is a short-hand notation for "anti-cooperative" and "coop" stands for "cooperative". Free energies in kcal mol$^{-1}$, surface area in Å$^2$.

between the plates in the absence of the binding methane so that part of the surface area buried by the binding methane was already buried by the other particles. For the extreme cases of systems L and M, which are most hydrophobic, the SASA model strongly underestimates the binding affinity, because the buried SASAs are smaller for these two systems than in systems J and K′. On the other hand, the MSA model to some extent predicts the right order of binding affinity among these systems. A similar analysis of systems [D, E, F, F′, G, G′, G″, H, H′, H″, I, I′, I″], yields similar conclusions.

### Comparison of MSA and SASA model predictions of the non-additivity effect

The non-additive contributions to the binding affinities of methane to all of the enclosures are listed in Table 1. We see from the table that while only small non-additive effects (±0.2 kcal mol$^{-1}$) were observed for the methane trimers by the

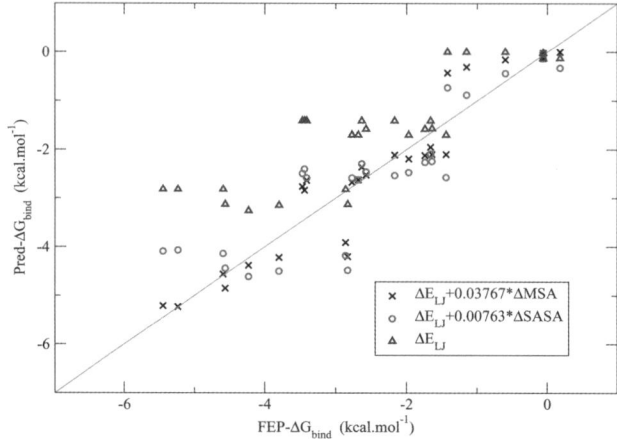

**Fig. 3** Buried surface area/molecular mechanics prediction of methane-enclosure binding affinities. The surface tension coefficients were chosen to minimize the MAE (mean average error) between the predicted and FEP results for the binding affinities. Predictions based on the MSA (molecular surface area) model performed better than those based on the SASA (solvent accessible surface area) model, indicated by a higher $R^2$ value (0.89 vs. 0.76) and smaller MAE (0.40 vs. 0.57). Both of these models performed much better than predictions based on the van der Waals surface model which takes into account the direct LJ interaction (giving $R^2 = 0.70$, and MEA = 0.94 kcal mol$^{-1}$). SASA can not differentiate the hydrophobicity among systems [J, K, K′, K″, L, L′, M, M′] (predicting a similar binding affinity of about −4.1 kcal mol$^{-1}$) nor among systems [D, E, F, F′, G, G′, G″, H, H′, H″, I, I′, I″] (predicting a similar binding affinity of about −2.2 kcal mol$^{-1}$), while the MSA based model performed much better for these systems and predicted the right order of hydrophobicity among these systems to some extent.

Chan and Scheraga groups,[4,9,12,22] large cooperative effects (as large as −1.14 kcal mol$^{-1}$ for system L′) and anti-cooperative effects (as large as 0.45 kcal mol$^{-1}$ for system J) were observed for these systems.

The data shown in Table 1 allows us to phenomenologically identify a connection between the sign of the non-additivity effect and the curvature of the molecular surface. To wit, for all the enclosures that exhibit an anti-cooperative effect, the molecular surfaces of the enclosures were convex without any concave or saddle parts, whereas for all enclosures that exhibit a cooperative effect, the molecular surfaces had concave or saddle parts, and the methane binds close to the saddle or concave part of the surface.

Consider systems E and G for example. The molecular surface of enclosure E has no concave part, and the binding affinity of methane is anti-cooperative, (less favored than pairwise additivity predicted). In the presence of another methane in enclosure G, the molecular surface has a saddle part, and when the methane binds to this saddle part, it shows a cooperative effect. Similar analyses can be done on systems J and K, K′ and M′, K′ and K″. Systems I′, H′, and F have the same enclosure whose molecular surface exhibits a saddle region, but the inserted methane binds at different locations of the enclosure. Comparing the methane binding affinities for these systems, we found that as the methane moves closer and closer to the saddle part (from systems I′ to H′ to F), the binding affinity gets more and more favorable (from −1.74 kcal mol$^{-1}$ to −1.97 kcal mol$^{-1}$ to −2.63 kcal mol$^{-1}$), and the cooperative effect changes from relatively weak (−0.05 kcal mol$^{-1}$ for system I′) to strong (−0.51 kcal mol$^{-1}$ and −0.37 kcal mol$^{-1}$ for systems H′ and F, respectively). Similar analysis can be done for systems M′, L′, and K.

The cooperative effects in systems H′ and L′ were found to be stronger than for systems F and K respectively. This might seem to be at odds with intuition because systems F and K are more compact than systems H′ and L′ respectively. This is not surprising though: the distance between the two methane molecules in systems H′ and L′ corresponds to the de-solvation barrier on the PMF curve of two methanes molecules in bulk water. In the presence of the plate(s), the de-solvation barrier between the two methanes may not exist, so that the binding affinities might be much more favorable than would be predicted by the pairwise additivity approximation.

In addition to the data for the non-additive contribution to the changes of SASA and MSA, we can investigate whether either of these implicit solvent models can predict the non-additivity in the hydrophobic interactions. Multiplying the non-additive part of the changes in MSA and SASA in methane complexation by the surface tension coefficient obtained in the previous section, we can determine the MSA and SASA predicted non-additive contributions to the hydrophobic interactions. Fig. 4 depicts the relationship between the SASA/MSA predicted non-additive contributions and the FEP results of the non-additive hydrophobic effects. It can be seen that the MSA predicted results have a strong correlation with the FEP results while SASA predicted results anticorrelate with the FEP results.

At first glance at Fig. 4, it might seem that MSA anticorrelates with SASA, in contradiction to our common understanding of these models. To better understand this "strange" behavior, the non-additive contribution of SASA and MSA for a model methane trimer system was investigated. With the two methanes kept at their contact distance, the position of the third methane can be specified by two coordinates $(\theta, d)$. (See the top right corner of Fig. 5). Fig. 5 shows the predictions of the non-additive part of hydrophobic interactions given by calculating the changes in the MSA and the SASA each multiplied respectively by their corresponding surface tension coefficients (as in the previous section) as a function of the distance $d$ when $\theta = 0$. Actually, this figure is representative of what we found for all of the angles $\theta$. (See Fig. 9 of ref. 4) We see from this figure that for $d = 3.23$ Å, which corresponds to the configuration where the third methane is in contact with the other two methanes, the predictions of MSA and SASA have opposite signs. For all of the systems studied in this paper, the binding methane is in contact with the other methane(s) and/or plate(s) in the enclosures, thus it is not surprising to find the anticorrelation between MSA and SASA predictions observed in Fig. 4.

In addition, it can be seen from Fig. 5 that while the MSA model can predict additive, anti-cooperative, and cooperative effects, the SASA model only predicts additive and anti-cooperative effects. This conclusion seems to be generally valid. In the

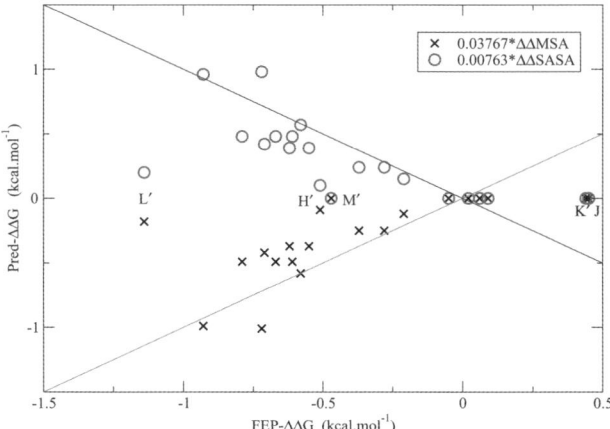

**Fig. 4** Relationship between the surface area models predicted non-additivity of hydrophobic effects and the corresponding FEP results. There is a strong correlation between MSA (molecular surface area) predictions and FEP results, but the SASA (solvent accessible surface area) predicted results anticorrelate with the FEP results.

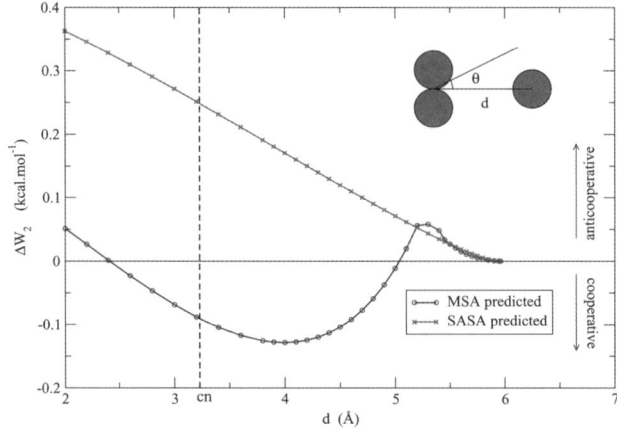

**Fig. 5** The MSA/SASA (molecular surface area/solvent accessible surface area) predicted non-additivity effects of hydrophobic interactions (non-additive part of the change in MSA and SASA multiplied by the corresponding surface tension coefficient) for the methane trimer system as a function of the distance $d$ for the configuration depicted in the top right corner when $\theta = 0$. The distance corresponding to the configuration where the third methane is in contact with the remaining two were indicated by the dotted line labeled "cn".

SASA model, the surface considered is formed from overlapping spheres, each of whose radii is the sum of the corresponding atomic van der Waals radii plus the radius of water.[40] The buried surface area between a sphere and a cluster of spheres, when the cluster of spheres has no overlaps, is always equal to the sum of the buried surface areas between the sphere and each individual sphere in the cluster, and smaller than this when the cluster of spheres has overlaps. This means that the SASA model can never predict the cooperative effects. However, in the MSA model, the surface is traced by the inward-facing surface of the probe sphere,[41] and it has the potential to predict all of the non-additivity effects. This partially explains why the MSA based model performs better than the SASA based model both for the binding affinity and for the non-additivity effects.

It should be noted from Fig. 4 that while the MSA model successfully predicts how the non-additivity effects correlate with the FEP results, there were a few outliers. We will interpret these outliers case by case.

In system M′, the region between the plates is enclosed on four different sides by hydrophobic moieties. This causes the density of water in the enclosed region to be much smaller than in bulk (1/4 of the bulk value), close to a hydrophobic dewetting condition. Part of the time there was one water molecule in the enclosed region and part of time it was empty, indicating an interface different from the normal nonpolar solute–liquid water interface. It is well known that in hydrophobic dewetting there is a strong driving force to bring the hydrophobic particles together,[1] which explains the strong cooperative effect observed in system M′.

For systems H′ and L′, as previously mentioned, the distance between the two methane molecules corresponds to the de-solvation barrier on the PMF curve of two methane molecules in bulk water, and in the presence of the plate(s), the de-solvation barrier may not exist at all. Because of the barriers in the pair potential of mean force, the pairwise additivity approximation predicts that this configuration will be very unfavorable, but the full calculation shows that in fact they are not unfavorable. In fact, the binding affinities in these particular configurations are much more favorable than would be predicted by assuming pairwise additivity. This may be the reason for the strong cooperative effects observed for these systems. In addition, the dewetting argument discussed in connection with systems M′ also applies to system L′, which further validates the strong cooperative effect observed for system L′.

To investigate the reason for the strong anti-cooperative effects observed in systems J and K′, we analyzed the structure of water between the two plates and found that, at the surface of one hydrophobic plate (system D), water breaks one hydrogen bond on average, which caused the average interaction energy between water at the surface and the rest of the system to be higher than that in bulk water by 1.12 kcal mol$^{-1}$. However, between two hydrophobic plates (system J), water breaks less than two hydrogen bonds on average because of its flexibility in making hydrogen bonds. This is supported by the fact that the average interaction energy between water molecules located between the two plates and the rest of the system is higher than in bulk water by only 2.05 kcal mol$^{-1}$, less than twice 1.12 kcal mol$^{-1}$ (2 × 1.12 = 2.24) expected from doubling the effect of one plate. It is well known that the hydrophobic effect for large scale systems is enthalpy driven,[1,44] because of breaking hydrogen bonds at the surface of the hydrophobic plates. A large contribution of the methane-enclosure binding affinity comes from de-solvation of solvent between the plates.[45] The anti-cooperative effect observed for systems J and K′ may be due to the fact that the excess of the interaction energy of water located between two hydrophobic plates (system J) over that of bulk water is found to be smaller than twice the value of the water at one hydrophobic plate (system D).

**Non-additivity effect at wetting-dewetting transition**

In the previous sections, we have shown how the non-additivity effects of methane binding affinities in enclosures with different topologies correlate with the MSA measurements. It is well known that the hydrophobicity of the enclosures depends not only on the topologies but also on the LJ parameters for atoms making up the enclosures.[1] So it will be interesting to study how the non-additivity effects depend on the LJ parameters for particles making up the enclosure. In this section, we will explore this effect by changing the LJ $\varepsilon$ parameter for particles making up the plates for one representative enclosure, enclosure J.

Fig. 6 depicts how the binding affinities of methane in enclosure J ($\Delta G_J$) and in enclosure D ($\Delta G_D$) changes as a function of the LJ $\varepsilon$ parameter for particles making up the plate(s) from FEP simulations. The pairwise additivity predicted binding affinities for enclosure J, which is two times that for enclosure D, are also depicted in the figure. As can be seen from this figure, while the binding affinities for enclosure D decreases (or alternatively the free energy becomes more positive) monotonically with increasing values of $\varepsilon$, the binding affinities for enclosure J first increases (the free energy becomes more negative) and then decreases, and the non-additivity effect goes from slightly anti-cooperative at very low $\varepsilon$ regions (smaller than 0.06 kcal mol$^{-1}$), to cooperative at intermediate $\varepsilon$ regions (between 0.06 and 0.23 kcal mol$^{-1}$), back to anti-cooperative at high $\varepsilon$ regions (higher than 0.23 kcal mol$^{-1}$).

The solvation free energy of methane in the enclosure can be decomposed into two components: the free energy to create a cavity with the size of methane in the enclosure and the free energy to turn on the attractive part of interactions between methane and the rest of the system. For enclosure D, when increasing the $\varepsilon$ parameter, the free energy to create the cavity will become more unfavorable because the solvent will become denser at the surface of the plate; however, the free energy to turn on the attractive part of interactions between methane and the plate will become more favorable with increasing values of $\varepsilon$. These two factors have opposite effects, but the first component dominates, so the overall binding affinity will decrease slightly. For enclosure J, at low values of $\varepsilon$, the region between the plates dewets, so the free energy to create the cavity is almost zero and changes slightly with increasing values of $\varepsilon$, but the free energy to turn on the attractive interactions between the methane and the plates becomes more negative, so the binding affinity will increase in this dewetting region. The critical value of $\varepsilon$ corresponding to the

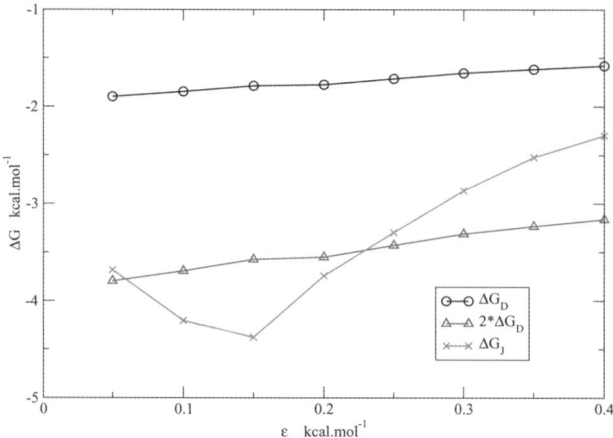

**Fig. 6** The methane enclosure binding affinities for enclosure D and J as a function of the LJ $\varepsilon$ parameter for atoms making up the plate(s). The binding free energy increases monotonically for enclosure D with increasing values of $\varepsilon$, while it decreases at the lower $\varepsilon$ region and increases at the higher $\varepsilon$ region for enclosure J. The non-additivity effect for enclosure J goes from anti-cooperative in the lower $\varepsilon$ region, (smaller than 0.06 kcal mol$^{-1}$), to cooperative in the intermediate $\varepsilon$ region, (between 0.06 and 0.23 kcal mol$^{-1}$), and then to anti-cooperative again in the higher $\varepsilon$ region (larger than 0.23 kcal mol$^{-1}$).

wetting–dewetting transition is $\varepsilon \approx 0.15$ kcal mol$^{-1}$. At this point, the probability for observing a dry inter-plate region is 50%. For $\varepsilon$ larger than this, the free energy to create the cavity grows rapidly with increasing values of $\varepsilon$, becoming the dominant effect, so that the overall binding affinity decreases rapidly with increasing values of $\varepsilon$. At the critical value of $\varepsilon$, that is at the wetting–dewetting transition, there is a large cooperative non-additive effect on the binding of the methane between the plates. With increasing values of $\varepsilon$, the two plates affect the density fluctuation of solvent by more than twice what one plate does, and the slope of the binding affinity versus $\varepsilon$ is therefore much larger for enclosure J than that for enclosure D. For sufficiently large $\varepsilon$ there is a large anti-cooperative deviation from additivity. At $\varepsilon = 0.23$, these two effects balance each other, so the free energy is additive at this point. The $\varepsilon$ value for system J studied in the previous section is $\varepsilon = 0.294$ kcal mol$^{-1}$ which is higher than 0.23, so we observed an anti-cooperative effect there. As $\varepsilon$ is decreased below the critical value of $\varepsilon$, the binding affinity increases for enclosure D and decreases for enclosure J and becomes anti-cooperative in this very low $\varepsilon$ region.

### Higher order multi-body interactions

In the previous section, we investigated the non-additive effects manifested when methane is inserted into different model enclosures. For systems consisting of three components, the non-additive effect corresponds to three-body interactions; for systems consisting of more than three components, the non-additive effect is the summation of all higher-than two-body interactions involving the insertion methane, corresponding to $\delta W(1, 2,..., n-1;n)$, defined in eqn(5). We now investigate the contributions beyond three body interactions defined recursively in eqn(3).

Table 2 lists the total PMF, the two-body contribution to the PMF, $W_2$, the three-body contribution to the PMF, $W_3$, and the subsequent higher order contributions for all the systems consisting of more than three components. The deviations found by truncating the series up to order $n$, $\Delta W_n$, equivalent to the GKSA approximation to the $n$-th order, is also listed in the table. (For example, $\Delta W_2 = W(1, 2,..., n) - W_2$.) From this table, we see that the error arising from truncation of the total PMF at the pairwise term can be as large as 1.43 kcal mol$^{-1}$ (system L), indicating

**Table 2** Multi-body PMF (potential of mean force) calculated from FEP, (free energy perturbation) and the contribution from two-, three-, four-, five-body interactions[a]

| Systems | $W$ | $W_2$ | $\Delta W_2$ | $W_3$ | $\Delta W_3$ | $W_4$ | $\Delta W_4$ | $W_5$ | $\Delta W_5$ |
|---|---|---|---|---|---|---|---|---|---|
| G | −4.90 | −4.37 | −0.52 | −0.47 | −0.06 | −0.06 | 0.00 | — | — |
| H | −7.04 | −5.95 | −1.08 | −1.46 | 0.38 | 0.38 | 0.00 | — | — |
| I | −6.85 | −6.19 | −0.65 | −0.74 | 0.09 | 0.09 | 0.00 | — | — |
| K | −7.42 | −7.19 | −0.24 | 0.14 | −0.37 | −0.37 | 0.00 | — | — |
| L′ | −7.07 | −6.36 | −0.71 | −0.14 | −0.56 | −0.56 | 0.00 | — | — |
| M′ | −6.64 | −6.60 | −0.04 | 0.78 | −0.82 | −0.82 | 0.00 | — | — |
| L | −12.31 | −10.88 | −1.43 | −1.40 | −0.03 | −0.53 | 0.51 | 0.51 | 0.00 |
| M | −12.09 | −11.12 | −0.98 | −0.21 | −0.73 | −1.39 | 0.63 | 0.63 | 0.00 |

[a] The PMF between two plates was set to zero when $D = 7.46$ Å, which corresponded to the configuration for all the systems including two plates. The choice of base line for the PMF between the two plates does not affect the multi-body contributions. Free energies in kcal mol$^{-1}$.

the importance of non-additivity effects in these systems. In addition, the higher order multi-body contributions can be either positive or negative, suggesting that hydrophobic interactions can be quite complex.

Fig. 7 and 8 gives the deviations of the PMF predicted by the GKSA from the total PMF, when the latter is truncated to order $n$, as a function of $n$. For most of the cases, the higher the order of multi-body interactions considered, the smaller the magnitude of the error (systems G, H, I, L′, L, M). However, this is not generally true since for systems K and M′, inclusion of the three-body interactions increases the error.

## 5 Conclusion

In this paper, the binding affinities between a united-atom methane and various model hydrophobic enclosures were determined using high accuracy FEP molecular dynamics simulations. Comparisons were made between the binding affinities from FEP and predictions based on assuming pairwise additivity, and through this it was possible to investigate the non-additive contributions of the hydrophobic interactions. Small non-additivity effects were found in the methane trimer systems by

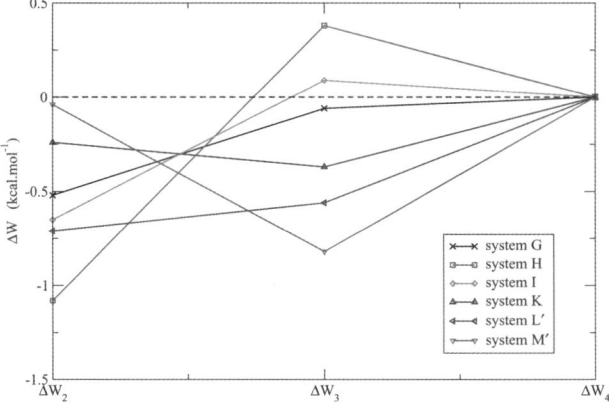

**Fig. 7** The deviation of GKSA (generalized Kirkwood superposition approximation) predicted PMF (potential of mean force) from the total PMF by truncating the total PMF to the second-, third-, and fourth- order as a function of the order.

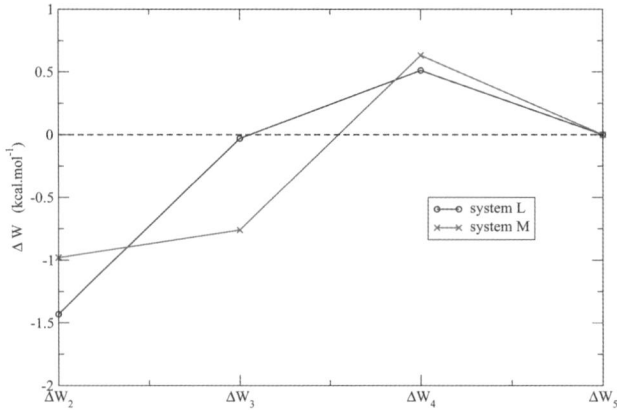

**Fig. 8** The deviation of GKSA (generalized Kirkwood superposition approximation) predicted PMF (potential of mean force) from the total PMF by truncating the total PMF to the second-, third-, fourth- and fifth- order as a function of the order.

the Chan and Scheraga groups,[4,9,12,22] but we find large cooperative effects (as large as $-1.14$ kcal mol$^{-1}$) and anti-cooperative effects (as large as 0.45 kcal mol$^{-1}$) for our relatively larger systems. Although approximations based on the pairwise additivity of the PMF have been used to study the transition state of protein folding[46] and the force-extension behavior of protein,[47] simulations done in this paper indicate that higher order correlations may be very important in real biological systems such as protein folding, protein–ligand binding and other relevant fields. This should not be surprising since the Kirkwood superposition approximation fails in dense simple fluids.

Phenomenologically, the sign of the non-additive contributions to the binding affinities of methane in the enclosures was found to be correlated with the curvature of the enclosures. To be specific, anti-cooperative effects were observed only in enclosures whose molecular surfaces were convex without any concave or saddle part, and cooperative effects were observed in enclosures whose molecular surfaces had concave or saddle parts, in which case the methane was found to bind close to the saddle or concave part of the surface. Such observations might be useful for further development of models to incorporate the non-additivity effect.

We also investigated whether two kinds of implicit solvent models are consistent with the observed binding affinities and non-additivity effects. In these models the solvent induced free energy for particle insertion was computed from the product of a "surface tension" and the area of the buried surface upon methane complexation. The area of the buried surface was computed by using the solvent accessible surface area (SASA) and by using the molecular surface area (MSA) as described in the text. We found that the MSA based model performed much better than the SASA based model in predicting the binding affinities, an observation consistent with previous findings.[4,8,10,11] In addition, the SASA based model always predicts non-additive effects which anticorrelate with the FEP reference data, whereas the MSA based model performed reasonably well in predicting the non-additive effects, except for a few outliers which were explained in the text. Further analysis indicated that the MSA based model predicts all cooperative and anti-cooperative non-additivity effects, and the SASA based model exhibits an intrinsic problem in that it can never predict cooperative effects. Furthermore, because of the correlation we observed between the non-additive contributions and the curvature of the molecular surface in the MSA based model, we believe that the MSA based model is a far better descriptor of the curvature of simple solutes than the SASA based model.

The non-additivity effect depends not only on the topology but also on the LJ parameters for atoms making up the enclosure. By changing the LJ $\varepsilon$ parameter for atoms making up the plates for enclosure J, we observed a wetting–dewetting transition in the inter-plate region, and the non-additivity effect changes from anti-cooperative in the low $\varepsilon$ region, to cooperative in the intermediate $\varepsilon$ region, and then to anti-cooperative again in the higher $\varepsilon$ region. This complicated re-entrant behavior of the non-additivity effect results from the competition between the two factors contributing to the solvation free energy: the free energy to create the cavity and the free energy to turn on the attractive interactions between the solute and the rest of the system. While the first factor dominates in the higher $\varepsilon$ region (wetting region), the second factor dominates in the lower $\varepsilon$ region (dewetting region).

The decomposition of the PMF for cluster formation can be expressed as a sum of increasing orders of multi-body interactions. We found that the multi-body correlations can be either negative or positive, implying that the hydrophobic interaction will depend on the topology of the surfaces enclosing the particle. In addition, increasing the order of multi-body interactions included can usually improve the accuracy, but this was not generally true.

## Note added at Proof

In this paper, what we call "dewetting" is really "capillary evaporation" induced by confinement; and what we call "wetting" means there is no "capillary evaporation" induced by confinement.

## Acknowledgements

We thank Dr Robert Abel and Prof. Tom Young for many helpful discussions. BJB acknowledges support by NIH GM 43340. RAF acknowledges that this work was supported in part by the National Science Foundation through TeraGrid resources provided by NCSA and ABE, (MCA08X002), and NIH GM 52018.

## References

1 B. J. Berne, J. D. Weeks and R. Zhou, *Annu. Rev. Phys. Chem.*, 2009, **60**, 85–103.
2 W. Kauzmann, *Adv. Protein Chem.*, 1959, **14**, 1–63.
3 K. A. Dill, *Biochemistry*, 1990, **29**, 7133–7155.
4 S. Shimizu and H. S. Chan, *Proteins: Struct., Funct., Genet.*, 2002, **48**, 15–30.
5 X. Li and J. Liang, *Proteins: Struct., Funct., Bioinf.*, 2005, **60**, 46–65.
6 G. Némethy and H. A. Scheraga, *J. Phys. Chem.*, 1962, **66**, 1773–1789.
7 F. Brugè, S. L. Fornili, G. G. Malenkov, M. B. Palma-Vittorelli and M. U. Palma, *Chem. Phys. Lett.*, 1996, **254**, 283–291.
8 J. A. Rank and D. Baker, *Protein Sci.*, 1997, **6**, 347–354.
9 S. Shimizu and H. S. Chan, *J. Chem. Phys.*, 2001, **115**, 1414–1421.
10 C. Czaplewski, S. Rodziewicz-Motowidlo, A. Liwo, D. R. Ripoll, R. J. Wawak and H. A. Scheraga, *Protein Sci.*, 2000, **9**, 1235–1245.
11 C. Czaplewski, D. R. Ripoll, A. Liwo, S. Rodziewicz-Motowidlo, R. J. Wawak and H. A. Scheraga, *Int. J. Quantum Chem.*, 2002, **88**, 41–55.
12 C. Czaplewski, A. Liwo, D. R. Ripoll and H. A. Scheraga, *J. Phys. Chem. B*, 2005, **109**, 8108–8119.
13 C. Czaplewski, S. Rodziewicz-Motowido, A. Liwo, D. R. Ripoll, R. J. Wawak and H. A. Scheraga, *J. Chem. Phys.*, 2002, **116**, 2665–2667.
14 S. Shimizu and H. S. Chan, *J. Chem. Phys.*, 2002, **116**, 2668–2669.
15 S. Shimizu, M. S. Moghaddam and H. S. Chan, *J. Phys. Chem. B*, 2005, **109**, 21220–21221.
16 C. Czaplewski, S. Kalinowski, A. Liwo, D. R. Ripoll and H. A. Scheraga, *J. Phys. Chem. B*, 2005, **109**, 21222–21224.
17 C. Czaplewski, S. Kalinowski, A. Liwo and H. A. Scheraga, *Mol. Phys.*, 2005, **103**, 3153–3167.
18 S. Shimizu and H. S. Chan, *J. Chem. Phys.*, 2000, **113**, 4683–4699.
19 M. S. Moghaddam, S. Shimizu and H. S. Chan, *J. Am. Chem. Soc.*, 2005, **127**, 303–316.
20 T. Ghosh, A. E. Garcia and S. Garde, *J. Phys. Chem. B*, 2003, **107**, 612–617.
21 T. Ghosh, A. Kalra and S. Garde, *J. Phys. Chem. B*, 2005, **109**, 642–651.

22 C. Czaplewski, S. Rodziewicz-Motowidlo, M. Dabal, A. Liwo, D. R. Ripoll and H. A. Scheraga, *Biophys. Chem.*, 2003, **105**, 339–359.
23 D. A. Mcquarrie, *Statistical Mechanics*, Sausalito, CA: University Science Books, 2000.
24 I. Z. Fisher and B. L. Kopeliovich, *Dokl. Akad. Nauk SSSR*, 1960, **133**, 81–83.
25 H. Reiss, *J. Stat. Phys.*, 1972, **6**, 39–47.
26 J. G. Kirkwood, *J. Chem. Phys.*, 1935, **3**, 300–313.
27 J. G. Kirkwood, *J. Chem. Phys.*, 1939, **7**, 919–925.
28 J. G. Kirkwood and E. M. Boggs, *J. Chem. Phys.*, 1942, **10**, 394–402.
29 K. J. Bowers, E. Chow, H. Xu, R. O. Dror, M. P. Eastwood, B. A. Gregersen, J. L. Klepeis, I. Kolossvary, M. A. Moraes, F. D. Sacerdoti, J. K. Salmon, Y. Shan and D. E. Shaw, *SC '06: Proceedings of the 2006 ACM/IEEE conference on Supercomputing*, New York, NY, USA, 2006, p. 84.
30 W. L. Jorgensen, J. D. Madura and C. J. Swenson, *J. Am. Chem. Soc.*, 1984, **106**, 6638–6646.
31 Schrodinger, LLC, New York, NY, *Maestro, version 8.5*, 2008.
32 H. Berendsen, J. Postma, W. van Gunsteren and J. Hermans, in *Intermolecular Forces*, ed. B. Pullman, Reidel: Dordrecht, 1981, pp. 331–342.
33 T. C. Beutler, A. E. Mark, R. C. van Schaik, P. R. Gerber and W. F. van Gunsteren, *Chem. Phys. Lett.*, 1994, **222**, 529–539.
34 S. Nosé, *J. Chem. Phys.*, 1984, **81**, 511–519.
35 W. G. Hoover, *Phys. Rev. A: At., Mol., Opt. Phys.*, 1985, **31**, 1695–1697.
36 G. J. Martyna, D. J. Tobias and M. L. Klein, *J. Chem. Phys.*, 1994, **101**, 4177–4189.
37 T. Darden, D. York and L. Pedersen, *J. Chem. Phys.*, 1993, **98**, 10089–10092.
38 C. H. Bennett, *J. Comput. Phys.*, 1976, **22**, 245–268.
39 B. Roux and T. Simonson, *Biophys. Chem.*, 1999, **78**, 1–20.
40 B. Lee and F. M. Richards, *J. Mol. Biol.*, 1971, **55**, 379–380.
41 M. L. Connolly, *Science*, 1983, **221**, 709–713.
42 M. L. Connolly, *The molecular surface package, version 3.9.3*, 2006.
43 D. J. Tannor, B. Marten, R. Murphy, R. A. Friesner, D. Sitkoff, A. Nicholls, B. Honig, M. Ringnalda and W. A. Goddard, *J. Am. Chem. Soc.*, 1994, **116**, 11875–11882.
44 K. Lum, D. Chandler and J. D. Weeks, *J. Phys. Chem. B*, 1999, **103**, 4570–4577.
45 R. Abel, L. Wang, R. A. Friesner and B. J. Berne, *J. Chem. Theory Comput.*, 2010, submitted.
46 C. Clementi, H. Nymeyer and J. N. Onuchic, *J. Mol. Biol.*, 2000, **298**, 937–953.
47 F. Grater, P. Heider, R. Zangi and B. J. Berne, *J. Am. Chem. Soc.*, 2008, **130**, 11578–11579.

# Water reorientation, hydrogen-bond dynamics and 2D-IR spectroscopy next to an extended hydrophobic surface

Guillaume Stirnemann,[*a] Peter J. Rossky,[b] James T. Hynes[ac] and Damien Laage[*a]

*Received 8th December 2009, Accepted 10th February 2010*
DOI: 10.1039/b925673c

The dynamics of water next to hydrophobic groups is critical for several fundamental biochemical processes such as protein folding and amyloid fiber aggregation. Some biomolecular systems, like melittin or other membrane-associated proteins, exhibit extended hydrophobic surfaces. Due to the strain these surfaces impose on the hydrogen (H)-bond network, the water molecules shift from the clathrate-like arrangement observed around small solutes to an anticlathrate-like geometry with some dangling OH bonds pointing toward the surface. Here we examine the water reorientation dynamics next to a model hydrophobic surface through molecular dynamics simulations and analytic modeling. We show that the water OH bonds lying next to the hydrophobic surface fall into two subensembles with distinct dynamical reorientation properties. The first is the OH bonds tangent to the surface; these exhibit a behavior similar to the water OHs around small hydrophobic solutes, *i.e.* with a moderate reorientational slowdown explained by an excluded volume effect due to the surface. The second is the dangling OHs pointing toward the surface: these are not engaged in any H-bond, reorient much faster than in the bulk, and exhibit an unusual anisotropy decay which becomes negative for delays of a few picoseconds. The H-bond dynamics, *i.e.* the exchanges between the different configurations, and the resulting anisotropy decays are analyzed within the analytic extended jump model. We also show that a recent spectroscopy technique, two-dimensional time resolved vibrational spectroscopy (2D-IR), can be used to selectively follow the dynamics of dangling OHs, since these are spectrally distinct from H-bonded ones. By computing the first 2D-IR spectra of water next to a hydrophobic surface, we establish a connection between the spectral dynamics and the dynamical properties that we obtain directly from the simulations.

## I. Introduction

While a significant and fruitful effort has been devoted over the past years to the understanding of water thermodynamics next to hydrophobic interfaces and of phenomena such as dewetting,[1,2] a comparable understanding of the water dynamics in these hydrophobic environments has remained elusive. Yet, the dynamics of water next to hydrophobic interfaces is critical for several biochemical processes such as

[a]*Dept. of Chemistry, Ecole Normale Supérieure, rue Lhomond, Paris, France. E-mail: guillaume. stirnemann@ens.fr; damien.laage@ens.fr*
[b]*Dept. of Chemistry and Biochemistry, University of Texas, Austin, TX, USA*
[c]*Dept. of Chemistry and Biochemistry, University of Colorado, Boulder, CO, USA*

protein folding,[1,3] where water expulsion from the hydrophobic core was suggested to be rate-limiting, and amyloid fiber agreggation,[4] which possibly involves protein "smooth spots" with highly labile hydration layers.[5]

Several of us recently characterized the effect of small hydrophobic solutes on the surrounding water dynamics.[6] We showed that the water hydrogen(H)-bond and rotational dynamics are only moderately slowed down by these solutes, due to an excluded volume effect at the H-bond acceptor exchange transition state.

However, this picture cannot be straightforwardly applied to extended hydrophobic surfaces, since the water structural and thermodynamic properties have been evidenced to change dramatically with the curvature of the hydrophobic interface.[1] Next to extended hydrophobic surfaces, water molecules cannot maintain their H-bond network. They reorganize in *anti*-clathrate-like arrangements with water molecule orientations inverted compared to a clathrate hydration structure, where molecules in the first layer sacrifice some of their H-bonds.[7–9] Such changes in the local structure are thus expected to strongly affect the water dynamics.

An improved understanding of water dynamics next to extended hydrophobic surfaces will be first relevant for numerous biomolecular systems such as melittin[10] or other membrane-associated proteins which exhibit large hydrophobic patches. But it will also be of great interest in the context of nanosciences, where water can be conveyed through nanoscopic hydrophobic channels.[11] For these systems, application of macroscopic notions (*e.g.* contact angle, boundary conditions of a fluid flow, friction forces) becomes questionable when only a few thousand of water molecules are involved next to a nano-sized object, *e.g.* a droplet at a flat surface.[2,12] At this scale, the fraction of interfacial waters is extremely important, and the dynamics of individual molecules at the interface can play an important role in the overall dynamics. Molecular studies should therefore provide insights that are not available from macroscopic hydrodynamic approaches.

In this paper, we study an atomically-detailed and uncharged silica surface as a model system. This surface has been shown to be hydrophobic (contact angle of 110°[12]) and has been extensively studied before, mainly from a structural and thermodynamic point of view.[7,13,14] Based on (classical) Molecular Dynamics simulations and analytical modeling, we show how this surface affects water reorientation, and interpret it within the theoretical framework several of us have previously proposed to explain water reorientation in the bulk[15,16] and around small solutes.[6,17–19] We also calculate the two-dimensional infrared (2D IR) vibrational echo spectra, and show that this technique which has recently emerged as a very powerful probe of the H-bond dynamics[20–22] can provide significant new insights about water dynamics at this interface.

The remainder of the paper is organized as follows. We first describe the simulation methodology in section II. We then examine in section III the structure of water molecules at the surface focusing on their orientation, in order to define sub-ensembles of water OHs whose dynamics is discussed in section IV. We then establish a connection with the results that can be obtained from vibrational spectroscopy experiments in section V, before ending with some concluding remarks in section VI.

## II. Methodology

### Simulation details

The model surface has the structure of the (111) face of β-cristobalite, with no silanol group at the surface.[13,14] The surface is rendered hydrophobic by removing all partial charges. The surface atoms thus interact with water only *via* a Lennard-Jones potential, whose parameters were taken from ref. 13. The plate is constituted of 624 atoms distributed over 4 layers of silica, and its resulting size is approximately 3.2 nm × 3.2 nm × 0.9 nm. Following previous work,[13] the plate is maintained rigid. The effect of this approximation on the water dynamics under study is expected to be very limited, since the plate only interacts with the water molecules through a short-ranged, Lennard-Jones interaction, and the amplitude of individual displacements of the plate atoms in a solid would be less than a fraction of an Ångström.[13] Water is described by the rigid SPC/E force-field.[23] The simulation box contains the plate, oriented perpendicularly to the $z$ direction, and 1530 water molecules (for the calculation of 2D-IR spectra, the system is slightly different, as detailed below); the dimensions of the simulation box in the parallel $x$ and $y$ directions are larger than the plate size. Our analysis of water dynamics is performed of a restricted region at the center on the plate to avoid finite size effects due to the plate edges.

Molecular Dynamics simulations are performed with the NAMD software.[24] After a first NPT equilibration with Langevin dynamics during 500 ps at $T = 298$ K and $P = 1$ bar, a typical trajectory is propagated in the NVE ensemble for 1.5 nanoseconds. The average temperature is $298 \pm 1$ K. Long-range electrostatic interactions are described by the Ewald-summation and periodic boundary conditions are applied. The distance between two images in the perpendicular $z$ direction is approximately 3.5 nm; a previous study of the same hydrophobic surface showed that such a separation is sufficient to avoid any confinement effect on water dynamics.[13]

**Spectra**

Most experimental 2D IR studies of water dynamics employ an isotopic mixture of dilute HOD in either $H_2O$ or $D_2O$ to avoid complications due to intermolecular vibrational energy transfer.[20,21,25] We chose to consider the OD stretch of dilute HOD in $H_2O$ rather than the OH stretch of HOD:$D_2O$ because of the greater relevance of $H_2O$ as a solvent, and of the longer vibrational population lifetime of the OD stretch which allows measurement of experimental spectra at longer delays.[26] The 2D IR vibrational echo spectra were calculated based on the 3rd order nonlinear response functions, including the non-Condon effects and without assuming Gaussian dynamics, following the procedure detailed in ref. 27 and 28. We simulate spectra of dilute HOD in $H_2O$ from our MD trajectories.[29] The OD stretch vibrational frequency was determined from the local electric field using the frequency map from ref. 27. While in our simulations all waters are $H_2O$ molecules and not dilute HOD in $H_2O$ in order to increase the sampling, the effects of this approximation have been verified to be negligible.[28]

**Calculation of jump times**

The jump time, or H-bond exchange time, of a water OH from one H-bond acceptor to another is defined as the average time to go from a stable H-bond with the initial acceptor to a stable H-bond with the new acceptor[6,16] (this time will be related to a reorientation time in Sec. IV.2). Following previous work,[16] the stable H-bond states are defined by tight geometric H-bond conditions ($R_{OO} < 3.0$ Å, $R_{OO} < 2.0$ Å, $\theta_{HOO} < 20°$). The jump time $\tau_{jump}$ is calculated through the cross time-correlation function between the initial (I) and final (F) states as $\langle p_I(0) p_F(t) \rangle = 1 - \exp(-t/\tau)$, where $p_{I,F}$ is 1 if the system is in state I (F respectively) and 0 otherwise. States I and F are defined within the Stable States Picture[30] to remove the contributions from barrier recrossing[15,16] and absorbing boundary conditions in the product state ensure that the forward rate constant is calculated. Key geometric quantities of these jumps and their amplitude can then be determined from simulations.

## III. Orientation of water molecules next to the surface

The structural arrangement of water molecules next to a hydrophobic surface has been extensively studied before[7,8,10,14,31] on various types of surfaces, including the same surface as the one presently studied.[13] We therefore only summarize the key features of the water structure that will be used to define the different configurations between which water molecules undergo dynamic exchanges.

The distribution function of water oxygens along the $z$ direction perpendicular to the plate displays a well-defined first layer of water molecules next to the hydrophobic surface (Fig. 1a). This layer is approximately 3.5 Å-thick, and contains water molecules with different orientations, as revealed by the distance–angle density probability $P(z,\theta)$ to find a water O–H bond whose oxygen lies at a distance $z$ from the plate center, and whose axis forms an angle $\theta$ with respect to a vector normal to the surface, pointing outwards. This probability is then normalized by $\sin(\theta)$, removing the bias resulting from the angular variation of the solid angle.[7,32] The probability distribution shown in Fig. 1b. evidences that water molecules within the broad first hydration shell lie along given orientations, leading to separated peaks corresponding to distinct sub-ensembles. These different configurations are schematically represented in Fig. 2.

The first population corresponds to the water OHs pointing toward the surface and lying very close to the surface (Fig. 1b). These OHs, usually referred to as "dangling", and denoted **D** in the following are not engaged in any H-bond. They result from the strain imposed by the extended hydrophobic surface on the water H-bond network, which is forced to "sacrifice" some H-bonds to accommodate the interface[1] (Note that an H-bond can also be sacrificed by pointing a water lone pair toward the surface,[7] a situation similar in energy to a dangling OH). Such dangling OHs have been evidenced experimentally next to solid surfaces,[9,31,33] water/organic liquid interfaces[33,34] and even very recently around small hydrophobic solutes,[35] mostly by sum-frequency vibrational spectroscopy[9,31,33] which can specifically probe molecules at interfaces. As expected in the absence of any H-bond acceptor, their OH vibration frequency is strongly blue-shifted compared to the average bulk value.[9,31,34,35]

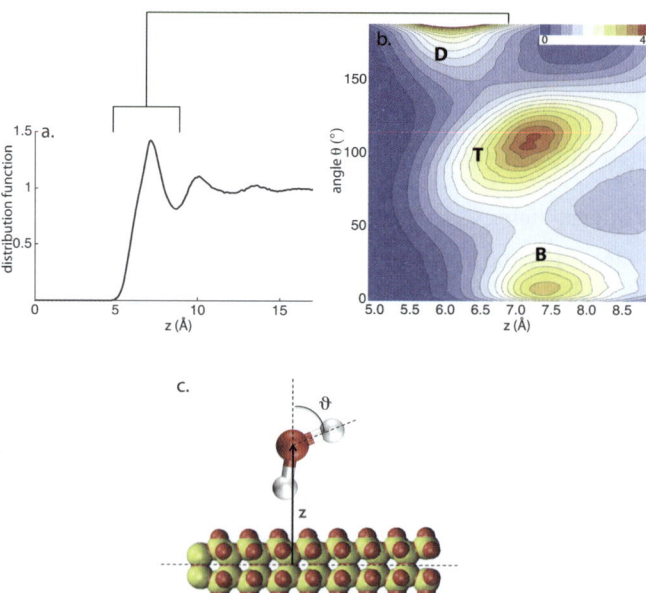

**Fig. 1** (a) Distribution function of water oxygens along the direction perpendicular to the surface, normalized by the bulk density value. (b) Distance–orientation distribution function of water OH bonds as a function of the distance $z$ and the angle $\theta$, normalized by $\sin(\theta)$. **D** (dangling), **T** (tangent) and **B** (bulk) refer to the populations of OHs as defined in the main text. (c) Definition of the $z$ and $\theta$ coordinates.

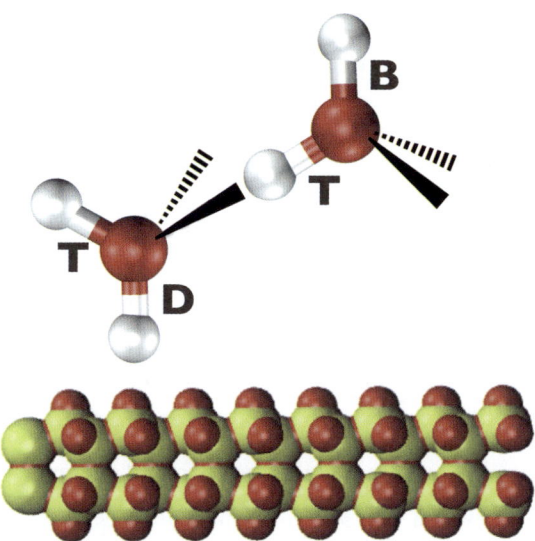

**Fig. 2** Schematic representation of the water structure next to the surface (using different scales for the surface and water molecules). Each oxygen atom is at the center of a tetrahedron formed by its two hydrogen atoms (explicitly represented) and its two lone pairs (black solid and dashed lines). Within the first hydration layer, molecules in direct contact with the surface point one vertex toward the surface, leading to *anti*-clathrate like arrangements (left hand side water on the scheme), with one OH or one lone pair dangling. These molecules are H-bonded to molecules of a second sub-layer, possessing a clathrate-like structure (right hand side). Letters between parentheses indicate to which population each OH belongs.

The second, more broadly spread, OH population corresponds to a configuration tangent (denoted **T**) to the surface (Fig. 2), and represents the major fraction of the first layer total population. These OHs donate an H-bond to other waters within the first layer. The broad peak in the distribution actually results from the contributions of two sub-populations, corresponding to two sub-layers within the first hydration layer. This situation can be analyzed through the local tetrahedral environment surrounding these water molecules. As shown in previous studies,[7] water molecules within the first hydration layer lie in a tetrahedral arrangement resembling that of ice $I_h$, although less structured and much more rapidly fluctuating. In an idealized ice structure $I_h$, each water tetrahedron can be found in one of two configurations. In the first of those, the tetrahedron can either point its apex to the surface, in an *anti*-clathrate configuration, as for the waters whose other OH is dangling (Fig. 2): the tangent OH then points slightly away from the surface, leading to the shoulder within the broad peak at short distances ($\sim$ 7 Å) and for angles around 80° (this configuration is energetically favored, as the loss of H-bond interactions is limited[7]). In the second of the configurations, the water tetrahedron can lie with one face flat on the surface, in a clathrate-like arrangement (Fig. 2): the tangent OHs then point slightly toward the surface, donating an H-bond to an *anti*-clathrate water, which leads to a contribution to the broad peak at just slightly longer distances from the surface and at angles around 110°. Such clathrate/*anti*-clathrate structures have been observed through MD simulations of water at a broad range of hydrophobic surfaces (realistic surfaces defined at an atomic scale or flat surfaces defined through a single distance dependent repulsive potential; disordered or structured; inorganic or biological[7,10,14]).

Finally, the third population for small angles with the normal vector corresponds to OHs pointing away from the surface, toward the bulk, for water molecules in a clathrate-like configuration. It will be denoted **B** (for bulk) in the following.

Based on the three distinct peaks within the first layer, we define distance and angle criteria to characterize these three sub-ensembles (see Table 1). These different states will be used to determine the rotational dynamics of each population and the exchange kinetics between these states. To eliminate the contribution from transient excursions to the exchange kinetics between states, we use the Stable States Picture[30] to define the three states. This implies that instead of defining only two "transition-state" separatrices between the three populations, each state is defined by boundaries along the distance and angle coordinates close to the corresponding distribution local maximum. A side effect of our definitions is that at any given time, a non-negligible fraction of the total first layer OH population, approximately 55%, does not fall within any of the **D**, **T** or **B** states; however, these systems are short-lived, and very rapidly return to one of the three defined states, on a 0.35 ps timescale. The respective populations in the stable states are 39% in **T**, 3% in **D**, and 3% in **B**, which shows that the vast majority of the OHs next to the hydrophobic surface lie in a tangent configuration.

## IV. Dynamics of water molecules next to the surface

In the following, we examine the reorientational dynamics of the **D**, **T**, and **B** subsets of the first layer OHs we defined in the previous part (see Table 1). The results are interpreted by considering the dynamics of H-bond acceptor exchanges, within and between the different states (see section II for calculation details). An analytic kinetic model describing the exchanges between the different states is then proposed.

### IV.1. Reorientation dynamics

Water reorientation can be followed using the second order Legendre polynomial time-correlation function

$$C_2(t) = \langle P_2[\vec{u}(0) \cdot \vec{u}(t)] \rangle \tag{1}$$

where $\vec{u}(t)$ designates the orientation of the water OH bond at time $t$. It is of particular interest since it can be related to experimental observables in NMR[36] and ultrafast infrared spectroscopy.[37,38] In bulk water at ambient temperature, $C_2$ decreases mono-exponentially with a 2.5 ps time constant, after a small fast initial drop (<200 fs) due to librations.[39] Bulk simulations based on the SPC/E model[16] lead to an excellent agreement with the experimental data, both from NMR[36] and ultrafast IR[37,38] spectroscopies.

We now compare the reorientation dynamics for OHs according to their initial situation, either in one of the three states **D**, **T**, and **B** within the first shell, or within the second shell, or beyond. We thus consider the $C_2(t)$ decays for OHs which lie in a given state at time $t = 0$, but without imposing any constraint on the future evolution of the system, and thus allowing the system to leave its initial state.

The anisotropy decays starting respectively from the **B** state of the first shell, from the second shell and from the further shells are almost identical (data not shown). They decay mono-exponentially with respective time constants of 2.7 ± 0.1 ps, 2.8 ± 0.1 ps and 2.6 ± 0.1 ps. Since these time scales are the same within error bars,

**Table 1** Definition of the OH populations as defined in the main text, in terms of distances to the surface (0 corresponds to the first non-zero value of the distribution function) and angles with respect to a normal vector pointing outwards (cf. Fig. 1)

| State | Dangling **D** | Tangent **T** | Pointing outwards **B** | Bulk |
|---|---|---|---|---|
| Distance to the surface/Å | 0–2.7 | 0–2.7 | 0–2.7 | >3.5 |
| Angle with respect to the normal vector (°) | 150–180 | 70–120 | 0–20 | Any |

this illustrates that the influence of the hydrophobic surface on the water dynamics is very short-ranged. The reorientation dynamics starting in the **B** state can therefore be approximated as bulk-like. In the following, **B**′ will be used to collectively designate the OHs within the **B** state or within the second layer and further from the surface.

For an OH initially tangent to the surface (**T**), the reorientation is slower than in the bulk, and is non mono-exponential (see inset of Fig. 3 on a logarithmic scale). A single exponential fit on the 2–10 ps interval leads to an effective relaxation time constant of 4.8 ps, thus approximately twice as long as in the bulk. We note that similar retardation factors in the 1.5–2.5 range have been measured by NMR for the water reorientation dynamics at the interface with small hydrophobic peptides[40] or with larger proteins such as BPTI and ubiquitin,[41] whose solvent-exposed surfaces include hydrophobic patches.

For OHs initially dangling (**D**), the $C_2(t)$ anisotropy decay is dramatically different from the bulk decay and exhibits very unusual features (Fig. 3). The initial drop is very pronounced (more than 50% *versus* ∼ 20% in the bulk, *i.e.* anisotropy values of respectively 0.38 and 0.79 after 100 fs, see Fig. 3). This corresponds to very large amplitude librational motions, due to the absence of any surface-fixed H-bond acceptor creating a restoring force; the OH orientation is mainly determined by the three other H-bonds to the water molecule. Following this fast initial decay, the anisotropy decreases on a ps timescale to reach *negative* values, before slowly returning to zero on a much longer time-scale, on the order of a few tens of picoseconds. Such negative anisotropy values are in stark contrast with the bulk behavior where the anisotropy always remains positive.

Several of us recently showed that the reorientation dynamics of water in the bulk[15,16] and around various solutes[17–19] reflects the dynamics of H-bond acceptor exchanges through large angular jumps. We therefore now analyze these acceptor exchanges in each OH population in order to rationalize the observed anomalous orientational relaxations next to the surface.

### IV.2. Hydrogen-bond exchange

In contrast with the traditional rotational diffusion picture, it was recently shown that beyond the initial fast librational decay, water reorientation mainly occurs

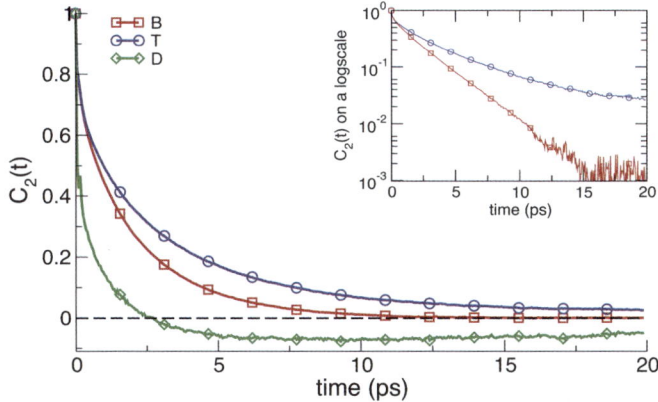

**Fig. 3** Second-order orientational time correlation function $C_2(t)$ eqn (1). The different curves correspond to distinct initial conditions: OHs initially in the first layer and respectively dangling **D** (green diamonds), tangent **T** (blue circles) or pointing to the bulk **B** (red squares). The decays for OHs initially in the second layer or beyond are superimposed on the **B** decay and are not represented for clarity. The horizontal dashed line corresponds to the vanishing anisotropy value. Inset: **B** and **T** decays on a logarithmic scale; the **D** decay is not represented since it becomes negative.

through large amplitude jumps due to the exchange of H-bond acceptors.[15,16,45,46] It is only once the environment has reorganized to offer a new stable H-bond acceptor that the water OH bond quickly executes a large-amplitude angular jump from its former H-bond partner to this new acceptor.[15,16] The characteristic time of the resulting orientational relaxation was shown to be well described by an analytic molecular jump model, whose two key ingredients are the jump time $\tau_{\text{jump}}$ (*i.e.* the inverse jump rate constant) and the jump average amplitude $\Delta\theta$.

We now use our simulations to characterize and determine $\tau_{\text{jump}}$ and $\Delta\theta$ for all the possible jumps occurring next to the interface, for each type of initial and final situations. For this purpose, in addition to the previous geometric conditions defining the stable **D**, **T**, and **B′** states, we add another set of conditions on the H-bond in order to consider only exchanges between stable H-bond states. These correspond to $R_{\text{OA}} < 3.0$ Å, $R_{\text{HA}} < 2.0$ Å, and $\theta_{\text{HOA}} < 20°$, where A is the H-bond acceptor.

### IV.2.a D state.

We first consider OHs which are initially dangling. These cannot form any stable H-bond since there is no initial H-bond acceptor. However, by extension, the initial stable state is defined here with the same criteria as detailed above, considering as a fictitious acceptor the point on the surface lying at the vertical of the donor oxygen. For an initially dangling OH, the jumps exclusively lead to a tangent **T** situation, as evidenced by the computed jump amplitude distribution (data not shown), which does not contain any noticeable contribution beyond 130° which would have led to a **B** state, and whose average is approximately 90°. This average amplitude is significantly greater than in the bulk[16] ($\sim 68°$), due to the different local water arrangement imposed by the surface. The jump time $\tau_{\text{jump}}$ is about twice as fast as in the bulk (1.6 ps *vs.* 3.3 ps in the bulk[16]). This acceleration results from the absence of any initial H-bond. Indeed, two of us showed[16] that a significant contribution to the H-bond acceptor exchange activation free energy comes from the free energy cost to elongate the initial H-bond to reach the bifurcated H-bond transition state configuration. Here this contribution vanishes and the jump rate constant accelerates accordingly.

The fast jump from **D** to **T** is also evidenced by the time dependent probability for an initially dangling OH to form an angle $\theta$ with respect to its initial orientation (Fig. 4a). After a transient initial delay where the average angle rapidly increases, jumps to a tangent situation lead to a distribution centered around an average value of approximately 90°, which very slowly increases with time because of subsequent reorientation once in state **T**. This latter process occurs on a much longer time scale, as shown by the probability distribution for an initially **T** OH (Fig. 4b), and is as discussed below in more detail.

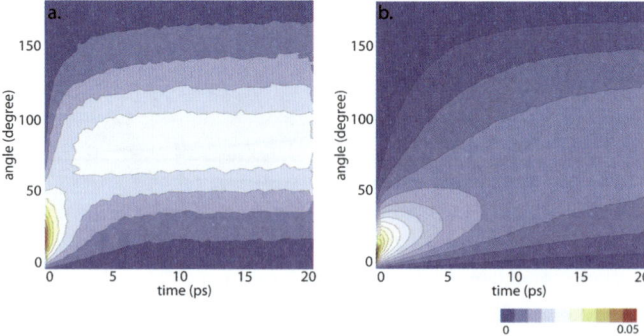

**Fig. 4** Probability density distributions for an OH at time $t$ to form an angle $\theta$ with respect to its initial orientation, starting from state **D** (a) or from state **T** (b). For the sake of clarity, only the data after $t = 0.05$ ps is shown to avoid the initial peaked value of 1 at $\theta = 0$.

**IV.2.b T state.** OHs that are initially tangent can jump to form one of three states: (i) either a dangling **D** OH, or (ii) a new H-bond with a water within the first hydration layer, corresponding to a different **T** configuration, or (iii) a new H-bond with a molecule lying in the bulk, leading to a **B'** configuration. The overall jump rate constant (inverse jump time) is the sum of the three rate constants, since these are independent processes.

The **T** → **D** jump corresponds to the reverse reaction with respect to the **D** → **T** jump discussed above. It thus has the same amplitude (∼ 90°), but the jump time is much longer (21 ps). The ratio $\tau_{\text{jump}}^{D \to T}/\tau_{\text{jump}}^{T \to D}$ is indeed equal to the equilibrium constant between the **D** and **T** populations (which is approximately equal to 13) reflecting the large excess of tangent OHs with respect to the dangling OHs.

We now focus on the jump to a new H-bond acceptor, either from the first hydration layer or from further out from the surface. For both cases, the exchange mechanism is the same as in the bulk, as evidenced by the same value of the average jump amplitude (68°). As for the kinetics, considering both types of final acceptor together yields a global jump time of 5.9 ps to form a new H-bond. The exchange mechanism is therefore identical to the bulk case, but retarded by a 1.8 factor with respect to the 3.3 ps bulk time.[16]

This retardation phenomenon is completely similar to what we recently evidenced around small hydrophobic solutes,[6] where several of us showed that it stems from an excluded volume effect on the possible transition state locations for the new acceptor. The presence of the hydrophobe (solute or surface) hinders the approach of a new H-bond partner, thus slowing down the exchange rate. The retardation factor $\rho_V$ is defined as

$$\rho_V = \frac{\tau_{\text{jump}}^{\text{hydrophobic}}}{\tau_{\text{jump}}^{\text{bulk}}} \quad (2)$$

in which $\tau_{\text{jump}}^{\text{hydrophobic}}$ designates the jump time to a new H-bond partner in the presence of the hydrophobic solute/surface (5.9 ps here). The associated decrease in the transition state (TS) entropy can be estimated in the simulations through the determination of the fraction $f$ of the transition state locations forbidden by the solute/surface excluded volume.[6] This leads to a transition state excluded volume (TSEV) factor, which provides a quantitative description[6] of the jump time retardation factor $\rho_V$ through

$$\rho_V = \frac{1}{1-f} \quad (3)$$

Next to convex surfaces (as is the case around small solutes), less than half of the space is forbidden in the TS, and the resulting retardation factor is less than 2; for a wide range of small hydrophobic solutes, the TSEV factor was computed to be approximately 1.4, in good agreement with experimental data.[6]

In the present case of an OH tangent to an extended flat hydrophobic surface, a crude estimate of the excluded volume fraction would be $f = ½$, leading to a slow-down factor of 2. However, the OH lies slightly above the surface, and the computed TSEV factor is $\rho_V = 1.8$, in perfect agreement with the observed retardation on the jump rate given above.

Among the possible new H-bond partners for an initially **T** OH, we now distinguish between the different localizations of final partners. These can first come from the first hydration layer ($\tau_{\text{jump}} = 9.6$ ps) leading to a final **T** situation where the OH stays tangent to the surface. In this case, the angular jump occurs roughly in a plane parallel to the surface and the rotating OH remains in the first shell. Final partners can alternatively originate from the second shell ($\tau_{\text{jump}} = 15.2$ ps); the jump arc is therefore perpendicular to the surface and then leads to a final **B'** situation where the OH points away from the surface and interacts with second shell.

It is relevant for the jump times scales that H-bond exchange events induce translational displacements of the water molecules.[17] The slower **T** → **B′** jumps perpendicular to the surface relative to the faster **T** → **T** jumps parallel to the surface are consistent with the previous observation of a larger translational diffusion constant for water along a surface than perpendicular to it (see *e.g.*[47] for water next to protein surfaces). As most of the OHs in the first hydration shell are in a tangent **T** state, the above results can also provide a tentative explanation for the well-known decoupling between translational and rotational motions that has been observed next to solutes,[48] hydrophobic surface,[49] and close to the glass transition.[50] The translational dynamics becomes markedly retarded while the rotational dynamics is little affected by the surface (as shown in section IV.1, the $C_2$ decay for **T** is only retarded by a factor of approximately 2).

**IV.2.c B′ state.** Within the **B′** situation, two types of jumps can occur: either a **B′** → **T** jump, which is the reverse event with respect to the **T** → **B′** jump described above, or a **B′** → **B′** jump, which is identical to the jumps characterized in bulk water in previous work ($\tau_{jump} = 3.3$ ps and $\Delta\theta = 68°$).[16]

All the jump characteristic times and amplitudes are summarized in the scheme shown in Fig. 5.

**IV.2.d Connection to the orientational relaxation.** The above analysis has shown that the jump rate constants differ between each of the **D**, **T** and **B′** states. As reorientation mainly occurs through these large angular jumps, this implies that the reorientation times are different within each state. Within the Extended Jump Model (EJM),[15,16] the reorientation time within one state is given by

$$\frac{1}{\tau_{reor}} = \frac{1}{\tau_{jump}} \left[ 1 - \frac{1}{5} \frac{\sin(5\Delta\theta/2)}{\sin(\Delta\theta/2)} \right] + \frac{1}{\tau_{frame}} \quad (4)$$

The different terms are now discussed. It is important to note here that the first contribution, the jump reorientation time, differs from the $\tau_{jump}$ jump time by an angular factor[15,16] accounting for the reorientation amplitude each time a jump occurs. In the case of different final partners, as is the case starting from **T**, the overall jump contribution will simply be the sum of each individual jump contribution. An additional minor contribution to the reorientations times arises due to the orientational relaxation of intact H-bonds between jump events, whose time constant is denoted $\tau_{frame}$.[15,16] We now briefly discuss how $\tau_{frame}$ is modified for OHs belonging to the one of the three considered states **D**, **T** or **B**.

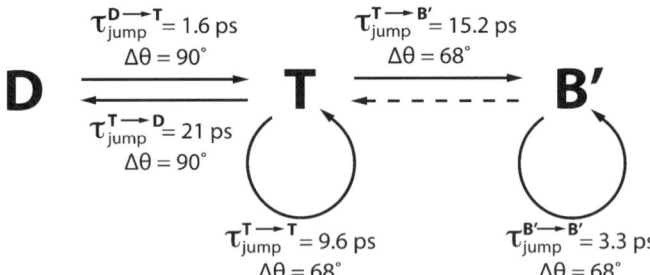

**Fig. 5** Schematic representation of the different H-bond exchanges considered here. Jump times $\tau$ and jump amplitude $\Delta\theta$ are given for each type of exchange. When different products are involved, the global rate constant is equal to the sum of the rate constants corresponding to the formation of each product.

Starting from **D**, the $\tau_{\text{frame}}$ contribution simply vanishes because no H-bond is formed until a jump event occurs. Starting from **T**, the calculated $\tau_{\text{frame}}$ is 13 ps, twice as long as in the bulk (6 ps[16]) due to the presence of the immobile surface. Finally, for an OH in state **B** (in the first shell but pointing away from the surface), $\tau_{\text{frame}}$ is the same as in the bulk within error bars (this provides further support for the definition of a single state **B'** including state **B** and the second layer). Surprisingly, the observed retardation of $\tau_{\text{frame}}$ in both cases is close to the one for the jump time ($\rho_V$, eqn (2)); as a consequence, $\tau_{\text{frame}}$ always remains sufficiently larger than $\tau_{\text{jump}}$, such that its contribution to the overall orientational relaxation remains minor.

A further consequence of a jump, which we have not yet elucidated, is the changing of the OH configuration and the induction of transitions between the **D**, **T** and **B'** states. To model the anisotropy decays detailed in section IV.1, we therefore have to determine not only the characteristic reorientation times within each state, but also the exchange kinetics between these states. This is the focus of the following sub-sections.

### IV.3. Kinetic model and interpretation of the anomalous anisotropy decays

We consider the exchanges between the **D**, **T** and **B'** states, where **B'** includes all the first layer OHs pointing to the bulk (**B** state) and all the OHs lying in the second shell and further.

The resulting kinetic scheme is

$$\mathbf{D} \underset{1/\tau_{\text{jump}}^{\text{T}\rightarrow\text{D}}}{\overset{1/\tau_{\text{jump}}^{\text{D}\rightarrow\text{T}}}{\rightleftharpoons}} \mathbf{T} \underset{k_{\text{b}}}{\overset{1/\tau_{\text{jump}}^{\text{T}\rightarrow\text{B}'}}{\rightleftharpoons}} \mathbf{B}' \quad (5)$$

The rate constants for transitions between the **D** and **T** states, and from **T** to **B'** correspond to the inverse jump times between the corresponding states. A proper description of the exchange from **B'** to **T** would include the effect of translational diffusion within **B'** away from the interface with **T** (see *e.g.* 51). Here we approximate this back reaction by a rate equation, which was shown to yield a satisfactory description at delays shorter than the back reaction inverse rate constant.[52] This rate constant was determined to be $k_B \approx 1/(45 \text{ ps})$ from the equilibrium constant between the **T** state and the second layer population, which contains all the OHs susceptible to jump to a tangent state.

This simple kinetic scheme (eqn (5)) can then be solved analytically for a given distribution of initial populations, providing the time evolution of each population.[54] Fig. 6 shows the respective populations in each state, starting either from a dangling state **D** (top) or from a tangent state **T** (bottom). These analytic results only require the rate constants determined from the numerical simulations. They can be compared with the direct calculation of the time evolution for the respective populations in the simulations. As shown in Fig. 6, the agreement is extremely good and thus supports this simple kinetic model (the discrepancy for **D** at very short delays corresponds to fast **D** → **T** librational exchanges, not described in our kinetic model).

We now analyze the time-evolutions of the populations for the different initial states to interpret qualitatively the anomalous anisotropy decays shown in Fig. 3.

We first consider the fate of an initially dangling **D** OH (top panel of Fig. 6). The anisotropy decay (Fig. 3) was shown to decay much faster than in the bulk and to reach negative values before slowly returning to an isotropic distribution. This can now be interpreted through the kinetics, which shows that an initial **D** state rapidly jumps within a few ps to a tangent **T** configuration within the first hydration layer. This transiently populated **T** state subsequently decays to populate the **B'** state. Since the average angle between the **D** and **T** configurations is 90°, a crude estimate for the

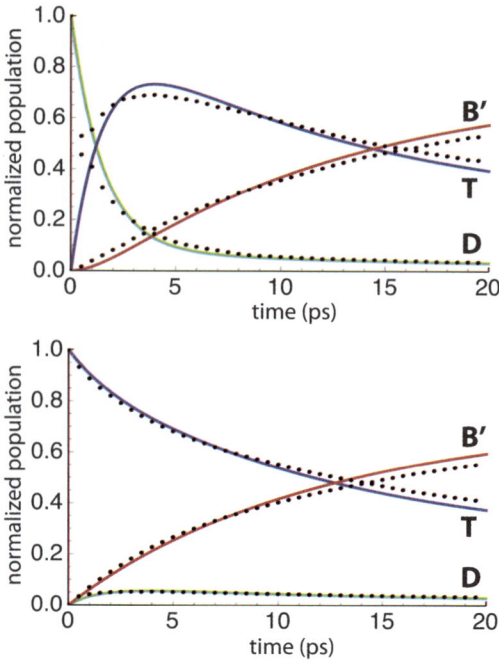

**Fig. 6** Evolution of the normalized population in each state (dangling (green), tangent (red) and bulk (blue)) as a function of time, with different initial conditions: starting from a dangling OH (top panel) or from a tangent OH (bottom panel). Plain lines represents results from our simple kinetic model, and black dots represent the corresponding data, directly extracted from simulations.

$C_2$ anisotropy while in the **T** state is $P_2[\cos(90°)] = -1/2$.[55] This explains the negative values (the actual $C_2$ is less negative than in this simplified picture, mainly because at any given delay, some systems are not in **T** and bring positive contributions to $C_2$); these negatives values are reached on the timescale of the **D** → **T** jump time, *i.e.* 1.6 ps. The slow return of the anisotropy to a vanishing value is due to the **T** → **B'** jumps, whose timescale was shown above to be much slower (15.2 ps).

We now consider the systems which are initially tangent **T** (Fig. 6, bottom panel). For these, the anisotropy decay had been shown (Fig. 3) to decay more slowly than in the bulk, and with a pronounced multi-exponential behavior. This can be explained by examining the time evolution of the populations, which shows that state **T** populates the **B'** state on the same timescale as the reorientation (while state **D** never becomes significantly populated). The anisotropy decay is thus the sum of two contributions. The first one comes from systems which have already switched to the **B'** state and reorient in a bulk-like fashion. The second one originates from systems remaining in state **T**, where the reorientation is slower due to the retarded jump time (slowdown factor of 1.8) and occurs mostly in two dimensions instead of three, due to the dominant contribution of jumps between **T** configurations; the asymptotic anisotropy value is then ¼[56] and not zero as in the three-dimensional case. The sum of the **B'** and **T** contributions thus leads to an overall anisotropy decay which is slower than in the bulk and not mono-exponential.

## V. Vibrational spectroscopy

We now examine whether the anomalous water dynamics induced by the hydrophobic surface that we have discussed above can be evidenced experimentally.

Most experimental studies of water next to an interface rely on vibrational sum-frequency generation (SFG) spectroscopy, which can specifically probe waters at the interface due to the lack of local symmetry in their environment.[9,31,33,57,58] While this technique yields valuable information about the water structure at the interface, it does not provide any direct information about the dynamics.

Other techniques, although not selectively sensitive to the interface waters, have already been successfully used to investigate water dynamics around solutes. NMR can determine an average water reorientation time within a hydration shell through a concentration-dependence study of the global reorientation time in dilute solutions and the assumption of a two-state bulk/hydration shell model (see *e.g.*[59]). However, in the present case of an interface, it is not clear that such a concentration study could be performed, although measurements of water dynamics around extended hydrophobic systems such as carbon nanotubes have been reported.[60]

Ultrafast infrared spectroscopy can selectively excite the waters within a hydration shell when their vibrational frequency is shifted with respect to the bulk due to the interaction with the solute (see *e.g.* 61,62). As shown experimentally by SFG next to hydrophobic interfaces:[9,31,33,57,58] this is the case for the dangling water OHs whose frequency is blue-shifted. We now examine two types of ultrafast infrared spectroscopy techniques and assess what future studies based on these techniques can reveal about water dynamics next to a hydrophobic interface.

### V.1. Pump–probe spectroscopy

Ultrafast infrared pump–probe spectroscopy can measure the time-resolved anisotropy decay of an initial vibrational excitation, which is proportional to the $C_2(t)$ orientational time correlation function defined in eqn (1).[37,63] By tuning the frequency of the pump pulse, it is possible to selectively follow the reorientation of waters according to the initial strength of their donated H-bond.[64]

In order to examine the possibilities for this type of spectroscopy, we have calculated the linear infrared spectrum of water next to the hydrophobic surface. As explained in section II, we consider an isotopic mixture of dilute HOD in $H_2O$ in order to decouple the two water stretching modes and to avoid intermolecular vibrational energy transfer. Fig. 7.a shows the spectrum in the OD stretch frequency region. The dangling ODs lead to a visible peak in the 2700–2750 $cm^{-1}$ region. However, this peak overlaps with the broad main peak, whose intensity is much higher. Indeed, some waters from the bulk display very blue-shifted frequencies when they experience a transient H-bond breaking. Even if such events are very brief (< 200 $fs^{15,22,65}$), they are very frequent (more than once per picosecond[53]) and they give a non-negligible contribution to the spectrum on the blue edge. Using a pump frequency in that range will therefore also select these systems in addition to the dangling waters.

Fig. 8 shows the $C_2(t)$ anisotropy decay calculated for two ranges of initial frequencies, pumping respectively the center of the band (2300–2600 $cm^{-1}$) or the dangling peak region (2700–2750 $cm^{-1}$). The marked initial librational decay for the blue-shifted OHs is due to the large angle of the librational cone in which these systems librate, not only in the **D** state but also in the bulk, where the blue-shifted OH frequency corresponds to a weak H-bond.[39] The subsequent reorientation is faster for systems pumped on the blue edge than in the center of the band, but $C_2(t)$ does not reach negative values as does the decay from state **D** (see data of Fig. 3 also reported on Fig. 8), due to the large contribution of the bulk OHs. Using pulses linearly polarized along the direction perpendicular to the wall does not sufficiently increase the fraction of dangling OHs in the signal to observe negative anisotropy values. We also note that pumping at the center of the band leads to an anisotropy decay which is intermediate between the ones previously obtained for bulk and tangent OHs, whose initial frequencies are similar. This demonstrates

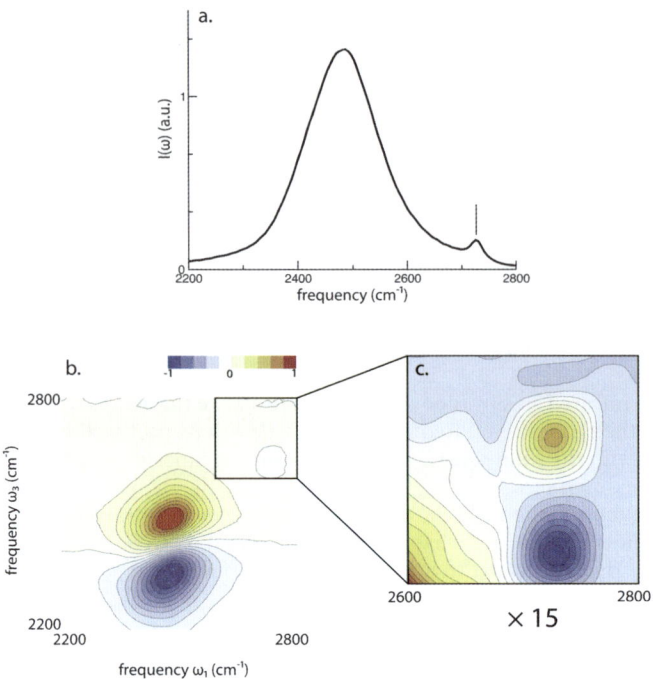

**Fig. 7** (a) Calculated linear spectrum of HOD in $H_2O$ for the system containing the plate and water molecules, polarized along the direction perpendicular to the surface. Bottom: calculated 2D-IR spectra at a delay of $T = 0.5$ ps, in the whole frequency range (b) and focusing on the peaks corresponding to dangling OHs (c), not visible on the whole spectrum. Each spectrum is normalized with respect to the positive peak height.

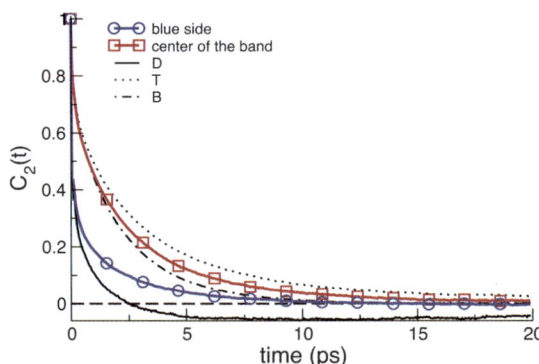

**Fig. 8** Second-order orientational time correlation function $C_2(t)$ eqn (1), by selecting the initial frequency range. The different curves correspond to distinct initial conditions: initial frequency between 2300 and 2600 $cm^{-1}$, corresponding to the center of the band (red squares); initial frequency between 2700 and 2750 $cm^{-1}$, corresponding to the small peak region (blue circles). Thick black lines (plain, dot-dashed and dashed) represent the anisotropy decay from the geometrically-defined **D**, **T** and **B** states, as in Fig. 3. No particular polarization is used here.

that while pump–probe spectroscopy reveals an anisotropy decay significantly different from that in the bulk, it cannot disentangle the different contributions and selectively follow the dynamics of dangling OHs.

## V.2. Two-dimensional vibrational echo spectroscopy

There is a way to selectively follow the dangling OHs without simultaneously probing the transiently non-H-bonded bulk OHs. This is to take advantage of the much longer lifetime of the dangling state (1.6 ps as shown in section IV.2) with respect to that of the transiently broken H-bonds (< 200 fs[15,22,65]); for transient breakings, the blue-shifted frequency rapidly returns to the center of the band. This can be done *via* two-dimensional infrared (2D-IR) vibrational echo spectroscopy. 2D-IR has recently emerged among ultrafast spectroscopy techniques as a very powerful tool to probe H-bond dynamics.[20,21,25] It measures the time dependence of the water vibrational stretch frequency, which reflects the local environment fluctuations, and has already been successfully applied to study H-bond dynamics in pure water and aqueous salt solutions.[22,26,65–67] 2D-IR spectra can be seen as a correlation map between the excitation frequency $\omega_1$ and the detection frequency $\omega_3$ of a given oscillator after a delay $T$. It is a Fourier transform technique using broadband excitation pulses, and it is therefore not limited by the time resolution of transient hole burning experiments, due to the required trade-off between time and frequency resolutions.

2D-IR spectra are computed from MD trajectories using the methodology described in Section II. A typical spectrum of HOD in $H_2O$ is shown in Fig. 7b. It exhibits two main peaks: a positive peak corresponding to the 0–1 OD transition (bleaching), centered around 2450 cm$^{-1}$, *i.e.* the linear spectrum maximum, and a negative peak corresponding to the 1–2 transition (induced absorption). In the following, we will focus only on the positive peaks, even if similar results can be obtained by studying negative peaks, that behave similarly.[62]

An additional diagonal peak on the blue edge reflects the presence of dangling ODs, but it is barely visible on the scale of the main peak. A magnification of the blue region (Fig. 7c) evidences the presence of a system of 0–1 and 1–2 peaks around 2720 cm$^{-1}$. The amplitude of these peaks is very small for two reasons. First, the fraction of dangling ODs is small with respect to the bulk and second, the transition moments are dramatically smaller in the blue region because of non-Condon effects.[27] This last effect is even more pronounced in the 2D spectra than in the linear spectra, because the volume of the peaks is proportional to $\mu^4$ [62] (compared to $\mu^2$ for the linear spectra). The relative height of the dangling peak can be increased by a) reducing the distance between the hydrophobic interfaces, thus increasing the fraction of interfacial waters, and b) using pulses linearly polarized along the direction perpendicular to the surface, thus lowering the isotropic contribution of the bulk compared to that of the dangling ODs.

We now discuss the 2D-IR spectral dynamics along the waiting time $T$. The time-evolution of both the main peak and the dangling peak results from three effects: spectral diffusion, vibrational population relaxation, and chemical exchange. We now detail each of these different effects.

At short delays (< 100 fs), the frequency memory between excitation and detection is largely preserved, and the 2D correlation spectrum is elongated along the diagonal. In contrast, for longer delays, the frequency diffuses within the spectrum due to the fluctuating forces on the oscillator exerted by the surrounding molecules,[22,66] and the positive-going band in the spectra becomes increasingly symmetrical and round, due to the loss of frequency correlation.

Vibrational population relaxation to the ground state occurs on the vibrational lifetime timescale, *i.e.* 1.45 ps for the OD stretch,[26] and results in an exponential decay of the signal.

The third, chemical exchange, effect appears when the OHs can lie in different states with distinct average frequencies, such as two states with different H-bond strengths.[19,62,68] Systems having undergone an exchange between two states during the waiting time have different excitation and detection frequencies, and lead to off-diagonal cross peaks in the 2D IR spectra. The growth kinetics of these off-diagonal peaks provides the exchange time between the two states.[62] Here, when an initially dangling (thus blue-shifted) OD jumps to a tangent situation, *i.e.* **D** → **T**, its final frequency falls within the OD spectrum main broad peak. However, the large size of the main peak hides the cross peak, which is not discernable in the spectra.

The exchange kinetics between the **D** and **T** states therefore cannot be measured *via* the appearance of the cross peak, but we will now show it can be determined experimentally from the decay of the diagonal blue-shifted peak, which is shown on Fig. 9 at different delays. We illustrate this approach on our calculated spectra to demonstrate how the exchange time between **D** and **T** can be determined from the 2D IR spectra exclusively, and show that this leads to a timescale very close to that obtained from the direct calculation.

We consider the total volume of the diagonal dangling peak and not its peak height in order to free our approach from spectral diffusion effects.[62] This peak volume is proportional to the population in the vibrationally excited **D** state, and its time evolution with the waiting time $T$ is

$$V(T) \sim \exp[-T(1/\tau_{\rm vib} + 1/\tau_{\rm jump}^{\bf D \to T})] \qquad (6)$$

where $\tau_{\rm vib}$ designates the vibrational lifetime and $\tau_{\rm jump}^{\bf D \to T}$ is the **D** → **T** jump time. In our calculated 2D spectra, we employ the same $\tau_{\rm vib}$ value (1.45 ps) for all the ODs, even though vibrational lifetimes at interfaces can be different from the bulk.[69] However, note that such vibrational lifetimes can be experimentally measured through a separate set of pump–probe experiments,[62] or they can be calculated[70] but such calculations are beyond the scope of the present study. Therefore, the decay of the peak volume due to vibrational lifetime can be experimentally determined (although we fix it to the bulk value here) and once it is known, an exponential fit of the peak volume decay provides the jump time. Fig. 9 shows the evolution of the blue diagonal peak with time.

The time decay of the blue diagonal peak volume determined from our calculated spectra is shown in Fig. 10. It exhibits an initial (< 200 fs) fast drop of the peak volume, which is due to the blue-shifted OHs from the bulk whose frequency very rapidly returns to the center of the band.[65] Beyond this initial decay, only the **D** OHs remain, and the decay is monoexponential. Using eqn (6) and the vibrational lifetime, a decay time of 1.9 ± 0.2 ps is determined, in good agreement with the 1.6 ps value from the direct determination detailed in section IV.2. We have checked that this time is only weakly sensitive to the definition of the blue peak boundaries.

This clearly demonstrates that 2D-IR will provide experimental information on H-bond dynamics next to a surface that until now have been only accessible through simulations. Current developments in SFG-2D-IR[71] also appear extremely promising in combining the surface-sensitivity of SFG with the time-correlation information provided by the 2D-IR approach.

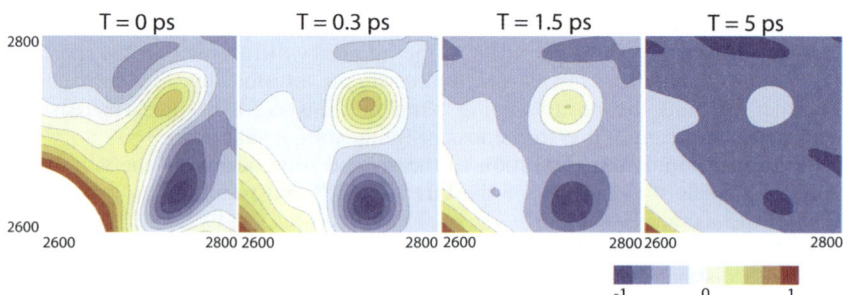

**Fig. 9** 2D-IR spectra in the region of dangling OHs at several delays. The horizontal and vertical axes correspond respectively to the excitation and detection frequencies. To evidence the peak decay due to exchange, each spectrum is normalized by a factor $e^{-T/1.45ps}$ corresponding to the global decay of the spectra because of vibrational population relaxation.

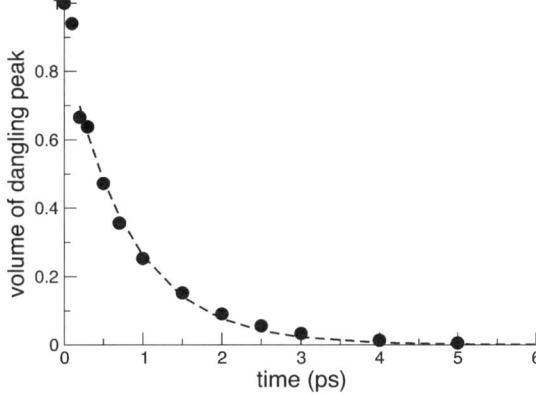

**Fig. 10** Evolution of the dangling peak volume as a function of time, normalized with respect to its initial value (filled circles). The dashed line represents the fit of these points after 200 fs, allowing determination of the jump time.

## VI. Concluding remarks

Based on simulations and analytic modeling, we have presented an extensive study of water dynamics next to a hydrophobic surface. While previous studies have mostly considered the dependence of the water dynamics on the distance from the water oxygen to the surface,[7,14,49] we have shown that considering the orientation of water OHs allows a more effective rationalization of the behavior. Indeed, depending on its orientation, a water OH experiences different environments, which play a critical role in its dynamics.

Most OHs in the first hydration layer are tangent to the surface. We showed that their reorientational dynamics is retarded by a factor close to 2 with respect to the bulk. We quantitatively explained this slowdown, within the Extended Jump Model, by a transition state excluded volume effect on the H-bond exchange mechanism, the presence of the surface hindering the approach of a new H-bond partner. The effect is very similar to the case of small hydrophobic solutes,[6] but is more pronounced because of the larger excluded volume induced by the extended surface.

Some other OHs within the first layer lie in a dangling configuration pointing to the hydrophobic surface, due to the sacrifice of some H-bonds imposed by the surface. They reorient much faster than bulk OHs. The second order orientational correlation function for these dangling OHs takes negative values during its decay, which were shown to be due to an exchange with the more stable tangent configuration.

Finally, we demonstrated that the exchange timescale between dangling and tangent configurations can be measured by 2D-IR experiments, which can selectively follow the spectrally distinct dangling OH population. We trust that future experiments will be able to confirm the very fast dynamics of a sub-ensemble of OHs next to the hydrophobic surface.

The present analysis should find numerous extensions. First, this framework could be applied to the water/organic liquid interface.[72] This liquid interface is expected to be similar to the present solid interface; the organic liquid is also hydrophobic and capillary waves have little influence on the water dynamics.[72] Recent simulation studies on such interfaces[72,73] have shown that water reorientation is strongly anisotropic next to the interface and depends on the initial OH orientation, and, further, that on average the H-bond exchanges within the first layer are slowed down by a factor of 1.4 with respect to the bulk and occur through larger amplitude jumps (84°). Such conclusions are completely consistent with our description here. Relative considerations should prove useful for the water–air interface.

The perspective of our present work can also be extended to surfaces with an increasing hydrophilic character, in connection with a recent simulation study that has evidenced that next to a surface gradually changed from hydrophobic to hydrophilic,[49] the average water reorientation time within the first layer exhibits a non-monotonic character. This time is first retarded by a factor of approximately 2 with respect to the bulk for the most hydrophobic case; it then first decreases with increasing hydrophilicity to nearly reach the bulk reorientation time, before increasing again for even more hydrophilic surfaces to eventually reach a value retarded by a factor of approximately 3 with respect to the bulk.[49] While this behavior was explained through transient ice-like structures,[49] our present work can offer an alternative explanation which emphasizes consideration of the total H-bond environment in analyzing the water dynamics; this would be the aim of future investigations.

## VII. Acknowledgments

This work was supported in part by NSF grants CHE-0750477 (JTH) and CHE-0910615 (PJR).

## References

1 D. Chandler, *Nature*, 2005, **437**, 640–647.
2 B. J. Berne, J. D. Weeks and R. Zhou, *Annu. Rev. Phys. Chem.*, 2009, **60**, 85–103.
3 T. Kimura, A. Maeda, S. Nishiguchi, K. Ishimori, I. Morishima, T. Konno, Y. Goto and S. Takahashi, *Proc. Natl. Acad. Sci. U. S. A.*, 2008, **105**, 13391–13396.
4 A. Fernandez and R. S. Berry, *Proc. Natl. Acad. Sci. U. S. A.*, 2003, **100**, 2391–2396.
5 A. De Simone, G. G. Dodson, C. S. Verma, A. Zagari and F. Fraternali, *Proc. Natl. Acad. Sci. U. S. A.*, 2005, **102**, 7535–7540.
6 D. Laage, G. Stirnemann and J. T. Hynes, *J. Phys. Chem. B*, 2009, **113**, 2428–2435.
7 C. Y. Lee, J. A. McCammon and P. J. Rossky, *J. Chem. Phys.*, 1984, **80**, 4448–4455.
8 M. C. Gordillo and J. Marti, *Phys. Rev. B: Condens. Matter Mater. Phys.*, 2008, **78**, 075432.
9 Y. R. Shen and V. Ostroverkhov, *Chem. Rev.*, 2006, **106**, 1140–1154.
10 Y. K. Cheng and P. J. Rossky, *Nature*, 1998, **392**, 696–699.
11 J. C. Rasaiah, S. Garde and G. Hummer, *Annu. Rev. Phys. Chem.*, 2008, **59**, 713–740.
12 N. Giovambattista, P. G. Debenedetti and P. J. Rossky, *J. Phys. Chem. C*, 2007, **111**, 1323–1332.
13 N. Giovambattista, P. J. Rossky and P. G. Debenedetti, *Phys. Rev. E: Stat., Nonlinear, Soft Matter Phys.*, 2006, **73**, 041604.
14 S. H. Lee and P. J. Rossky, *J. Chem. Phys.*, 1994, **100**, 3334–3345.
15 D. Laage and J. T. Hynes, *Science*, 2006, **311**, 832–835.
16 D. Laage and J. T. Hynes, *J. Phys. Chem. B*, 2008, **112**, 14230–14242.
17 D. Laage and J. T. Hynes, *Proc. Natl. Acad. Sci. U. S. A.*, 2007, **104**, 11167–11172.
18 F. Sterpone, G. Stirnemann, J. T. Hynes and D. Laage, *J. Phys. Chem. B*, 2010, **114**, 2083–2089.
19 G. Stirnemann, J. T. Hynes and D. Laage, *J. Phys. Chem. B*, 2010, **114**, 3052–3059.
20 J. Zheng, K. Kwak and M. D. Fayer, *Acc. Chem. Res.*, 2007, **40**, 75–83.
21 Y. S. Kim and R. M. Hochstrasser, *J. Phys. Chem. B*, 2009, **113**, 8231–8251.
22 J. J. Loparo, S. T. Roberts and A. Tokmakoff, *J. Chem. Phys.*, 2006, **125**, 194522.
23 H. J. C. Berendsen, J. R. Grigera and T. P. Straatsma, *J. Phys. Chem.*, 1987, **91**, 6269–6271.
24 J. C. Phillips, R. Braun, W. Wang, J. Gumbart, E. Tajkhorshid, E. Villa, C. Chipot, R. D. Skeel, K. Laxmikant and K. Schulten, *J. Comput. Chem.*, 2005, **26**, 1781–1802.
25 S. Mukamel, *Principles of Nonlinear Optical Spectroscopy*, Oxford University Press, New York, 1995.
26 J. B. Asbury, T. Steinel, C. Stromberg, S. A. Corcelli, C. P. Lawrence, J. L. Skinner and M. D. Fayer, *J. Phys. Chem. A*, 2004, **108**, 1107–1119.
27 J. R. Schmidt, S. A. Corcelli and J. L. Skinner, *J. Chem. Phys.*, 2005, **123**, 044513.
28 J. R. Schmidt, S. T. Roberts, J. J. Loparo, A. Tokmakoff, M. D. Fayer and J. L. Skinner, *Chem. Phys.*, 2007, **341**, 143–157.
29 The simulations used to calculate the spectra involved a smaller number of water molecules (500 instead of 1500) and infinite plates (obtained by adjusting the box size to the plate lattice parameters and periodic boundary conditions). The resulting distance between the

two plates is 2.1 nm, and the water layer is approximately 1.2 nm thick. Although this confinement very slightly slows down the water dynamics, we verified that the vibrational spectra and their relaxation dynamics are not affected.

30. S. H. Northrup and J. T. Hynes, *J. Chem. Phys.*, 1980, **73**, 2700–2714.
31. M. Maccarini, *Biointerphases*, 2007, **2**, MR1–MR15.
32. K. R. Gallagher and K. A. Sharp, *J. Am. Chem. Soc.*, 2003, **125**, 9853–9860.
33. G. L. Richmond, *Chem. Rev.*, 2002, **102**, 2693–2724.
34. L. F. Scatena, M. G. Brown and G. L. Richmond, *Science*, 2001, **292**, 908–912.
35. P. N. Perera, K. R. Fega, C. Lawrence, E. J. Sundstrom, T.-P. J. and D. Ben-Amotz, *Proc. Natl. Acad. Sci. U. S. A.*, 2009, **106**, 12230–12234.
36. R. Ludwig, F. Weinhold and T. C. Farrar, *J. Chem. Phys.*, 1995, **103**, 6941–6950.
37. Y. L. A. Rezus and H. J. Bakker, *J. Chem. Phys.*, 2005, **123**, 114502.
38. S. Park, D. E. Moilanen and M. D. Fayer, *J. Phys. Chem. B*, 2008, **112**, 5279–5290.
39. D. Laage and J. T. Hynes, *Chem. Phys. Lett.*, 2006, **433**, 80–85.
40. J. Qvist and B. Halle, *J. Am. Chem. Soc.*, 2008, **130**, 10345–10353.
41. C. Mattea, J. Qvist and B. Halle, *Biophys. J.*, 2008, **95**, 2951–2963.
42. D. Laage, *J. Phys. Chem. B*, 2009, **113**, 2684–2687.
43. J. Teixeira, M.-C. Bellisent-Funel, S. H. Chen and A. J. Dianoux, *Phys. Rev. A: At., Mol., Opt. Phys.*, 1985, **31**, 1913–1917.
44. A. Luzar, *Faraday Discuss.*, 1996, **103**, 29–40.
45. It was recently shown[42] that interpretation of experimental neutron scattering results[43] with our jump model leads to reorientation times in agreement with those obtained through other techniques, while their original interpretation with the Debye rotational diffusion picture lead to times twice shorter.
46. Our jump reorientation mechanism contrasts with previous suggestions, such as ref. 44, as detailed in ref. 16.
47. A. R. Bizzarri and S. Cannistraro, *J. Phys. Chem. B*, 2002, **106**, 6617–6633.
48. S. L. Lee, P. G. Debenedetti and J. R. Errington, *J. Chem. Phys.*, 2005, **122**, 204511.
49. S. R. V. Castrillon, N. Giovambattista, I. A. Aksay and P. G. Debenedetti, *J. Phys. Chem. B*, 2009, **113**, 1438–1446.
50. M. G. Mazza, N. Giovambattista, F. W. Starr and H. E. Stanley, *Phys. Rev. Lett.*, 2006, **96**, 057803.
51. N. Agmon, E. Pines and D. Huppert, *J. Chem. Phys.*, 1988, **88**, 5631–5638.
52. S. H. Northrup and J. A. McCammon, *J. Chem. Phys.*, 1980, **72**, 4569–4578.
53. A. Luzar, *J. Chem. Phys.*, 2000, **113**, 10663–10675.
54. Approximating the back reaction from $B'$ by a rate equation leads to a simple system of linear differential equations, in contrast to the explicit consideration of diffusion which requires a resolution through Laplace transforms.51,53.
55. This was already observed and discussed in section IV.2. by examining the time dependent probability for an OH to form an angle $\theta$ with respect to its initial orientation, as shown in Fig. 4a.
56. K. Seki, B. Bagchi and M. Tachiya, *Phys. Rev. E: Stat., Nonlinear, Soft Matter Phys.*, 2008, **77**, 031505.
57. A. Morita and J. T. Hynes, *J. Phys. Chem. B*, 2002, **106**, 673–685.
58. A. Morita and J. T. Hynes, *Chem. Phys.*, 2000, **258**, 371–390.
59. H. G. Hertz and M. D. Zeidler, *Ber. Bunsen-Ges. Phys. Chem.*, 1964, **68**, 821–837.
60. K. Matsuda, T. Hibi, H. Kadowaki, H. Kataura and Y. Maniwa, *Phys. Rev. B: Condens. Matter Mater. Phys.*, 2006, **74**, 073415.
61. A. W. Omta, M. F. Kropman, S. Woutersen and H. J. Bakker, *Science*, 2003, **301**, 347–349.
62. D. E. Moilanen, D. Wong, D. E. Rosenfeld, E. E. Fenn and M. D. Fayer, *Proc. Natl. Acad. Sci. U. S. A.*, 2009, **106**, 375–380.
63. A. Szabo, *J. Chem. Phys.*, 1984, **81**, 150–167.
64. H. J. Bakker, Y. L. A. Rezus and R. L. A. Timmer, *J. Phys. Chem. A*, 2008, **112**, 11523–11534.
65. J. D. Eaves, J. J. Loparo, C. J. Fecko, S. T. Roberts, A. Tokmakoff and P. L. Geissler, *Proc. Natl. Acad. Sci. U. S. A.*, 2005, **102**, 13019–13022.
66. J. J. Loparo, S. T. Roberts and A. Tokmakoff, *J. Chem. Phys.*, 2006, **125**, 194521.
67. M. L. Cowan, B. D. Bruner, N. Huse, J. R. Dwyer, B. Chugh, E. T. J. Nibbering, T. Elsaesser and R. J. D. Miller, *Nature*, 2005, **434**, 199–202.
68. S. Park, M. Odelius and K. J. Gaffney, *J. Phys. Chem. B*, 2009, **113**, 7825–7835.
69. A. Eftekhari-Bafrooei and E. Borguet, *J. Am. Chem. Soc.*, 2010, **132**, 3756–3761.
70. R. Rey, K. B. Moller and J. T. Hynes, *Chem. Rev.*, 2004, **104**, 1915–1928.
71. J. Bredenbeck, A. Ghosh, H.-K. Nienhuys and M. Bonn, *Acc. Chem. Res.*, 2009, **42**, 1332–1342.
72. J. Chowdhary and B. M. Ladanyi, *J. Phys. Chem. B*, 2009, **113**, 4045–4053.
73. J. Chowdhary and B. M. Ladanyi, *J. Phys. Chem. B*, 2008, **112**, 6259–6273.

# General Discussion

**Dr Stillinger** opened the discussion of the paper by Professor Widom: The two portions of Fig. 1 that graphically distinguish first-order and second-order wetting transitions on the basis of the way that $\Delta\sigma$ vanishes lead naturally to some questions seeking additional details about these phenomena.

(A) In the first-order case, Fig. 1(a), the possibility of a metastable extension formally associated with $\Delta\sigma < 0$ has been indicated by a dashed line. As $\Delta\sigma$ approaches zero from the positive side (thermodynamically stable states), the angle between the 1,2 and 2,3 interfaces continuously approaches zero. However that angle seems not to be defined for the metastable extension $\Delta\sigma < 0$. Is there a simple description of the resulting geometric arrangement present in the predicted three-phase metastable state that would allow its unambiguous experimental identification?

(B) As Fig. 1(b) points out, the signature of a second-order transition in the mean-field approximation is that $\Delta\sigma$ vanishes quadratically upon approaching the wetting transition. But as one has learned from critical point studies for bulk phases in two and three spatial dimensions, the neglect of large critical point fluctuations by the mean field approximation affects the predictions of critical exponents. In the present case, the analogous issue is whether one can anticipate *a priori* how restoration of the influence of critical fluctuations would shift the exponent-two quadratic prediction. In particular, can one say whether the exponent describing vanishing of $\Delta\sigma$ is actually greater than, or less than, two?

**Professor Widom** replied:

(A) The angle could in principle be observed and measured. It would not be 0, which would have been its equilibrium value. The metastability could have arisen, for example, by the film's spreading having been arrested by some asperity on the substrate (easier to imagine on a solid than on a liquid substrate). In the meantime the three interfacial tensions would continue to evolve normally, with $\Delta\sigma$ passing through 0 and becoming negative. The relations that, at equilibrium, determine the contact angles from the tensions, would then formally yield for the magnitude of the cosine of one of the angles a value greater than 1 — thus telling us that the system is not in true thermodynamic equilibrium when $\Delta\sigma < 0$.

(B) You are right that the fluctuations that are ignored in this mean-field density-functional model could affect the exponent that defines the order of the transition. Interestingly, even in mean-field models of this kind, one can find higher-order transitions in which the exponent can be anywhere between 1 and $\infty$, and even be non-universal; *i.e.*, a function of the parameters in the free-energy density or of the system's thermodynamic state. This is seen, for example, in ref. 4[1] of our paper, and likewise in our own work in ref. 2.[2]

1 E. H. Hauge, *Phys. Rev. B: Condens. Matter*, 1986, **33**, 3322.
2 K. Koga, J. O. Indekeu and B. Widom, *Phys. Rev. Lett.*, 2010, **104**, 036101.

**Professor McCarthy** opened the discussion of the paper by Dr Willmott: Does the relative humidity affect the drop mobility?

**Dr Willmott** answered: We attempted to control humidity by carrying out experiments within a small container filled with water (see our Experimental section). No dependence on humidity was noted in the experiments.

**Mr Boreyko** opened the discussion of the paper by Mr Wu: To what degree are the nanoparticles present on the side walls of the micropillars, and is wetting *via* the nanoparticles on the side walls how you account for the preferential wetting of the nano-scale even at the base of the surface for the superhydrophobic samples?

**Mr Wu** answered: Due to the nature of the coating process, in this case, spin-coating, the nanoparticles would preferentially deposit at the base of the micropillars rather than the sidewalls. While we believe there exists nanoparticles on the sidewalls of the micropillars, they are negligible (Fig. 6) in terms of their contribution to the X-ray scattering data.

At this stage, we cannot definitively account for the preferential wetting at the nano-scale for our superhydrophobic samples. If there were nanoparticles on the sidewalls, we wouldn't expect there to be a 'wicking' effect where the test fluid would travel preferentially to the base of the substrate. Thus, at this point, we do not believe it would affect the preferential wetting order between micro and nanostructures.

**Dr Henderson** asked Professor Widom: Is it possible to use the two parameters of your model together with temperature to generate a three-dimensional wetting phase diagram of the class of Fig. 2(a) of H. Nakanishi and M. E. Fisher?[1] If so, the transition between first-order and second-order wetting transitions occurs at a unique tri-critical point separating the two classes.

1 H. Nakanishi and M. E. Fisher, *Phys. Rev. Lett.*, 1982, **49**, 1565.

**Professor Widom** answered: Yes, mean-field density-functional models of this kind, but necessarily more general than the two I referred to here, can indeed have a state (a "tricritical" point) separating a manifold of first-order wetting transitions from another of higher-than-first order. The paper in our Reference 2 had not yet been published at the time this one was submitted but has since appeared.[1] There we treat a model like that in Eq. (8) of the present paper but with an additional parameter in it (we call it "$a$") that becomes the present ratio $a_1/a_2$. We then find in the $a,b$ plane a locus of wetting transitions on which the point $a = 1$, $b = 0.681$ separates a range of higher-than-first order (including infinite order) wetting transitions from one of first-order transitions.

Since $a_1/a_2$ at this point is less than 2, it is in accord with the necessary condition for higher-order transitions and the sufficient condition for first-order transitions presented in the present paper.

1 K. Koga and J. O. Indekeu and B. Widom, *Phys. Rev. Lett.*, 2010, **104**, 036101.

**Professor Evans** asked: It is remarkable that your simple square-gradient density functional treatment of fluid interfaces yields such rich wetting phase transition behaviour. But given that this richness does occur in what is arguably the simplest theory it is tempting to believe that these wetting transition scenarios might also occur in more sophisticated treatments.

I would be tempted to relate your parameters $a_1$ and $a_2$ to bulk correlation lengths in the relevant fluid phases. Is there a direct way of doing this? One knows that from general studies of wetting transitions based on effective interfacial potentials (treatments which write the excess grand potential as a function of the relevant film thickness) that competition between the different correlation lengths, which in turn determine the decay of density profiles into bulk phases, is crucial in determining the order of the wetting transition and whether, for continuous transitions, the critical exponents are universal. The paper by Aukrust and Hauge that you cite is an excellent example. Are you and your colleagues able to take your theory and construct the appropriate interface potential? I suspect that this would involve exponential terms decaying with different length scales. One might then be able to take over directly the results of Aukrust and Hauge and Hauge and Olaussen, for wetting a wall with an exponentially decay wall-fluid potential, to ascertain the wetting characteristics and how the fluctuations at continuous transitions renormalize the wetting behaviour.

**Professor Widom** responded: It is indeed the case that the $a_1$ and $a_2$ here are the exponential decay lengths for the decay of $\rho_1$ and $\rho_2$ to their values (0 and b, respectively) in the bulk phase 2. Thus, on the trajectory in Fig. 2 that passes through 0,b, on approaching that point, b-$\rho_2$ vanishes proportionally to $|\rho_1|^{(a_1/a_2)}$ as long as this exponent is less than 2. This immediately gives the criterion for first-order vs. higher-order wetting transition, since on the non-wetting trajectory, at higher-than-first-order wetting, the second derivative of $\rho_2$ with respect to $\rho_1$ diverges as $\rho_1, \rho_2 \to 0,b$, so that must be true also on the wetting trajectory. You are right that this can also be associated with non-universal wetting exponents. While the $F$ in the present Eq. (8) leads only to a second-order wetting transition, with an additional parameter in (8) it can yield a manifold of wetting transitions of continuously varying order. In the present ref. 2 we introduce a parameter "$a$" that is the present $a_1/a_2$. Then for $0 < a < 1$ the wetting transition occurs at $b = 0$ and $\Delta\sigma$ vanishes proportionally to $b^{[2/(1-a)]}$. No doubt one could construct a related interface-potential model that would yield analogous results.

**Professor Craig** asked Professor Willmott: Geoff, I like your experimental set-up it is simple and elegant. I am therefore a little surprised about the level of noise in your data. Could you please comment on the source of variation in your results.

If the noise is due to dynamic effects associated with the drop meeting the capillary perhaps you could eliminate them by creating a drop by applying pressure to the capillary. Or perhaps it is the dynamic effects that you wish to explore.

**Dr Willmott** answered: A number of experimental effects could contribute to scatter in the data, as discussed in Section 4 of our paper. The most significant of these is likely to be roughness and associated contact angle hysteresis within the tube. The idea of controlling the pressure is interesting. There remains considerable interest in studying the dynamics as the drop comes into contact with the capillary, because this process is comparable to the classical capillary uptake case and reflects what might happen in a real 'microfluidics' environment.

**Dr Vollmer** asked Mr Wu: The patterned silicon substrate showed a fixed pillar height of 2.5 micrometres. The authors determined a maximum penetration depth between 2 and 2.3 micrometres. What causes the difference between the penetration depth and the pillar height?

**Mr Wu** replied: Freeze fracture microscopy involves the formation of vitreous ice to 'imprint' the water/surface interface. Even with crystallinity mismatch between the substrate and the ice, there exist flaws and imperfections that can contribute to the premature cracking of the vitreous ice along the bulk rather than at the interface.

In other words, while the data presented in the paper reflects the limitation of precision in this technique, it is still sufficient for us to 'track' the progression of wetting at the microscale

**Ms Mishchenko** asked Dr Willmott: I notice signs of contact angle hysteresis on some of your images. Are you using different superhydrophobic surfaces in some of your experiments? Is this affecting your results and conclusions?

**Dr Willmott** replied: The same superhydrophobic surface is used in all experiments. As noted in the Experimental section (and associated references), some hysteresis was recorded in contact angle measurements. The surface fabrication method involves polishing the copper surface to achieve the desirable level of roughness — the surface texture is therefore not precisely controlled and the presence of imperfections is possible. Additionally, drops are preferentially located at imperfections, leading to the hysteresis observed in our images. This effect should not affect our major results, because the droplet pressure is determined by drop radius (which we measure), and is independent of contact angle on the substrate. Variable adherence to the surface is noted with respect to the five identified modes of interaction — see, for example, the first paragraph of Section 3.2.

**Professor McCarthy** said: This polymer is likely not PTFE (polytetrafluoroethylene) because it is transparent it is likely a copolymer.

**Dr Willmott** replied: Note our reply to Ms Hecht.

**Ms Hecht** said: Can you give more exact information about the capillary you used? What influences the attraction of the drop to the capillary, *i.e.* is the drop attracted to the capillary material or only to the reduction in energy experienced when entering the capillary? I wonder how the answer to this question could be used to predict the behavior of a similar capillary with thinner walls or of capillaries of other materials.

**Dr Willmott** responded: The tubing used was obtained from Cole-Parmer, product reference EW-06417-70. During the discussion, it was noted that although the product is called 'PTFE tubing', this is not the same as pure PTFE. We now think that the attraction between the drop and the capillary (before contact) is probably driven by surface charge. Minimization of surface energy becomes important following contact.

Qualitatively, we can expect drops to be more stable prior to contact when the charge is reduced. The influence of thick walls is not usually accounted for in theoretical work, although non-zero thickness is unavoidable in practice — so this is an area of some interest. With less charge and/or thinner walls, the instability which inhibits filling of a dry capillary (see Fig. 8) should be less significant.

**Professor Debenedetti** asked: Have you done dimensional analysis of equation 2 to identify the relevant dimensionless groups? This could help you organize your data in the form of a "phase diagram" showing the ranges of dimensionless groups where each of the 5 regimes that you have identified occurs.

**Dr Willmott** responded: Such analysis would represent a good extension to this work. With regard to separating the regimes, we note that their occurrence overlaps in terms of the variables we have measured; this work identifies further important factors (for example, contact angle hysteresis inside the tube) that are not included in Eq. 2.

**Professor Evans** asked Professor Widom: If one considers real fluid mixtures then at first sight one might say that a square-gradient treatment is not appropriate as this does not incorporate the London dispersion (van der Waals) forces that are omnipresent. However, we know that in certain circumstances it might be possible to weaken the dispersion forces, *e.g.* in colloidal fluids, so that the system under consideration is effectively governed by short-ranged forces. Do you think that real systems might be found that comply to your model so that the exciting prediction of non-universal critical exponents would be pertinent?

**Professor Widom** responded: First an incidental remark before addressing your main question. There is a sense in which the square-gradient theories do, or may, incorporate the dispersion forces. The coefficients of the square-gradients, which we here took to be "1/2", are in reality proportional to the second moments of the intermolecular potentials — finite numbers for $1/r^6$ potentials. But you imply the more important point that the interfacial composition profiles decay to their bulk-phase values exponentially in the square-gradient theories and so do thereby miss the longer ranged tails from the dispersion forces. I am heartened by your remark that there may be colloidal systems in which this may not be relevant. Also, for wetting transitions near a critical endpoint they are believed not to be relevant (see, *e.g.*, Section IX B of Dietrich's review "Wetting Phenomena" in the Domb–Lebowitz series), but there, by the Cahn argument, one may expect the wetting transition to be first order. So maybe colloidal systems, as you suggest, is where to look for non-universal higher order wetting exponents. Two-phase equilibria in such systems are seen, *e.g.*, in the beautiful experiments in the Lekkerkerker group in Utrecht. Asking for three-phase equilibria in such systems, to be analogous to the systems considered here, may be a major challenge but a most interesting prospect.

**Dr Christenson** asked Dr Willmott: I propose that the jumping of droplets that you have observed, *e.g.* Figure 8, is due to electrostatic charging of the capillary tubes. Dry fluorocarbons would easily build up a charge that would induce an opposite charge in the droplets and cause them to jump onto the side of the tubes. The problem may go away if a radioactive source, a β-emitter, is placed in the vicinity of the tube.

**Dr Willmott** replied: Charge seems to be the most plausible explanation for the 'jump to contact' behaviour, and should be controlled or investigated in future experiments. We note that the velocity of the drops when they reach the tube, while likely to affect the initial dynamics of the fluid, does not materially affect the conclusions of the present experiments — that is, the relationship between drop size and the position of the arrested meniscus.

**Professor Debenedetti** asked Professor Widom: What is known about the mechanisms by which wetting transitions such as you described can occur? In view of the fact that these transitions can be very weakly first-order, are there reasons to expect deviations from conventional nucleation?

**Professor Widom** replied: The analogous mechanisms are at work as in bulk-phase transitions. For example, the metastable wetting film that must ultimately transform to an equilibrium thin film ("de-wetting") can do so by the analog of homogeneous nucleation or spinodal decomposition. These mechanisms are discussed by de Gennes, Brochard, and Quéré, with references, in their book "Capillarity and Wetting Phenomena".

I would not expect qualitative differences when the wetting transition is only weakly first order, although there may well be significant quantitative differences since the free energy of metastability would then be small.

**Dr Hummer** commented: The coefficients $a_1$ and $a_2$ in Eq. 6 appear to be related to both bulk compressibility and interfacial properties. Is there a possible connection to the Egelstaff–Widom length?[1]

1 P. A. Egelstaff and B. J. Widom, *Chem. Phys.*, 1970, **53**, 2667–2669.

**Professor Widom** replied: There might be some connection. The parameters $a_1$ and $a_2$ are two slightly different measures of the width of the 1–2 interface (differing only according to which component's spatial profile is used to measure it).

The length to which you referred is of the order of magnitude of that width, and relates the interfacial tension to it and to the mechanical compressibility of the liquid phase. The relation was originally proposed for the liquid–vapor interface in a one-component system. It could be of interest to see if an analogous relation exists for the interface between liquid mixtures and their osmotic compressibilities.

**Ms Mishchenko** asked Mr Wu: How accurate are your contact angle measurements, especially ones showing 0 to 20° contact angle? Also, does one get any unwanted bubbles or other artificial features in the cryofixation portion of your sample prep? Do you feel those affect your conclusions?

**Mr Wu** replied: The accuracy of our contact angle measurements was governed by the intrinsic error of contact angle measurements, which in this case, is $\pm$ 5°. Of course, this error margin drastically increases with decreasing contact angle, tending towards 0°. Nevertheless, the point we wish to highlight here is the fact that despite the observed transition between classical Cassie state towards Wenzel state (from CA > 150 to CA < 30), wetting at the nanoscale was observed to continue beyond CA reaching 0°.

Careful consideration and preparations were made to minimize any unwanted artifacts during the rapid cryofixation process. Unwanted bubbles may appear in two distinct areas: one within the bulk of the drop, which would cause the fracturing process to occur prematurely along the bulk of the fluid rather than at the interface, and one at the water/surface interface, which would affect the penetration depth measurements. In each case, the experiment was repeated several times to ensure such artifacts were identified, and more importantly, their sources so that we were able to systematically avoid them.

We were diligent in calibrating the system such that the sources and symptoms of errors were identified and removed. By subjecting a series of well-defined surfaces to the same process using fluids that would induce nanobubbles *in situ*, we were able to identify and observe the effect of unwanted bubbles on the fracturing and imprinting process.

While it is true that these errors, should they exist, would affect our conclusions, we are confident that they have been minimized to negligible levels.

**Professor Luzar** asked Professor Widom: Your answer to Dr Hummer's question was positive, namely that your model can be extended to higher component mixtures. When you talk about density fluctuations in the mixture, are you referring to concentration fluctuations, that experimentalists measure in liquid mixtures for example? What I mean by density is the total number density of all species present in the system. What I mean by concentration is the number density of a specified component, *i.e.* concentration pertains to a density of specified component. In a mixture, fluctuations in densities of either species can be much greater than total density fluctuations. Such spatial heterogeneities, *i.e.* concentration fluctuations on length scales slightly above the intermolecular distances can be tested experimentally with small angle scattering. The distinction between density and concentration fluctuations seems to be important and I can give one example: a proof that water is not a mixture of amorphous ice (LDA) clusters inside an ideal liquid water comes from small angle neutron scattering (SANS) experiments. These experiments show that density fluctuations alone are sufficient to explain the scattered intensity (*i. e.* the extrapolated value of the zero-angle scattered intensity, $S(0)$, is proportional to the isothermal compressibility).

**Professor Widom** responded: Thanks for your interesting question and remarks. The "density" fluctuations I referred to are fluctuations in the chemical composition, controlled by the osmotic compressibility, which, as you said, can be large even in dense liquid solutions. The fluctuations in total density are slight (although measurable, as you noted) because of the near incompressibility of the dense liquid solvent.

**Dr Christenson** asked: Is there anything about the wetting transitions illustrated in Figure 1 that limits the discussion to liquid–liquid interfaces? One might expect that the proposed metastable non-wet state, where the expected wetting transition does not actually occur, has experimental counterparts in quite a few systems. Lifshitz theory often predicts wetting by long-range van der Waals forces, *e.g.* by most low-density liquids on mineral surfaces, but this is not always observed. Examples might be water and some organic liquids on mica.[1–4]

To study this experimentally measurements should be carried out both by thickening films on adsorption from vapour, and by thinning films, *e.g.* by pressing a bubble up against a surface. T. D. Blake carried out such a study of the wetting of alumina by n-octane and n-decane,[5] but found good agreement between the two methods.

1 D. Beaglehole, E. Z. Radlinska, B. W. Ninham and H. K. Christenson, *Phys. Rev. Lett.*, 1991, **66**, 2084–2087.
2 H. K. Christenson, *J. Phys. Chem.*, 1993, **97**, 12034–12041.
3 H. K. Christenson, *Colloids Surf. A*: *Physicochem. Eng. Asp.*, 1997, **123–124**, 355–367.
4 T. E. Balmer, H. K. Christenson, N. D. Spencer and M. Heuberger, *Langmuir*, 2008, **24**, 1566–1569.
5 T. D. Blake, *J. Chem. Soc.*, *Faraday Trans.*, *1*, 1975, **71**, 192–208.

**Professor Widom** responded: Thanks for your remarks and references. The kind of mean-field density-functional model I outlined here is not at all restricted to liquid–liquid interfaces and has in fact been more often applied to the interface between a liquid and a rigid solid substrate. In that case there is only one contact angle in question, while with three fluids (say, a vapor and two liquids, as in this paper), there are two independent ones, so in that respect it is slightly more general.

**Dr Vollmer** asked Mr Wu: What might cause 40% of the area close to the bottom of the pitches to remain dry, although the droplets are in the Wenzel state? Might it be caused by the method the nanoparticles are fixed or by an inhomogeneous distribution of SDS in the nanopores? SDS has a high electron density contrast and therefore a high scattering contrast.

**Mr Wu** responded: Since it is well known that SDS possesses the tendency to physisorb onto silica surfaces, we agree that the SDS could have potentially been 'locked' into the nanopores, resulting in an inhomogeneous distribution of SDS. X-ray scattering caused by SDS molecules typically occurs when the SDS molecules are arranged in a micelle. In other words, as long as the concentration of SDS in solution is below the CMC, the distribution of SDS is dilute enough to cause negligible X-ray scattering.

**Professor Quéré** asked Dr Willmott: In a drop/tube geometry, the way the liquid first penetrates the tube must be quite affected by the geometry, compared to the classical bath/tube case. Do you have evidences for that, in short time regimes of rise?

**Dr Willmott** responded: The present experiments lacked the resolution (temporal and spatial) to study this regime, although it is of high interest for our future studies. In this initial regime, the present experiments differ from the classical bath/tube geometry in several interesting ways: the filling height is not zero, so viscous effects may be significant; meniscus reorientation is affected by the presence of a second meniscus, due to fluid within the tube; the surface of the drop is significantly curved; and the tube surface is non-wetting. A relevant topic of particular interest is the dynamic contact angle for a non-wetting surface, which has not been widely studied.

**Mr Peter** asked Mr Wu: A question on the freeze fracture atomic force microscopy measurements:

Was there any damage observed during the freeze procedure? Did you perform AFM topography measurements before and after the freeze procedure?

How big was the water drop used for this technique?

**Mr Wu** responded: There was indeed damage during the freeze procedure, and thus we were diligent in calibrating the technique by subjecting damage-prone samples to the same procedure to help identify the appearance of imperfections. Through this, we were able to reproduce the same results with photolithographic samples, and subsequently, applied this technique to our superhydrophobic samples.

AFM topography of the water/surface interface must be done *in situ* and is often done on smooth surfaces. The nature of our samples pertains roughness that is too great for such a sensitive technique. This was the reason why the freeze-fracture process was required for our surfaces.

The water droplet used in this technique was approximately 1.5 mm in diameter.

**Mr Kang** commented: In Fig. 7 (b) of your pre-print, the percentile change of invariant increases above 60% CMC for the case of 1 μm pitch + nano and 0 to 60% CMC for 5 μm pitch + nano while, for all other cases, it generally decreases as the %CMC decreases.

Could you comment on these two cases?

**Mr Wu** responded: The slight increases/decreases in the measured invariant $Q$ is within experimental error, which is the reason we interpreted the observed transition between 0 to 20% CMC as constant or negligible change in nano-wetting behaviour.

**Professor Law** remarked: Equation (3) must only provide a hand waving explanation of Figures 2 and 3. If one inserts approximate values into this equation then one gets penetration depths of meters rather than microns.

**Mr Wu** replied: It is indeed true that equation (3) is only an approximation to explain our findings in Figures 2 and 3. The reason being, equation 3 assumes a fully enclosed capillary of radius $r$, whereas our surface is essentially an 'open grid' structure, where there are only two sidewalls, defined by the pitch.

**Professor Luzar** opened the discussion of the paper by Professor Berne: This is a good time and place to define what is meant by "dewetting". If you mean that all water goes out from the confinement, as you suggest, you should call it capillary evaporation[1,2] to be consistent with the accepted nomenclature. See *e. g.* Evans' work of 20 years ago.[3] Capillary evaporation (through cavitation) is a macroscopic phenomenon, but there is no reason to use a different language for the same phenomenon at the nanoscale. One only needs to keep in mind that for evaporation in any system to take place, the usual (textbook) condition of hydrophobicity for surfaces to attract water less than water attracts itself is not sufficient.[4] In naturally occurring systems we cannot expect to observe spontaneous evaporation except from molecular-sized confinements in which the barrier to evaporation becomes thermally accessible.[4]

Let me summarize the important distinction between one and two wall confinement scenarios: liquid next to a single wall wets or de-wets. Liquid confined between two walls evaporates (or condenses) according to the well-known thermodynamic argument.[5] In prototypical biophysical confined systems one can make further distinctions between two possible scenarios: (1) Cavitation effect in which water density is decreased and a vapor bubble is formed. Laplace pressure pulls solutes together and squeezes the remaining water out of the hydrophobic gap. This effect can drive hydrophobic assembly. (2) Water expulsion scenario in which water is gradually expelled form the collapsed interior when solutes approach each other. This effect accompanies hydrophobic collapse and protein folding.

So in systems you study, one should call your "dewetting" capillary evaporation (pathway through cavitation) if water goes out from the confinement and undergoes liquid-to-vapor transition. Alternatively, if water evacuates from the confinement but remains in the liquid state, one should call this process spontaneous expulsion of water.

1 A. Luzar and K. Leung, *J. Chem. Phys.*, 2000, **113**, 5836; K. Leung and A. Luzar, *J. Chem. Phys.*, 2000, **113**, 5845.
2 A. Luzar, D. Bratko and L. Blum, *J. Chem. Phys.*, 1987, **86**, 2955.
3 R. Evans, in Liquids at Interfaces, Proceedings of the Les Houches Summer School in Theoretical Physics, Les Houches, 1988, J. Charvolin, J. F. Joanny and J. Zinn-Justin ed., North Holland, Amsterdam, 1990.
4 A. Luzar, *J. Phys. Chem. B*, 2004, **108**, 19859.
5 K. Lum and A. Luzar, *Phys. Rev. E.*, 1997, **56**, R6283.

**Dr Patel** commented: Fig. 6 shows that for large $\varepsilon$, when enclosure $J$ (methane between two plates) is in a wet state, the binding of methane displays anti-co-operative behavior. In the wet state, the binding affinity is dominated by the free energy of cavity formation, which is typically entropic in nature for a methane-sized cavity. On the other hand, for small $\varepsilon$, when dewetting is observed in the enclosure, there is co-operativity. In this case, the free energy of cavity formation is small and the binding affinity is dominated by attractions between methane and the enclosure, which is enthalpic. At first glance, it thus seems that there may be a correlation between co-operative behavior and the binding affinity being dominated by the enthalpic contribution to the free energy. Could you please comment on this?

**Professor Berne** responded: Generally there is no correlation between the sign of the non-additivity, $\Delta\Delta G$, (whether it is cooperative or anti-cooperative) and the enthalpic and entropic contributions to the binding affinity. For example in Fig. 6, in the capillary wetting region, cooperative effects were observed in the region $\varepsilon$ = [0.15, 0.23], and anti-co-operative effects were observed in the region $\varepsilon$ = [0.23,0.40]. There is a crossover from cooperative to anti-co-operative effects with increasing $\varepsilon$ all in the wetting region. In the capillary drying region, a similar crossover behavior was also observed but in the opposite direction. In general, the enthalpic and entropic contributions to the solvation free energy are length-scale dependent: for small length scale solutes, like methane, its entropy dominates over enthalpy; and for large length scale solutes, enthalpy dominates over entropy.

**Dr Jamadagni** commented: The results you have presented on co-operativity and anti-co-operativity in hydrophobic binding, and how they are related to the curvature of the surface are very interesting. Although the rationale for why SASA can never predict co-operative behavior is presented in the pre-print, I felt it was not very clear. Could you please elaborate on it, and perhaps provide some illustrations to better make the point.

**Professor Berne** responded: See the following figure (Fig. 1) for an explanation of the case where two particles form a contact dimer, and a third particle binds to it. When the third particle is far away, and the probe ball can pass through the center region of the three, it is clear that both the SASA and MSA models will predict that the binding affinity will be additive. When the third particle gets closer to the dimer, a probe ball can not pass through and if there is no three-particle overlap in the SASA model, as in the top left figure, the SASA model still predicts that the binding affinity will be additive. If there are three-particle overlaps, the actual buried surface area (dashed line in the top middle of the figure) is smaller than the pairwise additive buried surface area, where, in the top right of the figure, the dashed triangle is

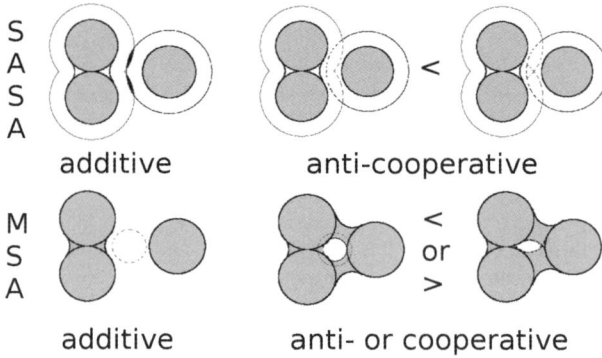

Fig. 1

counted twice. Thus the SASA model predicts that the binding will be non-additive or anti-cooperative. These are the only two possibilities, and under no condition will the SASA model predict cooperative effects. For the MSA model, when the probe ball can not pass through the center of the three particle region, the actual exposed surface area (middle bottom of the figure where the white circle represents the part of the probe ball exposed to water) can be larger or smaller than the pairwise-additive exposed surface area (bottom right of the figure) depending on the specific configuration of the three particles. So the MSA model can predict all three kinds of effects; anti-co-operative, co-operative, and additive. The conclusions for the three-particle case can be easily generalized to the $n$-particle case, and these conclusions are also found by explicit quantitative calculations of the buried surface area given in our paper.

**Dr Hummer** remarked: You find a remarkably good correlation between calculated free energies of binding and molecular surface area (MSA) estimates (in Figs. 3 and 4). As I understand, this correlation is seen for the case of the "standard" strong water–hydrophobe interactions. What happens for weakened interactions, in particular for cases where water recedes from the space between the plates, as in configuration $J$ ($\varepsilon < 0.15$ kcal mol$^{-1}$ in Fig. 6)? Does the MSA approximation break down?

**Professor Berne** replied: In general the MSA approximation does break down in the capillary evaporation regime. In the case where the atoms of the plates and the methane particles have the same LJ parameters, the MSA model with constant surface tension coefficient fails to predict the non-additivity effects when there is capillary evaporation. For example, in Fig. 4 (correlations between the MSA model predictions and FEP results for the non-additivity effects), systems L' and M' are outliers because of dewetting (capillary evaporation). As mentioned in our paper, a dry enclosure presents a different interface than the normal apolar solute/water interface so that the MSA model will fail there. Also, when the plate atoms have different LJ parameters than methane, for example, for system J with plate-atom energy parameter $\varepsilon < 0.15$ kcal/mol, corresponding to the capillary drying regime in Fig. 6, the MSA model with a constant surface tension coefficient isn't able to predict the non-additivity effect, because the surface tension coefficient itself should depend on the parameter $\varepsilon$. Different solutes have different local water densities and different effective solute/water surface tensions. The reason we use a constant surface tension coefficient in this paper is that all the solute atoms have the same interaction parameter except for the systems in Fig. 6.

If we set the binding affinity predicted by the MSA and SASA model to be equal to the FEP results, (*i.e.*, $\gamma(\varepsilon)\Delta S + \Delta E_{LJ} = \Delta G_{FEP}$) then we can get the effective

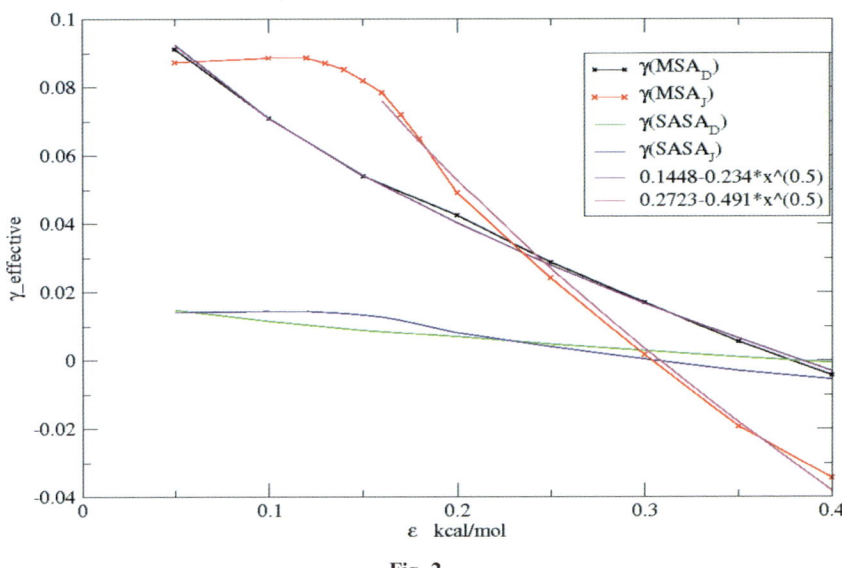

**Fig. 2**

microscopic surface tension coefficient as a function of $\varepsilon$ for systems D and J. These are displayed in Fig. 2.

For system D, the effective surface tension coefficient decreases with increasing $\varepsilon$, which makes sense because the larger the $\varepsilon$, the stronger will be the interaction between the plate and water, the less hydrophobic will be the plate, and the smaller will be the effective surface tension. Quantitatively, the decrease of the effective surface tension coefficient is approximately proportional to $\sqrt{\varepsilon}$. We expect this because macroscopically the surface tension is determined by the work that must be done to separate the plate from water and this scales as $\sqrt{\varepsilon}$. For system J, in the capillary drying region, the surface tension is approximately constant because the free energy to create the cavity is almost zero, so that the binding affinity difference comes from the direct LJ interactions. In the capillary wetting region, the effective surface tension coefficient has the same kind of behavior as that for system D, and the coefficient of the square root dependence term is approximately twice that for system D. The effective surface tension coefficients for the SASA model have the same kind of behavior as the MSA model for these two systems.

**Professor Bratko** said: The work you describe illustrates the nonadditivity of (multi-particle) potentials of mean force (PMF) manifested in surface interactions. In typical simulations, empirical molecular force fields routinely imply integration-out of intramolecular degrees of freedom, hence molecular potentials themselves also have the character of context-dependent PMFs and are only approximately additive. In one example, DFT calculations for neat water reveal up to 20% reduction of typical dipole moments of interfacial water molecules compared with those in the bulk phase.[1] Can context-dependence of intermolecular potentials be important in comparison with nonadditivity effects of molecular cluster PMFs you discuss?

1 C. J. Mundy and I. F. W. Kuo, *Chem. Rev.* 2006, **106**, 1282.

**Professor Berne** answered: It is indeed true that the parameters used in pairwise additive liquid state force fields are fitted to known properties of the liquid and therefore already include some of the many-body effects that are integrated out such as electronic degrees of freedom (polarizabilities), and intramolecular degrees of freedom *etc.* Nevertheless, as we show, even for these pairwise additive force fields, there are significant non-additive contributions to the free energies of binding as well as to the third and higher order PMFs, which arise from integrating out the solvent degrees of freedom. It is of interest to compare the non-additivity in two models of water, one with fixed charges and the other polarizable. The latter could be considered as an example of the context-dependent potential you mention. Two such models having similar structures for neat water (almost the same OO, O-H, and H-H radial distribution functions) are the fixed charge TIP4P-OPLS and the polarizable TIP4P-FQ models. Steven W. Rick[1] has studied the potential of mean force between two non-polar solutes in TIP4P-OPLS and in TIP4P-FQ water, and found that there is no significant difference in the PMFs for these two model potentials (see Table 1 of his paper). Thus we do not think the context-dependent potentials will significantly affect the non-additivity or our conclusions, at least for simple systems like those studied in our paper. Nevertheless it would be interesting to further study the effect of many-body potentials on binding and non-additivity. For example it is well known that even in simple atomic fluids, the Axelrod–Teller three-body dispersion interaction can contribute as much as 50% to the gas–liquid surface tension at the triple point. Unfortunately it is generally difficult to devise and simulate many-body potentials. *Ab initio* molecular dynamics can deal effectively with polarizability, but has problems with two and many-body dispersion interactions.

1 Steven W. Rick, *J. Phys. Chem. B*, 2003, **107**, 9853–9857.

**Ms Mooney** opened the discussion of the paper by Mr Stirnemann: Figure 6 shows at long times a divergence between the analytic kinetic model and the results from MD simulations. This divergence suggests that the simple kinetic model of eqn 5 is not capturing the long time distribution of **B'** and **T** states observed in the MD simulations. Within the proposed model, it is notable that the authors assume the independence of the reorientational dynamics of the O–H with regard to the other O–H on the same molecule.

What appears to be missing is that, when an O–H converts from a **D** to a **T** state, simultaneous reorganization of the other O–H may makes the "back reaction" more favorable than conversion to a **B'** state. This leads to a different rate than if the **T** state was reached by means of conversion from a **B'** state, implying that the reorientation process may not be Markovian.

How might such a modification modify the kinetic model result?

Because the authors conclude their work by expressing an interest in extended, network reorientational dynamics, can they comment on how they might incorporate reorientational dependencies into their model(s), as well as into their experiments where dilute HOD in $H_2O$ gives information on the marginal reorientational probability for **D** → **T** states while the information regarding the orientation of the other O–H on the molecule is inaccessible.

**Mr Stirnemann** responded: Our simplified jump reorientation mechanism indeed assumes that the two water OHs reorient independently. This simplified description was shown to provide a satisfactory description in all the systems we have studied so far (bulk water,[1,2] water around ions,[3] small hydrophobes,[4] amino-acids[5]). In the bulk situation, the jump probability has been verified to be Markovian.[2]

Regarding the divergence at very long delays (> 15 ps) between the populations from the kinetic model and from the simulations, it is due to the description of the **B'** → **T** exchange through a rate equation. At very long delays, a water molecule within the **B'** bulk state can diffuse away from the interface with the surface first hydration layer. A proper description of the **B'** → **T** exchange should therefore include the effect of diffusion. While such treatment has already been presented (see *e.g.* ref. 6), it has also been shown[6] that an effective description through a rate constant provides a good approximation for delays shorter than the inverse of this rate constant.

Concerning experimental evidence of water dynamics at such interfaces, it is certain that the dynamics of tangent OHs is not accessible *via* ultrafast infrared spectroscopy (fsIR) since their OH stretch frequency is indistinguishable from that of bulk water molecules. To isolate the contribution from water molecules in the hydration shell of small solutes, concentration studies are usually performed in NMR or fs IR experiments. Such an approach cannot be easily generalized to the case of surfaces. As mentioned in the concluding remarks, ongoing developments in two dimensional sum frequency generation may be a promising technique to specifically access the dynamics of water in the first hydration layers.

1 D. Laage and J. T. Hynes, *Science*, 2006,**311**, 832–835.
2 D. Laage and J. T. Hynes, *J. Phys. Chem. B*, 2008, **112**, 14230–14242.
3 D. Laage and J. T. Hynes, *Proc. Natl. Acad. Sci. U. S. A.*, 2007, **104**, 11167–11172.
4 D. Laage, G. Stirnemann and J. T. Hynes, *J. Phys. Chem. B*, 2009,**113**, 2428–2435.
5 F. Sterpone, G. Stirnemann, J. T. Hynes and D. Laage, *J. Phys. Chem. B*, 2010, **114**, 2083–2089.
6 N. Agmon, E. Pines and D. Huppert, *J. Chem. Phys.*, 1988, **88**, 5631–5638.

**Professor Luzar** commented: I have several comments on your otherwise nice work. The paper presents essentially a static view of water at the vicinity of a hydrophobic surface. The molecules are classified into three different classes, which can be acceptable from the methodological point of view, but is certainly a strong assumption as it appears very much like the study of the rotational motions in an orientational crystal. Diffusion process almost does not appear in your picture and molecules are only rotating going eventually from one class to another. This view is not very realistic. Thus the dynamics presented in your work concerns solely molecular reorientations. Indeed, the numbers you report (between 2 and 10 ps) are much longer than a hydrogen bond lifetime, and have the right order of magnitude for molecular reorientations. However, one should not relate such orientational dynamics to each hydrogen bond. Moreover, and this is important to note, the three classes are defined from structure, *i.e.* the orientation of the whole molecule. In each class there are automatically a variety of hydrogen bond environments.

The absence of mentioning neutron scattering data along with NMR and IR (2D or more) is rather surprising. The latter two techniques, powerful as they are, are more model dependent than QENS for example. Any relation of "molecular reorientations" with hydrogen bond dynamics can be deduced from IR only from a model, which is not the case for QENS that measures the positions of individual hydrogen atoms.

**Mr Stirnemann** replied: The focus of our paper is actually the *dynamics* of water next to an extended hydrophobic surface. This dynamics is very sensitive to the structure of the local environment surrounding a water molecule. We have therefore performed a rapid structural analysis to classify the different types of local environments which can be found at the interface with a hydrophobic surface. The key point of our study is that the orientation of each water OH (pointing respectively toward the surface, parallel to it, or away from it) determines how easily a new water will approach to exchange hydrogen-bond partners and allow a stable reorientation. Through our transition state excluded the volume picture, we can relate the water *structure* to the water orientation and hydrogen-bond *dynamics*.

The classification of the water orientations in three states is not arbitrary. It relies on the well-established clathrate/anticlathrate arrangement of water next to a hydrophobic surface,[1] and we verified that these states are stable configurations with well-defined exchange rate constants between them (see Sections III and IV). We also verified that within each state, the distribution of local environments is sufficiently narrow to lead to a unique exchange time and not to a broad distribution of exchange rates. This was confirmed through the mono-exponential character of the cross time-correlation function (see section II) in contrast with the stretched-exponential behavior expected for broad distributions.

Several definitions have been suggested for the hydrogen-bond lifetime,[2,3] and each of them reflects a different aspect of the hydrogen-bond dynamics. The continuous lifetime, *i.e.* the time before a hydrogen-bond breaks for the first time, is determined by transient breakings, which are followed in most cases by a fast return to the same initial acceptor; this lifetime thus mostly reflects the hydrogen-bond strength. In contrast, the intermittent lifetime includes the contribution from bond reformations and is sensitive to the diffusion of the pair. We showed in the bulk situation that the exchange time between stable hydrogen-bond acceptors that we consider has a value between the continuous and intermittent extremes and is the relevant quantity for the orientational dynamics.[4] It is thus legitimate to connect this stable hydrogen-bond exchange dynamics with the orientational dynamics.

The focus of our present study was on hydrogen-bond exchanges and their effect on water rotational dynamics. However, water translational diffusion obviously also requires a reorganization of the hydrogen-bond network through hydrogen-bond exchanges, and our study has implications for the translational motion. Our prior analysis of the jump mechanism in bulk water evidenced that rotational and translational motions are coupled through the hydrogen-bond exchange mechanism.[4] As discussed in section IV.2.b, the consideration of the hydrogen-bond exchange times in the tangent state provides a rationalization of the different values of the translational diffusion constant measured for water along the surface or perpendicular to it. However, it is certainly the case that all these translational diffusion issues deserve further investigation.

The confrontation of the results from the simulations and from the model with experiments is a critical point. We agree that the three main techniques presently available to probe water dynamics are femtosecond infrared spectroscopy (fs IR), NMR spectroscopy and quasielastic neutron scattering (QENS). Of these, the technique which provides the greater wealth of information is certainly the fs IR spectroscopy since it is the only one to be time-resolved. This technique yields the time-resolved orientational time-correlation function for the water OH bond, averaged over the sample.[5] In contrast, NMR provides the integrated value of this correlation function, *i.e.* the average reorientation time. Concerning QENS, it is indeed sensitive to the locations of hydrogen atoms. However, the interpretation of the (frequency-domain) QENS spectra in terms of molecular translations and rotations requires two critical assumptions: (a) the translational and rotational motions are assumed to be decoupled, which is not fully correct since *e.g.* hydrogen-bond exchanges couple these motions,[4] and (b) a model for the rotational motion needs to be assumed. All prior interpretations of water QENS spectra employed the Debye rotational diffusion picture, which led to reorientation times faster by a factor of approximately two compared with the times obtained through other techniques such as fs IR spectroscopy, NMR or simulations (see ref. 6 and references within). Replacing the diffusion picture by the angular jump model leads to reorientation times in agreement with those of other techniques.[6] We therefore believe that fs IR and NMR spectroscopy results offer a more stringent test for our model.

1 C. Y. Lee, J. A. McCammon and P. J. Rossky, *J. Chem. Phys.*, 1984, **80**, 4448–4455.
2 D. C. Rapaport, *Mol. Phys.*, 1983, **50**, 1151–1162.
3 A. Luzar, *J. Chem. Phys.*, 2000, **113**, 10663–10675.
4 D. Laage and J. T. Hynes, *J. Phys. Chem. B*, 2008, **112**, 14230–14242.
5 H. J. Bakker and J. L. Skinner, *Chem. Rev.*, 2010, **110**, 1498–1517.
6 D. Laage, *J. Phys. Chem. B*, 2009, **113**, 2684–2687.

**Mr Ritchie** asked: Please rewrite or explain your orientation of water electrostatic hydrogen linking in terms of a reasonable mechanism.

**Mr Stirnemann** responded: The "water electrostatic hydrogen linking" is usually called a "hydrogen bond". Most of the peculiar properties of water originate from these special bonds. The covalent *vs.* electrostatic nature of the hydrogen bond, or its description through the Mulliken charge-transfer picture has been a subject of great debate. For further information, you may refer to the following references.

1 G. C. Pimentel and A. L. McCellan, *The Hydrogen Bond*, WH Freeman and Company, San Francisco and London, 1960.
2 D. Eisenberg and W. Kauzmann, *The Structure and Properties of Water*, Clarendon Press, Oxford, 1969.
3 H. Ratajczak, *J. Phys. Chem.*, 1972, **76**, 3000-3004.
4 R. Z. Khaliullin, A. T. Bell, M. Head-Gordon, *Chem.–Eur. J.* 2009, **15**, 851-855.

**Mr Acharya** asked: What fraction of water in the first hydration layer is in the **T** or the **D** configuration?

**Mr Stirnemann** responded: We summarize here the detailed discussion of the populations in the different states that can be found in section III. The different states are defined such that the exchange kinetics between them does not include any contribution from transient excursions. We therefore adopt the Stable States Picture and each state is defined by a set of tight geometric criteria. Consequently, the first layer populations in these states are respectively 39% in T, 3% in B and 3% in D. Thus in an instantaneous configuration, 55% of the first layer water OHs do not fall within any of these states; however, we verified that the lifetime of such configurations is short compared to the exchange time between the states, showing that these are unstable configurations which quickly return to one of the defined stable states.

**Mrs Seyed-Yazdi** commented: Why in the kinetic model did you not consider D-jump to a new D-jump that will be the result of rotation of a water molecule around its dangling bond? Can the neglect of this mechanism contribute to the discrepancy between the results of the kinetic model and direct calculation from simulation at very short times (Fig. 6a)?

**Mr Stirnemann** replied: We actually did consider the direct jumps from one **D** situation to another **D** situation. Our simulations showed that such jumps can be neglected since an initial **D** state first jumps intermediately to a **T** state before being able to jump to a new **D** state. We have therefore not included such direct **D** → **D** jumps in our kinetic model, but the **D** → **T** → **D** is fully described.

Regarding the discrepancy at very short times (<1 ps) between the populations from the kinetic model and from the simulations (Fig 6a), as mentioned in section IV.3, it is due to water librational motions which cause rapid but unstable population exchanges. Our kinetic model relies on a description through stable states and does not account for this transient effect.

**Professor Evans** opened a general discussion of the papers with: This is a general comment addressed to all the authors and participants. It concerns nomenclature — something that I hope *Faraday Discussions* would see as pertinent.

There was discussion after Professor Berne's paper that brings this to the fore. I believe that one should distinguish between phenomena that are concerned with confinement of a fluid between say two parallel substrates or walls at a finite separation $H$ and those that relate to a single substrate (or wall) $H = \infty$. For the former the bulk phase boundaries are usually shifted by confinement, and the effects of the substrate fields on the fluid particles, from their values in bulk. These shifted transitions are termed in the general adsorption literature: (i) capillary condensation — condensation occurs at a chemical potential that lies below the saturation value $\mu < \mu_{sat}(T)$ (ii) capillary evaporation where the transition from liquid to gas occurs for $\mu > \mu_{sat}(T)$. It is straightforward to show from macroscopic arguments that (i) occurs for large $H$ provided the contact angle for identical substrates has $\theta < \pi/2$ and (ii) occurs provided $\theta > \pi/2$. Similarly capillary freezing refers to confinement whereby the freezing transition in the capillary, for finite $H$, occurs at a chemical potential below the bulk freezing value.

Wetting and drying are terms that refer to a single substrate or wall, $H = \infty$. In the adsorption community complete wetting is associated with the equilibrium contact angle $\theta = 0$ where a macroscopically thick film of liquid intrudes between the substrate and the (bulk) gas. Complete drying is the opposite case; this is wetting by the gas phase. Here a macroscopically thick film of gas intrudes between the (solvophobic) substrate and the bulk liquid so that the equilibrium contact angle $\theta \to \pi$. For the majority of fluids and substrates $0 < \theta < \pi$. However, the terms partially wet $0 < \theta < \pi/2$ and partially dry $\pi/2 < \theta < \pi$ are often used. Note also that in the adsorption community the term 'wetting transition' has a very specific meaning. Singularities in surface excess thermodynamic quantities occur only as $\theta \to 0$ (wetting) or as $\theta \to \pi$ (drying). These transitions can be induced by changing the strength of substrate-fluid attractive field or by changing the temperature at $\mu = \mu_{sat}$. It is important to recognise that such singular behaviour only occurs for $H = \infty$, i.e. for a single substrate in contact with bulk fluid. Unfortunately the notion of 'drying' in confined geometry has become a little ambiguous. There is no reason for this. To be specific: capillary evaporation will occur for $\theta > \pi/2$ and finite $H$, whereas complete drying occurs only for $\theta = \pi$ and for $H = \infty$. Only for purely repulsive or very weakly attractive substrate-solvent potentials will complete drying occur. An example of a continuous transition to complete drying, brought about by increasing the strength of wall-fluid attractive interactions, is given in Fig. 1 the paper by M. Stewart and R. Evans,[1] where density profiles corresponding to a range of contact angles approaching $\pi$ are plotted.

1 M. Stewart and R. Evans, J. Phys. Condens. Matter, 2005, **17**, S3499.

**Dr Henderson** answered: There are equally interesting issues of nomenclature and physics associated with the term capillary condensation. Many experimentalists refer to capillary condensation in the context of open wedges. Such as the annular wedge generated in the Surface Forces Apparatus. However, in a physicists model, complete condensation (or evaporation) in a wedge occurs *at* bulk saturation; *i.e.* the wedge is open to bulk fluid (which would have to be deliberately held off bulk coexistence to prevent complete filling). For wedge geometry, physicists have introduced the term "filling" in a technical sense — to distinguish both from wetting and from condensation. All these issues are ultimately concerned with the defining role played by geometry.

# The search for the hydrophobic force law

Malte U. Hammer, Travers H. Anderson, Aviel Chaimovich, M. Scott Shell and Jacob Israelachvili*

*Received 11th December 2009, Accepted 22nd January 2010*
DOI: 10.1039/b926184b

After nearly 30 years of research on the hydrophobic interaction, the search for the hydrophobic force law is still continuing. Indeed, there are more questions than answers, and the experimental data are often quite different for nominally similar conditions, as well as, apparently, for nano-, micro-, and macroscopic surfaces. This has led to the conclusion that the experimentally observed force–distance relationships are either a combination of different 'fundamental' interactions, or that the hydrophobic force-law, if there is one, is complex – depending on numerous parameters. The only unexpectedly strong attractive force measured in *all* experiments so far has a range of $D \approx 100-200$ Å, increasing roughly exponentially down to $\sim 10-20$ Å and then more steeply down to adhesive contact at $D = 0$ or, for power-law potentials, effectively at $D \approx 2$ Å. The measured forces in this regime (100–200 Å) and especially the adhesive forces are much stronger, and have a different distance-dependence from the continuum VDW force (Lifshitz theory) for non-conducting dielectric media. We suggest a three-regime force-law for the forces observed between hydrophobic surfaces: In the first, from 100–200 Å to thousands of ångstroms, the dominating force is created by complementary electrostatic domains or patches on the apposing surfaces and/or bridging vapour cavities; a 'pure' but still not well-understood 'long-range hydrophobic force' dominates the second regime from $\sim$150 to $\sim$15 Å, possibly due to an enhanced Hamaker constant associated with the 'proton-hopping' polarizability of water; while below $\sim$10–15 Å to contact there is another 'pure short-range hydrophobic force' related to water structuring effects associated with surface-induced changes in the orientation and/or density of water molecules and H-bonds at the water–hydrophobic interface. We present recent SFA and other experimental results, as well as a simplified model for water based on a spherically-symmetric potential that is able to capture some basic features of hydrophobic association. Such a model may be useful for theoretical studies of the HI over the broad range of scales observed in SFA experiments.

## Introduction

The phenomenon of the low solubility of non-polar moieties or their strong mutual attraction in water is known as the hydrophobic interaction (HI). The HI is arguably the most important non-specific interaction in biological systems and is responsible for the creation of enclosed compartments by proteins and lipid bilayers in water, which was fundamental for the evolution of cells and therefore life. Molecular mechanisms of protein folding and adsorption are also regulated by the HI. Despite its importance, the underlying mechanisms of HI and its fundamental force law, if there is one, are still unresolved. Even though the first direct measurement of the

*Departments of Chemical Engineering, University of California, Santa Barbara, CA, 93106, USA. E-mail: Jacob@engineering.ucsb.edu*

attraction between two nominally hydrophobic surfaces was done nearly 30 years ago, the experimental and theoretical research that followed has left more puzzles than solutions. Puzzling aspects are the strength and range of the attractions measured between hydrophobic surfaces. Experiments report attractive ranges from ~100 to ~6500 Å.[1,2] The range of a few thousand ångstroms can hardly be explained by structured water alone. For rough (super)hydrophobic surfaces attractive forces are measured for separations up to 3.5 μm strongly indicating the influence of roughness.[3] To solve the puzzle it is helpful to differentiate between a general HI, involving all types of forces observed in experiments, and the pure HI involving just forces unique to hydrophobic moieties (molecules, molecular groups and extended surfaces) that are still not well understood but arise from changes in the properties of water near hydrophobic surfaces. The first step in understanding the hydrophobic phenomena is to understand the 'pure HI' which itself may have more than one distance regime. Review of the 'old' data suggested that there are three regimes that could be attributed to hydrophobic interactions: one – an 'extended long-range' regime, extending out to many thousands of ångstroms – that is *indirectly* produced by hydrophobic interactions, for example, the overturning of monolayers into charged bilayer patches or domains that then interact *via* long-range electrostatic forces, or bridging cavities; and two – a long-range and short-range hydrophobic interaction (HI), extending out to a few hundred ångstroms – that are *directly* related to hydrophobic effects and which we call the 'pure' HI. The new data with chemisorbed surfactant monolayers show that the long-range interaction appears both with physiorbed and chemisorbed surfactants, thereby showing it to be part of the pure (long-range) HI.

In addition to the lack in theory there are experimental problems creating suitable hydrophobic surfaces. Recently, it was shown that very few of the reported results are directly related to the intrinsic hydrophobicity of the surfaces in question.[1] Many seemingly contradictory or inconclusive results from studies concerning the effects of electrolyte concentration,[4–12] temperature,[13–15] and dissolved gases[7,8,10,16,17] can be attributed to surface or solution preparation techniques. To determine how various parameters affect hydrophobic interactions quantitatively, fundamental studies concerning temperature and electrolyte effects must be carried out on smooth, stable hydrophobic surfaces, such as octadecyl-tri-ethoxysilane (OTE) monolayers chemisorbed on activated mica. One of the few aspects that appear to be clear in the hydrophobic puzzle is the fact that the measured range of attraction decreases dramatically if great care is taken to eliminate electrostatic and capillary forces, for example, by rigorously deaerating the solutions. Deaeration has been found to significantly decrease the range of the extended long-range interaction between physiorbed monolayers even though cavities no longer exist while bilayer domains still exist. Does this suggest that deaeration somehow removes the long-range electrostatic interaction? Deaeration has also been found to reduce the attraction between surfactant-free oil droplets in water[18] as well as the thickness of the vapour-depleted layer at hydrophobic–water interfaces[1] but not the value of the hydrocarbon–water interfacial tension.[1] Thus, the effects of deaeration on both the long- and short-range (adhesion) forces between hydrophobic surfaces in water are not currently understood.

Direct measurement of the attractive forces between two nominally hydrophobic surfaces in aqueous solution was first published nearly 30 years ago[19,20] resulting in an attractive force acting as far as 80–100 Å from contact. During this time, the range was considered extremely large and was referred to as a "long-ranged" attraction. In the following years these first measurements, the experimental situation became murkier rather than clearer with different experiments producing different magnitudes and ranges for the 'hydrophobic force'. As experimental methods improved surface hydrophobization became more diverse, and the range of the attraction between hydrophobic surfaces was reported to be considerably longer than had been reported initially. The initially reported range of 100 Å is now considered to be part of the "short-ranged" hydrophobic interaction, with "long-range" forces extending out to separations greater than 200 Å, and even as far as 3000 Å or incredibly 6500 Å.[2]

The variable and surprisingly long-ranged nature of the force reported between hydrophobic surfaces used in experiments has been particularly vexing from a theoretical standpoint. It is clear from even a cursory look at the three typical types of force curves measured in these systems (*e.g.*, Fig. 2 from ref. 1) that no simple theory could be expected to account for all of the observed behaviours. Furthermore, any credible theory must be independent of the surface preparation technique, yet different surface hydrophobization techniques have resulted in force curves that are not even qualitatively similar,[10,21] indicating that secondary factors just beyond the surface hydrophobicity contribute to the measured forces. To find a simple theory describing just the pure HI acting between all hydrophobic moieties in water, we have to find suitable conditions for experiments to avoid the secondary effects associated with hydrophobic forces. One would expect the forces associated with the pure HI to be present between all hydrophobic surfaces. Unfortunately, even the search for the pure HI has proven difficult as it is often accompanied by other forces that are difficult to avoid in typical experiments.

While one major source of experimental confusion comes from the apparent dependence of the attraction on the hydrophobic surface preparation technique, another comes from the geometry of the surfaces used in the experiments, *e.g.*, whether macroscopic and of low curvature as in surface force apparatus (SFA) experiments, or macroscopic but with nanoscopic contact of tip and surface with high curvature as in atomic force microscopy (AFM) experiments. Most experiments have studied only one type of surface under a single or limited range of solution conditions. In many cases, it has become clear that the long-range attraction observed in so many experiments or systems is not in fact due to the direct pure hydrophobic attraction at all, but to originally hydrophobic monolayers overturning into charged patchy bilayers, giving rise to a long-ranged attractive electrostatic or double-layer force,[17,22] or to bridging of pre-existing nanobubbles,[23–28] giving rise to attractive capillary forces, again depending on the technique used to hydrophobize the surfaces. For the attractive electrostatic force, lateral motion of the bilayer patches on the surfaces is required. We have seen this in our AFM imaging (the patches move and rearrange both during and between scans) as well as the finite time (seconds) the attractive force takes to build up as the two surfaces are brought together. This assumption is in agreement with the literature, which describes evidence that dynamic interactions between polarized domains of lipid layers on both surfaces may result in attraction with a great range between hydrophobized mica surfaces.[29]

In fact, the only force present between *all* the hydrophobic surfaces so far studied is the "short-range" ($D < 100$–$200$ Å) force which, for biological interactions, would be considered as a long-range force. There is therefore a need to carry out unambiguous and systematic experiments of the forces between smooth, stable, hydrophobic surfaces under different solution conditions, and to closely coordinate the experiments with theoretical modeling. To date, OTE chemisorbed on activated mica has proved to be the only surface that satisfies all of the above requirements, making it a suitable surface for rigorous investigation of the hydrophobic interaction.[1,30,31,32] Knowledge that the pure HI should have a range of < 100–200 Å, and becomes particularly strong below 10 Å,[33] opens the door to several theoretical explanations that were previously dismissed by some due to their inability to explain the observed "long-range" force.

We present here recent SFA measurements done on smooth stable OTE surfaces and compare them with published measurements on different surfaces. Based on this data we would like to discuss a concept for the hydrophobic force law based on three different distance regimes. But, first, a few words about theoretical modelings of the HI, which have so far mainly focused on the shortest of the three distance regimes.

**Theoretical background**

Molecular simulations are frequently invoked in order to provide a better theoretical understanding of the physics underlying the HI, yet as this interaction spans multiple spatial scales, it becomes too computationally demanding to examine it with conventional fully-atomistic simulations based on semi-empirical force-fields. One possibility in overcoming this challenge might lie in multiscale simulations. Such simulations utilize simplified, coarse-grained molecular models extracted from detailed, fully-atomistic ones. Ideally, multiscale simulations employ models with a sufficient amount of information as required to describe the pertinent phenomena at the scale of interest. Such modeling strategies hold promise for examining the HI over its entire range of length scales.

Recently, Shell introduced a fundamental thermodynamic framework for multiscale simulations that can be applied to any arbitrary system of arbitrary Hamiltonian.[34] The key concept in this approach is the relative entropy, $S_{rel}$, given by,

$$S_{rel} = \sum_v p_v \cdot \ln\left(\frac{p_v}{\tilde{p}_v}\right) \qquad (1)$$

where $p_v$ denotes the probability of a particular configuration $v$ in a detailed, fully-atomistic molecular system while $\tilde{p}_v$ is the corresponding value in a simplified, coarse-grained molecular system. $S_{rel}$ measures the extent the configurational ensemble of the simplified system reproduces the correct one in the detailed system, measured, in at least one context, by a log-likelihood.[34] It is bounded below by zero (indicating perfect ensemble duplication), with higher values indicating decreasing adequacy of replication. Given a reference detailed model, a novel simplified model can be optimally determined by minimizing the relative entropy.[34]

Shell and coworkers used this approach to examine a particular spherically-symmetric model of water.[35] Its potential, $u(r)$, is constructed by the superposition of a Lennard-Jones with a Gaussian, given by,

$$u(r) = 4\varepsilon\left[\left(\frac{\sigma}{r}\right)^{12} - \left(\frac{\sigma}{r}\right)^{6}\right] + B\exp\left(\frac{-(r-r_o)^2}{\Delta^2}\right) \qquad (2)$$

The usual Lennard-Jones parameters, $\varepsilon$ and $\sigma$, define the energy and length scales of this interaction; the parameter $B$ sets the strength of the Gaussian while $r_o$ and $\Delta$ govern its center and spread, respectively. Consequently, Shell and coworkers evaluated the optimal values of these parameters for a grid of state points by correspondingly performing $S_{rel}$ minimization.[35] Surprisingly, the optimized models appeared to manifest an effective hydrogen bonding interaction in the form of an energetic variation of the region in the spherically-symmetric potential corresponding with water's nearest-neighbor distance.[35] Importantly, this variation in state contributes significantly in the ability of the spherically-symmetric model in reproducing aspects of waterlike behavior (e.g., water's anomalous pairwise structure along with its unique coordination number of 4).[35]

This coarse-grained model for water is a candidate for analyzing the multiscale nature of the HI. In this work, we do not yet intend to simulate the experiments discussed above; instead, we examine the ability of spherically-symmetric water in describing the HI on the molecular scale. Rather than examining the HI between infinite surfaces, we focus here on a simpler study, the association of a pair of methane-like hydrophobes. By this analysis, we evaluate the ability of the (optimized) spherically-symmetric model in capturing some basic features of the HI on the molecular scale, an issue which must be addressed before proceeding to use such models in the examination of the multiscale phenomena observed with the SFA.

## Materials and methods

### SFA measurements

The hydrophobic surfaces were prepared by a modified chemical vapor deposition based on a method by ref. 32 on back-silvered mica usually used for SFA measurements. The mica surfaces were treated with argon water plasma for 10 min at 450 mTorr. 0.5 ml OTE were placed together with mica discs into a glass Petri dish and OTE vapor was deposited during 4 h in a rough vacuum at 70 °C. The discs were rinsed with $CHCl_3$ and nitrogen dried before mounting to a SFA box. The SFA chamber was filled with water (Milli-Q A-10 water purification system (Millipore)). The dynamic force measurements[36] were done with a SFA 2000[37] at room temperature.

### Computer simulations

Monte Carlo simulations are used to examine the ability of spherically-symmetric water in describing the HI on the molecular scale. The simulation constructs a (periodic) box with 216 water molecules; its size is chosen so that water's density is 1.00 kg L$^{-1}$. One hydrophobic molecule rests at the center while another one moves along a single dimension. The corresponding potential of mean-force, a mathematical formulation of the HI, is evaluated *via* the multicanonical algorithm;[38] we perform 5 iterations to attain a sufficient flat-histogram of 50 bins. To ensure adequate statistics, 5 replicates are done. This protocol is performed for a few temperatures near ambient conditions.

We employ the spherically-symmetric water model discussed above, invoking the $S_{rel}$-optimized parameters for eqn (2).[35] Importantly, we examine two separate scenarios in modeling the medium: in one scenario, the potential for the water model is variable with state conditions while in the other scenario, it is constant. The parameters for the water model, along with their corresponding use in the two scenarios, are summarized in Table 1. The hydrophobes are described by methane-like molecules, expressed *via* a Lennard-Jones interaction.[39] These solutes interact with the solvent *via* Lorentz-Berthelot mixing rules; the parameters of these interactions, involving the hydrophobes, are summarized in Table 2.

## Results and discussion

### Experimental results

The measured interactions as a function of distance are shown in Fig. 1 in a semi-log and a log plot for variously prepared hydrophobic surfaces together with data from Wood and Sharma[31] and Meyer *et al.*[1]

As shown in Fig. 1, the shape of the measured force curve can hardly be explained by one curve. It seems that there are three different regimes, first a regime up to

**Table 1** The optimal parameters of the spherically-symmetric water model (as defined by eqn (2)) are given below for a range of temperatures, $T$, at a density of 1.00 kg L$^{-1}$.[35] Assuming infinite-dilution, these are invoked for the waterlike bath in the two different scenarios: in the variable case, the spherically-symmetric water model takes on the optimal parameters at each temperature, while in the constant case, the spherically symmetric model fixes the optimal parameters at ambient temperature (300 K)

| $T$/K | $\sigma$/Å | $\varepsilon$/kJ mol$^{-1}$ | $B$/kJ mol$^{-1}$ | $r_0$/Å | $\Delta$/Å |
|---|---|---|---|---|---|
| 280 | 2.42 | 22.4 | 25.8 | 2.45 | 1.10 |
| 300 | 2.43 | 20.4 | 23.3 | 2.46 | 1.09 |
| 320 | 2.43 | 19.7 | 22.9 | 2.44 | 1.11 |

**Table 2** The Lennard-Jones parameters associated with the (vacuum) interactions of the methane-like hydrophobes are given below.[39] Importantly, the mixing rules invoke the Lennard-Jones parameters of the fully-atomistic water model utilized in the optimization of the spherically-symmetric water model.[35]

| Hydrophobes interacting | $\sigma$/Å | $\varepsilon$/kJ mol$^{-1}$ |
| --- | --- | --- |
| with themselves | 3.73 | 1.23 |
| with water | 3.45 | 0.89 |

~10 Å, followed by an exponential regime up to ~100 Å, beyond which the forces between hydrophobic surfaces become more scattered depending on the surface and solution preparation but nevertheless are stronger than expected for van der Waals forces.

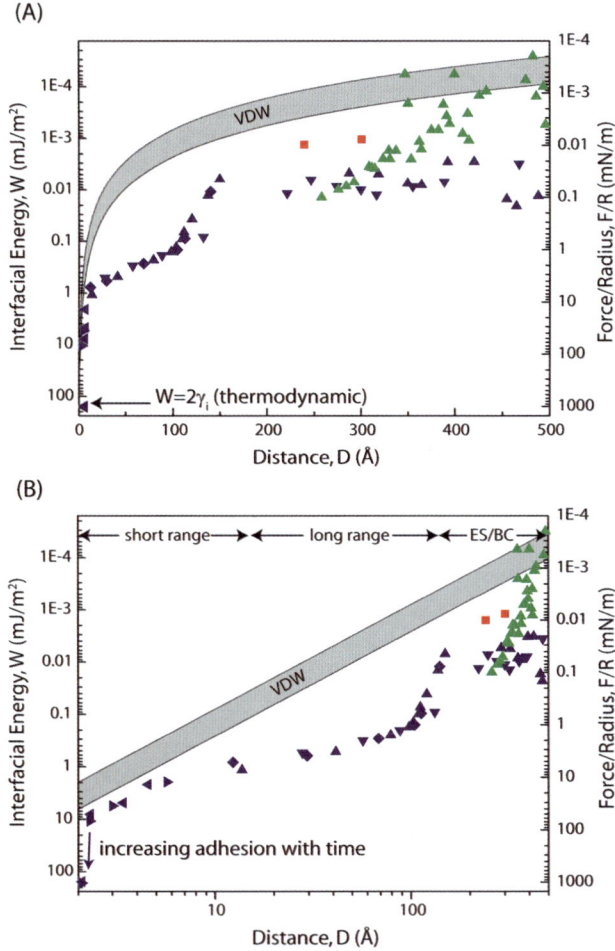

**Fig. 1** Results of experimentally measured forces between hydrophobic surfaces under different conditions presented as a semi-log (A) and a log-log (B) plot are indicated as followed: red squares: DMDOA, LB-deposited, deaerated, Wood and Sharma;[31] green triangles: OTE, chemical vapor deposition, deaerated; all data points in blue were obtained by Meyer et al.,[1] using OTE and DODA surfaces prepared by LB-deposition. Suggested regimes are marked as the short range, long range, and ES/BC, and are discussed below. The shaded VDW force band corresponds to Hamaker constants between 3 and $10 \times 10^{-21}$ J.

Up to now, no model exists that is capable of describing the attractive interaction between two hydrophobic surfaces for all described experimental data. The only force observed in all experiments is an attractive force below distances of 200 Å. Several reports suggest that the reported attraction above several hundred ångstroms results from different attraction forces, but not directly from HI between the surfaces.[33,40–42] Possible mechanisms for this attractive force are mainly based on local charge fluctuations created by imperfect, patchy preparation of hydrophobic surfaces resulting in attractive interaction between oppositely charged patches of the surface as well as bridging cavities.[1,42] Deaerating of water results in a dramatic decrease in the range of the attractive force.[1] There is an apparent correlation between contact angle hysteresis, indicating molecular rearrangements of the hydrophobic surface, and the existence of a force in the hundreds to thousands ångstrom range.[41] These surface defects can also act as nucleation centers for bubbles. The typical Debye length of 1500 Å for pure water suggests a long range electrostatic interaction. It is worth noting that the force curve resulting from differences between experiments with and without monovalent salt, has a decay length correlating to the expected Debye length of the salt used.[1] This indicates that the observed force in this regime is not directly related to the HI and is due to imperfect hydrophobic surfaces and is mainly due to electrostatic attraction and/or bridging cavities. Therefore, we suggest the name 'electrostatic and/or bridging cavitation' (ES/BC) for this regime.

Ederth and Liedberg[33] observed an attractive interaction above 200 Å that was apparently the result of bridging bubbles and concluded that the pure HI has a range of <200 Å. Within this regime, the measured force distance relationship can be fitted by an exponential function from 10 Å up to 100 or 200 Å[1] with a decay length of 10–20 Å[43] but not for distances below 10 Å. Interestingly in this range the expected VDW attraction for typical hydrocarbon surfaces with Hamaker constants of $3–10 \times 10^{-21}$ J is not negligible as shown in Fig. 1. Due to jump-in instabilities associated with the very strong attractive forces, and difficulties in distance resolution, experimental data for approaching surfaces below 10 Å are rare and have a large error. The maximum adhesion forces measured by Meyer, et al.[1] was determined to be $F_{ad}/R \approx 500$ mN m$^{-1}$, corresponding to an adhesion energy (based on the JKR theory) of $W = 2\gamma_i = 2F_{ad}/3\pi R \approx 100$ mJ m$^{-2}$, or $\gamma_i \approx 50$ mJ m$^{-2}$ (mN m$^{-1}$), which is the value for hydrophobic surfaces in water. As shown by Fig. 1, the VDW attraction is not negligible in the regime below ~10 Å and surely plays a part in the strength of the interaction, but is clearly too weak to account for the whole of the force. Published results show that the purely attractive forces measured are greater than any conceivable van der Waals force.[44] Other explanations related to water structuring effects associated with surface-induced changes in the orientation and/or density of water molecules and H-bonds at the water–hydrophobic interface[45] may explain the existence of the unusually strong attraction in the short-range regime.

## Theoretical results

Interestingly, the spherically-symmetric water bath appears to yield an effective HI that is comparable with the HI of fully-atomistic water media. For ambient conditions, Fig. 2 evaluates our computed potentials of mean-force with those utilizing fully-atomistic models for water.[46,47] On comparison, our HI has the same first-order oscillatory behavior (defined by hydration shells of water); in the spherically-symmetric case, the frequency is slightly higher and the amplitude is moderately lower. Importantly, it is encouraging that the binding energy is of the same order of magnitude.

The relation of HI with temperature is depicted in Fig. 3 for both scenarios (the water model is variable in one case and constant in the other). Importantly, the variable spherically-symmetric water attains an aspect of the HI that appears to be one of its signatures: the binding energy increases considerably with increasing

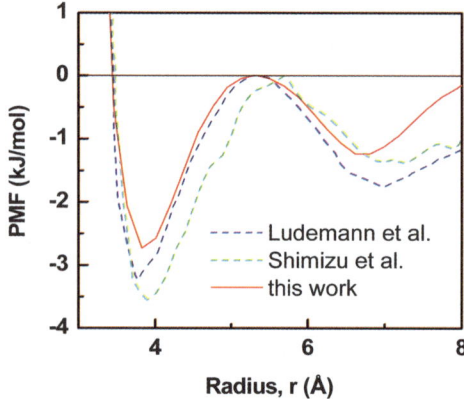

**Fig. 2** The potential of mean-force, PMF, in terms of the radial distance, $r$, between the pair of hydrophobes, is given above. The solid curve is obtained through our Monte Carlo simulations utilizing the spherically-symmetric water model. The dotted curves roughly depict the reported results of Ludemann et al.[46] and Shimizu et al.;[47] these two employ conventional semi-empirical force-fields for water, SPC[48] and TIP4P,[49] respectively. Our PMF has a slightly higher frequency and a moderately lower amplitude. The curves are shifted up–down so as to fix the "barrier" at zero energy.

temperature near ambient conditions.[46,47] Conversely, the constant spherically-symmetric water fails to describe this facet of the HI. These results reiterate observations made by Shell and coworkers: the spherically-symmetric water model must manifest a particular variation with state in the contact energy with its nearest-neighbors in order to capture waterlike properties; this allows for an effective hydrogen bonding interaction that can produce a number of signatures of water's anomalous bulk properties.[35]

The approximate description of the HI with the use of spherically-symmetric water suggests that water's molecular directionality can be effectively averaged for the "pure" HI, at least for the methane-like case study investigated here. In this simple case of the HI, the medium can be simply thought as composed of spheres, having the energy between nearest-neighbors subtly vary in state. This also suggests

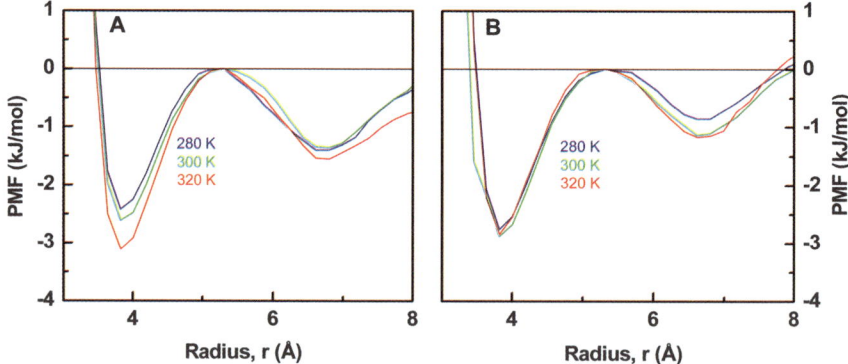

**Fig. 3** The potential of mean-force, PMF, in terms of the radial distance, $r$, between the pair of hydrophobes, is given above for the variable (A) and constant (B) scenarios. Each curve represents a different temperature. The variable case but not the constant case notably exhibits the following trend: with increasing temperature, the wells deepen, especially the one corresponding to binding. The curves are shifted up–down so as to fix the "barrier" at zero energy.

that it is water's unique pairwise structure that dominates on these scales of the "pure" HI.

## Conclusions

Based on this data we suggest a concept for the hydrophobic force law based on three regimes created by different forces dominating in specific distances. The longest range with an effective range of several thousand ångstroms is not a pure hydrophobic force. Instead it is based mainly on electrostatic effects due to heterogeneously charged surfaces and/or bridging effects (capillary forces) of cavities. The second force regime involved in our concept is the 'long range' force and although not well understood is likely a pure hydrophobic force with a range of 10–20 Å up to 100–200 Å. This force is possibly due to an enhanced Hamaker constant associated with the proton-hopping polarizability of water (*cf.* Grotthuss Effect) and may be important in explaining rapid protein-folding rates, faster then predicted by simple kinetic theories. Proton-hopping (the Grotthuss mechanism) rather than dipolar reorientations is the cause of the high dielectric constant (high polarizability) of liquid water and ice. Protons move over large distances and create giant dipoles. This effect on the van der Waals forces has not been considered, but may be expected to enhance the magnitude and/or range of the interaction. Since hydrophobic surfaces do not restrict the motion of water adjacent to them, we may expect that the enhanced polarizability of water at and between two hydrophobic surfaces could enhance the Lifshitz van der Waals type attraction to a magnitude and/or range that is comparable to that between conductors, *i.e.*, by up to two orders of magnitude. This could be the origin of the long-range 'pure' HI, which would be very different from the short-range interaction due to water structuring effects.

The 'short range' force is the overwhelming force in the distance from zero (contact) up to 10 Å and is likely related to water structuring effects associated with surface-induced changes in the orientation and/or density of water molecules and H-bonds at the water–hydrophobic interface.

Finally, we show that a simplified model for water, based on a *spherically-symmetric* potential, is able to describe some features of the water-mediated association of molecular-sized hydrophobes (*e.g.*, their binding energy increases with increasing temperature). Consequently, this model may eventually allow for multiscale simulations of our experiments, in turn, providing a molecular picture of the physics governing each regime of the HI. Nevertheless, other coarse-grained models for water should be examined with $S_{rel}$ methodology, as it presents a powerful approach for multiscale simulations.

## Acknowledgements

We would like to thank the Camille and Henry Dreyfus Foundation, the National Science Foundation (Award No. CBET-0845074), and the National Institutes of Health (Award No. R01 GM076709) for supporting our work.

## References

1 E. E. Meyer, K. J. Rosenberg and J. Israelachvili, *Proc. Natl. Acad. Sci. U. S. A.*, 2006, **103**(43), 15739–15746.
2 X. Y. Zhang, Y. X. Zhu and S. Granick, *J. Am. Chem. Soc.*, 2001, **123**(27), 6736–6737.
3 S. Singh, J. Houston, F. van Swol and C. J. Brinker, *Nature*, 2006, **442**(7102), 526–526.
4 H. K. Christenson, P. M. Claesson, J. Berg and P. C. Herder, *J. Phys. Chem.*, 1989, **93**(4), 1472–1478.
5 H. K. Christenson, P. M. Claesson and J. L. Parker, *J. Phys. Chem.*, 1992, **96**(16), 6725–6728.
6 H. K. Christenson, J. F. Fang, B. W. Ninham and J. L. Parker, *J. Phys. Chem.*, 1990, **94**(21), 8004–8006.

7 R. F. Considine, R. A. Hayes and R. G. Horn, *Langmuir*, 1999, **15**(5), 1657–1659.
8 V. S. J. Craig, B. W. Ninham and R. M. Pashley, *Langmuir*, 1998, **14**(12), 3326–3332.
9 P. Kekicheff and O. Spalla, *Phys. Rev. Lett.*, 1995, **75**(9), 1851–1854.
10 L. Meagher and V. S. J. Craig, *Langmuir*, 1994, **10**(8), 2736–2742.
11 J. L. Parker, P. M. Claesson and P. Attard, *J. Phys. Chem.*, 1994, **98**(34), 8468–8480.
12 J. L. Parker, P. M. Claesson, J. H. Wang and H. K. Yasuda, *Langmuir*, 1994, **10**(8), 2766–2773.
13 Y. H. Tsao, S. X. Yang, D. F. Evans and H. Wennerstrom, *Langmuir*, 1991, **7**(12), 3154–3159.
14 H. K. Christenson, J. L. Parker and V. V. Yaminsky, *Langmuir*, 1992, **8**(8), 2080–2080.
15 Y. H. Tsao, S. X. Yang and D. F. Evans, *Langmuir*, 1992, **8**(4), 1188–1194.
16 J. Mahnke, J. Stearnes, R. A. Hayes, D. Fornasiero and J. Ralston, *Phys. Chem. Chem. Phys.*, 1999, **1**(11), 2793–2798.
17 E. E. Meyer, Q. Lin, T. Hassenkam, E. Oroudjev and J. N. Israelachvili, *Proc. Natl. Acad. Sci. U. S. A.*, 2005, **102**(19), 6839–6842.
18 N. Maeda, K. J. Rosenberg, J. N. Israelachvili and R. M. Pashley, *Langmuir*, 2004, **20**(8), 3129–3137.
19 J. Israelachvili and R. Pashley, *Nature*, 1982, **300**(5890), 341–342.
20 J. N. Israelachvili and R. M. Pashley, *J. Colloid Interface Sci.*, 1984, **98**(2), 500–514.
21 H. K. Christenson and P. M. Claesson, *Adv. Colloid Interface Sci.*, 2001, **91**(3), 391–436.
22 S. Perkin, N. Kampf and J. Klein, *Phys. Rev. Lett.*, 2006, **96**(3), 038301.
23 P. Attard, *Langmuir*, 1996, **12**(6), 1693–1695.
24 N. Ishida, T. Inoue, M. Miyahara and K. Higashitani, *Langmuir*, 2000, **16**(16), 6377–6380.
25 N. Ishida, M. Sakamoto, M. Miyahara and K. Higashitani, *Langmuir*, 2000, **16**(13), 5681–5687.
26 P. Attard, M. P. Moody and J. W. G. Tyrrell, *Phys. A*, 2002, **314**(1–4), 696–705.
27 J. W. G. Tyrrell and P. Attard, *Phys. Rev. Lett.*, 2001, **87**(17), 176104.
28 J. W. G. Tyrrell and P. Attard, *Langmuir*, 2002, **18**(1), 160–167.
29 F. Pincet, E. Perez, G. Bryant, L. Lebeau and C. Mioskowski, *Phys. Rev. Lett.*, 1994, **73**(20), 2780–2783.
30 J. Wood and R. Sharma, *Langmuir*, 1994, **10**(7), 2307–2310.
31 J. Wood and R. Sharma, *J. Adhes. Sci. Technol.*, 1995, **9**(8), 1075–1085.
32 H. Sugimura and N. Nakagiri, *J. Vac. Sci. Technol., B*, 1997, **15**(4), 1394–1397.
33 T. Ederth and B. Liedberg, *Langmuir*, 2000, **16**(5), 2177–2184.
34 M. S. Shell, *J. Chem. Phys.*, 2008, **129**(14), 144108.
35 A. Chaimovich and M. S. Shell, *Phys. Chem. Chem. Phys.*, 2009, **11**(12), 1901–1915.
36 D. Y. C. Chan and R. G. Horn, *J. Chem. Phys.*, 1985, **83**(10), 5311–5324.
37 J. Israelachvili, Y. Min, M. Akbulut, A. Alig, G. Carver, G. W. Greene, K. Kristiansen, E. Meyer, N. Pesika, K. Rosenburg and H. Zeng, *Rep. Prog. Phys.*, 2010, **73**, 036601.
38 B. A. Berg and T. Neuhaus, *Phys. Rev. Lett.*, 1992, **68**(1), 9–12.
39 W. L. Jorgensen, J. D. Madura and C. J. Swenson, *J. Am. Chem. Soc.*, 1984, **106**(22), 6638–6646.
40 M. Hato, *J. Phys. Chem.*, 1996, **100**(47), 18530–18538.
41 H. K. Christenson, *Colloids Surf., A*, 1997, **123–124**, 355–367.
42 S. Ohnishi, V. V. Yaminsky and H. K. Christenson, *Langmuir*, 2000, **16**(22), 8360–8367.
43 Y. Liang, N. Hilal, P. Langston and V. Starov, *Adv. Colloid Interface Sci.*, 2007, **134–135**, 151–166.
44 E. Kokkoli and C. F. Zukoski, *Langmuir*, 1998, **14**(5), 1189–1195.
45 E. Ruckenstein and N. Churaev, *J. Colloid Interface Sci.*, 1991, **147**(2), 535–538.
46 S. Ludemann, H. Schreiber, R. Abseher and O. Steinhauser, *J. Chem. Phys.*, 1996, **104**(1), 286–295.
47 S. Shimizu and H. S. Chan, *J. Chem. Phys.*, 2000, **113**(11), 4683–4700.
48 H. J. C. Berendsen; J. P. M. Postma; W. F. van Gunsteren; J. Hermans., *Intermolecular Forces*. Reidel: Dordrecht, 1981; p 331.
49 W. L. Jorgensen, J. Chandrasekhar, J. D. Madura, R. W. Impey and M. L. Klein, *J. Chem. Phys.*, 1983, **79**(2), 926–935.

# The effect of counterions on surfactant-hydrophobized surfaces

Gilad Silbert,[a] Jacob Klein*[a] and Susan Perkin*[b]

*Received 4th December 2009, Accepted 11th January 2010*
DOI: 10.1039/b925569a

A common method for creating hydrophobic monolayers on charged surfaces is by self-assembly of ionic surfactants from solution. Several factors are important in controlling the structure and properties of such layers: the hydrophobic interactions between adjacent chains, the electrostatic interactions between adjacent headgroups, and electrostatic interactions between the headgroups and the surface charges. We have discovered that the surfactant counterions can have a remarkable effect on the hydrophobicity and hydrophobic interactions of a self-assembled layer. The experimental system was stearoyl($C_{18}$)trimethylammonium surfactant with iodide, bromide or chloride counterion (STAI, STABr, and STACl respectively) self-assembled onto mica substrates. Changing the surfactant counterions alters the wetting properties of hydrophobic monolayers on mica. Using a surface force balance we have carried out direct measurements of the interaction force between two surfactant-coated surfaces across water, revealing a strong effect of counterion on the normal interactions. Paradoxically, STAI-coated mica has both the highest water contact angle (is 'most hydrophobic') at the same time as having the highest surface charge relative to STABr and STACl. We use measurements of interfacial tension, asymmetric force measurements, and XPS to lead us towards an interpretation of these results and an understanding of the effect of counterion on the structure of self-assembled monolayers.

## 2. Introduction

Since the earliest investigations of the hydrophobic interaction between non-polar surfaces across water, surfactants self-assembled or deposited onto charged substrates have been used as a convenient method of creating model surfaces.[1–6] It has been suggested that some apparently 'hydrophobic' interactions between surfactant coated surfaces may in fact be due to lateral inhomogeneity leading to attractions of electrostatic origin.[7–9] More recent experiments have provided some evidence for this mechanism, showing that surfactants can rearrange on a surface to give positive and negative patches and that the resulting interaction is long-range and attractive.[10–13] The self-assembly of surfactants is also used in many industrial and technological processes to alter the wetting, lubricating or aggregation properties of materials, so much effort has been expended to understand the factors controlling the structure and properties of these surface layers.[14]

The adsorption of cationic surfactants onto negatively charged surfaces immersed in surfactant solution below the critical micelle concentration (CMC) has been

[a]*Department of Materials and Interfaces, Weizmann Institute of Science, Rehovot, 76100, Israel. E-mail: jacob.klein@weizmann.ac.il*
[b]*Department of Chemistry, University College London, 20 Gordon Street, London, WC1H 0AJ, UK. E-mail: susan.perkin@ucl.ac.uk*

shown to occur *via* nucleation of islands.[15,16] Adsorption from solution at concentrations above the CMC can involve the adsorption of both micelles and individual monomers,[17] and the resulting surface structure can include hemi-micelles, cylindrical micelles, and bilayers.[18] When removed from surfactant solution and rinsed in pure water to remove unbound molecules, a smooth hydrophobic monolayer remains coating the surface.[10,19] The surfactant counterion can have an effect on this process because of their different affinity for the surfactant headgroup. Micelles in solution have a lower degree of ionisation, $\alpha$, and thus a lower surface charge when the counterion is $I^-$ than for $Br^-$ or $Cl^-$ counterions due to the less favourable hydration energy for $I^-$.[20,21] In addition, it has been shown that trimethylammonium headgroups have a particularly strong binding for $I^-$ counterions, further enhancing the ion specific effects for these headgroups.[22] These counterion specific effects for micelles could also extend to the structure of monolayers in a number of ways. Firstly, the extra headgroup–headgroup repulsion for surfactants with higher $\alpha$ could lead to less dense monolayers; an effect which has been observed at the air–water interface.[23] Secondly, the inclusion of different fractions of surfactants with their counterions (ion pairs) in a monolayer could alter the interaction of the monolayer with a charged substrate.

This interaction of the monolayer with the charged substrate is important for determining the stability of the layer when immersed in water. A number of factors can affect the stability of a monolayer: charge–charge binding between negative surface sites and ionised surfactant ($STA^+$ in these experiments), dipole–dipole interactions between non-ionised surface sites and surfactant (STAX), and chain–chain van der Waals interactions and hydrophobic interactions which are dependent on the chain length and their packing density. For example, for the hexadecyltrimethylammonium surfactants it has been noted that changing the counterion from bromide (CTAB) to fluoride (CTAF) led to a significant increase in stability of the monolayer in pure water.[10]

It is known that the structure and stability of hydrophobic surface layers in water is crucial in determining the range and strength of hydrophobic interactions between them.[24] The true hydrophobic interaction due to disruption of water hydrogen bonding structure—'Class I' in Christenson and Claesson's analysis—is a relatively shorter ranged effect (< 20 nm) which occurs between stable, uniform, uncharged hydrophobic surfaces. Much longer ranged attractions between hydrophobic surfaces (up to $\sim$250 nm) can be caused by nanobubbles or water-vapour layers on the surfaces leading to cavitation and capillary attraction between the surfaces when they approach to a certain distance ('Class II'). A final class of interactions ('Class III') are the long-range attractions ($\sim$5–50 nm) which have been attributed to patches or heterogeneous layer characteristics leading to surface correlations and attractions of electrostatic origin.[6,8–10,12,13,25,26] The heterogeneity may arise from flipped molecules interdigitated with the monolayer; rearrangement of surfactants into patches of bilayer; tilting of chains to give domains; or collecting together of chains to give dense hydrophobic patches and less dense more hydrophilic patches. Patch formation has been observed for $C_{18}$ chain surfactants on silica,[11] and for $C_{16}$ chain surfactants on mica.[10] The surface correlation requires that the molecules be mobile on the surface, or at least that their hydrophobic tails be able to change orientation, under the influence of a field from the other surface. It has been shown that in this case the attraction between the surfaces should be of a range comparable to the heterogeneous domain size, or the Debye length, whichever is the shorter.[9,14]

In this study we investigated self-assembled monolayers of stearoyltrimethylammonium surfactants on mica with iodide, bromide or chloride counterions (called STAI, STABr, and STACl respectively, or 'STAX' in general). We compare direct force measurements of the interaction between two monolayer-coated surfaces using a surface force balance (SFB) to characterisation of the surface layers using contact angle goniometry, AFM and XPS. In particular we focus on the effect of the counterion on the monolayers for the homologous series STAI, STABr, STACl. We find that changing the counterion has significant effects on the stability, hydrophobicity, surface charge, and interactions of the layers in water. The trend in stability for STAX is opposite to that previously observed for CTAX,[10] which we interpret in terms of the balance between chain–chain interactions and headgroup–surface interactions. The observed long-range attractions between the STAX layers are discussed and interpreted in terms of the layer structure, density, and charge and how these vary with the counterion, X. Finally, a preliminary investigation of surface correlation through measurement of forces between laterally-moving surfaces reveals no effect within the scatter over the range of parameters examined.

## 3. Materials and methods

### Materials

Mica was highest grade (V1) muscovite mica from S&J Trading Inc., freshly cleaved on both faces before all contact angle measurements. For SFB measurements the mica preparation was as described previously.[27] Octadecyltrimethyammonium iodide (STAI) was custom synthesised according to the method of method of Kodama et al.[28] Octadecyltrimethyammonium bromide (STABr) was purchased from Sigma-Aldrich, purity >98%, Octadecyl(stearoyl)trimethylammonium chloride (STACl) was purchased from Tokyo Chemical Industries, catalogue number S0087, batch analysed as 99.3% purity. All surfactant solutions were freshly made soon before each experiment (they were not stored and re-used), and were all 0.75–1.00 mg ml$^{-1}$, which corresponds to concentrations to 6–8 × CMC, and were heated to 70 ± 5 °C before dipping the mica substrates. This ensured that all surfactant was dissolved, since this temperature is above the Krafft point for all three surfactants. NaCl was purchased from Sigma-Aldrich at 99.999% purity. Water was taken directly from a Barnstead Nanopure water purification system, with TOC monitoring, and had readings of 18.1–18.2 MΩ cm$^{-1}$ resistivity and < 1 ppb TOC.

### Contact angle measurements

Contact angle measurements were conducted using a goniometer purchased from First Ten Angstroms, model FTA200. Mica sheets were cut to ∼3 cm × 4 cm, cleaved on both sides, and dipped into surfactant solution (as described above) for 30 s before rinsing in pure water at 70 ± 5 °C for ∼30 s. They were then immersed in pure water for time 0 < $t$ < 30 h, withdrawn, and dried in a laminar flow cabinet before contact angle measurements were made. For each reading a new mica sample was used. A Hamilton syringe and straight needle were used to inject into, or withdraw water from, a droplet at a constant rate. Using a camera and image capture software, movies of the droplet advancing or receding on the mica surface were recorded at 2 frames per second. Each data point for advancing or receding contact angle, $\theta_A$ or $\theta_R$, represents the average value of the advancing or receding angle for one of these series of measurements.

### Surface force measurements

Interaction forces between the monolayer coated surfaces across pure water were recorded using a surface force balance (SFB) according to procedures described earlier.[29] Two mica sheets from the same original facet, of area ∼1 × 1 cm and of thickness 1–3 μm are mounted onto hemi-cylindrical lenses. Calibration of the surface separation between the un-coated mica surfaces in air and then in pure water is made at the start of each experiment. These substrates are then taken out of the SFB, dipped in a solution of the surfactant solution (either STAI, STABr, or STACl) at 70°–73° for 30 s (solutions as described above). The surfactant coated substrates were then rinsed in pure water also at 70°–73° for 30 s, before re-mounting inside the SFB which is filled with pure water. Force profiles (Force, $F$, between the surfaces as a function of surface separation, $D$) were recorded at 2–9 contact points on each pair of mica surfaces and the time after immersion in water was noted for each. Occasionally a different procedure was followed: the dry mica substrates were dipped directly into surfactant solution (*i.e.* no water calibration), rinsed, then put into the SFB and immersed in water. The results using this procedure were identical, within the scatter, to those recorded following the first procedure; demonstrating that the SFB water calibration step does not lead to subsequent differences in surfactant adsorption, relative to the procedure followed for contact angle measurements where the dry mica was dipped directly into surfactant solution.

In some experiments a high-speed video recording of the interference fringe pattern was used to analyse the interactions during the rapid jump into contact. In these cases the spectral interference fringes were recorded and later analyzed to obtain separation, $D$, as a function of time, $t$. The slope $dD/dt$ of the data points at large separation is constant and equal to the approach velocity of the surface due to the motor motion. Any deviation from this slope is due to the deflection of the spring connected to the lower lens. Using the spring constant and the deflection at each point the forces between the surfaces are obtained. From this force we subtract the hydrodynamic force given by the Taylor equation $F_D = \dfrac{6\pi\eta R^2}{D}$, (were $D$ is the separation between the surfaces, $\eta$ is the dynamic viscosity of the media, $\dot{D}$ is the approach velocity of the surfaces and $R$ is the mean radius of curvature of the surfaces) under the assumption of no slip boundary condition. (This assumption was tested by us for the STAI monolayers where we found that the hydrodynamic slip length is no more then 10 nm [unpublished data]; this can introduce a small error in the force at the last few nanometres[30]). To get the local velocity at each point the data is fitted to a polynomial and the differential at each point is taken. The area of the jump is fitted separately in order to achieve a more accurate profile. The acceleration term can also be calculated from the $D$ vs. $t$ plot using the second derivative, but is found to be negligible. Thus we remain only with forces arising from the interacting surfaces.

### AFM imaging

Images were recorded using a Digital Instruments AFM in tapping mode using a standard $Si_3N_4$ cantilever. All parts which contact the water were rinsed in ethanol and water and irradiated with UV light before use.

### X-ray photoelectron spectroscopy (XPS) measurements

Measurements were carried out on a Kratos AXIS ULTRA spectrometer, using monochromatized Al (K$\alpha$) X-ray source ($h\nu = 1486.6$ eV) at 10–75 W and detection pass energies ranging between 10 and 40 eV.

## 4. Results

### Wetting properties

We investigated the wetting properties of the self-assembled STAX monolayers ($X^- = I^-$, $Br^-$, $Cl^-$) after various immersion times of the surfactant monolayer in pure water. Mica substrates coated with monolayers of STAI, STABr, and STACl were immersed in pure water for varying times between $t = 0$ up to $t = 30$ h. They were then withdrawn from the water and dried in a laminar flow cabinet. Both advancing and receding contact angles of a pure water droplet were then measured on the surfaces. The results are shown in Fig. 1. In the case of STAI (Fig. 1A), the advancing contact angle, $\theta_A$, is $102° \pm 0.5°$ ($\bar{x} \pm \sigma/\sqrt{n}$), remaining almost unchanged

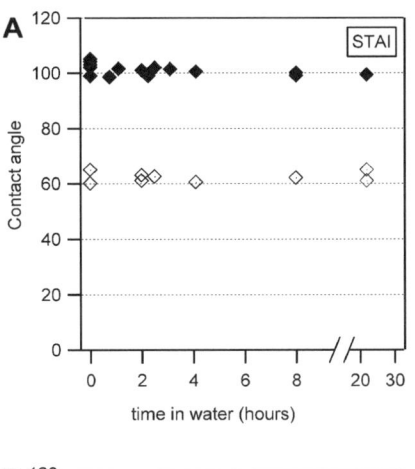
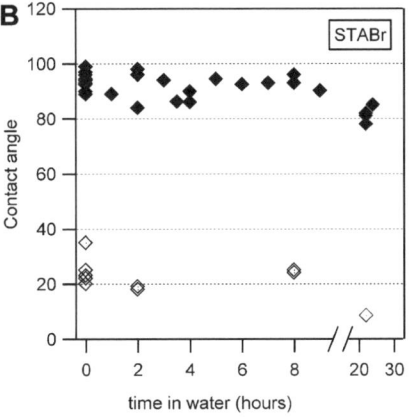
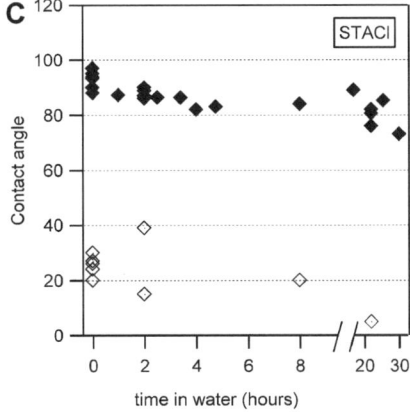

**Fig. 1** Advancing ($\theta_A$) and receding ($\theta_R$) contact angles of water droplets on surfactant monolayer coated mica sheets as a function of their immersion time in pure water after surfactant self-assembly (procedure described in Methods section). $\theta_A$ are shown with filled diamond symbols; $\theta_R$ are shown with open diamond symbols. Parts **A**, **B**, and **C** show the results for STAI, STABr, and STACl respectively. Each point represents the average value of a series of measurements once steady-state contact line motion has been reached.

over the 23 h timescale investigated. The receding contact angle, $\theta_R$, is $62° \pm 0.6°$, also remaining unchanged after periods of immersion in water up to $t = 23$ h.

STABr and STACl (Fig. 1B, C) appear similar to one another, but quite different to STAI. They both have advancing contact angles which decrease with immersion time in water from $\theta_A \sim 94°$ at $t = 0$ to $\theta_A \sim 80°$ at $t = 30$. Compared to STAI they have a significantly larger hysteresis between advancing and receding contact angles, with receding angles of $\theta_R = 23° \pm 2°$ for STABr and $\theta_R = 25° \pm 3°$ for STACl. For STABr and STACl there is also a larger degree of scatter of data.

We noticed that after long periods of time in water (*e.g.* 23 h) the droplet of water advances and recedes on STAI in a very smooth way, with a constant advancing or receding angle. In contrast, water droplets on STABr or STACl surfaces after the same long period in water advance in 'jumps', with the angle increasing and the line remaining 'stuck' until $\theta_A$ is reached at which point the line jumps to a new static contact angle and the cycle repeats. An example of this contrasting behaviour is shown in Fig. 2. The difference in angle between adv and static angles (peak to trough of the jumps) is 5–10°.

If the monolayer-coated surfaces are immersed in 0.1 M NaCl electrolyte solution instead of pure water, the difference between STAI and STABr was found to be even more pronounced. We observe a dramatic difference in the stability of the STABr layer after immersion in the electrolyte, whereas STAI remains stable in electrolyte (Fig. 3). When the STAI monolayer was immersed in 0.1 M NaCl solution for up to 24 h the advancing contact angle remains steady, at a slightly lower value than that seen after immersion in pure water. STABr, on the other hand, gives a sharp decrease in contact angle towards totally hydrophilic after only 15 min in the electrolyte solution. After this time $\theta_A$ was too small to measure (the droplet wetted the surface completely).

### 4.2 Force measurements between the hydrophobised surfaces

Force measurements between surface layers across water can inform us about several properties of the layers: their surface charge and surface potential, their 'hydrophobic' interaction, their stability over time, their interfacial energy, amongst other things. Using a SFB we measured the interaction force as a function of

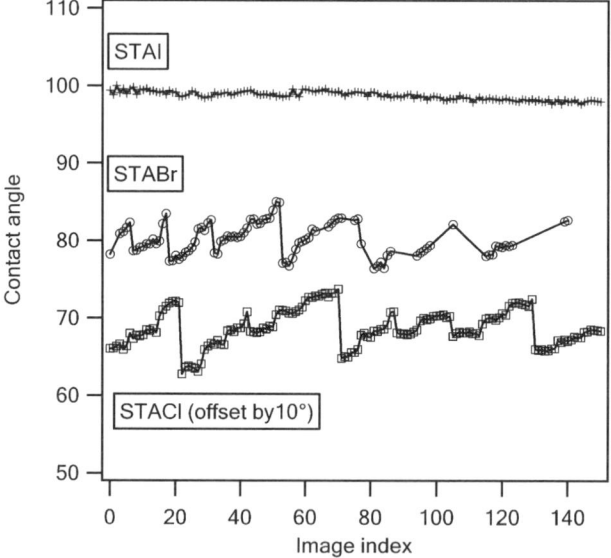

**Fig. 2** Advancing contact angles on STAI, STABr and STACl layers after 23 h immersion in pure water, recording one data point every 0.5 s during the advance of the droplet. The STACl data is offset by $-10°$ for clarity.

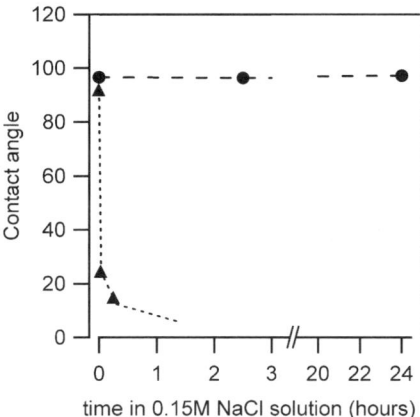

**Fig. 3** Advancing contact angle of water droplets on STAI monolayers (filled circles) and STABr monolayers (filled triangles) after varying immersion times in 0.15 M NaCl solution.

separation between two identical monolayer-coated surfaces for each of the three surfactants investigated: STAI, STABr, STACl (Fig. 4A–C). At least three separate experiments (separate mica sheets) were performed for each of the surfactants, and many (up to 9) contact points were investigated during each experiment. The data presented in Fig. 4 are representative traces from these experiments. STAI layers (Fig. 4A) show a long ranged repulsion from $D \sim 120$ nm, followed by a strong attraction from $\sim 15$ nm which pulls the surfaces into an adhesive contact. The repulsive part of the interaction can be fitted to a DLVO expression based on the non-linearized Poisson–Boltzmann equation,[31] using a surface potential value of $|\psi_0| = 70$ mV. (The sign of $\psi_0$ is determined using an asymmetric SFB experiment as shown in Fig. 6 and described later.) The attractive interaction is longer ranged than expected for simply van der Waals interactions, as highlighted in the inset to Fig. 4A. Thus we find that the STAI layers have a considerable surface charge, as well as attracting one another from a surface separation larger than can be explained as purely van der Waals interactions.

The interaction between STABr layers across water (Fig. 4B) is different to that seen for STAI: there is little or no interaction between them until they reach a separation of 20–30 nm at which point they are attracted together into adhesive contact. The attractive interaction appears to be longer-ranged than can be attributed to van der Waals interaction alone. The interaction between two STACl layers (Fig. 4C) is similar to those seen between STABr layers—there is little or no long-range electrostatic repulsion—although the attraction into contact is from a shorter distance of 8–20 nm. In some STABr and STACl experiments a small repulsion was recorded before this attraction (which could be fitted to $|\psi_0| = 5$–30 mV) during the first 2 h in water, always to be replaced by a purely attractive interaction after 2 h. An example of this is the STACl data in Fig. 4C, where a small repulsion is seen during the first 1–2 h in water, to be replaced by a purely attractive interaction after longer times in water. In many experiments no repulsion was observed, or was too small to be detected, at all times (as in Fig. 4B). We note that no force measurements could be made during the first $\sim 30$ min after addition of water (due to the necessary setup and thermal equilibration of the experiment), therefore the forces we describe are all recorded for $t > 30$ min.

### 4.3 Surface energy of the STAX: water interface

When the STAX layers reach contact in the SFB experiment, the interaction is adhesive in every case. By measuring the force required to pull the layers apart, and using JKR theory, we are able to calculate the STAX layer–water interfacial energy.[32] In Fig. 5 we compare this interfacial energy for STAI, STABr and STACl as a function

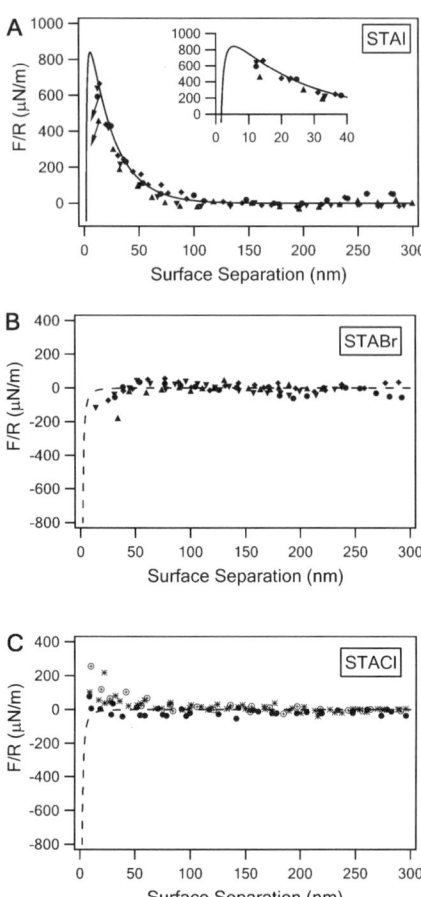

**Fig. 4** Interaction force profiles between two STAX coated mica sheets across pure water measured using the SFB. The results are presented as $F/R$ where $R$ is the radius of curvature of the substrate surfaces ($R \sim 1$ cm), in order that $F/R$ is directly proportional to the interaction energy between two parallel plates obeying the same force law (the Derjaguin approximation). A: interaction between STAI coated surfaces. B: interaction between STABr coated surfaces. C: interaction between STACl coated surfaces. The STAI force profiles were independent of immersion time in water, and always showed the repulsive interaction from $\sim$100 nm before attraction at $\sim$15 nm. The solid line in **A** shows a fit of the longer range repulsive interaction to DLVO theory using a non-linear Poisson–Boltzmann expression for the electrostatic repulsion and a surface potential of $|\psi_o| = 70$ mV and $c = 2 \times 10^{-4}$. The STABr data in **B** was recorded over a period of 30 h in water, and the STACl data in **C** was recorded over 8 h in water. The STACl data symbols include open circles for the first 1–2 h in water, stars for 3–4 h in water, and filled symbols for 5–8 h in water. Both STABr and STACl force profiles were in some experiments observed to have this small repulsive region after short times in water turning to purely attractive interactions after longer times in water (as seen in **C** for the STACl data here). In other experiments using STABr or STACl the small repulsive region at short times was not observed (as seen in **B** for the STABr data here). Dashed lines in **B** and **C** show the pure van der Waals component of the interaction.

of immersion times of each in pure water. The data shown represents 4–5 contact points for each surfactant. We see that STAI layers have the highest surface energy, of 55–60 mJ m$^{-2}$, and that this value remains unchanged with the time of immersion in water over the 23 h period. STABr and STACl have lower interfacial energies, and the values decrease further as the surfactant layers remain immersed in water to below 20 mJ m$^{-2}$ after 30 h.

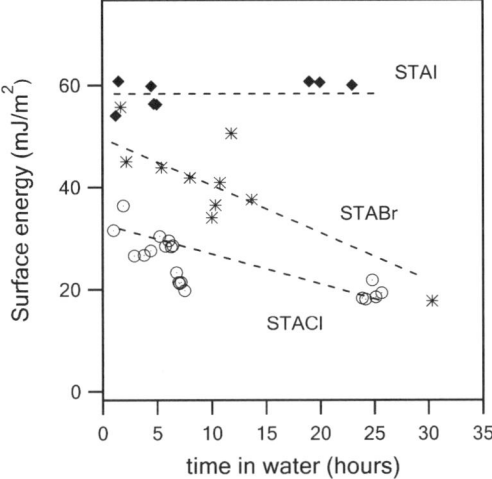

**Fig. 5** Surface energy measurements for STAI (filled diamonds), STABr (stars), and STACl (open circles) surfaces in pure water as a function of immersion time in the water. The measurements were made in the SFB by recording the force required to pull apart the adhered surfaces—the adhesion force, $F_{adh}$—in pure water. The surface energy is then calculated as $F_{adh}/3\pi R$.

### 4.4 Asymmetric force measurements between a STAX layer and a bare mica surface

By measuring the interaction between a STAI monolayer on one surface and a bare (uncoated) mica surface, which has a negative charge of known magnitude, it is possible to determine whether the substantial surface charge on the STAI layer is negative or positive in sign. Fig. 6A shows the interactions measured in the STAI-bare mica asymmetric case. There is no repulsion between the surfaces, only a long-ranged attraction with an onset between 25–50 nm. This attraction is expected for oppositely charged surfaces,[33] demonstrating that the STAI layer on mica is positively charged.

Using asymmetrical surface force measurements between an adsorbed surfactant layer and a surface with known constant charge, such as mica, is a sensitive way to probe the stability of the layer with time in water.[34] Fig. 6B shows the asymmetric experiment for STABr interacting with bare mica. There is no interaction until the surfaces reach 20–35 nm, at which point there is an attractive interaction and the surfaces jump into contact. The shorter range of attraction compared to the STAI:-bare mica case in Fig. 6A suggests a smaller positive charge or no net charge on the STABr layer. This interaction remains unchanged for 40 h immersed in water, showing that the STABr layers remain bound to the surface for long periods in water. This is in contrast to what has been seen in a similar experiment for $C_{16}$TABr, where the layer is gradually released into the water.[34] In addition, sequential force measurements at the same contact point show identical behaviour suggesting that the surfactant molecules are not transferred from one surface to another on reaching contact.

### 4.5 An experiment to investigate surface correlation

We see in the case of the STABr layer that after immersion in water the layer–water interfacial energy decreases—suggesting that it is more hydrophilic—and yet the attraction between two layers across water remains long ranged. Indeed, in some cases the attraction becomes stronger or more long ranged with time. In addition,

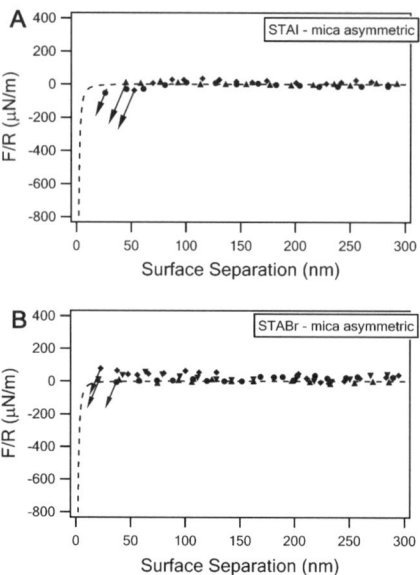

**Fig. 6** Interaction force profiles between one surfactant coated surface and one bare mica surface ('asymmetric' experiment). **A**: STAI–bare mica; **B**: STABr–bare mica, data recorded between 5 h and 42 h after addition of water. Arrows indicate the points from which the surfaces jump into contact. The dashed curve shows the form of the purely van der Waals component of the interaction.

we have seen, mainly from the asymmetrical surface force experiments, that no net negative charge develops on the layer during the time in water. This implies that in general the surfactant monomers do not desorb from the surface, a process which should lead to the development of net negative charge originating from the bare mica. In order to rationalise these results we must consider the possibility that the long range attraction in the symmetrical case is due to correlation between charge patches or other lateral heterogeneity in the surfactant layer caused by rearrangements of the surfactants on the surfaces driven by the lowering of surface energy with water.[6,10,12,13,24,35] We have attempted to test this correlation mechanism by applying a back-and-forth lateral motion to the top surface during the approach of the surfaces. In this way, the correlation between any structures on opposing surfaces is expected to be reduced or prevented as long as this lateral velocity is faster than the velocity at which charge patches may move laterally on the surface (so that they 'do not have time to correlate'). In this preliminary study, the amplitude of the shear motion used was only up to 120 nm peak to peak; this is comparable with the range of charge patch dimensions observed earlier for $C_{16}TAB$ surfactants[13] which could reduce the perturbing effect of the lateral motion of the top surface. To perform this experiment we used a high speed camera to detect the surface position as a function of time as described in the Methods section. Alternate force measurements on the same contact spot were made with shearing ('shear approach') and without shearing ('direct approach') for direct comparison. These were also compared to first approaches at different contact points carried out either as direct approach or shear approach. Fig. 7 shows two such interaction force profiles—one 'shear approach' and one as a 'direct approach'. Measurements similar to those shown in Fig. 7 were made at different contact points on the surfaces and after different immersion times in water, and in each case we found no systematic difference between the direct approach force profiles and the shear approach force profiles. It seems that at the lateral applied velocities and amplitudes used in these experiments there is no systematic effect on the normal forces at any value of $D$.

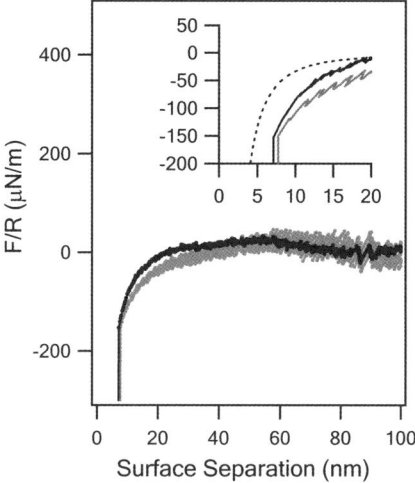

**Fig. 7** Interaction force profiles between two STABr coated surfaces recorded using high-speed video recording of the interference fringes to capture the details of the attractive part of the interaction. Two approach traces are shown, one a 'direct approach' (black line) and one a 'shear approach' (grey line) as described in the text. The additional noise on the shear approach is brought about by the shearing motion. The approach velocities are 4.7 nm s$^{-1}$ (direct approach) and 4.5 nm s$^{-1}$ (shear approach). They were measured 10 min apart, after approximately 8 h of immersion in water. Similar measurements have been carried out at a range of shear amplitudes up to 120 nm peak to peak, and frequencies up to 3 Hz, and no systematic effect of the shearing on the normal interaction force is detected in any of these cases. The inset shows the attractive region on an expanded scale, and also shows (dashed line) the van der Waals interaction expected between the surfaces at each separation using a Hamaker constant of $2 \times 10^{-20}$ J.

Therefore, if there is indeed a correlation occurring between any heterogeneities on the surfaces as a result of lateral movement within the surface, causing the attractive interaction, it is not perturbed within the range of amplitudes and velocities applied to the surface in our study.

### 4.6 XPS investigation of the ratio of STAI molecules to mica surface sites

Using XPS we have measured the ratio of surfactant molecules to mica surface sites. Atomic concentrations are presented in Table 1, normalised to the Si 2p signal. In order to determine the amount of negatively charged sites on the mica, which can bind electrostatically to the positively charged surfactant, we compared the K 2p signal for freshly cleaved mica to the K 2p signal for mica washed in H$^+$ solution, as first suggested by Herder[36] and used later to investigate adsorption of trimethylammonium surfactants on mica.[37] This comparison provides a reference for the removal of all surface K$^+$ ions by the acidic solution, exchanging them for H$^+$. The residual K 2p signal after acid washing is attributed to sub-surface K$^+$ ions (a full quantitative treatment will be given elsewhere). Thus the difference between the two K 2p signal intensities can be used as the normalised quantity of surface sites

**Table 1** XPS data for mica and STAI-coated mica

| K$^+$ mica K 2p/Si 2p | H$^+$ exchanged mica K 2p/Si 2p | mica sites = Δ (K 2p/Si 2p) | STAI mica N 1s/Si 2p | Ratio (N 1s)/ (mica sites) |
|---|---|---|---|---|
| 0.3537 ± 0.0002 | 0.2757 ± 0.0001 | 0.0780 ± 0.0002 | 0.12 ± 0.02 | 1.53 ± 0.05 |

on the mica. Indeed, measurements at different angles confirmed that all the potassium ions located on the exposed surface plane were removed by the acid washing.

Next, we use the N 1s signal as a measure of the STAI coating. The mica surface prepared in the same way as for the SFB and CA experiments. A single nitrogen atom is present in each STAI molecule (and none is present in the mica), hence it can be directly compared to the number of surface sites (explained above). By comparing the normalised N 1s peak to the measured number mica sites we calculate a 53% excess of surfactant molecules compared to the mica surface sites.

It should be mentioned that the halogen counter-ion could in principle be used for this analysis. However, it turned out that it was unstable under the XPS measurement, namely under vacuum and X-ray plus electron irradiation. We noticed that in successive scans the halide signals decreased rapidly, reflecting their release from the surface. This process is highly assisted by the continuous supply of electrons from the neutralizer (the flood gun), which tends to break the halogen bonds and allow the creation of diatomic molecules (*e.g.* $I_2$) that can easily desorb from the surface. The N 1s peak was stable during successive scans of the surface and, therefore, more reliable for the above estimation. Angle resolved XPS measurements confirmed the above results.

If, when the layer is immersed in water, every STAI molecule were ionised then the monolayer would hold a surface charge density of 1.06 e/nm$^2$. Comparing this to the fit to the electrostatic repulsion in Fig. 4A, we see that the actual surface charge density (derived from the effective potential *via* the Grahame equation[14]) is 0.013 e/nm$^2$; thus less than 2% of the 'excess' STAI molecules are ionised to STA$^+$ in water.

### 4.7 AFM imaging of STAI

A self-assembled monolayer of STAI on mica was imaged under water using AFM, and is shown in Fig. 1. This image, recorded 1 h after immersion in the water, shows a layer which is very smooth (r.m.s. roughness 0.35 nm) with no features. This was generally observed for STAI in pure water. We have also imaged STABr in pure water using AFM, however we do not present an image here because there was a large variation in the results and no single image could be said to be representative.

## 5. Discussion

It is clear from measurements of the force between STAX surfaces (Fig. 4), surface tension (Fig. 5) and CA measurements (Fig. 1–3) that the STAI layer is quite different to the STABr and STACl layers. This difference is both in the initial states of the layer and in the stability over time in water and salt solution.

The advancing contact angles, $\theta_A$, on the STAX layers (Fig. 1) suggest that each of the surfaces are initially 'hydrophobic' according to the basic requirement that $\theta_A > 90°$, indicating that in air the monolayers are arranged with cationic headgroup towards the mica surface and hydrocarbon tails exposed to the air. However the contact angle hystereses ($\theta_A-\theta_R$) are large, in particular for STABr and STACl where $\theta_R \sim 25°$. The hysteresis in the contact angle on monolayers may be due to a reversible overturning or rearrangement of molecules when immersed in water— driven by a reduction in interfacial energy—so that advancing angles are measuring a surface with less overturned molecules and receding angles are probing a surface with more overturned or rearranged molecules. This mechanism has already been suggested in 1938 by Langmuir.[2] The $\theta_A$ and $\theta_R$ results in Fig. 1 suggest that the STAI layer is more compact (giving a higher $\theta_A$) with less overturning and rearrangement (giving a smaller hysteresis) than for STABr or STACl. Based on the timescale of the $\theta_A$ and $\theta_R$ measurements we can say that the reversible rearrangement, which occurs to a larger extent for STABr and STACl, must occur faster than $\sim$1 s. This interpretation leads to the idea that, for surfaces which are able to rearrange or reorganise in water (perhaps including the majority of self-assembled surfactant layers), the advancing contact angle is not in itself a good measure of the hydrophobicity, or the interfacial energy of the hydrophobic surface in water, and will not be related in a simple way to the strength of their hydrophobic interactions.

The decreases in contact angle for STABr and STACl after incubation in water implies that a second irreversible change is taking place over a timescale of the order of hours. In the past such irreversible rearrangement in water has been seen to lead, for $C_{16}$ chain surfactant, to patches of bilayer surrounded by areas of bare mica and the gradual desorption of the layer from the surface.[10,34] The advancing contact angles now reported for STABr and STACl reach a steady state value of $\sim 80°$, suggesting that the irreversible change in the layer (perhaps involving some overturning of molecules and some loss of monomers to the bulk solution) occurs to a far lesser extent than for $C_{16}$TABr; the layers remain much more stable after long periods in water than for $C_{16}$TABr.

The CA measurements on STAI layers in water and in salt solution are stable over time, while on the STABr and STACl layers they show a sharp decrease after a short period of time in salt solution. The differences in the initial state and in the stability are also evident from the surface tension values (Fig. 5) and from the surface forces (Fig. 4). The large and long-ranged repulsion measured between STAI surfaces across water—an electrostatic repulsion due to a positive surface potential of 70 mV—is completely absent for STABr and STACl which are uncharged or almost uncharged. These differences suggest that the STAI surface is more compact and well organized, maximizing the chain–chain interaction, relative to STABr and STACl. The strong affinity of the iodide to the trimethylammonium headgroup increases the concentration of the counter ions between the charged head groups in the adsorbed layer, screening their mutual repulsion, leading to a tightly packed layer which is then further stabilised by hydrocarbon tail–tail interactions, leading to a more hydrophobic and stable layer. The excess charge on the STAI layer might, at first thought, seem unlikely since $I^-$ counterions are less favourably solvated than $Br^-$ or $Cl^-$ and so perhaps fewer STAI molecules should be dissociated. However this effect is outweighed by the much greater density of the STAI layer compared to STABr and STACl. Indeed, the XPS analysis for STAI reveals this high density of surfactant molecules adsorbed to the mica surface: about 1.5 molecules per negatively-charged site on the mica surface. From the measured surface potential we can estimate that less then 2% of these 'additional' molecules are ionised to $STA^+$, the rest are neutralized by the $I^-$ counter ion. The STABr and STACl layers, on the other hand, are less dense (due to the lower fraction of screening counterions in the layer) which leads to a gradual irreversible desorption of ionised surfactant molecules until the layer is charge neutral. The gradual desorption of charged surfactant molecules is seen in the slight decrease in $\theta_A$ with time in water (Fig. 1), the decrease in interfacial energy measured in the SFB with time in water (Fig. 5), and the decrease to zero of any repulsive electrostatic interaction forces (Fig. 4).

All of the STAX layers show attractive interactions across water which are longer ranged than expected from van der Waals interactions alone. In light of the contact angle hysteresis observed, it is tempting to suggest that these attractions may be due to rearrangements of the layer into a laterally heterogeneous structure and subsequent correlation attraction between the domains.[6,8–10,12,13,24] The asymmetric measurement between STABr and bare mica also demonstrates this attractive interaction, despite the STABr layer being net neutral in charge, in agreement with previous measurements of asymmetric interactions where domains were suggested to be the cause.[8] We have described an experiment to investigate whether correlation occurs on the surfaces as they approach (Fig. 7)—by shearing the surfaces during this approach to suppress such correlations. As explained earlier, within the range of parameters of the preliminary measurements no systematic effect could be observed.

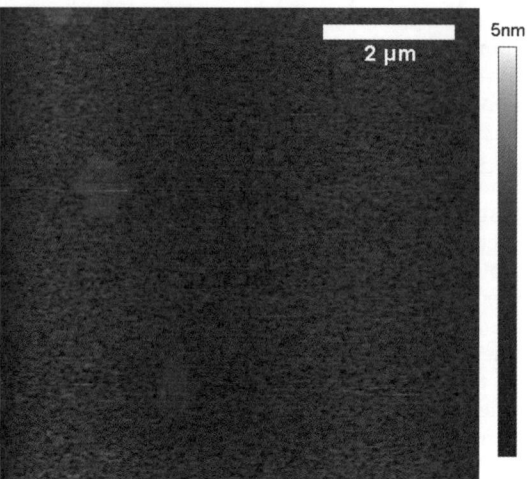

**Fig. 8** AFM image recorded under water of the STAI layer. The image is recorded after 1 h in water. The layer is smooth with no features other than some small aggregates, with an rms roughness over the whole image of 0.35 nm.

The uniform and smooth STAI layer revealed by AFM (Fig. 8) is in line with the contact angle and adhesion measurements which each indicate that the STAI layer is stable in water, and the XPS result that the STAI layer is densely packed. The fact that STAI is relatively insensitive to scanning parameters, repeatedly showing a smooth layer, whereas the STABr layer was very sensitive to scanning conditions may be related to the fact that STABr is found to be less stable than STAI. In the past, imaging the growing of STABr layer from surfactant solution below the c.m.c, has revealed solid-like domains surrounded by a more liquid like monolayer.[38]

Through our study of the series STAI, STABr, and STACl we find that STAI is dramatically different to STABr and STACl, which behave more similarly to one another. This is perhaps unsurprising since $I^-$ is less favourably hydrated, as well as having a particular affinity for the trimethylammonium headgroup.[22] The affinity of $I^-$ for the $STA^+$ ion leads to a high degree of screening between the $STA^+$ headgroups, a closer packing of molecules in the monolayer (an effect which is well known for micelles[21] but less considered for monolayers), which is also favoured by the tight packing of hydrocarbon tails in the dense layer. This density leads to both kinetic and thermodynamic barriers for overturning or desorption of molecules from the layer, with the result that the STAI layer is stable in water (over timescales of at least 24 h, as also shown in the past[39]). The dense layer is so favoured that it can support a small fraction (<2%) of ionised monomers, which leads to the excess surface charge and osmotic repulsion between the layers. The importance of this balance of driving forces (chain–chain, headgroup–surface, counterion hydration) is clear if we compare the results here for $C_{18}$ chains to previous results for $C_{16}$ hydrocarbon chains:[10,37] an opposite trend in stability with halide counterion is observed for $C_{16}$ compared to $C_{18}$ showing that for the $C_{16}$ chains the headgroup interactions with the mica surface—stronger for the layers with a greater degree of ionisation—is the overriding factor whereas for $C_{18}$ chains the dense packing is the overriding factor which is favoured by the $I^-$ ions which can screen headgroup repulsions and promote the dense packing.

## 6. Conclusions

We have studied the surface interactions in water, the wetting properties, and layer structure of a series of cationic surfactants self-assembled on mica substrates. STAI monolayers on mica are densely packed with 1.5 molecules per mica site, while STABr and STACl seem less dense and as a result less hydrophobic and less stable in water. When the STAI layer is immersed in water, the majority of 'excess' molecules in the layer are present as ion pairs (with the I- counterion bound to the headgroup); only a small fraction are present as STA$^+$. This small fraction is stable in the layer (they do not desorb over time), and gives rise to the net positive surface charge of approximately +70 mV. The dense chain packing for STAI—caused by I$^-$ screening of the headgroup repulsions and favoured by the long $C_{18}$ chains—leads to stability over long times in water, a large advancing contact angle (compared to other self-assembled layers) of 102°; and high STAI-water interfacial tension.

STABr and STACl monolayers on mica are less densely packed, leading to slightly lower advancing contact angles. These layers are uncharged (or almost uncharged) since there is no large excess of molecules compared to mica sites. This reduced density leads to a lower stability of the layer in water. As a result the layers, when exposed to water, can (a) rapidly and reversibly reorganise in order to lower the interfacial energy, and (b) slowly and irreversibly lose molecules to the bulk. This leads to the very low receding contact angles (the water droplet is moving over a hydrophilic surface as it recedes) and to the gradual decrease in contact angles and interfacial tension over time in water. The changes with time in these layers leads to the interactions between these layers being attractive and longer ranged than pure van der Waals interactions, which could possibly be due to correlations between surface domains. However an experiment designed to detect these correlations did not show any effect, from which we conclude that if correlations occur then they must be faster than the shearing velocity in the test experiment.

## 7. Acknowledgements

We would like to thank Bob Thomas and Sam Safran for useful discussions, Haggai Cohen and Tatyana Bendikov for performing the XPS measurements, and Tamar Yelin for contributing to the contact angle measurements. SP gratefully acknowledges Merton College, Oxford and the US Office for Naval Research (grant N00014-10-1-0096) for financial support. JK and SP thank the Weizmann-UK Joint Research Programme for financial support.

## 8. References

1 I. Langmuir, *Trans. Faraday Soc.*, 1920, **15**, 62.
2 I. Langmuir, *Science*, 1938, **87**(2266), 493.
3 J. Israelachvili and R. Pashley, *Nature*, 1982, **300**(5890), 341.
4 R. Pashley, P. M. McGuiggan, B. Ninham and D. F. Evans, *Science*, 1985, **229**, 1088.
5 P. M. Claesson, C. Blom, P. Herder and B. Ninham, *J. Colloid Interface Sci.*, 1986, **114**(1), 234.
6 Y.-H. Tsao, D. F. Evans and H. Wennerström, *Science*, 1993, **262**, 547.
7 P. Attard, *J. Phys. Chem.*, 1989, **93**(17), 6441.
8 Y.-H. Tsao, D. Fennell Evans and H. Wennerström, *Langmuir*, 1993, **9**, 779.
9 S. J. Miklavic, D. Chan, L. R. White and T. W. Healy, *J. Phys. Chem.*, 1994, **98**, 9022.
10 S. Perkin, N. Kampf and J. Klein, *J. Phys. Chem. B*, 2005, **109**(9), 3832.
11 J. H. Zhang, R. H. Yoon, M. Mao and W. A. Ducker, *Langmuir*, 2005, **21**(13), 5831.
12 E. E. Meyer, Q. Lin, T. Hassenkam, E. Oroudjev and J. N. Israelachvili, *Proc. Natl. Acad. Sci. U. S. A.*, 2005, **102**(19), 6839.
13 S. Perkin, N. Kampf and J. Klein, *Phys. Rev. Lett.*, 2006, **96**(3), 038301.
14 D. F. Evans and H. Wennerström, *The Colloidal Domain*, 2nd edn. 1999, New York: Wiley.
15 B. Li, Q. Lin, M. Fujii, F. Kazuhiro, T. Kato and T. Seimiya, *Thin Solid Films*, 1998, **312**, 20.
16 J. M. Mellott, W. A. Hayes and D. K. Schwartz, *Langmuir*, 2004, **20**, 2341.
17 Y. L. Chen, S. Chen, C. Frank and J. Israelachvili, *J. Colloid Interface Sci.*, 1992, **153**(1), 244.
18 W. Ducker and E. J. Wanless, *Langmuir*, 1999, **15**, 160.
19 M. Fujii, B. Li, K. Fukada, T. Kato and T. Seimiya, *Langmuir*, 2001, **17**(4), 1138.

20 L. Sepulveda and J. Cortes, *J. Phys. Chem.*, 1985, **89**(24), 5322.
21 N. Jiang, P. Li, Y. Wang, J. Wang, H. Yan and R. K. Thomas, *J. Colloid Interface Sci.*, 2005, **286**(2), 755.
22 L. Kellaway and G. G. Warr, *J. Colloid Interface Sci.*, 1997, **193**(2), 312.
23 M. M. Knock and C. D. Bain, *Langmuir*, 2000, **16**, 2857.
24 H. K. Christenson and P. M. Claesson, *Adv. Colloid Interface Sci.*, 2001, **91**(3), 391.
25 R. Podgornik, *J. Chem. Phys.*, 1989, **91**(9), 5840.
26 R. Podgornik and V. A. Parsegian, *Chem. Phys.*, 1991, **154**, 477.
27 S. Perkin, L. Chai, N. Kampf, U. Raviv, W. Briscoe, I. Dunlop, S. Titmuss, M. Seo, E. Kumacheva and J. Klein, *Langmuir*, 2006, **22**(14), 6142.
28 M. Kodama, K. Tsujii and S. Seki, *J. Phys. Chem.*, 1990, **94**(2), 815.
29 J. Klein and E. Kumacheva, *J. Chem. Phys.*, 1998, **108**(16), 6996.
30 O. Vinogradova, *Langmuir*, 1995, **11**, 2213.
31 D. Chan, R. Pashley and L. R. White, *J. Colloid Interface Sci.*, 1980, **77**(1), 283.
32 K. L. Johnson, K. Kendall and A. D. Roberts, *Proc. R. Soc. London, Ser. A*, 1971, **324**, 301.
33 N. Kampf, D. Ben-Yaakov, D. Andelman, S. A. Safran and J. Klein, *Phys. Rev. Lett.*, 2009, **103**(11), 118304.
34 G. Silbert, N. Kampf, S. Perkin, and J. Klein, (in preparation), 2010.
35 H. K. Christenson and V. V. Yaminsky, *Colloids Surf., A*, 1997, **129–130**, 67.
36 P. C. Herder, P. M. Claesson and C. E. Herder, *J. Colloid Interface Sci.*, 1987, **119**(1), 155.
37 B. Li, M. Fujii, K. Fukada, T. Kato and T. Seimiya, *J. Colloid Interface Sci.*, 1999, **209**, 25.
38 W. A. Hayes and D. K. Schwartz, *Langmuir*, 1998, **14**(20), 5913.
39 N. Kampf, J.-F. Gohy and J. Klein, *J. Polym. Sci., Part B: Polym. Phys.*, 2005, **43**, 193.

# Hydrophobic forces in the wetting films of water formed on xanthate-coated gold surfaces

Lei Pan and Roe-Hoan Yoon*

*Received 22nd December 2009, Accepted 25th January 2010*
DOI: 10.1039/b926937a

The kinetics of thinning of water films on hydrophobic gold substrates has been studied using the thin film pressure balance (TFPB) technique. The changes in the thickness of the wetting films have been monitored by recording the profiles of the dimpled films as a function of time using a high-speed video camera. It was found that the kinetics, measured at the barrier rim of a wetting film formed on a hydrophilic silica surface, could be predicted using the Reynolds lubrication approximation with the no-slip boundary condition. However, the wetting films formed on hydrophobized gold substrates thinned much faster, and the kinetics increased with increasing hydrophobicity. The data obtained with gold surfaces of different hydrophobicities have been fitted to the Reynolds approximation to determine the hydrophobic force constants ($K_{132}$) of a power law. $K_{132}$ increased with increasing contact angle and decreased with electrolyte (NaCl) concentration. It was also found that the $K_{132}$ values can be predicted from the hydrophobic force constants ($K_{131}$) for the interaction between hydrophobic surfaces and the same ($K_{232}$) for the foam films using the geometric mean combining rule that is frequently used to predict asymmetric molecular forces from symmetric ones.

## 1. Introduction

Properties of the thin liquid films between particles, bubbles and drops control the behavior of their suspensions and interactions with each other. In flotation, air bubbles collide with particles and create wetting films between them. If the films are unstable, particles can rupture the films, attach themselves to the bubbles, and float. If the films are stable, no flotation would be possible. Thus, control of the stability of wetting films is of critical importance in flotation. The key parameter controlling the stability of wetting films is the hydrophobicity of particles. Flotation is a rate process, and its kinetics increases with particle hydrophobicity.[1,2] Various reagents are used to render selected minerals hydrophobic. For sulfide minerals and precious metals, short-chain alkyl xanthates and thionocarbamates are commonly used as hydrophobizing agents (collectors). For the flotation of non-metallic minerals such as silica and iron oxides, long-chain high HLB surfactants are used for hydrophobization. Air bubbles have been in use for minerals flotation since 1905 when the process was first patented,[3] and yet the basic mechanisms involved in the rupture of wetting films are not well understood.

When an air bubble is pressed against a hydrophilic plate such as mica and quartz in water,[4-6] the intervening liquid drains until a stable film is formed. The stability of the wetting film is due to the disjoining (or 'wedging-apart') pressure, which was considered to arise from double-layer force, van der Waals-dispersion force, and

*Center for Advanced Separation Technologies, Virginia Tech, Blacksburg, Virginia, USA.*
*E-mail: ryoon@vt.edu*

structural force.[7] The first two were classical colloidal forces, while the third was attributed to the hydrogen bonding between the solid and water molecules in the film. At film thicknesses above 20 nm, double-layer force dominates, while at thicknesses below approximately 10–15 nm dispersion force also contributes to stabilizing the wetting films.[8,9] The film thickness decreases with electrolyte concentration and the valence of electrolyte.[10] It has also been reported that the wetting films on quartz rupture when the charge of the substrate was reversed by $Al^{3+}$ ions[11] or by a cationic surfactant.[12] Thus, the wetting films on hydrophilic surfaces behave just like a typical colloidal film, for which the DLVO theory may be useful.

Blake and Kitchener[8] used the bubble-against-plate technique to study the wetting films formed on both hydrophilic and hydrophobic silica plates. The hydrophobic silica was prepared by coating the surface with trichloromethylsilane (TMCS). The thicknesses of the aqueous films formed on both substrates were approximately the same, which was attributed to the observation that the methylation did not significantly change the $\zeta$-potentials of quartz. When the film thickness was reduced by KCl addition, however, the film on the hydrophobic surface ruptured spontaneously at a thickness of 64 nm, which was attributed to the presence of hydrophobic force in the wetting film. Tchaliovska et al.[13] studied the wetting films on mica hydrophobized with dodecylammonium hydrochloride (DAH) and suggested that both hydrophobic force and attractive electrostatic force are important in determining the film lifetime and the rates of expansion of the meniscus perimeter. More recently, Mahnke et al.[14] modeled the rupture of the wetting films formed on methylated glass plates using a long-range hydrophobic force with a decay length of 13 nm, while Churaev[15] discussed the role of hydrophobic force in the rupture of wetting films.

The first direct measurement of hydrophobic force was reported by Israelachvili and Pashley.[16] The measurements were made using the surface force apparatus (SFA) in cetyltrimethylammonium bromide (CTAB) solutions. Many investigators[17] conducted follow-up experiments using SFA and atomic force microscopy (AFM) with surfaces coated with various hydrophobizing agents and reported much longer-ranged and stronger hydrophobic forces than reported by Israelachvili and Pashley. However, the origin of the hydrophobic force is not yet known, and many investigators suggested various possible mechanisms. These include electrostatic interactions between the charged domains of adsorbed surfactants,[18] cavitation,[19,20] nanobubbles,[21,22] and others. Of these, the possibility of nanobubbles causing the long-range attractions has received much attention in recent years. It has been shown, however, that long-range attractions were still observed in degassed solutions,[23,24] although the attraction becomes weaker. Further, recent thermodynamic studies showed that macroscopic hydrophobic interactions entail entropy decrease, contrary to the case of molecular-scale hydrophobic interactions.[25,26] This finding suggests that the long-range hydrophobic force is due to the structuring of the water molecules in the confined spaces between hydrophobic surfaces.

The possibility of nanobubbles playing a role in the rupture of wetting films has been explored by some investigators. Stöckelhuber et al.[27] suggested that the thin liquid films formed between the nanobubbles nucleating on a hydrophobic surface and the air/water interface of a wetting film act like foam films, in which attractive van der Waals forces can cause rupture by the capillary wave mechanism.[28,29] According to Stöckelhuber et al.,[9] there are no attractive forces in wetting films; therefore, its rupture cannot be explained by the capillary wave mechanism.

Platikanov[30] monitored the kinetics of thinning of the wetting films on hydrophilic glass plates. The author found that the films formed dimples initially and produced flat films at equilibrium as predicted by Frankel and Mysels.[31] The dimples disappeared, however, when film radii became small. The author showed that the kinetics measured in the presence of 0.1 M KCl can be described by the Reynolds lubrication approximation with the no-slip boundary condition for both the solid/water and air/water interfaces. Thus, the author concluded that the dynamic method of using the Reynolds approximation can be used to determine the disjoining pressure if liquid films are flat and plane-parallel. Wang and Yoon[32] also used the Reynolds approximation to determine the contributions from the hydrophobic force to the disjoining pressures in single foam films. They found that air bubbles are hydrophobic and that the hydrophobic force in foam films decreases with increasing surfactant and NaCl concentrations.

Schulze and his co-workers[9,11] measured the critical rupture thicknesses of the wetting films formed on hydrophobic surfaces and compared the results with the film thinning kinetics predicted using the Reynolds equation. Without recognizing the presence of hydrophobic force, the authors suggested that the kinetics of film thinning should follow the same theoretical curve, because the surface charge did not change after hydrophobization with hexamethyldisilazane (HMDS). It was found that the critical rupture thicknesses plotted *vs.* film lifetime were randomly distributed around the theoretical thinning curve, which led to their conclusion that films rupture due to the presence of gas nuclei formed on heterogeneous surfaces and the hole formation mechanism suggested by Sharma and Ruckenstein.[33] More recent work of Sharma[34] suggested, however, that the hole formation is due to hydrophobic attraction.

In the present work, we monitored the kinetics of thinning of wetting films using the thin film pressure balance (TFPB) technique, which was originally designed to study foam films.[28,35] Wetting films of water were formed on gold plates hydrophobized with potassium amyl xanthate (KAX). The kinetics was monitored by means of a high-speed video camera, which allowed accurate measurements of film thicknesses changing with time at any point of a dimpled film. The results were analyzed using the Reynolds approximation with the non-slip boundary condition to determine the disjoining pressures. It was found that the kinetics of film thinning increased with increasing hydrophobicity, which was attributed to the increase in hydrophobic force in wetting films. The magnitudes of the hydrophobic forces measured in the wetting films were compared with those measured in foam films and in the films between hydrophobic solid surfaces.

## 2. Theoretical approach

When an air bubble is pressed against a flat solid surface in a horizontal orientation, the air/water interface deforms to produce initially a plane-parallel wetting film. The change in curvature associated with the deformation creates a pressure difference between the liquid in the film and the bulk and causes the film to thin. As the thinning continues, the film becomes a convex lens (or "dimple") with inverted curvature. The torus-shaped water film surrounding a dimple is referred to as a "barrier rim". The dimpled film is no longer plane parallel, but many investigators[36,37] modeled the thinning process using the Reynolds lubrication approximation, which has been derived using the boundary conditions that the two interfaces are parallel to each other and that the liquid velocity at the two interfaces are zero, *i.e.*, the film thins under non-slip conditions. The radii of the barrier rings observed in the present work were larger than 0.08 mm, while we were monitoring the thinning rate at film thicknesses below approximately 300 nm. The large difference in the length scales should satisfy the first boundary condition. Maali *et al.*[38] showed that the non-slip boundary condition is appropriate for the water flow on the hydrophilic surface. Also, Lin and Slattery[39] showed that the air/water interfaces are immobile in the presence of a trace of surfactant. Platikanov[30] and Frankel *et al.*[31] showed that the Reynolds lubrication approximation can be used to model the thinning of wetting films at the barrier ring.

The Reynolds lubrication approximation is usually presented in the following form:[35,40]

$$\frac{dh}{dt} = -\frac{2h^3 \Delta P}{3\mu R_f^2} \qquad (1)$$

where $h$ is the film thickness, $t$ drainage time, $\mu$ the liquid viscosity, $R_f$ the film radius, and $\Delta P$ is the pressure difference causing a film to thin. In dimpled wetting films, the $\Delta P$ may be expressed as follows,[41]

$$\Delta P = \frac{2\gamma}{R} - \frac{\gamma}{r}\frac{\partial}{\partial r}\left(r\frac{\partial h}{\partial r}\right) - \Pi \qquad (2)$$

where $R$ is the radius of the film holder (or bubble), $\gamma$ the surface tension of the liquid, $r$ the radial distance from the center of the barrier ring, $h$ the local film thickness, and $\Pi$ is the disjoining pressure. The first term represents the contribution from the Laplace pressure, the second term from the hydrodynamic pressure due to changes in curvature along the radial distance, and the third term represents the contribution from the disjoining pressure created by the surface forces in the film.

Thus, the thinning of wetting films is controlled by hydrodynamic forces initially and by surface forces during the latter stages. In flat films, the contribution from the changes in curvature may not be important as compared to the Laplace pressure. In this case, eqn (2) is reduced to:

$$\Delta P = \frac{2\gamma}{R} - \Pi \qquad (3)$$

The DLVO theory recognizes two surface forces, *i.e.*, a double-layer force and a van der Waals dispersion force. The latter is repulsive in wetting films,[42] while the former can also be repulsive in alkaline pH where both the solid/water and air/water interfaces are usually charged negatively. Despite the absence of an attractive force, wetting films formed on hydrophobic surfaces rupture. We, therefore, use the extended DLVO theory,[1]

$$\Pi_t = \Pi_d + \Pi_e + \Pi_h \qquad (4)$$

which includes disjoining pressures due to the van der Waals-dispersion force ($\Pi_d$), double-layer force ($\Pi_e$), and hydrophobic force ($\Pi_h$).

The disjoining pressure due to the van der Waals force can be given by,

$$\Pi_d = -\frac{A_{132}}{6\pi h^3} \qquad (5)$$

where $A_{132}$ is the Hamaker constant for the interaction between a solid **1** and an air bubble **2** in water **3**. $A_{132}$ can be obtained using the geometric mean combining rule from the values of the Hamaker constants for the interactions between solids in water ($A_{131}$) and between bubbles in water ($A_{232}$). In the present system, dispersion forces are much shorter-ranged than hydrophobic forces; therefore, the retardation effect has not been considered.

The electrostatic component of the disjoining pressure can be calculated using the Hogg–Healey–Fuerstenau (HHF) approximation,[43]

$$\Pi_e = -\frac{\varepsilon \kappa^2}{2\sinh(\kappa h)}\left[(\psi_1^2 + \psi_2^2)\mathrm{cosech}(\kappa h) - 2\psi_1\psi_2\mathrm{coth}(\kappa h)\right] \qquad (6)$$

in which $\varepsilon$ is the dielectric permittivity of water, $\psi_1$ and $\psi_2$ are the double-layer potentials at the solid/water and air/water interfaces, respectively, and $\kappa$ is the inverse Debye length.

Hydrophobic forces measured in experiments are usually represented in single- or double-exponential functions. It has been shown that a power law of the following form can also be used,[44,45]

$$\Pi_h = -\frac{K_{132}}{6\pi h^3} \qquad (7)$$

where $K_{132}$ is a constant representing the magnitude of the hydrophobic force in a wetting film of thickness $h$. Eqn (7) is of the same form as eqn (5), which makes it easier to compare $K_{132}$ directly with $A_{132}$.

## 3. Experiment

### Materials

Polished fused quartz plates (Technical Glass Inc.) and gold-coated glass plates (CA134, EMF) were used as substrates for wetting films. Potassium amyl xanthate (KAX, >90%, TCI, America) was purified twice by dissolution in acetone (HPLC, Fisher Sci.) and recrystallization in diethyl ether (99.999%, Sigma-Alrich). Sodium chloride (99.999%, Sigma-Aldrich) was roasted at 600 °C for 6 h to remove organic impurities. All solutions were prepared using the Millipore water of >18.2 MΩ/cm, which was obtained using a Direct-Q3 water purification system.

### Procedure

Both the fused quartz and gold plates were cleaned by boiling in piranha solutions (7 : 3 by volume of $H_2SO_4$ : $H_2O_2$) for 30 min, followed by rinsing with the Millipore water and drying in a nitrogen gas stream. After the cleaning, the gold plates were hydrophobized by immersing them in freshly prepared KAX solutions. The hydrophobicity was controlled by varying the concentration and immersion time. The treated surfaces were washed with Millipore water and dried by blowing high-purity nitrogen gas on the surface. The quartz was used as substrate for wetting films without hydrophobization.

The kinetics of film thinning was measured using the TFPB technique, Fig. 1 showing the experimental set-up. A flat substrate was placed on top of a film holder (2.0 mm radius) immersed in water. After making sure that no air bubbles were adhering on the surface, the assembly was placed on an inverted microscopic stage (Olympus IX51) to monitor the changes in film thickness with time. A halogen lamp (100 W, Osram) was used as a light source with a band-pass filter (NT46-053, Edmund Optics) to produce a monochromatic green light source ($\lambda = 526$ nm).

Initially, the thickness of a film was reduced by pulling the water out of the film holder by means of a piston. Once interference patterns (Newton rings) began to appear in the microscopic field of view, the film was allowed to thin spontaneously

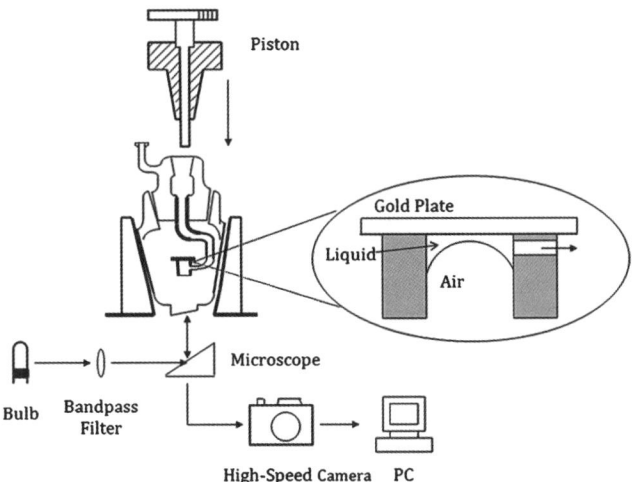

**Fig. 1** Schematics of the TFBC apparatus used for the study of wetting films.

while recording the images by means of a high-speed CCD camera (Fastcam 512PCI, Photron) at a speed of 60 frames per second. The camera was capable of taking the images at much higher speeds, but 60 FPS was found to be adequate. The interference patterns recorded were used to obtain timed profiles of a wetting film across the entire film holder, which in turn were used to monitor the changes in film thicknesses ($h$) as a function of time at any point of the film. The film thicknesses were calculated as described by Nedyalkov et al.[46]

The surface tensions of the solutions used in the present work were measured using the Wilhelmy plate method. Advancing and receding contact angles of the hydrophobized gold plates were measured using the sessile drop technique by means of a goniometer (Ramé-Hart, Inc.).

The $\zeta$-potentials of spherical gold power (1.5–3.0 μm, Alfa Aesar) were measured using the Zetasizer Nano. The gold power (99.96%) was hydrophobized in the same manner as the gold plate used for the film thinning kinetics measurements. It was assumed that the $\zeta$-potentials of the gold plate and powder were the same.

## 4. Result

Fig. 2 compares the timed profiles of the wetting films formed on gold substrates of two different hydrophobicities. Fig. 2a shows the profiles of the film formed on a freshly-cleaned gold plate that had not been hydrophobized. Its equilibrium contact angle ($\theta_e$) was 42° with 60° advancing ($\theta_a$) and 17° receding ($\theta_r$) angles. Initially, the film showed dimpled profiles. As the drainage continued, the film became flat and reached an equilibrium thickness ($h_e$) of 80 nm in 20 s, when the capillary pressure was equal to the disjoining pressure. Fig. 2b shows the timed profiles of the wetting film formed on a gold plate hydrophobized in a 5 × 10$^{-6}$ M KAX solution for 60 min to obtain $\theta_r = 79°$. The drainage rate became substantially faster than observed with the untreated gold plate; the film thickness decreased from 300 to 80 nm in 1.26 s and ruptured. The thickness at which the film fails catastrophically is referred to as the critical rupture thickness ($h_{cr}$). Note that the dimpling effect was more significant with the hydrophobic gold, indicating that the wetting film formed on the hydrophobic surface thins faster at the barrier rim than at the center. Note also that the curvature of the air/water interface on either side of the barrier rim was about the same, which would allow one to ignore the hydrodynamic pressure differential across the barrier rim.

Fig. 3 shows the changes in thickness of a wetting film formed on a hydrophilic fused quartz plate at its center and at the barrier rim. The measurements were conducted with a 0.1 M NaCl solution so that the $\Pi_e$ term of eqn (4) can be ignored.

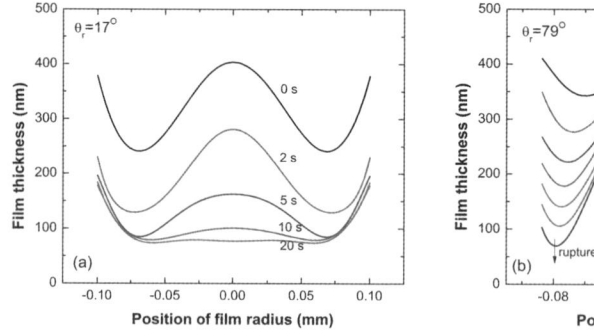

**Fig. 2** Comparison of the timed-profiles of the wetting films formed on a) an untreated gold plate with $\theta_r = 17°$ and b) a gold plate with $\theta_r = 79°$ after treatment with a 5 × 10$^{-6}$ M KAX solution for 1 h.

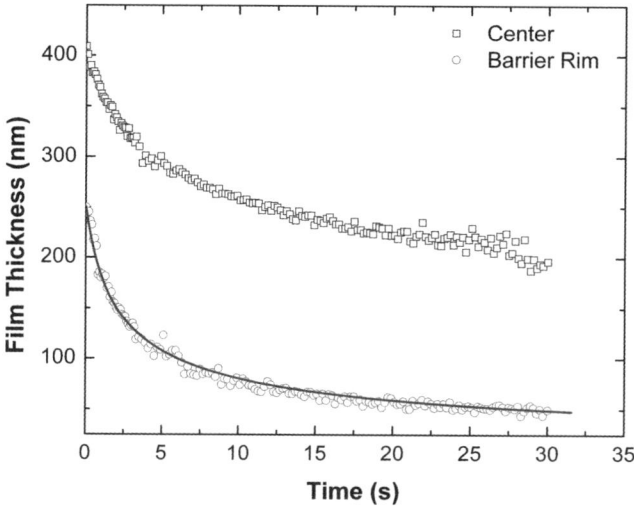

**Fig. 3** Changes in the thickness of a wetting film formed on a hydrophilic fused silica plate in a $10^{-1}$ M NaCl solution. The solid line represents a fit to eqn (1) with $R_f$ 0.084 mm, $A_{132} = -1.13 \times 10^{-20}$ J, and $\gamma = 72.4$ mN m$^{-1}$.

Since the substrate was hydrophilic, $\Pi_h$ was also ignored. Eqn (3) can then be reduced to,

$$\Delta P = \frac{2\gamma}{R} - \Pi_d = \frac{2\gamma}{R} + \frac{A_{132}}{6\pi h^3} \qquad (8)$$

By using the values of $\gamma = 72.4$ mN m$^{-1}$ at 0.1 M NaCl and $A_{132} = -1.13 \times 10^{-20}$ J,[47] the kinetics curve obtained at the barrier rim was fitted to eqn (1). The fit was excellent indicating that one can safely use the Reynolds approximation to describe the kinetics of wetting films. The results showed that the non-slip condition applies to both the surfactant-free quartz–water interface and the free air–water interface. The kinetics curve obtained at the center of the dimple thins slower than at the rim. The results presented in Fig. 3 are similar to the work of Platikanov.[30]

Fig. 4 compares the kinetics of film thinning on gold substrates with and without hydrophobization. After one hour of immersion time in a $5 \times 10^{-6}$ M KAX at open circuit and natural pH (= 7.3), the gold plate was rendered hydrophobic with $\theta_r = 79°$. The untreated gold plate showed only a slight hydrophobicity with $\theta_r = 17°$ possibly due to contaminants. When a gold surface was freshly cleaned in a piranha solution, the contact angle was zero. But the angle increased considerably after a short exposure to the atmosphere, during which time contaminants could adsorb on the surface from the air owing to the large Hamaker constant of gold. Despite the apparent but low-level hydrophobicity, its kinetics was much slower than on the treated surface, as shown in Fig. 4, and the film did not rupture. On the contrary, the wetting film formed on the treated gold surface ruptured at $h_{cr} = 80$ nm and thinned substantially faster, which may be attributed to the presence of a hydrophobic force in the wetting film.

In general, xanthate-coated mineral surfaces are negatively charged in water, and the $\zeta$-potentials do not change substantially at low concentrations.[48] Also, xanthate adsorption should not change the Hamaker constant of the gold plate ($A_{131}$) significantly. For the experiment conducted with the hydrophobic gold, the substrate was hydrophobized prior to forming a wetting film with pure water. Therefore, the chemistry of the air/water interface should be the same as that of the experiment

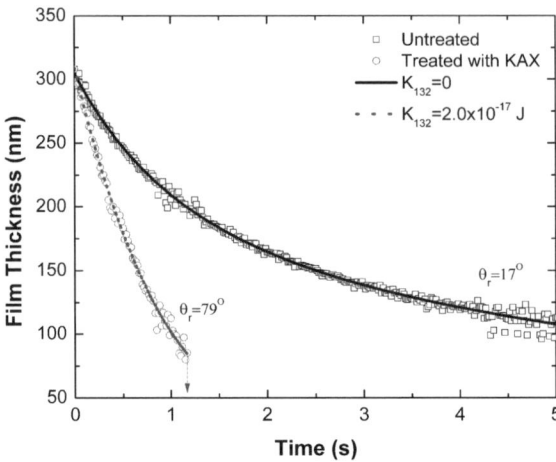

**Fig. 4** Changes in the film thicknesses measured at the barrier rims of the dimpled-wetting films formed on gold-coated silica plates with $\theta_r$ of 17° and 79°. The latter value was obtained after immersing a plate in a $5 \times 10^{-6}$ M KAX solution for one hour, and the former was for an untreated plate. The solid line represents the Reynolds approximation with $K_{132} = 0$, while the dashed line represents the same with $K_{132} = 2.0 \times 10^{-17}$ J, $R_f$ 0.075 mm, $\gamma = 72.5$ mN m$^{-1}$, $\Psi_1 = -45$ mV, $\Psi_2 = -80$ mV, $\kappa^{-1} = 243$ nm, and $A_{132} = -2.02 \times 10^{-20}$ J.

conducted with untreated gold, that is, the ζ-potential and the Hamaker constant of the air bubble ($A_{232}$) should essentially be the same as those of the untreated gold surface. It is, therefore, suggested that the fast kinetics of the film thinning on the hydrophobic surface was due to the hydrophobic force. We estimated the magnitude of the hydrophobic force by fitting the kinetics curves to the Reynolds equation (eqn (1)). $\Delta P$ was calculated using eqn (3)–(7). The values of the various parameters used for the fit are given in Fig. 4. In calculating the contribution from the hydrophobic force ($\Pi_h$), it was necessary to use the values of $K_{132} = 2.0 \times 10^{-17}$ J for the hydrophobized gold substrate and $K_{132} = 0$ for the untreated substrate. Note here that $K_{132}$ was positive and that its magnitude was much larger than that of the Hamaker constant ($A_{132} = -2.02 \times 10^{-20}$ J), which was negative.

Fig. 5 shows the results of the kinetics studies conducted with gold-coated silica plates by varying the immersion time in a $10^{-5}$ M KAX solution. The film thicknesses were measured at the barrier rims of the timed film profiles. It was found that the film thinning kinetics was the fastest after 10 min of immersion time and became slower at longer contact times. The results obtained after the 10-min contact time was fitted to the Reynolds equation with $K_{132} = 2.0 \times 10^{-17}$ J. After a 120 min immersion time, $K_{132}$ decreased to $7.0 \times 10^{-18}$ J and the drainage rate decreased, which may be attributed to a multilayer formation. It is well known that xanthate adsorption on sulfide minerals and precious metals results in the formation of a multilayer.[49–51] Xanthate adsorption results in the formation of chemisorbed xanthates in the first monolayer, followed by the adsorption of metal xanthates and/or dixanthogen on the top at higher electrochemical potentials.[52,53] Evidences for the multilayer formation in the gold-KAX system after long contact times and higher concentrations have been presented in another communication.[54] Note that $\theta_r$ was 82° after a 10 min contact time, which decreased to 76° after a 120 min contact time. Albeit small, the decrease in contact angle may be a reflection of the fact that the species adsorbing in the multilayer expose the head groups (-OCSSAu) toward the aqueous phase. Although the head group should be less hydrophobic than the end group (CH$_3$) of the chemisorbed xanthate in the monolayer, it may be substantially less polar than those of the high HLB number surfactants, providing an

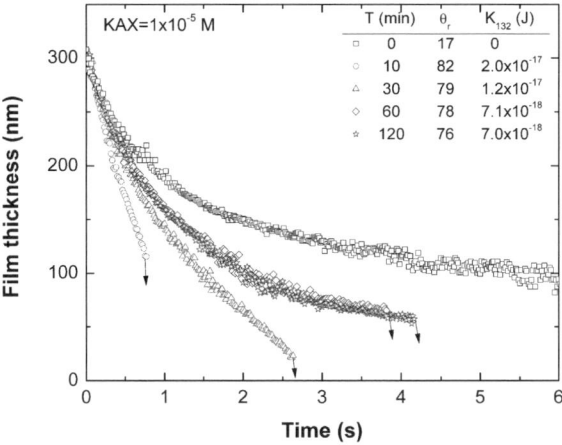

**Fig. 5** Effect of contact time between gold in a $10^{-5}$ M KAX solution on the kinetics of film thinning. The $K_{132}$ values were obtained by fitting the data to eqn (1) with $\gamma = 72.5$ mN m$^{-1}$, $\Psi_1 = -45$ mV, $\Psi_2 = -80$ mV, $\kappa^{-1} = 243$ nm, $A_{132} = -2.0 \times 10^{-20}$ J, and $R_f$ $0.075 \pm 0.004$ mm.

explanation for the relatively small decrease in $\theta_r$ observed in the present work. The decrease in $K_{132}$ with increasing contact time may thus be attributed to the decrease in the hydrophobicity of the gold substrate. This finding is consistent with the results of the AFM force measurements conducted between a xanthate-coated gold sphere and a gold plate.[54] It is also possible that the multilayer formation increases the roughness of the xanthate-coated gold plate, which should also contribute to the decrease in the drainage rate and hence the $K_{132}$ values estimated using the Reynolds approximation.

After the 10 min contact time, which was considered short enough to prevent the multilayer formation, a set of timed film profiles were obtained at different KAX concentrations ($10^{-6}$ to $10^{-4}$ M), and the changes in film thickness at the barrier rims were monitored and plotted in Fig. 6. The receding contact angle ($\theta_r$) increased from 43° at $10^{-6}$ M to 80° at $10^{-5}$ M KAX. As expected, the drainage rate increased with increasing KAX concentration and $\theta_r$, suggesting that the increased kinetics

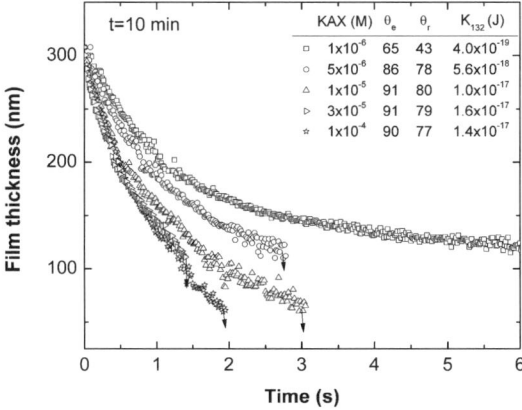

**Fig. 6** Effect of KAX concentration on the thinning of the wetting films formed on gold substrates. The contact time was 10 min, and the results were used to obtain the $K_{132}$ values using eqn (1), with $\gamma = 72.5$ mN m$^{-1}$, $\Psi_1 = -45$ mV, $\Psi_2 = -80$ mV, $\kappa^{-1} = 243$ nm, $A_{132} = -2.0 \times 10^{-20}$ J and $R_f = 0.091, 0.087, 0.084, 0.08$ and $0.076$ mm, respectively.

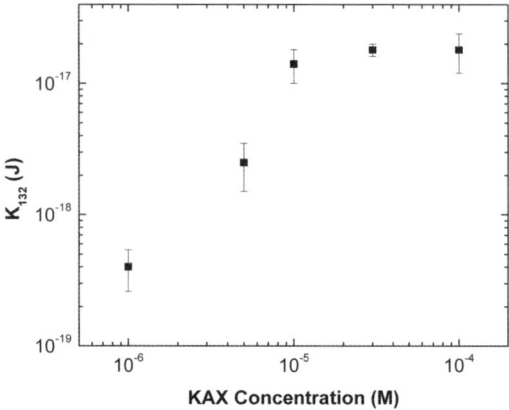

**Fig. 7** Changes in the hydrophobic force constant ($K_{132}$) with the concentration of KAX solutions, in which gold substrates were hydrophobized for 10 min.

was due to the increased hydrophobic force in the wetting films. Fig. 7 shows the $K_{132}$ values obtained by fitting the kinetics data to eqn (1) plotted *vs.* the KAX concentration. As shown, $K_{132}$ increased with increasing concentration and reached a maximum at approximately $3 \times 10^{-5}$ M. The data presented in Fig. 6 and 7 provides strong evidence that a hydrophobic force exists in the wetting films formed on the hydrophobic surfaces and serves as the major driving force for the film drainage and rupture.

Fig. 8 shows the effect of electrolyte (NaCl) on the kinetics of film thinning on hydrophobic surfaces. The experiments were conducted with gold plates treated in $5 \times 10^{-6}$ M KAX solutions for 60 min. Also shown for comparison is the result (dashed line) obtained with a wetting film formed with pure water on the surface of an untreated gold plate. In the absence of NaCl, the wetting film formed on a hydrophobized gold plate thinned substantially faster than that formed on an untreated gold plate, which can be attributed to the presence of a strong hydrophobic force in the former. The kinetics curve obtained with the hydrophobic

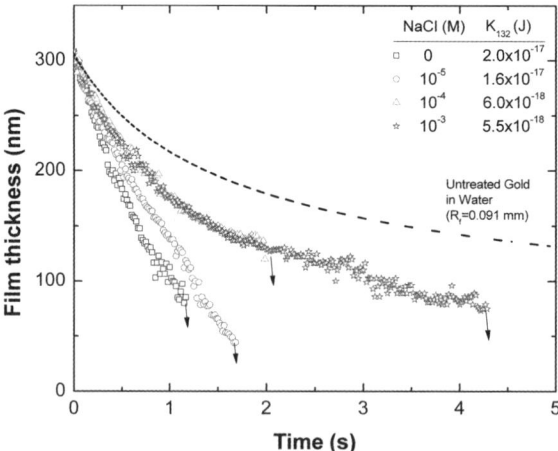

**Fig. 8** Effect of electrolyte (NaCl) on the kinetics of the wetting films formed on the surface of gold-coated glass plates hydrophobized in a $5 \times 10^{-6}$ M KAX solution for 60 min. The dashed curve represents the results obtained in pure water.

**Table 1** Model parameters used to fit the data in Fig. 8 to eqn (1)

| NaCl/M | $\gamma$/mN m$^{-1}$ | $R_f$/mm | $\Psi_1{}^a$/mV | $\Psi_2{}^b$/mV | $\kappa^{-1}$/nm | $K_{132}$/J |
|---|---|---|---|---|---|---|
| 0 | 72.3 | 0.075 | −40 | −80 | 241 | $2.0 \times 10^{-17}$ |
| $10^{-5}$ | 72.4 | 0.085 | −43 | −75 | 96.1 | $1.6 \times 10^{-17}$ |
| $10^{-4}$ | 72.3 | 0.091 | −47 | −50 | 30.4 | $6.0 \times 10^{-18}$ |
| $10^{-3}$ | 72.4 | 0.092 | −47 | −35 | 9.61 | $5.5 \times 10^{-18}$ |

$^a$ Present work. $^b$ Li and Somasundaran.[69,70]

gold in the absence of NaCl can be fitted to the Reynolds equation with $K_{132} = 2.0 \times 10^{-17}$ J. In the presence of $10^{-5}$ M NaCl, $K_{132}$ decreased to $1.6 \times 10^{-17}$ J, causing a decrease in the kinetics of film thinning. At $10^{-4}$ and $10^{-3}$ M NaCl, $K_{132}$ decreased further to $6.0 \times 10^{-18}$ and $5.5 \times 10^{-18}$ J, respectively, with a further decrease in the kinetics. Table 1 shows the various parameters used to fit the data presented in Fig. 8 to the Reynolds equation.

That the hydrophobic force in wetting films decreased in the presence of NaCl was consistent with the results from the AFM[55,56] and foam film[57] studies that the hydrophobic forces in the thin films confined between solid surfaces and between air bubbles decrease in the presence of electrolytes. It was suggested that electrolytes can break the hydrogen bonds between water molecules and, hence, cause a decrease in cohesive energy ($W_c$) and hydrophobic force.[57] The observation that the drainage rate of the wetting films formed on the hydrophobic surfaces decreased in the presence of NaCl appeared to be contrary to the DLVO theory, according to which the kinetics should actually increase due to double-layer compression. This apparent discrepancy simply indicates that the decrease in the attractive hydrophobic force was greater than the decrease in the repulsive electrostatic force due to double layer compression.

## 5. Discussion

We have shown that the wetting films formed on horizontal, planar surfaces begin to thin due to the capillary forces created by the changes in curvature. Its drainage rate can be predicted by the Reynolds lubrication approximation, with the capillary pressure serving as the driving force. As the film continues to thin, the solid/water and air/water interfaces interact with each other and create a disjoining pressure ($\Pi$), which also begins to affect the kinetics of film thinning.

Apart from the kinetics of film thinning, whether a film ruptures or not is determined by thermodynamics. When a wetting film ruptures, a new interface, *i.e.*, a solid–gas interface, is created at the expense of solid–liquid and solid–gas interfaces. Thus, the Gibbs free energy change ($\Delta G$) associated with the rupture can be given by the following relation:

$$\Delta G = \gamma_{12} - \gamma_{13} - \gamma_{23} < 0 \tag{9}$$

where $\gamma_{12}$, $\gamma_{13}$, and $\gamma_{23}$ represent the free energies at the interfaces between solid **1**, air **2**, and water **3** phases. Combining eqn (9) with the Young's equation

$$\cos\theta = \frac{\gamma_{12} - \gamma_{13}}{\gamma_{23}} \tag{10}$$

where $\theta$ is the water contact angle, one can find that wetting films rupture when

$$\Delta G = \gamma_{23}(\cos\theta - 1) < 0 \tag{11}$$

Eqn (11) suggests that wetting films can rupture when $\theta > 0$ at a critical rupture thickness ($h_{cr}$). Receding angles ($\theta_r$) may be relevant for the rupture of wetting films. If $\theta = 0$, a wetting film should stabilize at an equilibrium thickness ($h_e$). We have shown in the present work that the wetting films formed on gold substrates rupture when the surface is rendered hydrophobic by KAX. Earlier investigators showed that wetting films of water formed on quartz and mica, well-known hydrophilic substrates, rupture spontaneously at small film thicknesses. However, the lenses of water formed on the surface as a result of the rupture exhibited small contact angles in the range of 5–16°.[58,59]

From eqn (10) and (11), one can see that $\Delta G < 0$ when

$$\gamma_{12} - \gamma_{13} < \gamma_{23} \tag{12}$$

One can substitute the following relation into eqn (12),[60]

$$\gamma_{13} = \gamma_{12} + \gamma_{23} - 2\sqrt{\gamma_1^d \gamma_3^d} - 2\sqrt{\gamma_1^+ \gamma_3^-} - 2\sqrt{\gamma_1^- \gamma_3^+} \tag{13}$$

where $\gamma_1^d$ is the dispersion component of the solid surface tension, $\gamma_3^d$ is the same of liquid surface tension, $\gamma_1^+$ and $\gamma_1^-$ are the acidic and basic components of the solid surface tension, respectively, $\gamma_3^+$ and $\gamma_3^-$ are the acidic and basic components of the liquid surface tension, respectively, to obtain:

$$\left(2\sqrt{\gamma_1^d \gamma_3^d} + 2\sqrt{\gamma_1^+ \gamma_3^-} + 2\sqrt{\gamma_1^- \gamma_3^+}\right) < 2\gamma_{23} \tag{14}$$

Eqn (14) is equivalent to the following:

$$W_a < W_c \tag{15}$$

where $W_a$ is the work of adhesion of water on a solid and $W_c$ is the work of cohesion of water. In general, the dispersion component of $W_a$, i.e., $2\sqrt{\gamma_1^d \gamma_3^d}$, is smaller than $W_c$. It is, therefore, necessary to decrease the acid–base components of, i.e. $2\sqrt{\gamma_1^+ \gamma_3^-}$, and $2\sqrt{\gamma_1^- \gamma_3^+}$, by appropriate surface treatment. In the present work, we used KAX to satisfy eqn (15) and induce the first-order interfacial phase transition, a concept first discussed by Frumkin.[61]

Eqn (9) and (15) show that the spreading coefficient ($S$) becomes negative when $\theta > 0$, that is, a liquid film retreats and creates a finite solid/liquid interfacial area of contact. The larger the contact angle, the larger the area of contact between the bubble and the particle, and thereby help to minimize the probability of detachment of particles during flotation. (One should note here that advancing contact angles are relevant in detachment.) If a particle makes a point-to-point contact, it will be difficult to levitate coarse particles during flotation. On the other hand, ultrafine particles can be floated without film rupture, i.e., when $\theta = 0$. In dissolved air flotation (DAF), which is widely used for waste water treatment, fine particles are often floated without using hydrophobizing agents. Control of surface forces, e.g., double-layer forces by control of pH and coagulant addition, is sufficient.

Even if film rupture is thermodynamically favorable, the process can be kinetically hindered, e.g., by increasing $\Pi$, creating surface roughness to slow down drainage rate, increasing film elasticity, etc. We have considered that $\Pi$ consists of dispersion ($\Pi_d$), electrostatic ($\Pi_e$), and hydrophobic ($\Pi_h$) components, as shown in eqn (4). $\Pi_d$ is repulsive as $A_{132}$ is negative in wetting films, while $\Pi_e$ is also negative in the gold-xanthate system studied here and in many other systems. We found that the kinetics of film thinning increase with increasing xanthate concentration and contact angle ($\theta_r$), which suggests that a hydrophobic force is present in the wetting films formed on hydrophobic surfaces. Using the Reynolds lubrication approximation, we calculated the values of $K_{132}$ representing the magnitudes of $\Pi_h$ (and of hydrophobic forces), which have been found to increase with $\theta_r$. We found also that by recognizing the presence of hydrophobic forces in wetting films, it was not necessary to invoke the capillary wave models.[28,29] The wetting films ruptured spontaneously at $h_{cr}$. In the gold-KAX system, $\Pi_h$ was the only negative component of $\Pi$ and hence could bring the film thickness to $h_{cr}$ in a short time frame. Likewise, Manica et al.[41] found that when $\Pi_e$ is strongly negative, there was no need to invoke the capillary wave model to predict the thinning and rupture of the water films between mica and mercury.

That hydrophobic force is present in wetting films may be more readily acceptable if one can recognize that air bubbles in water are hydrophobic. In this case, the thinning and rupture of the wetting films formed on hydrophobic surfaces may be viewed as one of asymmetric hydrophobic interaction. van Oss et al.[62] suggested that the air side of the air–water interface is the most hydrophobic surface known and is about 30% more hydrophobic than octane and Teflon. The basis of this argument is that the tension at the air–water interface (72 mN m$^{-1}$) is substantially higher than those ($\sim$50 mN m$^{-1}$) at the hydrocarbon–water interfaces. The vibrational sum frequency (VSF) spectra of the water molecules straddling at the hydrophobic surface–water interfaces show sharp peaks at 3600–3700 cm$^{-1}$, which represent the characteristic non-hydrogen-bonded (free) OH stretch vibrations.[63,64] The high interfacial tensions at the hydrophobic surface/water interfaces are due to the presence of the free OH groups at these interfaces. Interestingly, the free OH peaks observed at the CCl$_4$–water and hexane–water interfaces are observed at 3669 $\pm$ 1 cm$^{-1}$, while the same for the vapor–water interface is observed at $\sim$3700 cm$^{-1}$. The red shift of the characteristic peak shows an attractive interaction between the free OH groups and the organic molecules at the interface. In fact, the binding energy for the CCl$_4$–H$_2$O dimer has been calculated to be $-1.4$ kcal mol$^{-1}$.[65] These reports are consistent with the fact that the dispersion components of $W_a$ at the hydrophobic liquid–water interfaces are $\sim$20 mJ m$^{-2}$, while it should be zero at the air–water interface.

We used the Reynolds lubrication approximation to estimate the $K_{132}$ values for hydrophobic disjoining pressure (eqn (7)). There are several questions to be raised in this approach. First, the approximation is useful for a film of fluid between nearly plane-parallel surfaces. Although we monitored the rate of film thinning at the barrier rim, which is a curved surface, the film thickness ($h$) was much smaller than the radius of curvature. Therefore, we have measured the thinning rates effectively between plane-parallel surfaces. Second, the roughness of the surface should affect the drainage rate and, hence, the $K_{132}$ values obtained using the Reynolds approximation. In the present work, we found that drainage rate decreased with increasing contact time in KAX solutions due to decreased hydrophobicity and the surface roughness created by multilayer formation.

Perhaps the most important question to be raised would be the question regarding the non-slip boundary condition employed in deriving the Reynolds approximation. There seems to be no doubt that it applies for the flows around hydrophilic surfaces.[38] For the flows around hydrophobic surfaces, however, some found that the non-slip condition may not be applicable. On the other hand, the correlation between slip lengths and contact angle was very poor.[66] As for the flow around air bubbles (or air–water interface), it was found that the liquid–gas interface becomes nearly immobile even with only a trace of surfactant present.[39] Although we have conducted experiments in the absence of surfactants, there is a possibility that traces of gold xanthates may be present at the air–water interface, making it possible to use the non-slip condition. If not, the $K_{132}$ values determined in the present work may have been overestimated to some extent and require appropriate corrections in future work. Nevertheless, it is highly unlikely that the correction would relinquish the possibility that a hydrophobic force is present in the wetting films formed on hydrophobic surfaces. It should be noted also that Horn and his co-workers found that the non-slip boundary condition applies for the surfactant-free mica–water–mercury systems.[41] Further, the drainage experiments conducted in the present work (see Fig. 3) and by Platikanov[30] with wetting films produced with surfactant-free air–water interfaces could be fitted to the Reynolds approximation with non-slip boundary conditions.

The geometric mean combining rule is used to predict the Hamaker constants for van der Waals interactions between unlike surfaces from those between like ones. It is based on the Berthelot relation derived originally for molecular interactions.[67] It has been reported previously that asymmetric hydrophobic force constants ($K_{132}$) can be predicted using the geometric mean combining rule as follows:[45]

$$K_{132} = \sqrt{K_{131} K_{232}} \qquad (16)$$

in which $K_{131}$ is the force constant for symmetric hydrophobic interaction between two surfaces of identical contact angle of $\theta_1$, and $K_{232}$ is the same for contact angle $\theta_2$.

In Fig. 9, the values of $K_{132}$ obtained using the Reynolds approximation have been plotted vs. the values of $K_{131}$ in logarithmic scales. According to eqn (16), the slope should be ½. From the intercept obtained numerically, we found that $K_{232} = 5.3 \times 10^{-17}$ J. This value is close to that estimated by extrapolating the $K_{232}$ values for the foam films stabilized at different concentrations of surfactants.[68] The $K_{131}$ values used in the plot shown in Fig. 9 include those by Wang and Yoon[54] and those measured specifically for the present work using the AFM force measurements conducted with KAX-coated gold surfaces in pure water. It is interesting that eqn (16) applies for the hydrophobic interactions between solid–liquid, gas–gas, and gas–solid interactions. This finding suggests that the hydrophobic force may be the molecular force representing the properties of the thin liquid films confined between hydrophobic surfaces, regardless of whether the interacting surfaces are solid, liquid, or gas. This finding is consistent with the results of the thermodynamic studies conducted by one of us showing that hydrophobic forces may originate from changes in the water structure as a thick film becomes a thin film.[26]

## 6. Conclusion

The results of the present investigation show that the kinetics of thinning for the wetting films of water formed on a hydrophilic silica surface can be fitted to the

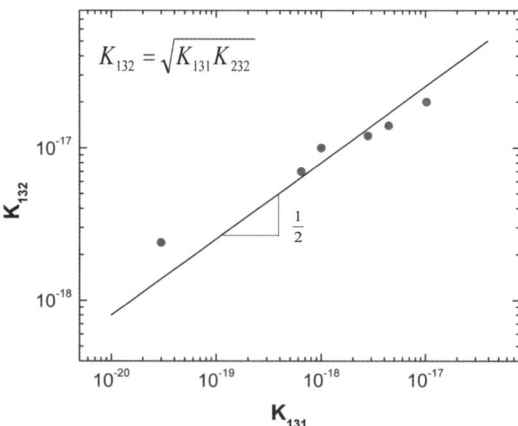

**Fig. 9** A plot of the asymmetric hydrophobic force constant ($K_{132}$) for wetting films vs. the square root of the symmetric hydrophobic force constant for thin films between hydrophobic solid surfaces. Under the condition that the slope is 0.5, one can determine the intercept of the plot numerically, which gives $K_{232} = 5.3 \times 10^{-17}$ J. This value is close to the value reported in ref. 68.

Reynolds lubrication approximation with non-slip boundary conditions. The same approach has also been used to study the kinetics of film thinning on the surface of gold substrates hydrophobized by KAX. The results show that the kinetics increase with increasing hydrophobicity. This finding suggests that hydrophobization of a substrate causes the disjoining pressure in the wetting films to decrease, which in turn can be attributed to the presence of a hydrophobic force in the wetting films. It has been found that the hydrophobic force constant ($K_{132}$) of the wetting film, as determined by fitting the kinetics data to the Reynolds approximation, increases with increasing receding contact angle of the substrate. It has also been found that $K_{132}$ decreases with increasing NaCl concentration and after an excessively long contact time between the gold substrate and KAX. The former can be attributed to the decrease in the cohesive energy of water ($W_c$) in the presence of the electrolyte, while the latter to the increase in surface roughness associated with a multilayer formation. Further, the values of $K_{132}$ can be predicted from the values of the hydrophobic force constants ($K_{131}$) for the interaction between solid surfaces of identical hydrophobicity and those ($K_{232}$) for soap films using the geometric mean combining rule.

## Acknowledgements

The authors would like to express sincere appreciation for the help from Ms. Zuoli Li for the AFM force measurements, permission from Dr Rick Davis to use the Zetasizer Nano for $\zeta$-potential measurements, and the financial support (DE-FC26-05NT42457) from the National Energy Technology Center, the U.S. Department of Energy, and FLSmidth Salt Lake City, Inc.

## References

1 R. H. Yoon and L. Q. Mao, *J. Colloid Interface Sci.*, 1996, **181**, 613–626.
2 L. Q. Mao and R. H. Yoon, *Int. J. Miner. Process.*, 1997, **51**, 171–181.
3 H. L. Sulman and H. E. Kirkpatrickpicard, US Patent 793,808, 1905.
4 B. V. Derjaguin and M. M. Kusakov, *Acta Physicochim. URSS*, 1939, **10**, 25–44.
5 H. J. Schulze, *Colloid Polym. Sci.*, 1976, **254**, 438–439.
6 B. V. Derjaguin and M. M. Kusakov, *Acta Physicochim. URSS*, 1939, **10**, 153–174.
7 B. V. Derjaguin and N. V. Churaev, *J. Colloid Interface Sci.*, 1974, **49**, 249–255.
8 T. D. Blake and J. A. Kitchener, *J. Chem. Soc., Faraday Trans. 1*, 1972, **68**, 1435–1442.
9 K. W. Stockelhuber, H. J. Schulze and A. Wenger, *Chem. Eng. Technol.*, 2001, **24**, 624–628.
10 A. D. Read and J. A. Kitchener, *J. Colloid Interface Sci.*, 1969, **30**, 391–398.
11 H. J. Schulze, K. W. Stckelhuber and A. Wenger, *Colloids Surf., A*, 2001, **192**, 61–72.
12 L. Alexandrova, R. J. Pugh, F. Tiberg and L. Grigorov, *Langmuir*, 1999, **15**, 7464–7471.
13 S. Tchaliovska, P. Herder, R. Pugh, P. Stenius and J. C. Eriksson, *Langmuir*, 1990, **6**, 1535–1543.
14 J. Mahnke, H. J. Schulze, K. W. Stckelhuber and B. Radoev, *Colloids Surf., A*, 1999, **157**, 1–9.
15 N. V. Churaev, *Adv. Colloid Interface Sci.*, 2005, **114–115**, 3–7.
16 J. Israelachvili and R. Pashley, *Nature*, 1982, **300**, 341–342.
17 H. K. Christenson and P. M. Claesson, *Adv. Colloid Interface Sci.*, 2001, **91**, 391–436.
18 E. E. Meyer, K. J. Rosenberg and J. Israelachvili, *Proc. Natl. Acad. Sci. U. S. A.*, 2006, **103**, 15739–15746.
19 H. K. Christenson and P. M. Claesson, *Science*, 1988, **239**, 390–392.
20 V. V. Yaminsky and B. W. Ninham, *Langmuir*, 1993, **9**, 3618–3624.
21 J. L. Parker, P. M. Claesson and P. Attard, *J. Phys. Chem.*, 1994, **98**, 8468–8480.
22 J. W. G. Tyrrell and P. Attard, *Langmuir*, 2002, **18**, 160–167.
23 E. E. Meyer, Q. Lin and J. N. Israelachvili, *Langmuir*, 2005, **21**, 256–259.
24 J. H. Zhang, R. H. Yoon, M. Mao and W. A. Ducker, *Langmuir*, 2005, **21**, 5831–5841.
25 J. Wang, Ph.D. Thesis, Virginia Tech, 2008.
26 R. H. Yoon, J. Wang and J. C. Eriksson, Science, submitted.
27 K. W. Stockelhuber, B. Radoev, A. Wenger and H. J. Schulze, *Langmuir*, 2004, **20**, 164–168.
28 A. Sheludko, *Adv. Colloid Interface Sci.*, 1967, **1**, 391–464.
29 A. Scheludko, *Proc. K. Ned Akad. Wet.*, 1962, **B65**, 76–87.

30 D. Platikanov, *J. Phys. Chem.*, 1964, **68**, 3619–3624.
31 S. P. Frankel and K. J. Myseis, *J. Phys. Chem.*, 1962, **66**, 190–191.
32 L. G. Wang and R. H. Yoon, *Colloids Surf., A*, 2005, **263**, 267–274.
33 A. Sharma and E. Ruckenstein, *J. Colloid Interface Sci.*, 1990, **137**, 433–445.
34 A. Sharma, *J. Colloid Interface Sci.*, 1998, **199**, 212–214.
35 A. Scheludko and D. Exerowa, *Kolloid-Z.*, 1959, **165**, 148.
36 D. S. Dimitrov and I. B. Ivanov, *J. Colloid Interface Sci.*, 1978, **64**, 97–106.
37 R. K. Jain and I. B. Ivanov, *J. Chem. Soc., Faraday Trans. 2*, 1980, **76**, 250–266.
38 A. Maali, Y. Wang and B. Bhushan, *Langmuir*, 2009, **25**, 12002–12005.
39 C.-Y. Lin and J. C. Slattery, *AIChE J.*, 1982, **28**, 147–156.
40 O. Reynolds, *Philos. Trans. R. Soc. London*, 1886, **177**, 157–234.
41 R. Manica, J. N. Connor, S. L. Carnie, R. G. Horn and D. Y. C. Chan, *Langmuir*, 2007, **23**, 626–637.
42 J. Laskowski and J. A. Kitchener, *J. Colloid Interface Sci.*, 1969, **29**, 670–679.
43 R. Hogg, T. W. Healy and D. W. Fuerstenau, *Trans. Faraday Soc.*, 1966, **62**, 1638–1651.
44 P. M. Claesson, C. E. Blom, P. C. Herder and B. W. Ninham, *J. Colloid Interface Sci.*, 1986, **114**, 234–242.
45 R. H. Yoon, D. H. Flinn and Y. I. Rabinovich, *J. Colloid Interface Sci.*, 1997, **185**, 363–370.
46 M. Nedyalkov, L. Alexandrova, D. Platikanov, B. Levecke and T. Tadros, *Colloid Polym. Sci.*, 2007, **285**, 1713–1717.
47 J. Israelachvili, *Intermolecular & Surface Forces* - 2nd Edition, Academic Press, San Diego, 1992.
48 M. B. M. Monte, F. F. Lins and J. F. Oliveira, *Int. J. Miner. Process.*, 1997, **51**, 255–267.
49 J. Leja, *Surface chemistry of froth flotation* Plenum Press, New York, 1982.
50 J. A. Mielczarski and R. H. Yoon, *J. Phys. Chem.*, 1989, **93**, 2034–2038.
51 J. A. Mielczarski and R. H. Yoon, *J. Colloid Interface Sci.*, 1989, **131**, 423–432.
52 R. Woods, C. I. Basilio, D. S. Kim and R. H. Yoon, *Colloids Surf., A*, 1994, **83**, 1–7.
53 J. O. Leppinen, R. H. Yoon and J. A. Mielczarski, *Colloids Surf.*, 1991, **61**, 189–203.
54 J. Wang and R. H. Yoon, Langmuir, submitted.
55 J. Wang and R.-H. Yoon, *Langmuir*, 2008, **24**, 7889–7896.
56 H. K. Christenson, J. Fang, B. W. Ninham and J. L. Parker, *J. Phys. Chem.*, 1990, **94**, 8004–8006.
57 L. G. Wang and R. H. Yoon, *Langmuir*, 2004, **20**, 11457–11464.
58 B. V. Deryagin and N. V. Churaev, *Langmuir*, 1987, **3**, 607–612.
59 B. V. Derjaguin, in *Theory of Stability of Colloids and Thin Films*, Consultants Bureau, New York, 1989, pp. 143–151.
60 C. J. van Oss, *Interfacial Forces in Aqueous Media* - 2nd Edition, CRC Press, Boca Raton, 2006.
61 A. Frumkin, *Zh. Fiz. Khim.*, 1938, **12**, 337–345.
62 C. J. van Oss, R. F. Giese and A. Docoslis, *J. Dispersion Sci. Technol.*, 2005, **26**, 585–590.
63 Q. Du, E. Freysz and Y. R. Shen, *Science*, 1994, **264**, 826–828.
64 L. F. Scatena, M. G. Brown and G. L. Richmond, *Science*, 2001, **292**, 908–912.
65 T. Chang and L. X. Dang, *J. Chem. Phys.*, 1996, **104**, 6772.
66 E. Lauga, in *Handbook of Experimental Fluid Dynamics*, ed. C. Tropea, A. Yarin and J. F. Foss, Springer, New York, 2007, pp. 1219–1240.
67 D. Berthelot, *Compt. Rend.*, 1898, **126**, 1703–1857.
68 R. H. Yoon and B. S. Aksoy, *J. Colloid Interface Sci.*, 1999, **211**, 1–10.
69 C. Li and P. Somasundaran, *Energy Fuels*, 1993, **7**, 244–248.
70 C. Li and P. Somasundaran, *J. Colloid Interface Sci.*, 1991, **146**, 215–218.

# Interfacial thermodynamics of confined water near molecularly rough surfaces

Jeetain Mittal[*a] and Gerhard Hummer[*b]

Received 8th December 2009, Accepted 26th January 2010
DOI: 10.1039/b925913a

We study the effects of nanoscopic roughness on the interfacial free energy of water confined between solid surfaces. SPC/E water is simulated in confinement between two infinite planar surfaces that differ in their physical topology: one is smooth and the other one is physically rough on a sub-nanometre length scale. The two thermodynamic ensembles considered, with constant pressure either normal or parallel to the walls, correspond to different experimental conditions. We find that molecular-scale surface roughness significantly increases the solid–liquid interfacial free energy compared to the smooth surface. For our surfaces with a water-wall interaction energy minimum of $-1.2$ kcal mol$^{-1}$, we observe a transition from a hydrophilic surface to a hydrophobic surface at a roughness amplitude of about 3 Å and a wavelength of 11.6 Å, with the interfacial free energy changing sign from negative to positive. In agreement with previous studies of water near hydrophobic surfaces, we find an increase in the isothermal compressibility of water with increasing surface roughness. Interestingly, average measures of the water density and hydrogen-bond number do not contain distinct signatures of increased hydrophobicity. In contrast, a local analysis indicates transient dewetting of water in the valleys of the rough surface, together with a significant loss of hydrogen bonds, and a change in the dipole orientation toward the surface. These microscopic changes in the density, hydrogen bonding, and water orientation contribute to the large increase in the interfacial free energy, and the change from a hydrophilic to a hydrophobic character of the surface.

## 1 Introduction

Water plays a central role in many biomolecular self-assembly processes, including the folding of proteins and the formation of lipid membranes.[1–10] In key steps of these biomolecular processes, water is often highly confined,[11] reduced for instance to a few layers of water molecules between the extended surfaces of large macromolecules. Considering the diversity in the chemical and physical topology of these surfaces, the interfacial behavior of water in a cellular environment will depend sensitively on the details of the confining surfaces. The wetting behavior of surfaces and its dependence on the physical and chemical nanostructure are also technologically relevant, for instance in the development of low-friction fluid flow channels.[11–13]

Quantitative experimental studies of water at interfaces are highly challenging,[14–17] and roughness can be a relevant parameter.[18] Recent advances in both theory and experiments[9] have re-invigorated the interest in the molecular underpinning of the

[a]Department of Chemical Engineering, Lehigh University, Bethlehem, Pennsylvania, 18015, USA. E-mail: jeetain@lehigh.edu
[b]Laboratory of Chemical Physics, National Institute of Diabetes and Digestive and Kidney Diseases, National Institutes of Health, Bethesda, Maryland, 20892-0520, USA. E-mail: hummer@helix.nih.gov

hydrophobic effect and the behavior of water near surfaces. Water confined between a "Janus interface" of adjoining hydrophobic and hydrophilic surfaces was found to fluctuate significantly during shear deformations.[19] This observation raises several interesting questions about water near heterogeneous surfaces and how local patchiness (wetting *versus* nonwetting) may affect hydrophobicity and microscopic density fluctuations.[20] To find answers to at least some of these questions, molecular simulations can complement the laboratory experiments because in the simulations surface details can be controlled precisely and molecular information can be obtained directly.

Previous theoretical and simulation studies have typically focused on the structure, thermodynamics, and dynamics of water confined between idealized smooth surfaces,[21-32] between atomistic surfaces,[33-37] within carbon nanotubes,[12] and between realistic protein-like surfaces.[38,39] These studies have vastly improved our molecular-level understanding of the expected changes in the water behavior due to confinement and the presence of hydrophilic *versus* hydrophobic surfaces.

In a laboratory experiment, one way to characterize surfaces as hydrophilic or hydrophobic is to measure the macroscopic contact angle of small water droplets on these surfaces. Recent molecular simulation studies have used a microscopic analog of the macroscopic contact angle to define the hydrophobicity of surfaces.[40-42] Giovambattista *et al.*[41] used contact angle data to show that water behavior can be a non-trivial function of the surface polarity. Godawat *et al.*[42] studied the behavior of water near self-assembled monolayer structures with a wide range of chemistries and showed that the hydrophobicities of such surfaces as measured by contact angles can be characterized well by measures related to the water density fluctuations. We similarly saw enhanced water-density fluctuations near extended non-polar surfaces,[43] associated with transitions between locally wet and dry regions and resulting in a broadened liquid–vapor like interfacial density profile.

It is well known in the surface wetting literature that the physical roughness of the surface can cause the contact angle to change significantly.[44] This change is typically described by either the Cassie[45] or the Wenzel[46] model. These models are widely used to describe the effects of the mesoscopic (micron-scale) roughness of superhydrophobic surfaces, such as Lotus leaves. Much less is known on how surface roughness of the order of a few molecular diameters,[47-49] which is relevant for instance to characterize protein surfaces, will affect the interfacial behavior of water.

This poor understanding of surface roughness effects on the interfacial thermodynamics also severely limits the applicability of surface-area based solvation models. Many widely used models for ligand binding and self-assembly rely on the interfacial free energy of surfaces to predict their solvation free energy.[50] However, at the molecular scale corrections should be made to the interfacial free energy to account for curvature and physical roughness.[51,52] Examples of such curvature corrections rely on the macroscopic concept of the Tolman length,[53] or on the microscopic observation of curvature dependent "cavity expulsion" of water.[54]

To explore how molecular-scale physical roughness affects the interfacial free energy, we perform molecular dynamics simulations of water confined between walls. One of the major difficulties in performing simulations of confined water is to define the thermodynamic state of the system and to make sure that water is present in a thermodynamically stable equilibrium. Here, we consider two thermodynamic ensembles, defined by constant normal and parallel pressures (Fig. 1) at a constant temperature and with a constant number of particles. We also monitor individual components of the pressure tensor to make sure that the system is in a stable state during the entire simulation time.

In our model system, the wall roughness can be changed in a controlled manner by modulating the local wall position with periodic functions of given amplitudes and wavelengths (see Fig. 2). We find that increasing wall roughness, as characterized by increasing roughness amplitude or decreasing roughness wavelength, can significantly increase the solid–liquid interfacial free energy. Roughness can even change

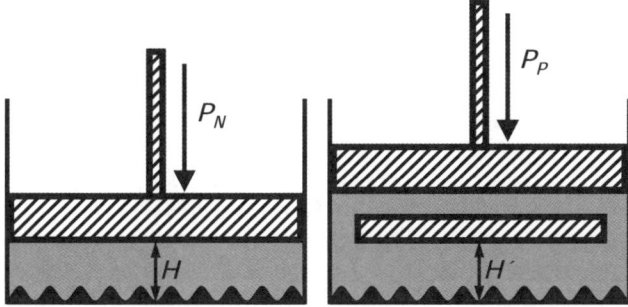

**Fig. 1** Thermodynamic ensembles. We simulate water confined between a lower rough wall and an upper smooth wall with two different constraints. (Left) Constant normal pressure $P_N$. Experimentally, the position of the upper surface would be controlled by a piston under constant pressure $P_N$. (Right) Constant parallel pressure $P_P$. Here the confined fluid is in equilibrium with bulk fluid at a constant pressure $P_P$. Note that in the two ensembles, the densities and thus the heights $H$ and $H'$ will be different in general for a given number of water molecules in a box of given projected area in the $xy$-plane.

the sign of the interfacial free energy from negative (hydrophilic) to positive (hydrophobic). We examine the microscopic origins of this change in wetting behavior in the context of hydration structure near rough walls and associated density and hydrogen bond distributions.

## 2 Models and methods

We use a system of $N_w = 1500$ SPC/E water[55] molecules in a rectangular box with periodic boundary conditions applied in two directions $x$ and $y$, and fixed boundary conditions applied in the $z$-direction. The walls are represented by a coarse-grained 9–3 Lennard-Jones potential,

$$V(z) = \varepsilon \left[ \frac{2}{15}\left(\frac{\sigma}{z}\right)^9 - \left(\frac{\sigma}{z}\right)^3 \right] \quad (1)$$

which is frequently used to represent interactions between a fluid particle and a wall made up of 12–6 Lennard-Jones particles. Here, $\varepsilon$ and $\sigma$ are the water-wall interaction energy and distance parameters, respectively. $z$ is the distance between the water oxygen atom and the wall surface. We use $\varepsilon = 1.15$ kcal mol$^{-1}$ and $\sigma = 0.346$ nm in this work. The resulting well depth of the wall–water interaction

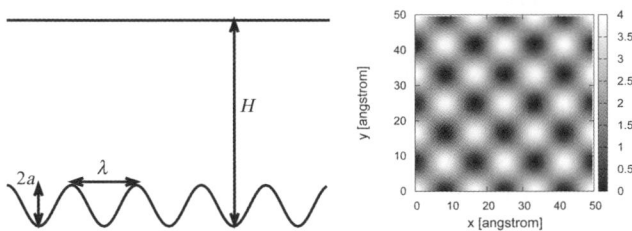

**Fig. 2** Model system. The two confining walls are placed normal to the $z$-direction and a distance $H$ apart. The physical roughness is represented by a cosine function, eqn (1) and 2, which gives rise to the wall pattern with peaks and valleys shown on the right hand side. The panel on the left shows a simplified one-dimensional cut through the peaks along the $x$ direction. $\lambda$ is the roughness wavelength, and $a$ is the amplitude of the roughness so that the lower wall position will vary from $z = 0$ to $2a$.

is ≈ −1.21 kcal mol$^{-1}$. To represent a rough wall as shown in Fig. 2, we use a periodic function of the following form,

$$z(x, y) = a\cos\left(\frac{2\pi x}{\lambda}\right)\cos\left(\frac{2\pi y}{\lambda}\right) \qquad (2)$$

where $z(x, y)$ represents the wall position as a function of $x$- and $y$-coordinates. $a$ is the roughness amplitude, and $\lambda$ is the roughness wavelength, which is an integer fraction of the simulation box length in the $x$ and $y$ direction. The resulting interaction $V[z_i - z(x_i, y_i)]$ of an oxygen molecule and the wall depends on all three oxygen coordinates ($x_i$, $y_i$, $z_i$).

The equilibrium behavior of our confined water system is obtained from molecular dynamics simulations using LAMMPS[56] in a canonical ensemble. The temperature is held constant at 300 K by using a Nosé-Hoover thermostat with a time constant of 0.5 ps. The simulation time step is 2 fs. For each state we simulate for a total time of 4 ns. The initial 2 ns are discarded as equilibration time, and the remaining 2 ns of the simulation runs are used for the final analysis. Electrostatic interactions are calculated with particle–particle particle–mesh solver. To account for non-periodicity in the $z$-direction we extend the simulation box to three times the height $H$ and apply the correction proposed by Yeh and Berkowitz.[57]

We calculate the solid–liquid interfacial free energy between the confined water and confining walls from the difference between normal and parallel pressures,[58]

$$\gamma \equiv \gamma_l + \gamma_u = \int_0^H (P_N - p_P(z))dz = (P_N - P_P)H \qquad (3)$$

where $P_P = H^{-1}\int_0^H p_P(z)dz$ is the average of the $z$-dependent parallel pressure $p_P(z)$. $\gamma_l$ and $\gamma_u$ are the interfacial free energies per unit area (of the surface projected onto the $xy$-plane) between confined water and the lower and upper wall, respectively, assuming separability.

## 3 Results and discussion

To calculate how the wall roughness may affect the interfacial free energy for the confined water system shown in Fig. 2, we first calculate the pressure equations of state as a function of wall distance $H$. The equations of state differ for the two ensembles under constant parallel and normal pressures. As our starting smooth wall system ($a = 0$), we use the wall–oxygen interaction parameter $\varepsilon = 1.15$ kcal mol$^{-1}$ and confinement length $H = 16.75$ Å to obtain a system close to a stable thermodynamic equilibrium and with about 4 water layers between the walls. To explore the effects of roughness, we then modulate the wall location of the lower surface according to the periodic function in eqn (1) and 2. Results for the interfacial free energy will be compared in states defined by conditions of 1 atm pressure in either the normal direction ($P_N$) or the parallel direction ($P_P$) (see Fig. 1).

Fig. 3 (left, insets) shows the normal pressure $P_N$ and the parallel pressure $P_P$ as a function of $H$ for roughness amplitudes $a$ between 0 and 10 Å (from left to right) of the lower wall. The simulation pressures are fitted accurately by a cubic polynomial in $H$. The fitted equations of state allow us to find the $H$ values for which $P_N$ or $P_P$ is equal to 1 atm. We then use the respective $H$ value to evaluate the pressure component that is not kept constant, i.e., $P_P(P_N = 1$ atm) and $P_N(P_P = 1$ atm). From the cubic fits, we also calculate the isothermal compressibility $\kappa_T = -H^{-1}(dH/dP)$ in both the normal and the parallel direction. The results are shown in Fig. 3 (right, bottom) where $dP_N/dH$ and $dP_P/dH$ are evaluated at $P_N = 1$ atm and $P_P = 1$ atm, respectively. We find that $\kappa_T$ increases with increasing $a$ for both constant $P_N$ and $P_P$. The confined water is thus more compressible near rough surfaces. We want to point out that this is the overall behavior of the confined system and not the local behavior near a surface as studied previously by Godawat et al.[42] and by us.[43]

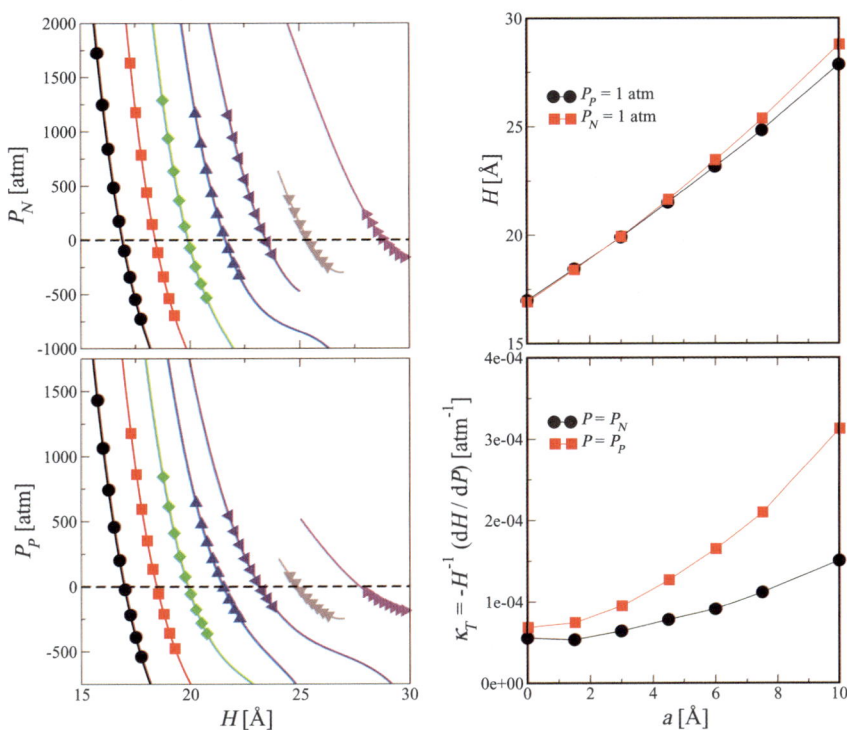

**Fig. 3** Equations of state. (Left) Normal pressure $P_N$ (top) and parallel pressure $P_P$ (bottom) as a function of the distance $H$ between the walls. The interfacial area $A$ of the simulation box in the $xy$-plane is kept constant. The water density is thus changing with $H$, effectively providing us with the pressure as a function of the density. The $P(H)$ data are shown for increasing wall roughness amplitudes $a = 0, 1.5, 3, 4.5, 6, 7.5,$ and 10 Å (from left to right) and for a roughness wavelength $\lambda = 11.59$ Å. The symbols are the simulation data and lines are fits to a cubic polynomial. The top and bottom insets show $P_P$ ($P_N = 1$ atm) and $P_N$ ($P_P = 1$ atm), respectively, as a function of the wall roughness amplitude $a$. (Right, top) Distance $H$ between the walls as a function of the roughness amplitudes $a$ at 1 atm pressure (squares for $P_N = 1$ atm, and circles for $P_P = 1$ atm). (Right, bottom) Isothermal compressibility $\kappa_T$ as a function of the wall roughness amplitude $a$ at 1 atm pressure (symbols as above).

A similar increase in $\kappa_T$ for a confined water system was found by Giovambattista et al.[41] with decreasing surface polarity and therefore increasing surface hydrophobicity. Also, the compressibility is higher for a system at constant $P_N$ as compared with a system at constant $P_P$ for a given $a$.

Thermodynamic stability is a possible concern in simulations of confined systems. For roughness amplitudes $a \geq 7.5$ Å, we find that the cubic fits to the $P$–$H$ data produce multiple roots for $P(H) = 1$ atm. For $a \geq 7.5$ Å, we thus expect phase co-existence (in a large system) between a liquid and vapor phase, or fluctuations (in a small system) between the two "phases." For these large roughness amplitudes $a \geq 7.5$ Å, the systems simulated are thus likely metastable.

The simulation systems can also be subject to mechanical instabilities. With one of the pressure components, $P_N$ or $P_P$, fixed at 1 atm, the other one can be negative. As shown in Fig. 3 (left panels), $P_P(P_N = 1$ atm) is negative for $a \geq 3$ Å, and $P_N(P_P = 1$ atm) is negative for $a \leq 1.5$ Å. To check whether the system underwent phase separation (through the formation of nanoscale bubbles), we track the full pressure tensor, making sure that the in-plane diagonal pressure components agree, $P_{xx} = P_{yy}$, and that the normal pressures on the lower and upper walls are equal, $P_{zz,l} = P_{zz,u}$. We do not find any signatures of such deviations for all the state points considered in this work, in particular not for conditions close to 1 atm pressure in either the normal or parallel direction. Also, the slope of the pressure equations of state (Fig. 3, left) is always negative, $\partial P/\partial H < 0$, which means that the thermodynamic stability condition is satisfied ($\kappa_T > 0$; Fig. 3 right bottom). These observations together provide a sufficient reason to believe that our systems of interest are in a thermodynamic equilibrium (or at least in a metastable equilibrium) during the entire simulation times.

Fig. 4 shows the interfacial free energy $\gamma$ as a function of wall roughness amplitude $a$ (left) and wavelength $\lambda$ (right) under the constraints of $P_N$ or $P_P$ being equal to 1 atm. We find that $\gamma$ changes sign with increasing wall roughness by either increasing $a$ (with $\lambda = 11.6$ Å fixed) or decreasing $\lambda$ (with $a = 3$ Å fixed). This change in sign means that the hydrophilic smooth surface ($\gamma < 0$) becomes hydrophobic ($\gamma > 0$) with increasing molecular scale roughness. We also find that $\gamma$ has a rather strong dependence on $a$ and this dependence is stronger for systems under the constraint of constant parallel pressure $P_P$. The dependence of $\gamma$ on the wavelength $\lambda$ for a fixed $a = 3$ Å is weaker than the dependence on the amplitude $a$ (Fig. 4, right). With increasing $\lambda$, the calculated $\gamma$ values drop toward the smooth-wall limit expected for $\lambda \to \infty$, albeit slowly. Importantly, physical roughness of the surfaces on the length scale of a single water molecule can change the wetting behavior of confined water qualitatively from hydrophilic ($\gamma < 0$) to hydrophobic ($\gamma > 0$).

If we assume that the interfacial free energy of the smooth wall, $\gamma_u$, is independent of the roughness of the lower wall, we can evaluate the interfacial free energy for the rough wall by subtracting the value for the smooth wall from the total free energy, $\gamma_l(a) \approx \gamma(a) - \gamma(a = 0)/2$. We find that the interfacial free energy for the rough wall calculated in this way varies from $-8.1$ mN m$^{-1}$ to $+89.1$ mN m$^{-1}$ for $P_P = 1$ atm and from $-3.6$ mN m$^{-1}$ to $+41$ mN m$^{-1}$ for $P_N = 1$ atm.

By assuming a solid–vapor interfacial free energy of $\gamma_{SV} \approx 0$ mN m$^{-1}$, we can use Young's equation based on mechanical equilibrium to find the contact angle $\theta$ of SPC/E water on our rough surfaces,

$$\cos\theta = \frac{\gamma_{SV} - \gamma_{SL}}{\gamma_{LV}} \qquad (4)$$

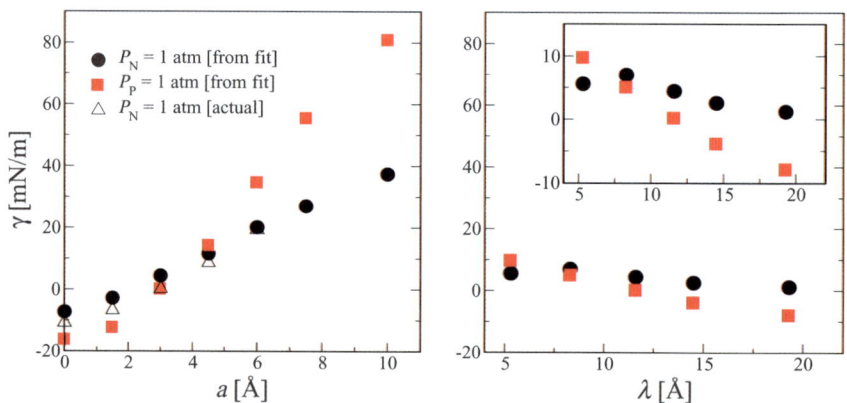

**Fig. 4** Effect of wall roughness amplitude and wavelength on interfacial properties. The combined solid–liquid interfacial free energy $\gamma$ between confined water and both the lower and upper walls is shown as a function of lower wall roughness amplitude $a$ for a fixed wavelength $\lambda = 11.6$ Å (left panel) and wall roughness wavelength $\lambda$ for a fixed amplitude $a = 3$ Å (right panel). The inset in the right panel shows the zoomed in version. The filled symbols are the $\gamma$ values calculated using eqn (3) and the cubic pressure equation of state fits shown in Fig. 3. The empty symbols are the data generated using additional simulations at $H$ values targeting $P_N = 1$ atm according to the equations of state.

where $\gamma_{SV}$, $\gamma_{SL}$, and $\gamma_{LV}$ are the solid–vapor, solid–liquid, and liquid–vapor interfacial tensions, respectively. With a liquid–vapor interfacial tension of $\gamma_{LV} = 61.3$ mN m$^{-1}$ for SPC/E water at 300 K,[59] and with $\gamma_l(a)$ assumed to be equal to $\gamma_{SL}$ and $\gamma_{SV} \approx 0$, we can use eqn (4) to estimate the contact angle. At constant $P_N = 1$ atm, the contact angle changes from 87° for a flat interface to 132° for $a = 10$ Å and $\lambda = 11.6$ Å. For constant parallel pressure, a contact angle of 180° is reached just above $a = 6$ Å, starting from $\theta \approx 82°$ for a flat surface. For larger amplitudes $a \geq 7.5$ Å, the calculated free energy of the liquid–solid interface exceeds that of the liquid–vapor and vapor–solid interface combined. It is thus thermodynamically favorable to separate the fluid phase from the rough surface by an intervening vapor layer, and eqn (4) does not have a solution for the contact angle $\theta$. This metastability is consistent with the equation of state at $P_P = 1$ atm having multiple roots for $a \geq 7.5$ Å (Fig. 3). We note, however, that in our model these drastic changes in the wetting behavior of microscopically rough surfaces result from the combined effects of wall roughness and confinement.

To understand the origin of the increased interfacial free energy of confined water near rough surfaces, we calculate various structural measures that characterize the spatial and orientational water distribution. We only perform these analyses for systems under the constraint of constant normal pressure, i.e., $P_N = 1$ atm.

Fig. 5(d) shows the water density profiles near the smooth upper and rough lower walls. The water density is plotted as a function of the local (x- and y-dependent) distance from the wall, $z_u - z$ and $z_l - z$, respectively, averaged over the entire system. Remarkably, neither at the smooth nor at the rough wall do we find any dramatic changes in the average density of the first water shell. Even for the restrictive confinement of our simulation setup, the presence of a rough surface does not appear to affect the density distribution near the opposite surface. However, as we will discuss below, the local changes in the density induced by roughness are pronounced.

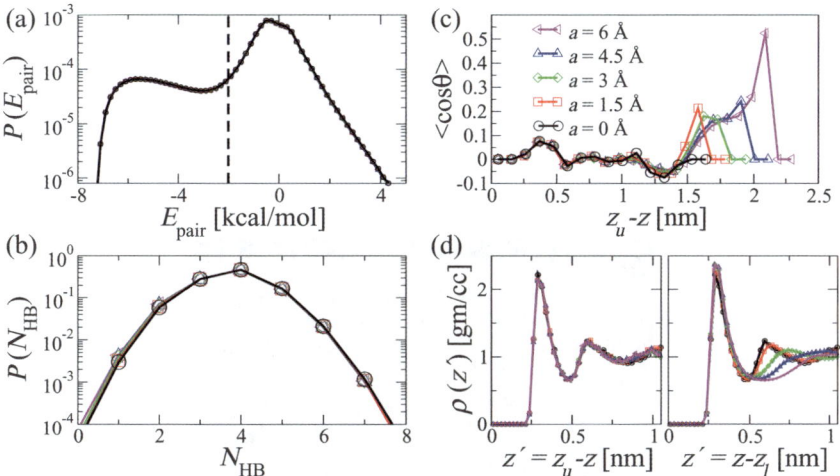

**Fig. 5** Water ordering between the walls. (a) Distribution of intermolecular pair-energy $E_{pair}$ between two water molecules for which the distance between the oxygen atoms is less than 6 Å. Two molecules are defined to be hydrogen bonded if this energy is less than $-2$ kcal mol$^{-1}$, indicated by the vertical dashed line. (b) Distribution of the number of hydrogen bonds per water molecule. (c) Average of cos $\theta$, where $\theta$ is the angle between the water dipole and the negative z-axis, shown as a function of the distance $z_u - z$ away from the upper smooth wall position. (d) Water density profile $\rho(z')$ shown as a function of distance with respect to the upper smooth wall position $z_u$ (left) and the lower rough wall position $z_l$ (right).

To describe the hydrogen bonding behavior of water, we use energetic criteria.[60] A pair of water molecules is defined to be hydrogen bonded if the intermolecular pair energy, which is the sum of Coulombic and van der Waals energies between all the atoms of the two molecules, is below some minimum value. We compute these pair energies for water molecules whose oxygen atoms are within 6 Å. Fig. 5(a) shows the resulting pair energy distributions for several roughness amplitudes $a$, with fixed $\lambda = 11.6$ Å. The peak at the lower energy values is due to molecules that are tightly held together, primarily by hydrogen bonds. All the curves are statistically indistinguishable from each other meaning that, on average, the wall roughness does not change the pair energy distribution. From this distribution, we also construct a histogram of the number $N_{HB}$ of hydrogen bonds per water molecule. Below an interaction energy cut-off of $-2$ kcal mol$^{-1}$, a molecule pair is considered to be hydrogen-bonded. As for the energy distribution, we find no major changes in the distributions $P(N_{HB})$ of the number of hydrogen bonds, averaged over the entire system, as the roughness amplitude is changed.

The dipolar orientation of water also depends sensitively on the roughness. Fig. 5(c) shows the average of $\cos \theta$ as a function of distance $z_u - z$ normal to the walls ($z$-direction), where $\theta$ is the angle between the water dipole and the negative $z$-axis, and $z_u$ is the location of the upper smooth wall. Near the upper smooth wall ($z_u - z \approx 0$), the dipole-moment vectors are on average lying in the $xy$-plane, with $\langle \cos \theta \rangle \approx 0$, consistent with the early simulations of Lee, McCammon, and Rossky.[61] In contrast, the dipoles of water molecules inside the rough grooves ($z_u - z > 1.5$ nm) are on average oriented partially toward the lower rough wall, with $\langle \cos \theta \rangle > 0$.

Upon closer inspection of individual structures, we find that the water molecules at the tips of the finger-like protrusions penetrating into the valleys tend to form peculiar hydrogen bonding structures. These water molecules are typically connected to the bulk-like water above by two hydrogen bonds, and have one or both of their OH bonds dangling, pointing toward the rough surface. An example of a water molecule accepting two hydrogen bonds and donating none is shown in Fig. 6 (indicated by a solid circle and a light blue arrow).

The locally-averaged number of hydrogen bonds per water molecule is consistent with this picture. With our definition, we have on average four hydrogen bonds in the bulk phase at the center of the slab (Fig. 7 right). Near the flat upper wall, the average number $\langle N_{HB} \rangle$ of hydrogen bonds drops to three. In the valleys of the rough interface, $\langle N_{HB} \rangle$ drops even further to about 2.5, 2, 1.5, and 1 for $a = 1.5, 3, 4.5$, and 6 Å.

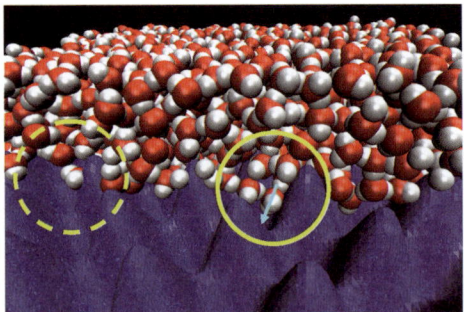

**Fig. 6** Snapshot of the water structure. A cut through the $xz$-plane shows water (O, red; H, white) penetrating deep into some of the valleys (solid yellow circle) of the rough surface (energy isocontour surface, blue; the smooth surface on top is not shown) but receding from other valleys (dashed yellow circle). The light blue arrow indicates the orientation of a typical water in a valley, with the dipole pointing toward the surface.

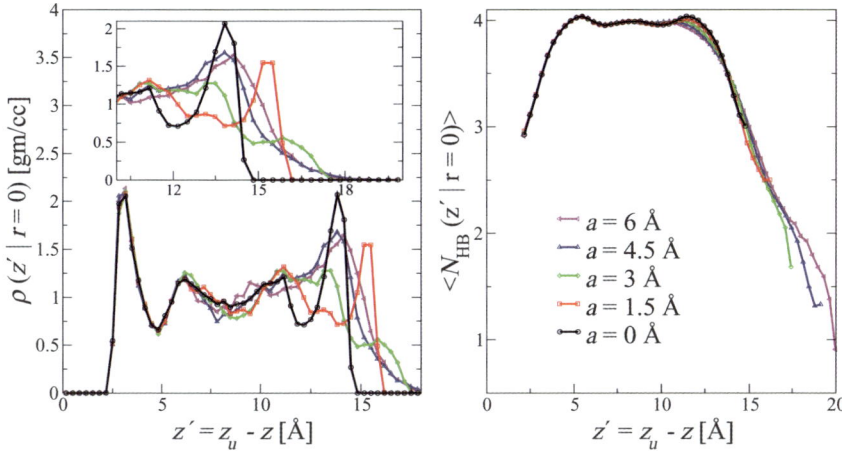

**Fig. 7** Local water ordering above the valleys of the rough surface. (Left) Local water oxygen density, averaged over cylindrical volumes of radius 0.29 Å above the valley centers and plotted as a function of the distance $z' = z_u - z$ from the upper smooth wall. The inset shows the zoomed in version near the rough wall surface to highlight the broadening of the interface with increasing $a$. (Right) Average number of hydrogen bonds per molecule $N_{HB}$ is shown as a function of $z'$.

This dramatic loss of hydrogen bonding interactions in the valleys of the rough surface is accompanied by a local depletion in water density. Whereas the average water density as a function of the distance from the rough surface is only weakly dependent on roughness (Fig. 5(d)), the local water-oxygen density at the center of the valleys is low. As shown in Fig. 7 (left), the water density directly above the center of the valleys drops only gradually toward the wall, in sharp contrast to the steep density profiles at the flat wall and at the center of the protrusions of the rough interface (not shown). These broadened density profiles in the valleys are reminiscent of the liquid–vapor interface and the interface near extended hydrophobic solutes.[43]

Consistent with the capillary-wave fluctuations in the density near extended hydrophobic solutes,[43] simulation snapshots show that some of the valleys are transiently filled by locally dense water "fingers" (solid yellow circle in Fig. 6), whereas other valleys are transiently empty (dashed yellow circle in Fig. 6). This analysis suggests that the valleys in the rough interface are fluctuating between locally "wet" and "dry" states. This local "dewetting" appears to be at least partially caused by the loss in hydrogen bonding in the fingerlike water structures able to penetrate into the valleys. Put together, our results suggest that the combination of loss in hydrogen bond energy, and local transient dewetting of the valleys are important factors contributing to the increase in interfacial free energy $\gamma$ with molecular-scale surface roughness.

## 4 Concluding remarks

We performed molecular dynamics simulations of SPC/E water confined between two surfaces, one perfectly flat and the other with sinusoidal roughness. Water has strong attractive interactions with both surfaces (well depth of $-1.2$ kcal mol$^{-1}$ in a 9–3 Lennard-Jones potential). From the difference between normal and parallel pressures, we determine the interfacial free energies. In the limit of infinite surfaces at infinite separations, these interfacial free energies would correspond to the surface tensions. Here, however, the two surfaces are close on a molecular scale, and we expect our interfacial free energies to be affected also by the confinement.

Despite having equal interaction strengths of water with both surfaces, we find that the interfacial free energy is negative only for smooth surfaces, but becomes positive when the roughness amplitude exceeds 3 Å at a wavelength of 11.6 Å. This sign reversal implies that the character of the surface changes from hydrophilic to hydrophobic as a result of relatively modest levels of roughness at molecular scales. In terms of contact angles, we find an increase from 82° to just under 180° upon going from a flat interface to a roughness amplitude of $a = 6$ Å at constant parallel pressure $P_P = 1$ atm.

At first sight, our results may seem counterintuitive: we start out with a surface with a favorable (negative) interfacial free energy, $\gamma(a = 0) < 0$. By making the surface rough but maintaining the interaction strength with water as a function of the height $z$, one in effect increases the surface area accessible to water at a fixed projected area. With the apparently favorable interactions between water and the surface, $\gamma(a = 0) < 0$, and the increased accessible area, one might naively expect the surface free energy to become even more negative with increasing roughness. However, this simplistic argument fails because it does not take into account that the interfacial water has to distort its hydrogen bond network to follow the bumpy surface. In our model system, the free energy penalty of the distorted hydrogen bond network appears to outweigh the energetic gain from the larger accessible area. As a consequence, we observe partial and transient dewetting of the valleys, and the rough surface changes its character from hydrophilic to hydrophobic.

This strong dependence of the interfacial free energy on the roughness may help in the design of superhydrophobic surfaces,[44] and the coating of high-flux nanofluidic and microfluidic channels.[11–13] Our results also have implications for the solvation thermodynamics and hydration-driven interactions of macromolecules, including proteins. The surfaces of these molecules are inherently rough on the molecular length scale. Improved surface area models of solvation could possibly be obtained by accounting for the strong effects of molecular roughness on the apparent interfacial free energy.

For protein molecules, the chemical heterogeneity of the surface adds another level of complexity to the solvation calculations.[39] MD simulations of water between partially hydroxylated silica surfaces (chemical patterning) showed that the boundary between a hydrophobic and a hydrophilic surface may become blurred because of chemical patchiness.[34] Willard and Chandler employed a coarse-grained lattice-gas model to investigate the interface between water-like solvent and chemically heterogeneous surfaces.[62] By varying the overall fraction and size of hydrophilic sites on a surface, they found that the interfacial fluctuations are large and spatially heterogeneous. Further, adding small uniform solvent–surface attractive interactions brought the interface closer to the surface while maintaining large density fluctuations close to the dewetting conditions.[62]

An important open question is the degree to which confinement affects the interfacial free energies. To explore this question, simulations of larger systems can be performed in which the separation of the two walls is increased. It will also be interesting to see if the contact angles inferred from $\gamma$ are consistent with simulations of water droplets on the rough surfaces.

## Acknowledgements

J.M. would like to thank Dr Vincent Shen (NIST) for helpful discussions. J.M. was supported by a start-up grant from the Lehigh University. G.H. was supported by the Intramural Research Program of the NIH, NIDDK. This study utilized the high-performance computational capabilities of the Biowulf PC/Linux cluster at the National Institutes of Health, Bethesda, MD (http://biowulf.nih.gov).

# References

1 W. Kauzmann, *Adv. Protein Chem.*, 1959, **14**, 1–63.
2 C. Tanford, *The Hydrophobic Effect: Formation of Micelles and Biological Membranes*, Wiley, New York, 1973.
3 A. Ben-Naim, *Water and Aqueous Solutions*, Plenum, New York, 1974.
4 F. H. Stillinger, *Science*, 1980, **209**, 451–457.
5 K. A. Dill, *Biochemistry*, 1990, **29**, 7133–7155.
6 K. Lum, D. Chandler and J. D. Weeks, *J. Phys. Chem. B*, 1999, **103**, 4570–4577.
7 L. R. Pratt and A. Pohorille, *Chem. Rev.*, 2002, **102**, 2671–2692.
8 B. Widom, P. Bhimalapuram and K. Koga, *Phys. Chem. Chem. Phys.*, 2003, **5**, 3085–3093.
9 D. Chandler, *Nature*, 2005, **437**, 640–647.
10 B. J. Berne, J. D. Weeks and R. Zhou, *Annu. Rev. Phys. Chem.*, 2009, **60**, 85–103.
11 J. C. Rasaiah, S. Garde and G. Hummer, *Annu. Rev. Phys. Chem.*, 2008, **59**, 713–740.
12 G. Hummer, J. C. Rasaiah and J. P. Noworyta, *Nature*, 2001, **414**, 188–190.
13 C. Cottin-Bizonne, J. L. Barrat, L. Bocquet and E. Charlaix, *Nat. Mater.*, 2003, **2**, 238–240.
14 H. K. Christenson and P. M. Claesson, *Adv. Colloid Interface Sci.*, 2001, **91**, 391–436.
15 P. Attard, *Adv. Colloid Interface Sci.*, 2003, **104**, 75–91.
16 S. Singh, J. Houston, F. van Swol and C. J. Brinker, *Nature*, 2006, **442**, 526.
17 E. E. Meyer, K. J. Rosenberg and J. Israelachvili, *Proc. Natl. Acad. Sci. U. S. A.*, 2006, **103**, 15739–15746.
18 R. R. Netz, *Curr. Opin. Colloid Interface Sci.*, 2004, **9**, 192–197.
19 X. Zhang, Y. Zhu and S. Granick, *Science*, 2002, **295**, 663–666.
20 S. Granick, *Science*, 2008, **322**, 1477–1478.
21 A. Wallqvist and B. J. Berne, *J. Phys. Chem.*, 1995, **99**, 2893–2899.
22 J. Forsman, B. Jö and C. E. Woodward, *J. Phys. Chem.*, 1996, **100**, 15005–15010.
23 T. M. Truskett, P. G. Debenedetti and S. Torquato, *J. Chem. Phys.*, 2001, **114**, 2401–2418.
24 D. Bratko, R. A. Curtis, H. W. Blanch and J. M. Prausnitz, *J. Chem. Phys.*, 2001, **115**, 3873–3877.
25 X. Huang, C. J. Margulis and B. J. Berne, *Proc. Natl. Acad. Sci. U. S. A.*, 2003, **100**, 11953–11958.
26 K. Leung, A. Luzar and D. Bratko, *Phys. Rev. Lett.*, 2003, **90**, 065502-1.
27 M. Ø. Jensen, O. G. Mouritsen and G. H. Peters, *J. Chem. Phys.*, 2004, **120**, 9729.
28 N. Choudhury and B. M. Pettitt, *J. Am. Chem. Soc.*, 2005, **127**, 3556–3567.
29 S. Vaitheeswaran, H. Yin and J. C. Rasaiah, *J. Phys. Chem. B*, 2005, **109**, 6629–6635.
30 P. Kumar, S. V. Buldyrev, F. Starr, N. Giovambattista and H. E. Stanley, *Phys. Rev. E: Stat., Nonlinear, Soft Matter Phys.*, 2005, **72**, 051503.
31 T. Urbic, V. Vlachy and K. A. Dill, *J. Phys. Chem. B*, 2006, **110**, 4963–4970.
32 P. Kumar, F. W. Starr, S. V. Buldyrev and H. E. Stanley, *Phys. Rev. E: Stat., Nonlinear, Soft Matter Phys.*, 2007, **75**, 011202.
33 N. Giovambattista, P. J. Rossky and P. G. Debenedetti, *Phys. Rev. E: Stat., Nonlinear, Soft Matter Phys.*, 2006, **73**, 041604-041604-14.
34 N. Giovambattista, P. G. Debenedetti and P. J. Rossky, *J. Phys. Chem. C*, 2007, **111**, 1323–1332.
35 J. Janeèek and R. R. Netz, *Langmuir*, 2007, **23**, 8417–8429.
36 M. J. Stevens and G. S. Grest, *Biointerphases*, 2008, **3**, FC13–FC22.
37 C. Sendner, D. Horinek, L. Bocquet and R. R. Netz, *Langmuir*, 2009, **25**, 10768–10781.
38 P. Liu, X. Huang, R. Zhou and B. J. Berne, *Nature*, 2005, **437**, 159–162.
39 N. Giovambattista, C. F. Lopez, P. J. Rossky and P. G. Debenedetti, *Proc. Natl. Acad. Sci. U. S. A.*, 2008, **105**, 2274–2279.
40 T. Werder, J. H. Walther, R. L. Jaffe and T. H. P. Koumoutsakos, *J. Phys. Chem. B*, 2003, **107**, 1345–1352.
41 N. Giovambattista, P. G. Debenedetti and P. J. Rossky, *J. Phys. Chem. B*, 2007, **111**, 9581–9587.
42 R. Godawat, S. N. Jamadagni and S. Garde, *Proc. Natl. Acad. Sci. U. S. A.*, 2009, **106**, 15119–15124.
43 J. Mittal and G. Hummer, *Proc. Natl. Acad. Sci. U. S. A.*, 2008, **105**, 20130–20135.
44 D. Quéré, *Annu. Rev. Mater. Res.*, 2008, **38**, 71–99.
45 A. B. D. Cassie and S. Baxter, *Trans. Faraday Soc.*, 1944, **40**, 546.
46 R. N. Wenzel, *Ind. Eng. Chem.*, 1936, **28**, 988.
47 S. Pal, H. Weiss, H. Keller and F. Müller-Plathe, *Langmuir*, 2005, **21**, 3699–3709.
48 J. T. Hirvi and T. A. Pakkanen, *Langmuir*, 2007, **23**, 7724–7729.
49 A. Wallqvist, E. Gallicchio and R. M. Levy, *J. Phys. Chem. B*, 2001, **105**, 6745–6753.
50 C. Chothia, *Nature*, 1974, **248**, 338–339.
51 D. Ben-Amotz, *J. Chem. Phys.*, 2005, **123**, 184504.

52 H. S. Ashbaugh and L. R. Pratt, *Rev. Mod. Phys.*, 2006, **78**, 159–178.
53 K. A. Sharp, A. Nicholls, R. M. Fine and B. Honig, *Science*, 1991, **252**, 106–109.
54 G. Hummer and S. Garde, *Phys. Rev. Lett.*, 1998, **80**, 4193–4196.
55 H. J. C. Berendsen, J. R. Grigera and T. P. Straatsma, *J. Phys. Chem.*, 1987, **91**, 6269–6271.
56 S. J. Plimpton, *J. Comput. Phys.*, 1995, **117**, 1–19.
57 I.-C. Yeh and M. Berkowitz, *J. Chem. Phys.*, 1999, **111**, 3155.
58 H. T. Davis, *Statistical Mechanics of Phases, Interfaces, and Thin Films*, VCH, 1996.
59 F. Chen and P. E. Smith, *J. Chem. Phys.*, 2007, **126**, 221101-1.
60 W. L. Jorgensen, J. Chandrasekhar, J. D. Madura, R. W. Impey and M. L. Klein, *J. Chem. Phys.*, 1983, **79**, 926.
61 C.-Y. Lee, J. A. McCammon and P. J. Rossky, *J. Chem. Phys.*, 1984, **80**, 4448–4455.
62 A. P. Willard and D. Chandler, *Faraday Discuss.*, 2009, **141**, 209–220.

# Mapping hydrophobicity at the nanoscale: Applications to heterogeneous surfaces and proteins

Hari Acharya, Srivathsan Vembanur, Sumanth N. Jamadagni and Shekhar Garde*

*Received 22nd December 2009, Accepted 10th February 2010*
DOI: 10.1039/b927019a

Approaches to quantify wetting at the macroscale do not translate to the nanoscale, highlighting the need for new methods for characterizing hydrophobicity at the small scale. We use extensive molecular simulations to study the hydration of homo and heterogeneous self-assembled monolayers (SAMs) and of protein surfaces. For homogeneous SAMs, new pressure-dependent analysis shows that water displays higher compressibility and enhanced density fluctuations near hydrophobic surfaces, which are gradually quenched with increasing hydrophilicity, consistent with our previous studies. Heterogeneous surfaces show an interesting context dependence – adding a single –OH group in a –$CH_3$ terminated SAM has a more dramatic effect on water in the vicinity compared to that of a single –$CH_3$ group in an –OH background. For mixed –$CH_3$/–OH SAMs, this asymmetry leads to a non-linear dependence of hydrophobicity on the surface concentration. We also present preliminary results to map hydrophobicity of protein surfaces by monitoring local density fluctuations and binding of probe hydrophobic solutes. These molecular measures account for the behavior of protein's hydration water, and present a more refined picture of its hydrophobicity map. At least for one protein, hydrophobin-II, we show that the hydrophobicity map is different from that suggested by a commonly used hydropathy scale.

## I. Introduction

Water molecules mediate a wide variety of colloidal and biological self-assembly phenomena in aqueous solutions.[1–3] In many cases, the assembly is driven by hydrophobic interactions between non-polar moieties (*e.g.*, between oily tails of surfactants[4]), or between hydrophobic patches present on more complex molecules such as proteins.[5] Understanding of protein phase behavior, protein–ligand binding, or design of biomolecular structures all require characterization of water-mediated interactions, and specifically of hydrophobicity at the molecular scale. The extent of hydrophobicity or philicity of macroscopic surfaces is characterized frequently by water droplet contact angle measurements. However, such measurements are not feasible for surfaces of nanoparticles or proteins,[6] which are many orders of magnitude smaller and contain chemical and topographical heterogeneities at the nanoscale.

For simple solutes, the oil to water partition coefficient (*i.e.*, solubility) may be used as a measure of hydrophobicity.[7] For molecules with heterogeneous chemistry and topography, solubility based measures provide only a coarse-grained

*The Howard P. Isermann Department of Chemical & Biological Engineering, Center for Biotechnology and Interdisciplinary Studies, Rensselaer Polytechnic Institute, Troy, New York, 12180, USA; Web: http://www.rpi.edu/~gardes. E-mail: gardes@rpi.edu*

quantification of hydrophobicity averaged over its entire surface. In principle, behavior of water molecules near an interface should allow characterization of hydrophobicity/philicity at subnanometer resolution. Recently we studied the hydration of flat, chemically homogeneous self-assembled monolayers (SAMs),[8,9] to identify molecular signatures of surface hydrophobicity.[10] For a range of chemistries we showed that the density of water in the vicinity of these surfaces provides a poor quantification of interface hydrophobicity. In contrast, the free energy of cavity formation, or correspondingly, the free energy of binding of probe hydrophobic solutes to an interface provide measures of hydrophobicity consistent with macroscopic expectations from contact angle studies.[10]

The enhanced probability of cavity formation near hydrophobic interfaces also implies larger density fluctuations and a higher compressibility of water relative to that in the bulk. The role of density fluctuations in interfacial and confined systems has been previously highlighted using coarse-grained models.[12,11] Molecular simulations of spherical solutes in water have also been employed recently to quantify interfacial density fluctuations.[13,14] SAMs using a range of head-group chemistries present excellent model systems for studies of interfacial water. Here, we use various measures of water density fluctuations (e.g., interfacial compressibility, water number distributions) to understand the differences between properties of water as a function of location (hydration layer/bulk) and chemistry. For homogeneous surfaces of different chemistries, we present new results on the pressure dependence of density to obtain compressibility of the water slab as a function of distance from the interface.

Prior studies using coarse-grained methods suggest that addition of a small number of hydrophilic sites can dramatically change the hydrophobicity of a given surface.[11,12] Such chemically heterogeneous systems can be constructed without introducing topographical complexities using self-assembled monolayers. At the most basic level, we are interested in quantifying hydrophobicity in the context of a given chemical background. For example, how does the introduction of a single –OH (or –CH$_3$) head-group in a sea of –CH$_3$ (or –OH) groups affect the local environment? How far does the perturbation due to that single group propagate? Can such nanoscale effects be detected using the aforementioned measures of hydrophobicity? How does this behavior evolve with increasing the lengthscale of chemical inhomogeneity (i.e., the size of the patch) or with surface composition? Our calculations highlight a large asymmetry in the ability of hydrophobic and hydrophilic inhomogeneities to affect the local compressibility of water at the interface, and the lengthscale of the perturbation such inhomogeneities cause.

Our results for heterogeneous systems are relevant to characterizing protein hydration. There is growing appreciation in the literature of the fact that computational prediction (or design) of ligand binding sites in proteins is not a solved problem.[15,16] Here we present preliminary results on the characterization of a protein, hydrophobin-II, by quantifying the behavior of water in its vicinity. Water density fluctuations, free energy of cavity formation, and binding of a hydrophobic probe solute to hydrophobin-II allow identification of hydrophobic and hydrophilic regions on its surface. This method represents a potentially powerful approach to characterize hydrophobicity (philicity) of complex and heterogeneous nanoscale surfaces.

## II. Simulation details

**Homogeneous interfaces**

The construction of model self assembled monolayer (SAM) surfaces has been described in detail previously;[10,17,18] we repeat the key information below. Our simulations include a bilayer presenting two monolayers to the water phase which fills the periodic system. Two alkane chains (C$_{10}$) terminated with a suitable head-group with chemistries from hydrophobic to hydrophilic (–CH$_3$, –OCH$_3$, –CONHCH$_3$, –CN, –OH, –CONH$_2$) are connected to a central sulfur atom to form the bilayer.

The sulfur atoms are position restrained consistent with the gold 111 lattice.[19] To prevent channeling of water into the SAM surface at high pressures, the seventh atom in the alkane chain (from the sulfur) was also position constrained in all simulations. The well-packed bilayer has a cross-sectional area of 3.46 × 3.5 nm² (with 112 total chains). The alkane chains were represented using the united atom model[20] and the head-groups were represented explicitly with the AMBER force-field.[21] For the –CN headgroup, the OPLS[22] force field was used.

### Patchy SAMs

To study the effect of chemical heterogeneity, we performed simulations by replacing $n$ contiguous headgroups in the –CH$_3$ SAM by –OH and *vice versa* for $n = 1, 2..., 9$, creating patches of different sizes (see Fig. 3 and 4). A ~4 nm thick slab of water containing about 1500 water molecules was used to solvate the homogeneous and patchy SAM systems, leading the z-dimension of ~7 nm in length. To obtain the spatial distribution of binding of hydrophobic probe solutes at the interface, additional simulations of the patchy SAMs in solutions containing 8 repulsive methane-sized Weeks–Chandler–Andersen[23] solutes ($\sigma = 0.373$ nm and $\varepsilon = 1.234$ kJ mol$^{-1}$) were also performed.

### Mixed SAMs

To study the effect of surface composition, 25%, 50% and 75% of the –CH$_3$ groups in a homogenous –CH$_3$ SAM were replaced with –OH groups (Fig. 6). For these simulations a larger SAM with a cross-sectional area 3.46 × 4.0 nm² with a total of 128 headgroups was used. The periodic system was solvated with a ~4 nm thick slab of water containing roughly 2200 water molecules. 20 ns long simulations were performed for all SAM systems with coordinates stored every ps for analysis.

### Protein systems

To quantify the overall compressibility of protein hydration shells we selected three proteins: hydrophobin-II (HYD, pdb-id 2b97), lysozyme (LYZ, pdb-id 1lyz), and staph nuclease (SNS, pdb-id 1snq). For hydrophobin-II only a single chain of the dimer was used in the simulations. All proteins were represented using the AMBER force field.[21] Proteins are solvated with ~10,000 water molecules in a periodic box of dimensions 7 × 7 × 7 nm³. Electroneutrality was maintained by addition of a suitable number of chloride counterions. To characterize the hydrophobicity of the protein surface, additional simulations of solutions containing methane-like Lennard-Jonesiums ($\sigma = 0.3855$ nm, $\varepsilon = 0.694$ kJ mol$^{-1}$) were performed for HYD, which is expected to have large hydrophobic patches.[24,25] Following 2 ns of equilibration, at least 20 ns of data were collected for all protein systems.

The SPC/E[26] water model was used in all simulations. Electrostatics were handled using the PME algorithm.[27] The Berendsen coupling schemes were used to maintain constant temperature (300 K) and pressure.[28] For the homogeneous and mixed SAMs, and proteins, simulations were performed at 1 bar, 1 kbar and 2 kbar to characterize the interfacial compressibility from the pressure response of density. For patchy SAMs, simulations were performed at 1 atm. For the SAM systems, anisotropic pressure coupling was used. All simulations were performed using the molecular dynamics package GROMACS[29–31]

## III. Results and discussion

### A. Homogeneous interfaces

Recently, we have shown[10] that the local density of water can not be quantitatively correlated with the wetting properties of the surface, and that water density fluctuations near a surface provide a better measure of the hydrophobicity of that surface. Density fluctuations are related to compressibility of the medium. To quantify the compressibility of water we performed MD simulations at pressures of 1, 1000, 2000, and 4000 bar. Calculations of the hydration layer compressibility for SAM surfaces have not been reported previously.

Fig. 1 shows water and SAM density profiles with increasing hydrostatic pressure. The SAM phase is crystalline, and its structure does not change appreciably with increasing pressure. In contrast, the response of local water density to pressure is clearly different near hydrophobic and hydrophilic surfaces. With increasing pressure layering is amplified near the $-CH_3$ surface, and the density profile moves closer to the SAM reducing the interfacial width. At the hydrophilic $-OH/-CONH_2$ surfaces, the water density is much less sensitive to pressure. These results suggest that compressibility of interfacial water may serve as a sensitive measure of surface hydrophobicity.

Fig. 2 shows the local compressibility, $\chi(z)$, obtained by taking the pressure derivative of local density, $\rho(z)$, in a 0.25 nm thick slab parallel to the surface.

$$\chi(z) = \frac{1}{\rho(z)} \left( \frac{\partial \rho(z)}{\partial P} \right)_{T,\Delta} \tag{1}$$

where $\Delta$ is the slab thickness, and $z$ is the location of the center of the slab. We note that $\rho(z)$ includes contributions from both water-oxygens and SAM heavy atoms, if their centers are located within the slab. In all $\chi(z)$ profiles, there is a maximum near the location where the SAM and water density profiles intersect, and the total heavy atom density (SAM + water oxygens) is a minimum. We define the compressibility at this point as the interfacial compressibility, $\chi_{int}$. Near the most hydrophobic $-CH_3$ surface, $\chi_{int}$ is ~12 times that in bulk water (Fig. 2b), indicating the high sensitivity of water density to pressure in the region closest to the SAM surface. With increasing surface hydrophilicity, the interfacial compressibility decreases, and near the $-OH$ SAM surface it becomes approximately equal to that of bulk water. Higher fluctuations and compressibility of interfacial water near a model hydrophobic surface and their gradual quenching/reduction with an applied field were observed previously.[32]

We also measured compressibility in a slab of width 0.25 nm centered at the first peak of $\rho_{water}(z)$ profile for different surfaces. This compressibility can be thought of

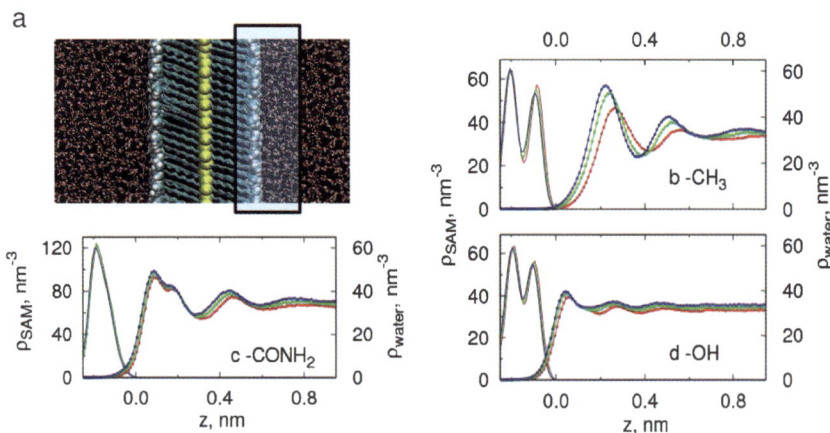

**Fig. 1** a: Snapshot of side-view of the SAM-water system. Sulfur (yellow) and headgroups (cyan) shown in spacefill representation. Alkane chains shown in licorice representation. Water (oxygen—red, hydrogen—white) shown in line representation. b–d: Heavy atom number density profiles of SAM (solid lines) and water (lines with points) for $-CH_3$, $-CONH_2$ and $-OH$ SAM systems in the boxed region of a. Color scheme: Red (1 bar), green (1 kbar), blue (2 kbar). At $z = 0$ the SAM number density is 0.05/nm$^3$ (~1% of the bulk SAM density).

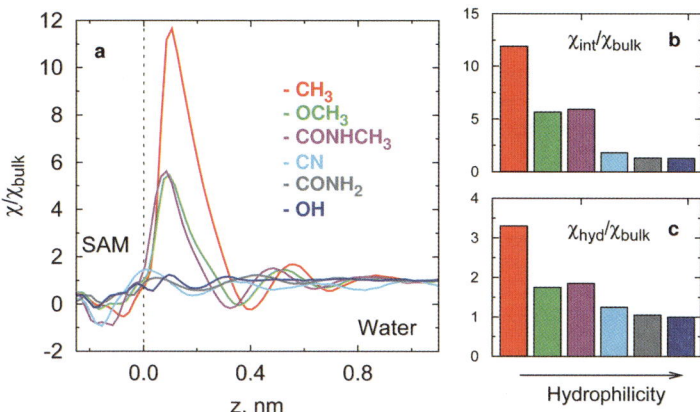

**Fig. 2** a: Normalized local compressibility, $\chi(z)/\chi_{bulk}$, in different SAM–water systems, along the surface normal in 0.25 nm thick slabs. Dashed line indicates where the SAM heavy atom density goes to ∼1% of the bulk density. **b**: $\chi_{int}/\chi_{bulk}$ (see text), and **c**: Normalized hydration-shell compressibility, $\chi_{hyd}/\chi_{bulk}$, at the first peak of $\rho_{water}(z)$. See the legend of panel **a** for color scheme.

as the "hydration shell compressibility", or $\chi_{hyd}$. Although compressibilities of the low density and high density regions, $\chi_{int}$ and $\chi_{hyd}$, are numerically different, the trend with chemistry is robust, with higher values near hydrophobic interfaces (Fig. 2c). Bratko et al.[33] and Giovambattista et al.[34] also found that the compressibility of water confined between hydrophobic plates is significantly higher than that of bulk water, especially at smaller separation between the plates.

Another measure of local compressibility based on fluctuations, $\chi_{fl}(z)$, can be obtained by monitoring the probability distribution of the number of heavy atoms, $N$, in an observation volume of interest, $V$, at different locations in the inhomogeneous system and using the compressibility equation,

$$\chi_{fl}(z) = \frac{V}{kT} \frac{\langle N(z)^2 \rangle - \langle N(z) \rangle^2}{\langle N(z) \rangle^2} \quad (2)$$

where $\langle \rangle$ denotes the ensemble average, $k$ is Boltzmann's constant, and $T$ is temperature. $\chi_{fl}$, depends on the size and shape of the observation volume (even in bulk systems) and equals $\chi$ only in the thermodynamic limit. For molecular-sized volumes near an interface the density fluctuations have previously been shown to correlate with hydrophobicity of the surface.[10] Here we find that $\chi$ and $\chi_{fl}$, calculated for considerably larger volumes (slabs of thickness 0.25 nm), track each other well, and support the picture that water at hydrophobic interfaces is more compressible (data not shown). The probability distributions obtained in such large slabs near hydrophobic surfaces also show non-Gaussian behavior in their low-N tails. The importance of fluctuations and their deviation from Gaussian behaviour has been highlighted previously.[35,36]

Collectively, our results highlight the utility of interfacial water compressibility in characterizing surface hydrophobicity. Below we explore the use of density fluctuations as well as interfacial compressibility as measures of hydrophobicity (philicity) to characterize chemically heterogeneous surfaces in model SAM systems as well as in proteins.

## B. Heterogeneous and patterned surfaces

Realistic interfaces, including those of biomolecules, contain chemical and structural heterogeneities. For smooth surfaces, when the lengthscale of chemical heterogeneities is large compared to the size of a water molecule, one may quantify interfacial free energy simply as a sum of different independent contributions weighted by composition.[37,38] Such linear descriptions are not expected to hold when heterogeneities are nano or subnanoscale in size.[39] To understand the effects of nanoscale heterogeneities, at the most basic level, we consider how a single –OH (or –CH$_3$) group affects the behavior of vicinal water when it is embedded in a sea of –CH$_3$ (or –OH) groups (Fig. 3a–b).

Fig. 3c–f show spatially resolved (inverse) normalized fluctuations, $[\langle N \rangle^2/\sigma_N^2]$, i.e., $V/(kT\chi_{fl})$, in methane-sized spherical observation volumes with a radius of 0.33 nm near these surfaces. Given that in small observation volumes number fluctuations are roughly Gaussian, the inverse normalized fluctuations are approximately equal to $2\mu^{ex}/kT$,[40] where $\mu^{ex}$ is the free energy of formation of a methane-sized cavity.[41] Thus, a higher value of $[\langle N \rangle^2/\sigma_N^2]$ at a given location indicates an unfavorable work of cavity creation at that location. Such a location would be termed hydrophilic. In contrast, density fluctuations are larger near hydrophobic surfaces and $[\langle N \rangle^2/\sigma_N^2]$ is correspondingly lower in magnitude.

Finding the signature for a single mutation in a large surface may appear challenging, similar to looking 'for a needle in a haystack'. However, our results show that (inverse) local density fluctuations, $[\langle N \rangle^2/\sigma_N^2]$, provide a sensitive measure to detect those single mutations (Fig. 3c–f). An interesting aspect is that the effects of implanting a single group are not symmetric. A single –OH group in the –CH$_3$ background quenches the local density fluctuations and increases the $\mu^{ex}$ value far more effectively than a single –CH$_3$ group in the background of –OH groups. The hydrophobic CH$_3$–SAM–water interface is soft, compressible, and accommodates larger fluctuations. These soft interfaces respond sensitively to solute–water attractions.[14] Thus, introduction of a hydrophilic –OH group which can hydrogen bond with water can effectively quench the local fluctuations. Oscillations in the $[\langle N \rangle^2/\sigma_N^2]$ values near the –OH surface with –CH$_3$ mutation (Fig. 5f) also resolves the underlying –OH pattern clearly.

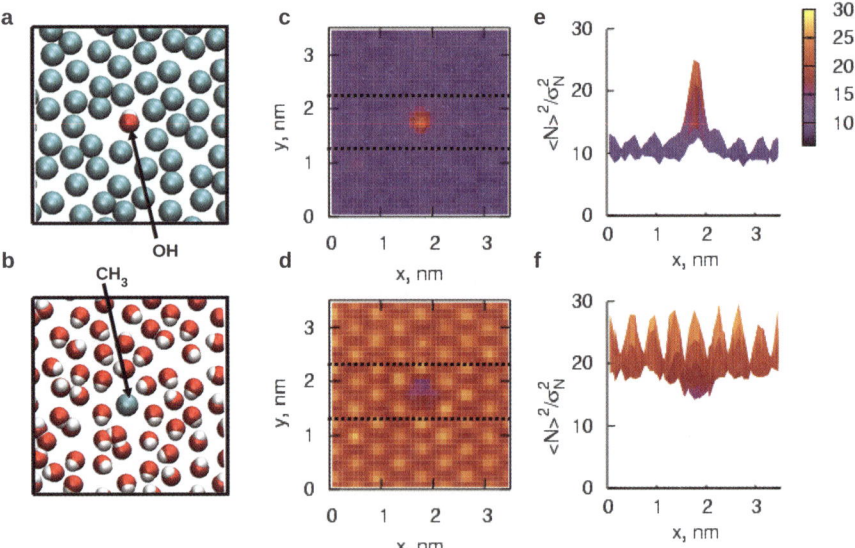

Fig. 3 a–b: Top view of the headgroup atoms of heterogeneous SAM surfaces containing one –OH group in a –CH$_3$ background and one –CH$_3$ group in an –OH background, respectively (red: oxygen, white: hydrogen, cyan: carbon). The hydrogens of the –CH$_3$ headgroup are not shown for clarity. c–d: Top views of the spatially resolved inverse fluctuations, $[\langle N \rangle^2/\sigma_N^2]$, in methane-sized spherical volumes (radius = 0.33 nm) at $z \sim 0$ nm are shown for the two heterogeneous SAMs shown in a–b (grid size of 0.1 nm). e–f: Side views of the data shown in c–d. For clarity, only the section within the dotted lines, 1.25 nm $< y <$ 2.25 nm, in c–d is shown.

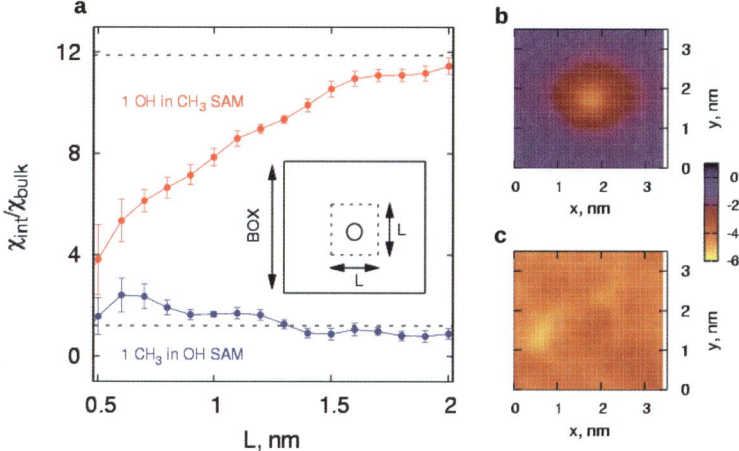

**Fig. 4** a: Lengthscale of the perturbation caused by a single mutation: Normalized interfacial compressibility, $\chi_{int}/\chi_{bulk}$, in cuboid observation volumes of size $L \times L \times \Delta$ ($\Delta = 0.25$ nm). In the $x$–$y$ plane, the observation volume is centered at the heterogeneous headgroup (see schematic in the inset of panel a). The dashed lines indicate $\chi_{int}/\chi_{bulk}$ of homogeneous –CH$_3$ and –OH SAM–water interfaces. b–c: Spatially resolved interfacial concentration of WCA solutes in arbitrary units (on a log scale) near surfaces containing one –OH group in a –CH$_3$ SAM and one –CH$_3$ group in an –OH SAM, respectively.

The asymmetry observed above can be further characterized by monitoring compressibility, $\chi_{int}$, in cuboid volumes ($L \times L \times \Delta$) centered on the location of the mutation. Near a homogeneous –CH$_3$ SAM, as shown in Fig. 2, $\chi_{int}/\chi_{bulk} \approx 12$. Fig. 4 shows that introduction of a single –OH group reduces $\chi_{int}/\chi_{bulk}$ to ~4 in cuboids of $L = 0.5$ nm. As the cuboid length increases, the compressibility increases gradually and reaches its limiting value of ~12 for $L \approx 2$ nm. In contrast, introduction of a –CH$_3$ group in an –OH background increases $\chi_{int}/\chi_{bulk}$ by a small amount in sub-nanometre length cuboids, and has negligible effects for cuboids of larger

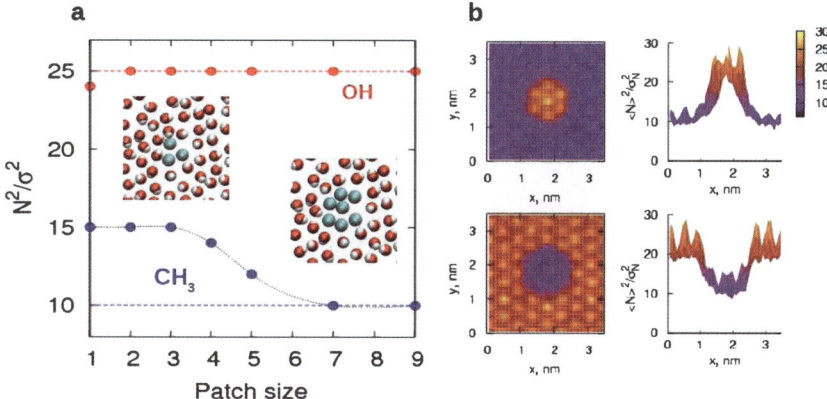

**Fig. 5** a: Inverse fluctuations, $[\langle N \rangle^2/\sigma^2_N]$, in a spherical observation volume of radius of 0.33 nm just above the center of a patch as a function of the number of head-groups in that patch. The horizontal dashed lines indicate the values near homogeneous –CH$_3$ and –OH SAM water interfaces. The black dashed line is a guide to the eye to highlight the sigmoidal transition as the hydrophobic –CH$_3$ patch size increases. Snapshots of surfaces containing 3 and 7 CH$_3$ groups in an –OH SAM are shown for reference. The representation is identical to Fig. 3a–b. b: The same as panels c–f of Fig. 3. The inhomogeneity now comprises a patch of 7 contiguous –OH groups in a –CH$_3$ SAM and *vice versa*.

lengths. This analysis again highlights a more significant effect of the introduction of a hydrophilic heterogeneity in the background of hydrophobic groups. The effect of this asymmetry can also be seen when the interfacial concentration of repulsive WCA probe solutes is spatially resolved (Fig. 4b–c). The probe solutes are excluded from the region around the central –OH headgroup in a –CH$_3$ background. In contrast, no local enhancement of probe solutes in the vicinity of the –CH$_3$ headgroup is seen in an –OH background.

We systematically explored how increasing the size of the mutated domain affects the vicinal fluctuations (Fig. 5). As the domain size increases, we expect the $\mu^{ex}$ value at the center of the domain to gradually approach that near a homogeneous surface. Fig. 5a shows that for a single –OH group in the –CH$_3$ background, the $\mu^{ex}$ in the vicinity is already close to that near the homogeneous –OH SAM. However, for a –CH$_3$ patch, the values of $[\langle N \rangle^2/\sigma_N^2]$ remain high even at the center of a patch containing three –CH$_3$ groups indicating boundary effects of surrounding –OH groups. There is a sigmoidal transition in $[\langle N \rangle^2/\sigma_N^2]$ values over patch sizes of four to seven. Thus, a contiguous patch containing at least seven –CH$_3$ groups is required before the value at the patch center approaches the homogeneous –CH$_3$ surface limit, as seen in Fig. 5a–b.

The asymmetry observed here implies that as hydrophilic heterogeneities are incorporated into a hydrophobic background surface, they will have significant effects even at relatively low concentrations. Further, at a given concentration, the effects will depend on the pattern of arrangement,[42] especially at low concentrations of hydrophilic groups. Lattice gas based simulations have predicted that the effects will be most significant when the hydrophilic heterogeneities are distributed uniformly.[12] To test this, we performed simulations of mixed –CH$_3$/–OH SAMs containing 25%, 50%, and 75% uniformly distributed mixtures, as shown in Fig. 6a–c. We monitored water compressibility as well as solvation chemical potential of hydrophobic probe solutes to characterize average hydrophobicity/philicity of these mixed surfaces.

As observed earlier in Fig. 2, the local compressibility is high near the hydrophobic –CH$_3$ SAM and low, bulk-like, near the –OH SAM. As expected from the asymmetry observed above, the compressibility of water near –CH$_3$ SAM drops dramatically when 25% of the groups are changed to –OH (Fig. 6d). Further increase in the concentration of –OH groups reduces the water compressibility but by a smaller magnitude, eventually approaching the homogeneous –OH value. Excess chemical potential of probe hydrophobic methane like solutes interacting with the system via purely repulsive WCA[23] potential are consistent with the compressibility data (Fig. 6e and f). A clear minimum is observed in the $\mu^{ex}$ value near homogeneous –CH$_3$ SAM surfaces, indicating that such a probe solute would preferentially bind the hydrophobic SAM surface. Addition of –OH groups to the –CH$_3$ SAM surface has the most dramatic effect at low concentrations. Increasing the surface hydrophilicity reduces the favorable hydration of probe hydrophobic solutes at the interface relative to that in the bulk, with homogeneous –OH surfaces effectively repelling the probe hydrophobic particles.

This asymmetry is likely important in understanding and engineering of protein–protein interactions. Our results indicate that single hydrophilic mutation in a hydrophobic patch can significantly reduce the strength of hydrophobically driven association and binding processes. Conversely, to strengthen hydrophobically driven binding significantly and to confer specificity one may need to mutate several hydrophilic groups to hydrophobic ones, leading to the formation of a contiguous patch.

## C. Proteins

Surfaces of proteins are chemically heterogeneous as well as rough. Traditionally hydrophobicity/philicity of different surface exposed residues (and the regions/patches they form) on proteins is assigned based on the so-called hydropathy

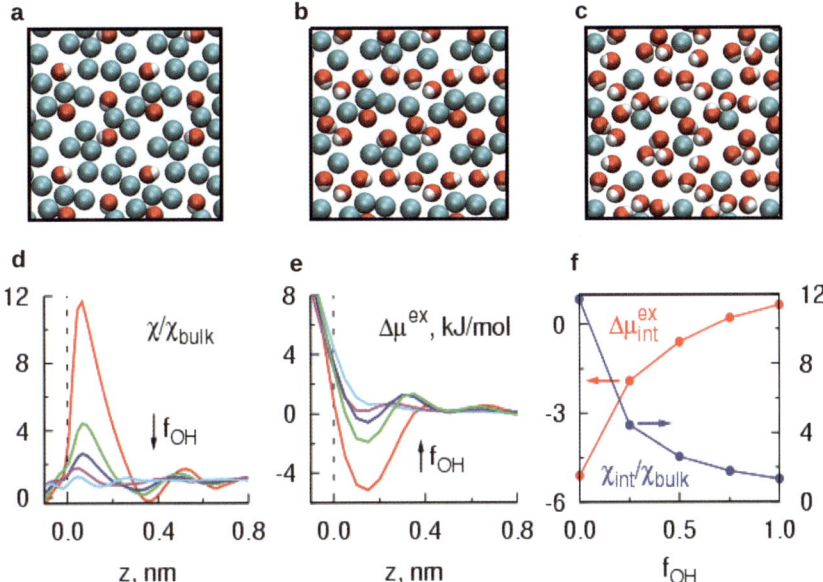

**Fig. 6** a–c: Snapshots of mixed SAM systems containing 25%, 50% and 75% –OH headgroups, respectively. **d**: Normalized local compressibility, $\chi/\chi_{bulk}$, in mixed systems, measured along the surface normal in slabs of thickness 0.25 nm. **e**: The excess chemical potential of a repulsive WCA methane relative to that in bulk water, $\Delta\mu^{ex}$, as a function of the distance from the mixed SAM surface. In **d** and **e**, the dashed line indicates the end of the SAM surface, and the arrow points in the direction of increasing –OH fraction or hydrophilicity. Colors: red (–CH$_3$ surface), green (25% –OH), blue (50% –OH), magenta (75% –OH), and cyan (–OH surface). **f**: $\chi_{int}/\chi_{bulk}$ and $\Delta\mu^{ex}$ at the interface as a function of the –OH group fraction. The values at the interface are taken to be those at the maxima and minima in the profiles shown in panels **d** and **e**, respectively.

scales.[43,44] Our study of homogeneous and heterogeneous model surfaces potentially presents a method for molecular level quantification of hydrophobicity that accounts for the behavior of vicinal water in the presence of chemical and topographical heterogeneities. Specifically, monitoring normalized density fluctuations, free energy of cavity formation, or the strength of binding of probe hydrophobic solutes to various regions on the protein surface would provide such a quantification. Similar ideas were employed previously by Siebert and Hummer[45] in their characterization of hydrophobicity maps of the surface of a HIV-1 gp41 peptide in water.

As a first step, we focused on quantifying the overall hydrophobicity of the entire protein surface by calculating the compressibility, $\chi_{hyd}$, of the protein hydration shell. Fig. 7 shows $\chi_{hyd}/\chi_{bulk}$ obtained from analysis of molecular dynamics simulations over a range of pressures for three different proteins. $\chi_{hyd}/\chi_{bulk}$ is ~1.7 for these proteins, and 3.3 and 1.3, respectively, for homogeneous –CH$_3$ and –OH surfaces, consistent with the fact that typical globular protein surfaces are mostly hydrophilic.

Fig. 7 also shows hydration shell compressibilities for two representative unfolded states of the protein staph nuclease, UF1 and UF2, generated using a water insertion algorithm.[14] Although both UF1 and UF2 expose more hydrophobic areas than the folded state of SNS, UF2 is a much more extended structure compared to UF1 (not shown). Hydration shell fluctuations for these conformations at low pressure were reported previously.[14] Our pressure-dependent calculations here again show that exposure of the buried hydrophobic residues in the unfolded states leads to higher hydration shell compressibilities for those states.[14,46–48] Interestingly, the hydration shell compressibility for the unfolded structures are even higher than that for

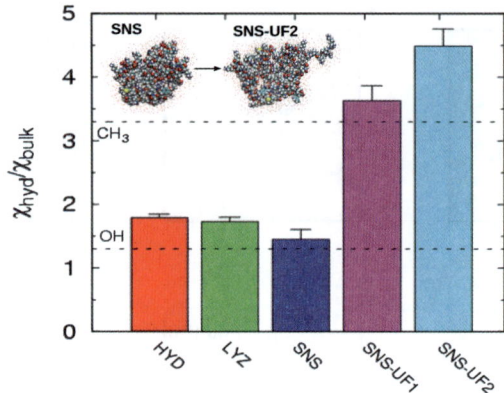

**Fig. 7** Normalized hydration shell compressibility, $\chi_{hyd}/\chi_{bulk}$, of three different proteins, hydrophobin-II (HYD), lysozyme (LYZ), and staph nuclease (SNS). We approximate $\chi_{hyd} \approx \frac{1}{N}\left[\frac{\partial N}{\partial P}\right]_T$ because the structure of the folded protein is largely unchanged over the range 1 bar–2 kbar and the hydration shell volume remains constant.[49] For SNS, data are also reported for two unfolded conformations, SNS-UF1 and SNS-UF2 that were generated by a water-insertion algorithm.[50,51] The dashed lines indicate $\chi_{hyd}/\chi_{bulk}$ of homogeneous –OH (1.3) and –CH$_3$ (3.3) SAM–water interfaces. The VMD[52] images at the top show representative conformations of SNS surrounded by hydration shell water in its folded and unfolded states.

a flat –CH$_3$ surface, which highlights the effects of surface roughness in increasing effective hydrophobicity of the surface.

Of the three folded proteins studied here, hydrophobin-II appears to be most hydrophobic, as its name would indicate! Hydrophobins are secreted by fungi in the extracellular environment.[24] They assemble in interfacial environments and in water to create hydrophobic coatings and form complex aerial structures and spores. To resolve hydrophobicity patterns of this protein, we performed simulations of the protein in an aqueous solution containing 75 probe hydrophobic solutes. The number density of probe solutes is small and they do not aggregate over the length of simulation. A sufficiently long simulation trajectory provides information on the average local density of probe particles in the vicinity of different parts of the protein, which can then be used to analyze the local excess chemical potentials of probe particles, and to quantify the hydrophobicity map of the protein. Our preliminary simulations were 20 ns long, and provide a qualitative picture based on analysis of probe densities. Fig. 8c–d show two views of the protein colored using the Kyte and Dolittle hydropathy scale[43] as well as our qualitative analysis based on probe binding studies (Fig. 8e–f). The hydropathy scale based mapping suggests large regions of the protein surface as hydrophobic (Fig. 8c). The probe-density based maps paint a slightly different/refined picture of the hydrophobicity of the hydrophobin-II surface with a few smaller regions identified as hydrophobic "hot spots".

To further explore the local hydrophobicity of the hydrophobin-II surface, guided by the probe binding simulations, we selected three regions, marked 1, 2, and 3 in Fig. 8e–f and monitored density fluctuations in their vicinity. Specifically, we selected three protein heavy atoms in these regions and calculated normalized heavy atom density fluctuations in methane-sized spherical observation volumes along an axis perpendicular to the plane of the atoms and passing through their centroid (Fig. 9a). Fig. 9b shows the inverse normalized fluctuations (or equivalently, $2\mu^{ex}/kT$ values) along that axis. A lower value of $\mu^{ex}$ indicates more favorable hydration of a hydrophobic probe solute and stronger binding. Indeed, we observe a minimum near the hydrophobic region identified by probe binding studies (region 1), whereas near the hydrophilic region (region 3), the $\mu^{ex}$ profile is slightly repulsive, consistent

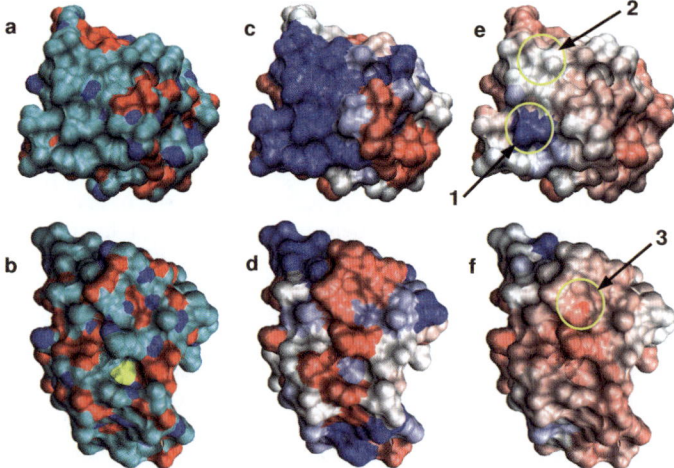

**Fig. 8** **a–b** Two different views of the protein hydrophobin with the atoms colored according to exposed heavy atom [carbon (cyan), oxygen (red), nitrogen (blue), and sulfur (yellow)] and represented using the "surface" option in VMD.[52] **c–d** Same as **a–b**, with the atoms being amino acids colored according to the Kyte and Dolittle hydropathy scale for amino acids.[43] Red indicates hydrophilic and blue indicates hydrophobic. **e–f** Protein atoms colored according to the frequency with which hydrophobic probe solutes are present within a cut-off distance of 0.4 nm. To smoothen the data, the value for the final coloring scheme for each atom is obtained by averaging over the values for all heavy atoms within a cut-off distance of 0.6 nm. The yellow circles highlight three regions (1—hydrophobic, 2—moderately hydrophilic, 3—hydrophilic) where we compare the results from hydropathy scales and probe binding simulations (see Fig. 9).

with weaker binding of probe solutes there. Region 2 displays an intermediate behavior consistent with the local density of probe solutes in its vicinity.

Although the preliminary results from probe binding and fluctuation-based analysis shown above may not be entirely unexpected, our approach can potentially identify regions on the protein surfaces whose hydrophobicity pattern is different

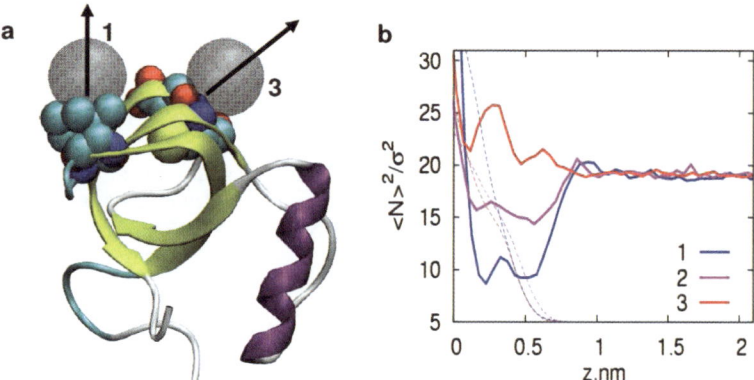

**Fig. 9** **a**: The protein hydrophobin is shown in cartoon representation. Atoms from region 1 (hydrophobic) and 3 (hydrophilic) of Fig. 8 are shown in van der Waals representation and heavy atom based coloring (see caption of Fig. 8). Typical observation volumes for monitoring density and density fluctuations are shown by grey spheres. We monitor fluctuations in such volumes along the direction shown by the arrows. **b**: Inverse density fluctuations calculated in spherical observation volumes of radius 0.33 nm for the three regions shown in Fig. 8e–f along the outward normal. Dashed lines indicate the protein heavy atom density for the three regions and are shown for reference to locate the interface.

from that suggested by hydropathy scales (*e.g.*, region 2 above). To this end, it would be ideal to develop a method to sample density fluctuations in the tails of the particle number distributions and for volumes that are larger than the size of methane and possibly have non-spherical shapes. Such studies are currently being pursued in our group.

## IV. Conclusions

Our simulations show that enhanced density fluctuations at hydrophobic–water interfaces also make those interfaces significantly more compressible than bulk water, whereas the compressibility of charge-neutral hydrophilic–water interfaces is bulk like. The density fluctuations are sensitive to interactions, and respond to small changes in the chemistry of the underlying surface. Specifically, the heterogeneous surfaces show an interesting context dependence – adding a single –OH group in a –$CH_3$ terminated SAM has a more dramatic effect on water in the vicinity compared to that of a single –$CH_3$ group in an –OH background. Also, the effect of a single –OH group extends roughly over an area of $\sim$4 $nm^2$. In contrast, the effect of a single –$CH_3$ group in a –OH SAM is weak and local. Consequently, for mixed –$CH_3$/–OH SAMS, hydrophobicity depends nonlinearly on the surface concentration. This asymmetry in how hydrophobic and hydrophilic chemistries affect fluctuations will have fundamental implications for controlling and manipulating surface hydrophobicity,[39] protein–protein[53] and protein–ligand interactions. We expect the asymmetric effects observed here to be amplified in the confined geometries again highlighting the context dependence of hydration phenomena.[54]

Finally, we presented results aimed at mapping the hydrophobicity of protein surfaces. Compressibility of protein hydration shells is only slightly more than that of bulk water, indicating the overall hydrophilic nature of the protein–water interfaces. The hydration shell compressibility of a protein provides only a coarse-grained measure of the surface hydrophobicity. The hydrophobic regions on the surface are expected to enhance density fluctuations or increase the probability of cavity formation in the vicinity. Indeed we observe a preferential binding of hydrophobic probe solutes to hydrophobic regions of the protein. We find that the regions suggested as hydrophobic by standard hydropathy scales and those identified by the probe binding simulations are not identical. This highlights a limitation of the context independent hydropathy scales which assign values independent of proximal environment in the protein. Our approach complements experimental methods for characterizing binding sites on protein surfaces using a variety of probe solutes.[55]

## V. Acknowledgments

We gratefully acknowledge discussions with Dr. Amish Patel, Dr. Gerhard Hummer, and Prof. David Chandler. We also acknowledge partial financial support of NSF-NSEC (DMR-0642573) and New York State NYSTAR program to Rensselaer Nanotechnology Center.

## References

1 K. A. Dill, *Biochemistry*, 1990, **29**, 7133–7155.
2 D. Chandler, *Nature*, 2005, **437**, 640–647.
3 C. Tanford, *J. Am. Chem. Soc.*, 1962, **84**, 4240.
4 J. N. Israelachvili, D. J. Mitchell and B. W. Ninham, *J. Chem. Soc., Faraday Trans. 2*, 1976, **72**, 1525–1568.
5 B. I. Giasson, I. V. J. Murray, J. Q. Trojanowski and V. M. Y. Lee, *J. Biol. Chem.*, 2001, **276**, 2380–2386.
6 S. Granick and S. C. Bae, *Science*, 2008, **322**, 1477–1478.
7 A. K. Ghose, V. N. Viswanadhan and J. J. Wendoloski, *J. Phys. Chem. A*, 1998, **102**, 3762–3772.
8 W. Mar and M. L. Klein, *Langmuir*, 1994, **10**, 188–196.

9 G. B. Sigal, M. Mrksich and G. M. Whitesides, *J. Am. Chem. Soc.*, 1998, **120**, 3464–3473.
10 R. Godawat, S. N. Jamadagni and S. Garde, *Proc. Natl. Acad. Sci. U. S. A.*, 2009, **106**, 15119–15124.
11 A. P. Willard and D. Chandler, *Faraday Discuss.*, 2009, **141**, 209–220.
12 A. Luzar and K. Leung, *J. Chem. Phys.*, 2000, **113**, 5836–5844.
13 J. Mittal and G. Hummer, *Proc. Natl. Acad. Sci. U. S. A.*, 2008, **105**, 20130–20135.
14 S. Sarupria and S. Garde, *Phys. Rev. Lett.*, 2009, **103**, 037803.
15 B. Schreier, C. Stumpp, S. Wiesner and B. Hocker, *Proc. Natl. Acad. Sci. U. S. A.*, 2009, **106**, 18491–18502.
16 M. R. Landon, R. L. Lieberman, Q. Q. Hoang, S. L. Ju, J. M. M. Caaveiro, S. D. Orwig, D. Kozakov, R. Brenke, G. Y. Chuang, D. Beglov, S. Vajda, G. A. Petsko and D. Ringe, *J. Comput.-Aided Mol. Des.*, 2009, **23**, 491–500.
17 N. Shenogina, R. Godawat, P. Keblinski and S. Garde, *Phys. Rev. Lett.*, 2009, **102**, 156101.
18 S. N. Jamadagni, R. Godawat and S. Garde, *Langmuir*, 2009, **25**, 13092–13099.
19 J. C. Love, L. A. Estroff, J. K. Kriebel, R. G. Nuzzo and G. M. Whitesides, *Chem. Rev.*, 2005, **105**, 1103–1169.
20 M. Mondello, G. S. Grest, E. B. Webb and P. Peczak, *J. Chem. Phys.*, 1998, **109**, 798–805.
21 W. D. Cornell, P. Cieplak, C. I. Bayly, I. R. Gould, K. M. Merz, D. M. Ferguson, D. C. Spellmeyer, T. Fox, J. W. Caldwell and P. A. Kollman, *J. Am. Chem. Soc.*, 1995, **117**, 5179–5197.
22 M. L. P. Price, D. Ostrovsky and W. L. Jorgensen, *J. Comput. Chem.*, 2001, **22**, 1340–1352.
23 J. D. Weeks, D. Chandler and H. C. Andersen, *J. Chem. Phys.*, 1971, **54**, 5237.
24 H. A. B. Wosten, *Annu. Rev. Microbiol.*, 2001, **55**, 625–646.
25 J. Hakanpaa, M. Linder, A. Popov, A. Schmidt and J. Rouvinen, *Acta Crystallogr., Sect. D: Biol. Crystallogr.*, 2006, **62**, 356–367.
26 H. J. C. Berendsen, J. R. Grigera and T. P. Straatsma, *J. Phys. Chem.*, 1987, **91**, 6269–6271.
27 U. Essmann, L. Perera, M. L. Berkowitz, T. Darden, H. Lee and L. G. Pedersen, *J. Chem. Phys.*, 1995, **103**, 8577–8593.
28 H. J. C. Berendsen, J. P. M. Postma, W. F. Vangunsteren, A. Dinola and J. R. Haak, *J. Chem. Phys.*, 1984, **81**, 3684–3690.
29 E. Lindahl, B. Hess and D. van der Spoel, *J. Mol. Model.*, 2001, **7**, 306–317.
30 H. J. C. Berendsen, D. Vanderspoel and R. Vandrunen, *Comput. Phys. Commun.*, 1995, **91**, 43–56.
31 J. P. Ryckaert, G. Ciccotti and H. J. C. Berendsen, *J. Comput. Phys.*, 1977, **23**, 327–341.
32 D. Bratko, C. D. Daub and A. Luzar, *Faraday Discuss.*, 2009, **141**, 55–66.
33 D. Bratko, R. A. Curtis, H. W. Blanch and J. M. Prausnitz, *J. Chem. Phys.*, 2001, **115**, 3873–3877.
34 N. Giovambattista, P. J. Rossky and P. G. Debenedetti, *Phys. Rev. E: Stat., Nonlinear, Soft Matter Phys.*, 2006, **73**, 041604.
35 A. J. Patel, P. Varilly and D. Chandler, *J. Phys. Chem. B*, In Press.
36 K. Lum, D. Chandler and J. D. Weeks, *J. Phys. Chem. B*, 1999, **103**, 4570–4577.
37 A. B. D. Cassie, *Discuss. Faraday Soc.*, 1948, **3**, 11–16.
38 J. R. Henderson, *Mol. Phys.*, 2000, **98**, 677–681.
39 J. J. Kuna, J. Voitchovsky, C. Singh, H. Jiang, S. Mwenifumbo, P. K. Ghorai, M. M. Stevens, S. C. Glotzer and F. Stellacci, *Nat. Mater.*, 2009, **8**, 837–842.
40 G. Hummer, S. Garde, A. E. Garcia and L. R. Pratt, *Chem. Phys.*, 2000, **258**, 349–370.
41 S. Garde, G. Hummer, A. E. Garcia, M. E. Paulaitis and L. R. Pratt, *Phys. Rev. Lett.*, 1996, **77**, 4966–4968.
42 M. Schneemilch, N. Quirke and J. R. Henderson, *J. Chem. Phys.*, 2003, **118**, 816–829.
43 J. Kyte and R. F. Doolittle, *J. Mol. Biol.*, 1982, **157**, 105–132.
44 S. D. Black and D. R. Mould, *Anal. Biochem.*, 1991, **193**, 72–82.
45 X. Siebert and G. Hummer, *Biochemistry*, 2002, **41**, 2956–2961.
46 J. L. Silva, D. Foguel and C. A. Royer, *Trends Biochem. Sci.*, 2001, **26**, 612–618.
47 V. M. Dadarlat and C. B. Post, *Biophys. J.*, 2006, **91**, 4544–4554.
48 G. Panick, R. Malessa, R. Winter, G. Rapp, K. J. Frye and C. A. Royer, *J. Mol. Biol.*, 1998, **275**, 389–402.
49 B. Pereira, S. Jain, S. Sarupria, L. Yang and S. Garde, *Mol. Phys.*, 2007, **105**, 189–199.
50 S. Sarupria, T. Ghosh, A. E. Garcia and S. Garde, *Proteins: Struct., Funct., Bioinf.*, 2010, **78**, 1641–1651.
51 A. Paliwal, D. Asthagiri, D. Bossev and M. Paulaitis, *Biophys. J.*, 2004, **87**, 3479–3492.
52 W. Humphrey, A. Dalke and K. Schulten, *J. Mol. Graphics*, 1996, **14**, 33–38.
53 N. Chennamsetty, V. Voynov, V. Kayser, B. Helk and B. L. Trout, *Proc. Natl. Acad. Sci. U. S. A.*, 2009, **106**, 11937–11942.
54 N. Giovambattista, P. G. Debenedetti and P. J. Rossky, *J. Phys. Chem. C*, 2007, **111**, 1323–1332.
55 C. Mattos and D. Ringe, *Nat. Biotechnol.*, 1996, **14**, 595–599.

# General Discussion

**Professor Bratko** opened the discussion of the paper by Dr Perkin: In your study, the increased stability of the STAI layer, associated with stronger binding of $I^-$ ions in the STAI layer is attributed to comparatively weaker hydration of free $I^-$, compared to $Br^-$ and $Cl^-$ ions. An alternative and likely dominant source of selective ion binding affinities is van der Waals binding of halide ions,[1] shown to increase in parallel with solvated ion polarizability[2,3] in the order $I^- > Br^- > Cl^-$ in agreement with present observations.

I also wish to add a remark on your observation of the insensitivity of intersurface attraction on rapid lateral motion of the surfaces. As you have shown, nonlinear Poisson–Boltzmann theory predicts attractions between neutral heterogeneous surfaces even in the absence of patch correlations. The proposed picture is reinforced by our calculations using Monte Carlo[4-6] and Hypernetted Chain[6,7] models where "the asymmetry in the strength and range of the interaction between colloidal entities with equal charges,[4,5] compared with oppositely charged pairs of the same absolute value,[6] suggests the presence of an overall attraction between electrostatically heterogeneous macroparticles or proteins with a nonuniform distribution of ionic groups."

1 B. W. Ninham and V. Yaminsky, *Langmuir*, 1997, **13**, 2097.
2 F. W. Tavares, D. Bratko, H. W. Blanch and J. M. Prausnitz, *J. Phys. Chem. B*, 2004, **108**, 9228.
3 M. Bostrom, F. W. Tavares, D. Bratko and B. W. Ninham, *J. Phys. Chem. B*, 2005, **109**, 24489.
4 J. Z. Wu, D. Bratko and J. M. Prausnitz, *Proc. Natl. Acad. Sci. USA*, 1998, **95**, 15169.
5 J. Z. Wu, D. Bratko, H. W. Blanch and J. M. Prausnitz, *J. Chem. Phys.*, 1999, **111**, 7084.
6 J. Z. Wu, D. Bratko, H. W. Blanch and J. M. Prausnitz, *Phys. Rev. E*, 2000, **62**, 5273.
7 D. Bratko, H. L. Friedman and E. C. Zhong, *J. Chem. Phys.*, 1986, **85**, 377.

**Dr Perkin** replied: The main point of our paper was to show how the anion selectivity affects the layer properties, probably this selectivity arises from several reasons which are all related one to another like hydration (solubility of the ions or in other words hydrophobicity or entropic effects), different charge density and thus different electrostatic interaction with the head groups, and also might be polarizability and van der Waals interactions. In the simulation and models mentioned, the solutions have high ionic strength so the pure electrostatic interaction becomes less dominant. Our case is more complicated, because the layer was adsorbed in surfactant solution (<5 mM) then later transferred to pure water where its stability and properties have been examined. Also the high packing in the case of $I^-$ which screens the headgroup repulsions suggests the importance of the electrostatic interactions between the head groups. Experimentally it has been shown that there is a dramatic preference of the iodide for the tri-methyl ammonium head group, much stronger than for the di-methyl ammonium head group, over bromide and chloride.[1] This was explained in terms of mutual de-hydration of the halide ion and the more hydrophobic headgroup.

Lastly, and regarding your remark on the interaction between heterogeneous surfaces: we see that your simulation work suggests an overall attraction between heterogeneous surfaces — without surface charge correlation effects — in line with the experimental evidence we present in our paper (and in a submitted paper by G. Silbert *et al.* on this specific topic). We refer to this also in our response to Professor Law's question above.

1 L. Kellaway and G. Warr, *J. Coll. Int. Sci.*, 1997, **193**, 312.

**Dr Christenson** opened the discussion of the paper by Professor Yoon: Your results are very interesting, but I am not sure how you can justify the fit to a van der Waals force with such a large Hamaker constant. The points in Figure 9 look as if they would fit a line of slope about 1/3 better than 1/2. Having said this, I must acknowledge that we did a similar thing with some results on the force between two fluorocarbon surfactant layers in water.[1] We used an "effective Hamaker constant" of $1.48 \times 10^{-18}$, which isn't all that different to what is used here. However, it was clear that we could get an equally good fit with two exponentials.

1 See Figure 9 of S. Ohnishi, V. V. Yaminsky and H. K. Christenson, *Langmuir*, 2000, **16**, 8360–8367.

**Professor Yoon** replied: Dr Christenson is questioning the validity of using the power law (Eq. 7) that is commonly used for weak van der Waals forces for the strong hydrophobic forces we have observed in the present work. We have been using both the power law and exponential force law interchangeably over the years to express hydrophobic forces. Most investigators use single- or double-exponential force laws with two or four fitting parameters, respectively, while the power law requires only one. We thought that in the absence of a force law based on a well-founded and generally accepted theory, it would be more meaningful to fit experimental data with a minimum number of parameters. An advantage of using the power law is that the values of $K_{131}$, $K_{232}$, and $K_{132}$ can be directly compared with the Hamaker constants. However, the main motivation for us using the power law was to see if we can use the geometric mean combining rule for hydrophobic interactions. We have shown previously that it can be used for asymmetric hydrophobic interactions between hydrophobic surfaces with different contact angles.[1,2] It appears that it can also be used for the asymmetric hydrophobic interactions in wetting films. We only have six data points in Fig. 9, and the point at the far left appears to be an outlier. If a line is drawn through this point, one might suggest that we cannot use the combining rule for wetting films as the slope is less than 1/2. We are currently in the process of conducting additional experiments to test the goodness-of-the-fit. We are happy to learn that Ohnishi *et al.*[3] also used the power law. Further, Professor Israelachvili[4] suggested that the "true" hydrophobic force at short separations below 1–2 nm is actually an extension of the van der Waals force due to protein hopping.

1 R.-H. Yoon, D. Flinn, and Y. Rabinovich, *J. Colloid Interface Sci.*, 1997, **185**, 363–370.
2 R.-H. Yoon and R. Pazhianur, *Colloid. Surf. A.*, 1998, **144**, 59–69.
3 S. Ohnishi , V. Yaminsky and H. K. Christenson, *Langmuir*, 2000, **16**, 8360–8367.
4 M. Hammer, T. Anderson, A. Chaimovich, M. Scott Shell and J. Israelachvili, *Faraday Discuss.*, 2010, **146**, DOI:10.1039/b926184b

**Dr Patel** opened the discussion of the paper by Professor Israelachvili: The spherically symmetric water model reproduces the PMF between two small hydrophobes. This implies that the model is likely to capture the solvation behavior of sub-nanometre sized solutes and their interactions. However, the solvation behavior and interactions between bigger extended solutes, such as the hydrophobic surfaces studied by the experiments are governed by the surface tension of water. Hence, it is crucial that the surface tension of the water model being used, agrees with the experimentally measured surface tension of water. Have you calculated the surface tension of the spherically symmetric water model?

**Mr Chaimovich** replied: We have not yet attempted to replicate the surface tension of water with various nonpolar surfaces. What we have presented thus far is only a first step in the assessment of the spherically-symmetric water model in describing the "pure" HI. Further examination of our model is necessary before we can make a comprehensive judgment regarding its overall capability in describing the true properties of the unique medium. In fact, we are currently extending our molecular simulations to allow for the computation of the HI between spheres of varying radii. Notably, we have already implemented an improved method for the determination of the HI based on the transition-matrix algorithm. Nevertheless, there is no guarantee that spherically-symmetric water can correctly reproduce the "pure" HI over all its length scales; however, the relative entropy optimization methodology at least provides us with a systematic approach to enhancing such models in this context.

**Professor Evans** addressed Professor Israelachvili and Mr Chaimovich: In the second part of Professor Israelachvili's paper Monte Carlo results are presented for the potential of mean force (PMF) between two hydrophobes immersed in a model of water developed recently by Shell and Chaimovich. The model is based on a (spherical) pair potential with parameters chosen to reproduce the properties of bulk water, using an approach that minimizes a quantity termed the relative entropy. Comparisons with results from fully atomistic models are made in Fig. 2 and there is qualitative agreement for the single (ambient) state considered. How well does the Shell model, *i.e.* the results in Fig. 3, describe the temperature variation of the PMF found for the atomistic models? It is implied in the text that the variable scenario Fig. 3(A) that yields a pronounced increase in the well depth with increasing temperature is consistent with results of ref. 1 and 2 (46 and 47 in the paper) but it would be valuable to see a detailed comparison.

The results in Figs. 2 and 3 are for hydrophobic separations $r$ out to 8 Å. Can the authors explain how results on these short length scales can shed new insight into the experimental SFA results that pertain to macroscopic surfaces and much larger separations $D$?

1 S. Ludemann, H. Schreiber, R. Abseher and O. Steinhauser, *J. Chem. Phys.*, 1996, **104**(1), 286–295
2 S. Shimizu and H. S. Chan, *J. Chem. Phys.*, 2000, **113**(11), 4683–4700.

**Mr Chaimovich** responded: Let us now describe the PMF curves in ref. 1 and 2, focusing in particular on the temperature trend of the energy difference between the innermost minimum and maximum. In ref. 1, this energy value changes from ~2.6 kJ mol$^{-1}$ at 250 K to ~4.6 kJ mol$^{-1}$ at 350 K. In ref. 2, this energy value changes from ~3.2 kJ mol$^{-1}$ at 278 K to ~4.2 kJ mol$^{-1}$ at 328 K. Assuming a linear trend, both of these suggest that this energy changes by about 0.8 kJ mol$^{-1}$ over a temperature drop of 40 K. For our spherically-symmetric water, this energy grows by ~0.7 kJ mol$^{-1}$ over our entire temperature change, 40 K. As this is only ~10% error, we deem the spherically-symmetric water model as a sufficient description of the medium for this particular phenomenon.

Nevertheless, we do not make any claims that the molecular simulations performed thus far can shed light on the SFA experiments. This computational work serves solely as a preliminary examination of spherically-symmetric water's ability in describing the 'pure' HI. Further molecular simulations are necessary before we can explain the phenomena observed *via* the SFA.

1 S. Ludemann, H. Schreiber, R. Abseher and O. Steinhauser, *J. Chem. Phys.*, 1996, **104**(1), 286–295
2 S. Shimizu and H. S. Chan, *J. Chem. Phys.*, 2000, **113**(11), 4683–4700.

**Mr Silbert** said: What are the evidences that support the idea of correlation between domains on opposing surfaces as the origin of the attraction between heterogeneously charged surfaces? (In the paper it was mentioned that the domains rearranged during AFM scans, is the force exerted by the AFM tip during scans comparable to the estimated force that might act between such patches at these separations?)

**Professor Israelachvili** answered: The AFM scan shows the presence of mobile domains, consistent with the cationic monolayers turning over to form positively charged bilayers floating on the negatively charged mica surface. That the domains are mobile during force measurements in the SFA can be sensed from the way the surfaces start to drift towards each other once they are closer than 100 nm — with the rate of drift noticeably accelerating as the surfaces get closer together. Equilibrium interaction forces do not show such drifts. The inwards drift implies an increasing attraction with time (at any given separation) due to, we suggest, the positioning of the domains into electrostatic registry, *i.e.*, with the positive domains facing negatively charged mica patches, so that at final contact there is only a single uniform bilayer between the surfaces. See also our response to Ms Mishchenko.

**Dr Christenson** commented: Given that the energetics of a small hydrophobic molecule in water, and of an extended hydrophobic surface in water are completely different, I do not see the point in attempting any linear extrapolation of the type described in Professor Israelachvili's 1982 paper with Pashley. Any agreement can only be purely coincidental.

**Professor Israelachvili** replied: I think it is important to at least try to establish whether there is some nanoscopic to microscopic (or intermolecular to intersurface) scaling of hydrophobic interactions, especially if we are to understand many biological processes such as the opening up of hydrophobic 'pockets' on protein or membrane surfaces leading to adhesion, binding or fusion, protein insertion into membranes, and protein folding. I agree that hydrophobic interaction potentials are unlikely to scale simply with the molecular or particle size (radius) as does the van der Waals force, although Pashley & Israelachvili did suggest a simple scaling relation for the adhesion energy that appears to be at least semi-quantitatively correct. When comparing the hydrophobic interactions between molecules with those between surfaces two important qualitative differences may be noted: (1) all the measured interaction forces so far measured between two hydrophobic surfaces appear to be monotonically attractive, exhibiting no intermediate energy barriers or maxima, whereas most simulations show an oscillatory energy–distance function with a peak that is sometimes above the $E = 0$ baseline. (2) For molecules the hydrophobic binding or adhesion energy increases with temperature (Tanford), while for surfaces it decreases, as manifested by the decreasing interfacial tension of the hydrocarbon–water interface with increasing temperature.

**Professor Craig** asked: Professor Yoon, with respect to Figure 2 in your paper it is apparent that the thin film of liquid drains much more rapidly on the surface with the higher contact angle. You have proposed that this is due to an additional attractive force. I suggest you consider another explanation. Much recent work (for example see the talk of Professor Charlaix at this meeting) suggests that on a surface with a contact angle similar to your treated gold surface the hydrodynamic boundary condition is that of partial slip — whereas on surfaces with lower contact angles (such as your untreated gold surface) the appropriate boundary condition is no slip. A partial boundary slip will result in a smaller hydrodynamic force and faster drainage. I'm wondering if you apply a reasonable slip length (*e.g.* 20 nm) to the solid–liquid boundary and employ the Vinogradova model[1] for drainage in the presence of slip whether this could account for the differences you see in drainage between the two cases in Figure 2. Perhaps the difference is too large to be explained in this manner. I note that the early work of Vinogradova in this area was partially motivated by the hydrophobic attraction — Vinogradova was suggesting that part of the 'attraction' in dynamic measurements may be due to an overestimation of the drainage force. So I guess my question is much the same for your system.

Also I note that recent work by Dagastine at the University of Melbourne and Parkinson at the University of South Australia show that unless the system is extraordinarily clean the no slip boundary condition is applicable to the air–water interface.

1 O. I. Vinogradova, *Langmuir*, 1995, **11**, 2213–2220.

**Professor Yoon** responded: The data shown in Fig. 2 for the gold surface with $\theta_r = 79°$ gives a thinning velocity, $v$, to be approximately 300 nm s$^{-1}$ at the barrier rim. Using this value as a characteristic velocity and the film thickness ($h$) of 100 nm as a characteristic length, one can obtain a shear rate of 3 s$^{-1}$. At a slip length of 20 nm as suggested by Professor Craig, one obtains a slip velocity ($u$) of 60 nm s$^{-1}$ at the solid–liquid interface. Assuming that the film drainage is driven purely by the hydrodynamic slip, $v$ may be estimated by the following relation, derived from the continuity equation for a flat disk, $v = -u(2h/R)$ where $R$ is the film radius. At $u = 60$ nm s$^{-1}$, $h = 100$ nm, and $R = 0.1$ mm, $v$ becomes 0.12 nm s$^{-1}$. This value is orders of magnitude smaller than the thinning velocity ($\approx 300$ nm s$^{-1}$) shown in Fig. 2 for the hydrophobic surface. It can be said, therefore, that hydrodynamic slip at the hydrophobic surface is not responsible for the expedited film thinning and rupture processes observed on hydrophobic surfaces. We thank Professor Craig for letting us know that the non-slip boundary condition is applicable for the air–water interface unless the system is extraordinarily clean. The results presented in Fig. 3 and the earlier work reported by Prof. Platikanov (ref. 30[1] in our paper) show the same.

1 D. Platikanov, *J. Phys. Chem.*, 1964, **68**, 3619–3624.

**Professor Law** commented: Your gold coated substrates exhibit a large contact angle hysteresis of 43°, which would imply that your substrates are very rough. What is the surface roughness of your substrates? The thin film kinetics, that you observe, are likely be influenced by the surface roughness of your substrates.

**Professor Yoon** answered: The contact angle hysteresis we have observed was not as large (43°) as indicated by Professor Law. Figure 6 shows equilibrium and receding angles only. Although not shown in the manuscript, the hysteresis was actually in the range of 20–25°. Also, the AFM images of the bare gold surfaces showed roughness in the order of 2–3 nm only, which did not change significantly after xanthate coating. In principle, roughness should affect the rate of film thinning and hence the hydrophobic disjoining pressures determined using the method described in the present work. However, its impact should not change the outcome of our conclusion, that is, the rate of film thinning is most significantly affected by hydrophobic force. If the surface had been perfectly smooth, the magnitude of the hydrophobic force would have been larger than we have reported in the present work. Reference 27[1] suggested that the thinning of wetting films is driven by the attractive van der Waals force in the foam films formed between the nano-bubbles adhering on the hydrophobic surface and the air–water interface of the wetting film. This is unlikely in the present system in a sense that our AFM images detected no bubbles on the surface. The failure to detect nano-bubbles may be attributed to the fact that the hydrophobicity of the xanthate-coated gold surfaces was substantially lower than those studied by many other investigators.

1 K. W. Stockelhuber, B. Radoev, A. Wenger and H. J. Schulze, *Langmuir*, 2004, **20**, 164–168.

**Dr Christenson** asked: I would like to make a comment with regard to Professor Law's question about equilibrium during force measurements, namely that the problem can usually be resolved by carrying out experiments at different rates of approach of the surfaces.

**Professor Law** responded: Even if one eliminates fluid exchange kinetics during the experiment, by conducting the experiment very, very slowly so that the bulk and surface composition of the confined binary mixture have time to adjust to the changing separation distance between two walls, it would appear that the hydrophobic force could still be caused by confinement effects of a binary mixture. Confined binary mixtures possess different bulk and surface compositions relative to systems where the two walls are infinitely far apart. Recall that in the surface forces apparatus experiment that we are discussing the confined binary mixture is in contact with an infinite bulk mixture reservoir, hence, the confined binary mixture can change it's composition *via* exchanges with this bulk reservoir.

**Professor Craig** responded: Professor Law's comments lead me to recall the late Dr Vasili Yaminsky's approach to understanding surface forces — That if it is an equilibrium force it must be described by the Gibb's adsorption isotherm. That is to say that you cannot have a force between two surfaces without a change in surface excess of at least one component and therefore in the presence of a force, as the surfaces approach you must have proximal adsorption or desorption occurring — and this can be viewed as the origin of the force rather than a consequence of it.

**Ms Mishchenko** asked Professor Israelachvili: When two mica plates with surfactant monolayers approach, why does one always end up with a continuous bilayer? Is this not a rate dependent end product and depend on whether or not correlation takes place? Also, considering the limitations of measurement, how does one know that this is truly a continuous bilayer?

**Professor Israelachvili** replied: We start with a continuous monolayer on each surface in air (resulting in a continuous bilayer when they are brought into contact in the SFA, as measured using the FECO optical technique) and end up with something very similar (in thickness and refractive index) when the surfaces finally come together in water. We cannot be certain that the bilayer is truly continuous or that it does not have defects or 'pinholes'. Concerning the rate effects see our response to Dr Perkin.

**Dr Perkin** addressed Professor Israelachvili and Ms Mishchenko: In response to Professor Law's earlier question, Professor Israelachvili mentioned that when measuring the interaction between charge-patched (heterogeneous) surfaces they see first a slow attraction then an accelerating attraction, and that they interpret this as evidence of correlation occurring during the approach. If this is the case, the interaction should be altered if the speed of the approach is changed (controlled); *e.g.* the attraction would be reduced if the surfaces were 'forced together' more quickly. Is this something that you have tested? I also have a comment following on from the question of Ms Mishchenko, relating to the surface separation after two charge-patch surfaces come together under the attractive interaction. When we perform these experiments, we find that the surfaces come to a closest distance (after the 'jump-in') equivalent to two contacting bilayers. After this the bilayer patches can slowly reorganise to allow the surfaces to come closer still. We discussed this in some detail in an earlier paper.[1]

1 Susan Perkin, Nir Kampf and Jacob Klein, *J. Phys. Chem. B*, 2005, **109**, 3832.

**Professor Israelachvili** replied: We have tried but cannot control the speed of approach once the inward drifting starts: it is difficult to control and too rapid to stop, and it determines the final speed of approach into contact which ends up being the same regardless of the initial (starting) speed of approach.

**Professor Debenedetti** addressed Professor Israelachvili and Mr Chaimovich: In discussing Figure 3, you state that "water's molecular directionality is not crucial for the 'pure' hydrophobic interaction". However, the parametrization of your spherically-symmetric water model leading to the interesting results shown in Figure 3(A) (binding energy increasing with temperature) was done by minimizing the relative entropy with respect to a reference that includes directionality. This suggests that your statement regarding the role of directionality needs to be qualified.

**Mr Chaimovich** responded: What we meant to suggest here is that an average over water's molecular directionality sufficiently describes the "pure" HI, as one basic aspect of the phenomenon is observed with spherically-symmetric water. This appears very perplexing taking into consideration the important role of the tetrahedral network of hydrogen bonds in water's unique properties; in the spherically-symmetric water model, the directional effect of hydrogen bonding is taken into account by a distinctive behavior of the model in state space. To clarify our idea, we intend to rephrase our statement in the publication: "The approximate description of the HI with the use of spherically-symmetric water suggests that water's molecular directionality can be effectively averaged for the "pure" HI, at least for the methane-like case study investigated here."

**Professor C. Chen** asked: Your paper seems to suggest that beyond 20 nm separation there is no "pure long-range hydrophobic force". There is a theory attempting to explain why 100 nm length scale is important for robust superhydrophobicity: liquid water spontaneously dewets hydrophobic plates held less than 100 nm apart (because of hydrophobic interactions). Can you comment on that theory in light of your experiments?

**Professor Israelachvili** replied: I am not sure which theory you are referring to. To my understanding, superhydrophobicity occurs only on certain rough or patterned surfaces where the macroscopic contact angle is greater than the microscopic (thermodynamic) angle $\theta$ on a flat surface. On the other hand, at true thermodynamic equilibrium, two flat parallel hydrophobic surfaces in water, where $\theta > 90°$, will have a vapour cavity or neck bridging them, which will be accompanied by a 'long-range' attractive force due to the Laplace pressure. This effect is true regardless of how far the two surfaces are apart.

**Professor Luzar** commented: It is worth noting during this discussion that hydrophobic interaction is a kinetic problem, in addition to a thermodynamic one. Hydrophobic interaction, solvent induced interaction, has been traditionally explained by equilibrium thermodynamics, *i.e.* as reversible work for changing areas of solutes exposed to water. Our own work[1,2] obtained new insights in the process by which two nonpolar surfaces come into contact in water. Because the activation barrier scales quadratically with surface separation[2] water confined between macroscopic hydrophobic surfaces separated by no more than 2 nm will persist in a metastable liquid state indefinitely.[1]

1 K. Leung, A. Luzar and D. Bratko, *Phys. Rev. Lett.*, 2003, **90**, 065502.
2 A. Luzar, *J. Phys. Chem. B*, 2004, **108**, 19859.

**Professor McCarthy** asked Dr Perkin: Can you show (in the discussion) the AFM image of STA-Br that was presented at the meeting. Could the white faceted structures be crystalline (crystallized) STA-Br?

**Dr Perkin** replied: In our paper we present experimental evidence which demonstrated significant differences between layer structure and stability for octadecyltrimethylammonium iodide (STAI) and octadecyltrimethylammonium bromide or chloride (STABr/STACl) monolayers on mica under water. We confirmed the uniformity and stability of the STAI layer using AFM, as shown in Figure 8 of our paper. We have now also optimized the scanning conditions for STABr (this can be more complicated for the less-stable surfactant layers) and now achieve reproducible images showing that the STABr layer reorganizes into patches, of approximately bilayer height and with patch diameters from 50–200 nm, on the mica surface after immersion in water. Two example images are shown in Fig. 1. We have proposed that the differences in contact angle hysteresis, long-range interactions across water, and time-dependent behaviour for STAI and STABr/STACl are due to the difference in stability of the layers in water. The AFM images clearly show that the STABr layer reorganizes into patches under water, in contrast to the STAI which remains as a stable monolayer on mica under water.

Regarding the structure of these patches, cross-section analysis of the AFM images of STABr show us that the height of the patches is ~3.5 nm; we propose this is due to bilayers of STABr molecules with headgroups of the lower leaflet contacting the mica surface (and electrostatically bound to it) and headgroups of the outer leaflet facing the water. The chains may be tilted or interdigitated. It is not possible from the AFM images to determine the liquid-like or solid-like nature of the hydrocarbon chains within the patches. Nonetheless, the non-uniform shapes of the bilayer domains do not give any indication of crystallinity, and probably it is unlikely to have crystallization in pure water (far below the solubility of the surfactant at this temperature). Similar bilayer domains at the mica–water interface were observed previously by us for a homologous surfactant with $C_{16}$ chain in the past.[1]

1 Susan Perkin, Nir Kampf and Jacob Klein, *J. Phys. Chem.*, 2005, **109**, 3832.

**Professor Ducker** remarked: Our experiments[1] show that when you raise the temperature above the chain melting temperature, domains of surfactant are transformed into a continuous layer. These experiments were performed in equilibrium with a surfactant solution.

1 J.-F. Liu and William A. Ducker, *J. Phys. Chem. B*, 1999, **103**, 8558.

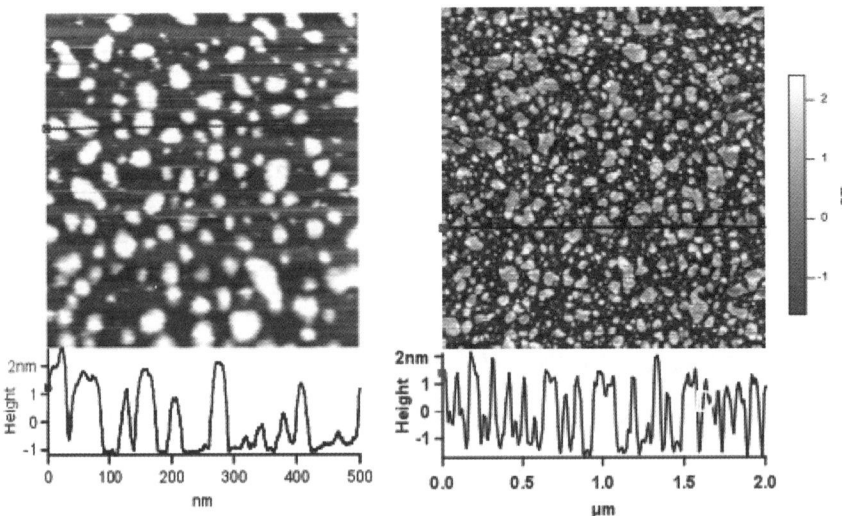

**Fig. 1** AFM images of STABr layers recorded under pure water (prepared in the same way as described in the paper): left, (0.5 × 0.5 m) recorded after 5.5 h immersion in water; right, (2 × 2 m) after approximately 2 h immersion in water.

**Dr Perkin** replied: In your article,[1] you have reported that for C18TAB surfactants on mica in pure water with no added salt (the system most relevant to our own results presented in this *Faraday Discussion*) there is no change in the flat surface morphology within the range of temperature studied, *i.e.* for $T/°C = 26$ to 50. In your experiments, where the surface layer is in equilibrium with the bulk solution of concentration above the CMC, there is always a bilayer or other structures with headgroup-outermost (micelles, cylinders) at the surface. Our experiments, in contrast, deal with monolayers in contact with pure water and so the situation is different: there is no equilibrium between surface and bulk surfactant and the layer is metastable.

The question about whether the hydrocarbon chains of the surfactants in the layer are solid-like or melted is related to a later question of Professor McCarthy and we discuss it further there.

1 J.-F. Liu and William A. Ducker, *J. Phys. Chem. B*, 1999, **103**, 8558.

**Professor Ducker** asked: Results presented by Prof. Yoon and Prof. Israelachvili show very large differences in both the range and magnitude of the measured hydrophobic force. Please comment on why these large differences might occur.

**Professor Israelachvili** responded: We are very careful not to claim that we are measuring the pure hydrophobic interaction unless we also measure the thermodynamic adhesion energy at contact ($D = 0$), *viz.* the interfacial tension of $\gamma_i \approx -50$ mJ/m$^2$. Prof. Yoon's adhesion energies are very low, or not shown in his paper presented at this meeting. I think it is fair to say that unless one measures the thermodynamic adhesion energy at contact one cannot say with certainty that one has a pristine hydrophobic surface or that the long-range attraction is due to the pure hydrophobic interaction. It could be due to bridging vapour cavities, long-range electrostatic forces, or even polymer bridging forces. There are also differences in the experimental setup and conditions (*e.g.*, deaeration): we measure the forces directly rather than try to determine Hamaker-like constants using a fitting equation (Eq. 7 in their paper). Pan & Yoon mentioned that their experimental forces "are usually represented in a single- or double-exponential function" but they nevertheless use a single power-law function to interpret their results. This could be an additional reason why their calculated values are different from our measured forces.

**Professor Yoon** replied: An easy answer to this question would be that our system is different from Prof. Israelachvili's. We used AFM with gold surfaces hydrophobized by a chemisrobing reagent, *i.e.*, potassium amylxanthate (KAX), in air-saturated water, while Prof. Israelachvili used SFA with mica surfaces hydrophobized by dimethyldioctadecylammonium (DODA) bromide and octadecyltriethoxysilane (OTE). The long-range attractions observed with OTE-coated mica surfaces as obtained by Wood and Sharma[1] are very weak. However, we have shown previously that silica surfaces coated with silanes exhibit much stronger and longer-range attractions.[2] Thus, the range of hydrophobic forces may vary depending on the type of surfaces, measurement methods and techniques, dissolved gas, *etc.*

In our view, there is one good reason that in general the hydrophobic forces in wetting films should be larger than those between hydrophobic surfaces. In the former, the hydrophobic force is the result of asymmetric interaction between air–water and solid–water interfaces. Thermodynamically, the former should be more hydrophobic than the latter in view of the large differences in interfacial tensions involved, *i.e.*, $\approx 70$ *vs.* 50 mN m$^{-1}$. We have shown in the present work that the hydrophobic force constant ($K_{232}$) between two air–pristine water interfaces is $5.3 \times 10^{-17}$ J, while those ($K_{131}$) for xanthate-coated gold surfaces are in the range of $4 \times 10^{-20}$ to $2 \times 10^{-17}$ J depending on the contact angles. Therefore, $K_{132}$ should be larger than $K_{131}$ according to the combining rule (Eq. 16). It has been shown, however, that $K_{232}$ decreases with increasing surfactant and NaCl concentrations.[2] Therefore, the values of $K_{132}$ should be lower in the presence of surfactant and/or electrolyte in the thin film of water. Note here that the critical rupture thicknesses (hcr) observed in the present work are in the range of 30 to 120 nm. These values are in general agreement with those reported by Schultze *et al.*[3] Note here that the values of hcr are well beyond the range ($h$ <10–20 nm) where Prof. Israelachvili[4] observed "true" hydrophobic forces, which may be another indication that the hydrophobic forces in wetting films are much longer ranged than those in colloid films.

1 J. Wood and R. Sharma, *J. Adhes. Sci. Technol.*, 1995, **9**(8), 1075–1085.
2 R.-H. Yoon and L. Wang, in *Colloid Stability: The Role of Surface Forces*, Part I, ed. T. F. Tadros, Viley-VCH, Weinheim, 2007, pp. 161–186.
3 H. J. Schulze, *Colloid Polym. Sci.*, 1975, **253**(9), 730–737.
4 M. Hammer, T. Anderson, A. Chaimovich, M. Shell and J. Israelachvili, *Faraday Discuss.*, 2010, **146**, DOI:10.1039/b926184b.

**Professor J. Klein** asked: In your paper you write that the long-ranged hydrophobic attraction in many experimental systems is due to "originally hydrophobic monolayers overturning into charged patchy bilayers, giving rise to a long-ranged attractive electrostatic or double-layer force.[17,22]" This is indeed one of the important realizations concerning "hydrophobic" interactions that have emerged in recent years: For this reason it would be helpful to cite the paper where it first appeared.[1] In this paper the authors propose (p. 3836): "The long-ranged attraction between surfaces comprising patches of positive and negative charge (as we infer from the contact angle, SFB and AFM results in the next section) is not easy to account for. It is possible that it originates as the charge patches shift across the surface to a correlated orientation, with positive bilayer rafts opposite negative mica surface areas". This idea of electrostatic attraction between surfaces coated with charge-patches, first suggested in Perkin 2005, was subsequently proposed also in ref. 17.[2]

1 Susan Perkin, Nir Kampf and Jacob Klein, *J. Phys. Chem. B*, 2005, **109**, 3832–3837.
2 E. E. Meyer, Q. Lin, T. Hassenkam, E. Oroudjer and J. N. Israelachvili, *Proc. Natl. Acad. Sci. USA*, 2005, **102**, 6839 and in J. Zhang, R. H. Yoon, M. Mao and W. A. Ducker, *Langmuir*, 2005, **21**, 5831.

**Professor Israelachvili** replied: The phenomenon of bilayer domains forming from overturning physisorbed LB-deposited monolayers (not SAMs) was fully described the previous year at the 6–11 June, 2004, Gordon Research Conference (Mount Holyoke College, MA, USA) on "Interfacial Water in Cell Biology" by J. Israelachvili in his talk "Measurements of Hydrophobic Forces". Unfortunately, GRC proceedings are not published.

**Professor Law** addressed Professor Israelachvili, Dr Perkin and Professor Yoon :It seems that the hydrophobic force law that you observe could be due to binary liquid confinement effects between two surfaces (of, for example, a Surface Forces Apparatus). Within a finite-size gap between the two surfaces, the equilibrium bulk and surface concentration of a binary mixture will be different than for an infinite gap separation. There must be composition exchanges between the fluid in the gap and the bulk fluid reservoir, as the gap separation is decreased. The kinetics of the changing surface and bulk composition will influence the measured hydrophobic force law between the two plates. If this is indeed the cause of the hydrophobic force then it should happen with any binary liquid mixture. Wouldn't it be easier, therefore, to get rid of water (which is a very complicated liquid) and instead study a simpler binary liquid mixture? If the effect is still present then this would eliminate theoretical suggestions, such as proton hopping, as a mechanism for the hydrophobic force law.

If kinetics determine the force law then it seems unlikely that there will be any universal force curve that will describe this effect.

**Dr Perkin** responded: In our experiments we can be confident that there is no redistribution between surface and bulk phases during confinement because sequential force measurements between the same surfaces (at the same 'contact point') show identical behaviour. This would not be the case if material were desorbed because the surface layers are metastable and any desorbed surfactant would be lost permanently to the bulk solution rather than re-adsorbing when the surfaces are separated to large distances. It is also important to note that in many cases long ranged attractive 'hydrophobic interactions' have been measured in systems where the surface layer cannot reorganise or desorb during approach due to chemical bonds to the substrate (this is discussed, for example, in Hugo Christenson's review article[1]) thus the attractive interaction between hydrophobic surfaces across water occurs independently of molecular exchange between the surface and the bulk.

A similar mechanism to the one you suggest which involves adsorption and desorption while surfaces approach in surfactant solution, or alternatively lateral mobility of monomers within a hydrophobic surfactant layer while surfaces approach in pure water, was suggested by Christenson and Yaminsky[2] and Yaminsky *et al.*[3] The advantage of this explanation is that it rationalises why surfaces with large contact angle hysteresis exhibit long-range attraction. Later it was showed by us and by other groups (Ducker, Israelachvili), that some of these surfaces with large hysteresis can undergo even more dramatic transformations, reorganising in pure water into heterogeneously charged surfaces. This broadens the question, to include the origin of the long range attraction between heterogeneously charged surfaces. Here there are several phenomena that might be involved, with different degrees of importance, such as correlation between the oppositely charged patches on the approaching surfaces. This mechanism is supported by theoretical work that predict long-range attraction between correlated surfaces. Another mechanism we proposed[4] involves the different nature of attraction and repulsion interaction in pure water, suggesting that the overall attraction is stronger than the overall repulsion between two randomly heterogeneously overall neutral charged surfaces. A remark by Prof. Bratko brought to our attention a simulation of oppositely charged macroions interacting in salt solution, which also shows that attractions are stronger then repulsion. The mean field view together with the simulations based on ion–ion pair potentials suggests that attraction between heterogeneously charged surfaces might arise due to the dominance of attraction over repulsions.

1 H. K. Christenson and P. M. Claesson, *Adv. Colloid Interface Sci.*, 2001, **91**, 391–436.
2 H. K. Christenson and V. V. Yaminsky, *Colloids Surf., A: Physicochem. Eng. Asp.*, 1997, **129–130**, 67–74.
3 V. V. Yaminsky, B. W. Ninham, H. K. Christenson and R. M. Pashley, *Langmuir*, 1996, **12**, 1936–1943.
4 G. Silbert *et al.* 2010, submitted for publication.

**Professor Israelachvili** replied: People have measured the forces between surfaces across other liquids, including binary mixtures (Kyle Vanderlick *et al.*) and have obtained short-range oscillatory (structural) forces, very much as expected, with no long-range monotonic attraction other than the expected van der Waals force. As mentioned in our discussion, we think that the observed forces in our setup deals with pure water, not a binary mixture. In our view, any effect due to the fluid being a binary system, including dissolved gases, would certainly modify the HI but not be its primary cause.

**Mr Stirnemann** communicated: If proton-hopping makes a significant contribution to the long-range hydrophobic force as you suggest, one would expect a significant isotope effect for the force, similar to that for proton/deuteron mobility. Is there any experimental evidence concerning such an effect?

**Professor Israelachvili** communicated in reply: Most properties that have to do with molecular interactions are not very different for normal and heavy (deuterated) water, $H_2O$ vs. $D_2O$. Thus, the surface tensions are very similar: 72.0 vs. 71.9 mJ/m$^2$, as are the dielectric constants: 78.4 vs. 78.1, although a possibly significant difference is found for the pH vs. pD values: 7.00 vs. 7.43. Thus, (i) $H_2O$ is slightly more strongly bound to itself than $D_2O$, and (ii) $H_2O$ is more dissociated than $D_2O$ by about a factor of 5. If the long range pure HI is driven by proton-hopping, considering the above information, we could expect a weaker long-range pure HI for $D_2O$ than for $H_2O$, but the magnitude and range of this effect is unknown both experimentally and theoretically. However, we shouldn't expect a change in the short-range pure HI which is more likely to be governed by properties like the surface tension and dielectric constant. Preliminary tests in our lab on the contact angles of water on OTE (a property determined by short-range forces) suggest that the advancing angle of $H_2O$ may be greater than that for $D_2O$.

**Dr Christenson** asked Professor Yoon: The results presented by Dr Perkin suggest that correlations between mobile charges are not responsible for the long-range attraction in their case. I have also noted that a poster by Gilad Silbert describes how a long-range attraction between overall neutral surfaces with a patchy charge distribution can give rise to a net attraction without the need for correlations. Is it not possible that there are several different mechanisms that may give rise to a long-range attraction? For example, the results of Kekicheff and Spalla[1] on forces between glass spheres in hexadecyltrimethylammonium bromide solution that show a long-range attraction with a decay length equal to half the Debye length. I am not sure how this fits with the results presented here. Nevertheless, to me there is something attractive about the correlation explanation for the force between the Langmuir–Blodgett films of dimethyldioctadecylammonium bromide (and a double-chain fluorocarbon surfactant) that Per Claesson I first studied over 20 years ago.[2,3] Just as Professor Israelachvili described a while ago, the surfaces could often be seen initially to remain stationary after having been brought closer, *e.g.* to within 50 nm or so of contact. They would then slowly gather speed and drift together. It was very much as if some time-dependent process was occurring.

1 P. Kekicheff and O. Spalla, *Phys. Rev. Lett.* 1995, **75**, 1851–1854.
2 H. K. Christenson, P. M. Claesson and R. M. Pashley, *Proc. Indian Acad. Sci.* (*Chem. Sci.*), 1987, **98**, 379–389.
3 P. M. Claesson and H. K. Christenson, *J. Phys. Chem.*, 1988, **92**, 1650–1655.

**Dr Perkin** communicated in reply: One of the interesting and also complicating factors of the investigation and discussion of 'hydrophobic interactions' (as you have pointed out very nicely in the past) is that there appears to be many different categories or types of behaviour, each occurring in different experimental systems, which cannot be explained using a single mechanism. For example, it is clear from the theoretical work done on correlations that this mechanism must lead to attractive interactions whenever a heterogeneous system is mobile on the relevant timescale. In cases where the correlation speed of the system is of the same order as the approach speed of the surfaces, the effect should be evident in ways such as you describe; and the mechanism can then be probed by measuring the effect of approach speed on the force law. In our particular experimental system we found that there appears to be no effect on the force law of shearing the surfaces rapidly during their approach, indicating that in this case there is no contribution of correlation on the interaction forces (unless the correlation is much faster than the shearing). Nonetheless, we find a strong and long-ranged attractive interaction.

As proposed in the work of Gilad Silbert (submitted for publication), random heterogeneously charged surfaces can be attractive across pure water even if there is no correlation of the charge patches. This mechanism must be acting between all heterogeneous surfaces whenever the conditions of surface charge and Debye length are satisfied, and so even in the cases where correlations do occur this interaction must also be contributing an additional attraction.

**Professor Yoon** communicated in reply: The hydrophobizing agent, potassium amyl xanthate (KAX), we used here is very different from any of the surfactants used in the references cited by Prof. Christianson. Although AX is anionic, it chemisorbs on gold via an electrochemical mechanism shown below:

Au$^o$ + AX$^-$ → AuAX + e ½O$_2$ + H$^+$ + e → OH$^-$ as reported in ref. 54. The spectroscopic evidence suggests that the xanthate adsorption results in the formation of gold amyl xanthate (AuAX), with smaller amounts of dixanthogen being found in multi-layers. According to the above mechanism, the negative charges of the xanthate ions are taken away by dissolved oxygen, while the gold substrate remains negatively charged. Therefore, the adsorption mechanism does warrant the presence of positively and negatively charged patches on the surface. Furthermore, it is unlikely that chemisorbed xanthate can have a sufficiently high degree of lateral mobility as two surfaces approach each other to separations below 50 nm. Thus, our results do not support the charged patch model. We have tested the charged patch model of Miklavic et al.[1] on a physisorbing system, i.e., alkyltrimethylammonium chloride (CnTACl) self-assembled on silica.[2] With C18TACl, the AFM force measurements were conducted in solutions of varying NaCl concentrations. The hydrophobic force decreased with increasing NaCl concentration, with the decay length ($D$) varying as one-half of Debye length ($\frac{1}{2}\kappa^{-1}$), which could be considered to suggest that hydrophobic force may be of electrostatic origin. In the absence of electrolyte, however, $D$ varied as $1/4\kappa^{-1}$, which cannot be explained by the model. It was found also that as $n$ increased from 12 to 18, the patch size increased from 18 to 172 nm. However, these values were substantially smaller than the experimental results reported by Fan et al.[3] It has been shown that alkylammonium bromide adsorbs on negatively-charged alumina with a normal mode of orientation, forming patches (or hemimicelles) and rendering the substrate hydrophobic.[3] Once the surface charge has been neutralized, however, the cationic surfactant begins to adsorb with inverse orientation, causing a decrease in hydrophobicity. Thus, the surface becomes most hydrophobic at the point of charge neutralization (p.c.n.). This mechanism is supported by the results of our AFM force measurements, which showed that the strongest hydrophobic forces are observed at the p.c.n.'s for the silica-CnTACl system.[2] For a patch to be charged, it is necessary that additional surfactants co-adsorb in the patch. The most thermodynamically favorable mode of orientation would be the inverse (or flip-flop) orientation, which should decrease hydrophobicity and hydrophobic force.

1 S. J. Miklavic, D.Y. Chan, L.R. White and T.W. Healy, *J. Phys. Chem.*, 1994, **98**, 9022–9032.
2 J. Zhang, R.-H. Yoon and J. C. Eriksson, *Colloid Surf., A*, 2007, **300**, 335–345.
3 A. Fan, P. Somasundaran and N. J. Turro, *Langmuir*, 1997, **13**, 506–510.

**Dr Christenson** asked Dr Perkin: I am puzzled by the very large pull-off forces measured between the STAI-coated surfaces shown in Figure 5. At first sight they would seem to correspond to an impossibly large interfacial energy, especially given that the interfacial free energy between water and a surface exposing mainly -CH$_3$ groups to the aqueous phase should be lower than between a surface of mainly -CH$_2$- groups. Surface roughness could explain why the adhesion between the other monolayers is low, but a surface energy of almost 60 mJm$^{-2}$ is difficult to rationalise. There are many "artefacts" that can reduce the adhesion, *e.g.* surface roughness, platinum particles,[1,2] adsorption of contaminants, *etc.* but an enhanced adhesion is more difficult to rationalise. If cavitation (capillary evaporation) occurs on separation it may not be visible, but the adhesion should then be given by $F/R = 4\pi R\gamma LV\cos\theta$, (where $\gamma LV$ is now the surface tension of water, *i.e.* the bubble–water interface — perhaps lowered by surfactant migrating to this interface). Presumably the advancing angle of 102° would be the correct one to use as the water has to move in to replace the bubble during separation of the surfaces. Possibly, a direct solid–solid contribution inside the cavity might also contribute. However, considering the numbers involved it would still be difficult to come up with such a large pull-off force. If these events are all very rapid it may be difficult to justify the use of any of the equations discussed. Do you have any ideas on this?

1 S. Ohnishi, M. Hato, K. Tamada and H. K. Christenson, *Langmuir*, 1999, **15**, 3312–3316.
2 M. M. Kohonen, F. C. Meldrum and H. K. Christenson, *Langmuir*, 2003, **19**, 975–976.

**Dr Perkin** communicated in reply: We agree that the measured surface energy for the STAI layers in water, which remains unchanged (within the scatter of the data) over many hours, is high compared to what is expected for a -$CH_3$ terminated hydrocarbon layer. One possible reason for the high pull-off force (we measure the surface energy from the force required to separate the surfaces) could be the tilted nature of the surfactant molecules in the monolayers, leading to exposed $CH_2$ groups as well as the possibility of interdigitation between molecules on the opposing layers when they are in contact. Evidence for tilt comes from the thickness of the bi-layer, which is less than expected for molecules standing perpendicular to the surface. Although our experiments do not reveal the origin of this high pull-off force (for example we did not observe in the interference fringes any evidence of cavitation or air bubbles, which might have been evident as discontinuities in the fringes, although there are limits to the time- and length-scale of our observation which prevent us ruling these out as a mechanism), the important point for us is that the surface energy remains constant over many hours and throughout the experiments. The constancy of (high) STAI-water surface energy after long periods of time provides evidence that the layer is stable, in contrast to the STABr and STACl layers which show a significant drop in surface energy over similar timescales.

**Mrs Seyed-Yazdi** opened the discussion of the paper by Dr Hummer: Do you know the ratio of empty and filled valleys in your simulated system? If not, is there a way to deduce it from your results?

**Dr Hummer** answered: The inset in the left panel of Figure 7 shows the average density profile as a function of $z$ at the center of the valleys. In the curves for large roughness amplitudes $a$ (green, blue, and magenta), one can see that the water density gradually drops to zero, over a range of about 3 Å. In contrast, the drop is very sharp at the flat surface. Under the assumption that the water "fingers" penetrating into the valleys have about the same density as water (consistent with our previous observations;[1] and with the theory of the liquid–vapor interface by Professor Weeks,[2] these curves would give us a rough estimate of the probability that a valley is filled with water up to a distance $z$.

1 J. Mittal, G. Hummer, *Proc. Natl. Acad. Sci.*, 2008, **105**, 20130–20135.
2 J. D. Weeks, *J. Chem. Phys.*, 1977, **67**, 3106–3121.

**Dr Sarupria** asked: Can you comment on the other reasons in addition to loss of H-bonds that play a role in the local 'dewetting' observed at your surfaces? Can concepts of cavity expulsion be applied here due to the different curvatures (convex *vs.* concave) involved in this surface? Can such a behavior be expected for a simple Lennard-Jones fluid?

**Dr Hummer** answered: The question is indeed interesting whether hydrogen bonds are required for the observed increase in the interfacial free energy. While we do not have any specific published results for, say, a Lennard-Jones fluid, I would expect qualitatively similar behavior for other strongly associating liquids, based on simulations by V. K. Shen, J. R. Errington, and J. Mittal (unpublished). For Lennard-Jones fluids at smooth hard surfaces, a crossover to drying has been reported.[1] With regard to the second part of your question, the concept of cavity expulsion[2] would indeed seem to capture aspects of the observed behavior at least qualitatively. Consistent with cavity expulsion, the loss of water–water attractive interactions in the concave surface-features is a major factor in the transition to hydrophobic behavior.

1 D. M. Huang and D. Chandler, *Phys. Rev. E*, 2000, **61**, 1501–1506.
2 G. Hummer and S. Garde, *Phys. Rev. Lett.*, 1998, **80**, 4193–4196.

**Dr Sarupria** opened the discussion of the paper by Professor Garde: Can concepts of co-operativity be applied to quantify the asymmetry in hydration observed in your work? Can such calculations then be used to predict the asymmetric behavior based on some linear combination of each pure component?

**Professor Garde** answered: I do not fully understand your question. However, our observation that properties of water (*e.g.*, density fluctuations or compressibility) at interfaces depend non-linearly on the composition of hydrophobic or hydrophilic groups in the interface indicates that those effects are non-additive. Developing a predictive, non-linear model to describe that dependence will be challenging.

**Professor Butt** addressed Professor Garde and Professor Bratko: Let me ask a question about the role of dissolved gas. It might adsorb at the interface and be enriched. Can it enhance fluctuations or serve as one of the inhomogeneities and thus influence the interaction significantly?

**Professor Garde** answered: I invite Prof. Dusan Bratko to address this comment, as his and Prof. Alenka Luzar's groups have done significant work on the role of dissolved gases and their adsorption at interfaces and in confined systems.[1,2] The solubilities of gases in water are rather small, and therefore, their enrichment at the interface is expected to be small as well. One may need to increase the pressure of the gas significantly (~1 kbar or more) before any effect will be visible. We note that increasing pressure will also quench local density fluctuations as shown in our paper.

1 D. Bratko and A. Luzar, *Langmuir*, 2008, **24**, 1247–1253.
2 A. Luzar and D. Bratko, *J. Phys. Chem. B*, 2005, **109**, 22545–22552.

**Professor Bratko** answered: Professor Butt asked about the influence of dissolved gases on interfacial water density fluctuations. We have addressed this issue in open ensemble simulations of confined aqueous slabs at ambient conditions and in supersaturated gas solutions.[1,2] The presence of dissolved gases invariably increased the extent of density fluctuations, however, the amount of dissolved gases ($N_2$, $CO_2$, $O_2$, Ar) corresponding to usual ambient conditions remains too low to produce significant changes in interfacial water behavior. Important differences, including the possibility of single-wall bubble nucleation were found[1] in supersaturated gas solutions.

1 D. Bratko and A. Luzar, *Langmuir*, 2008, **24**, 1247.
2 A. Luzar, D. Bratko, *J. Phys. Chem. B*, 2005, **109**, 22545.

**Professor Bratko** commented: The interesting result of the asymmetric response of water density fluctuations to changes in local hydrophobicity/philicity points to a more general behavior, also observed on uniform surfaces. In Fig. 2, we collect our Grand Canonical Monte Carlo ($\mu VT$) simulation results[1-3] for interfacial water compressibility, $\rho k_B T \kappa = (<N^2> - <N>^2)/<N>$, as a function of thermodynamic contact angle, $\theta_c = \cos^{-1}(-\Delta\gamma_{sv}/\gamma_{sl})$. Contact angle was varied by tuning surface chemistry,[1-3] or by applied electric field.[1,2] When gradually reducing the contact angle of a hydrophobic surface between 135° and ~90°, the compressibility is first suppressed rapidly. The sensitivity weakens at smaller contact angles, with apparent saturation observed in the hydrophilic regime (contact angle below 90°) where the compressibility settles close to that of bulk water.[1] The above behavior is consistent with the strong reduction in local density fluctuations when a hydrophilic moiety is introduced on a hydrophobic background, and helps interpret the minute effect of a -CH$_3$ group on strongly hydrophilic surfaces where the interfacial compressibility is much less sensitive to changes in surface/water interaction.

1 D. Bratko, C. D. Daub, A. Luzar, *Faraday Discuss.*, 2009, **141**, 55.
2 D. Bratko, C. D. Daub, A. Luzar, *J. Am. Chem. Soc.*, 2007, **129**, 2504.
3 D. Bratko, R. Curtis, H. W. Blanch, J. M. Prausnitz, *J. Chem. Phys.*, 2001, **115**, 3873.

**Professor Garde** replied: Thank you for sharing these very interesting results with us. Your results on the variation of reduced compressibility with droplet contact angles complement our results beautifully and possibly point to a more general connection between these quantities in interfacial environments.

**Professor Bratko** asked: (1) In anisotropic systems such as aqueous interfaces, the normal and lateral pressure components, $P_n$ and $P_l$, diverge when the surface contact angle deviates from 90° [see *e.g.* J. Mittal and G. Hummer[1]]. The tensorial nature of pressure cannot be ignored over the whole range of conditions covered in your study. Can you specify the meaning of the reported pressure $P$ in this work, and the function of the barostat at anisotropic situations considered herein. Similarly, the compressibility formula, eqn (1) in your article, given with scalar $P$ warrants clarification.

1 J. Mittal and G. Hummer, *Faraday Discuss.*, 2010, **146**, DOI:10.1039/b925913a.

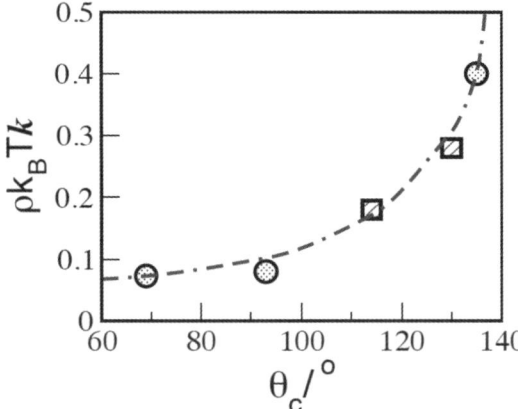

**Fig. 2** Reduced compressibility, $\rho k_B T \kappa$, of interfacial water in a planar nanopore of width $D =$ 1.64 nm as a function of contact angle, $\theta_c$, determined by Grand Canonical Monte Carlo simulations in SPC/E water. Contact angle of chemically homogeneous pore walls was varied through surface chemistry (circles at $\theta_c$ = 135, 93, or 69°),[2,3] or by applying an electric field across a hydrophobic pore (squares at $\theta_c$ =114 and 129°).[1] The line is a guide to the eye.

**Professor Garde** replied: This is a very interesting question. Indeed, in the vicinity of an interface, the normal and transverse components of the pressure tensor will be numerically different, depending on the interfacial tension. This creates ambiguity in the definition of local compressibility, which characterizes the pressure dependences of a scalar field variable, density. For simplicity, we have used definition of compressibility as a derivative of the local density with respect to the normal component of the pressure tensor. We consider this to be a simulation analog of a typical simple piston-cylinder experimental set up (left panel of Figure 1 in Mittal and Hummer's paper[1]). It may be worth pursuing your question further in future, where one might explore quantifying local compressibility using the response of density to different elements of the pressure tensor.

1 J. Mittal and G. Hummer, *Faraday Discuss.*, 2010, **146**, DOI:10.1039/b925913a.

**Professor Weeks** addressed Dr Hummer: You mention that the compressibility is higher for a system with constant normal pressure than for one with constant parallel pressure at the same roughness amplitude. Can you provide any physical insight into why this is true?

**Dr Hummer** replied: This is an interesting point for which I have, at best, a qualitative explanation, namely that in response to normal pressure the valleys of the rough surface are more easily filled. Note, however, that the state points at $P_N = 1$ atm and $P_P = 1$ atm are different.

**Professor Luzar** addressed Professor Garde: We are attacking similar problems in our lab. Our experience is that while it is easy to quantify local density fluctuations along synthetic surfaces (*e.g.* mixed $CH_3/NH_2$), several problems arise with protein surfaces due to pronounced surface roughness, irregular shapes, cut-off radii, threshold lengthscales are very hard to determine for each protein patch as well, defining hydration shells for each patch becomes a rather artificial game.....We describe an alternative approach of measuring protein hydrophobicity in terms of effective contact angles (as described in the poster presented by Jihang Wang, at this Discussion). At any rate, it seems to me that we all need to think harder in order to be able to report on exciting findings in this domain of research. Unpublished work presented in the introductory lecture given by Professor Peter Rossky looks like the right direction to follow.

**Professor Garde** responded: You are correct in highlighting the difficulties in measuring density fluctuations near complex heterogeneous and rough surfaces. Our recent work points to intimate relationships between density fluctuations, cavity formation, solute binding, and water–water correlations; suggesting that any one of them could, in principle, be used as a quantifier of hydrophobicity. Practically, however, we find that monitoring binding/local concentrations of probe hydrophobic solutes near complex surfaces is most efficient. Perhaps, as you suggest, some combination of our approach and that of Prof. Peter Rossky (focusing on atom-based hydropathy scales) will lead to a highly efficient (first pass) method that will work better than current hydropathy scales.

**Dr Hummer** addressed Professor Luzar and Professor Garde: It may be worth noting that protein hydrophobicity has been successfully mapped by X-ray crystallography experiments with added organic probe molecules.[1] Protein active sites, for instance, could be properly identified. We have also seen strong correlations between the experimentally determined locations of hydrophobic groups of inhibitors bound to HIV gp41 and locally enhanced water density fluctuations at the unliganded protein surface, as well as increased test-particle insertion probabilities.[2] These earlier results from both experiment and simulation suggest that the approach described by Professor Garde will be very powerful in the study of protein hydrophobicity.

1 D. Ringe, *Curr. Opin. Struct. Biol.*, 1995, **5**, 825–829.
2 X. Siebert, G. Hummer, *Biochemistry*, 2002, **41**, 2956–2961.

**Professor Garde** responded: This is a very good comment. The experiments of D. Ringe and coworkers that you cite are indeed seminal in that they highlight the utility of binding experiments in identifying binding sites and mapping hydrophobic regions on protein surfaces. Their method relies on the use of protein crystals and diffusion of probe organic molecules through the crystals to bind to accessible protein surface. We recently performed preliminary NMR experiments to study binding of ligands to protein in solutions complemented with information from MD simulations of probe binding,[1] with great success. From the theory/modeling point of view, recent work from our group shows that various measures, such as density fluctuations, solute binding (or cavity formation — as you point out), or water–water correlations can be used to quantify hydrophobicity of interfaces. For complex and chemically heterogeneous surfaces of proteins, studying binding of probe solutes presents computationally the most efficient approach.

1 C. J. Morrison, R. Godawat, S. A. McCallum, S. Garde and S. M. Cramer, *Biotechnol. Bioeng.*, 2009, **102**(5), 1428–1437.

**Mr Wang** communicated: It is amazing that you observed the connection between the hydrophobicity of the surface and the compressibility of water at the interface. From the figure in your 2009 PNAS paper, it seemed that the number density of water at the interface is Gaussian distributed, only the variance is different for surfaces with different hydrophobicity. It is true that the spontaneous density fluctuations in bulk water is Gaussian distributed according to fluctuation-dissipation theory, (which is the basis of information theory) but I did not see any reason why in such inhomogeneous systems the density fluctuation is still Gaussian distributed. Would you please give a physical explanation?

1 Rahul Godawat, Sumanth N. Jamadagni and Shekhar Garde, *Proc. Natl. Acad. Sci. USA*, 2009, **106**(36), 15119–15124.

**Professor Garde** communicated in reply: In our 2009 PNAS paper[1] we focused primarily on density fluctuations in relatively small volumes at interfaces. Those fluctuations are approximately Guassian in nature, similar to that in bulk water, as reported in our 1996 PNAS paper on information theory.[2] We note however, that the fact that the fluctuations are Gaussian in small volumes is not anticipated *a priori* (by fluctuation–dissipation or by any other theorem) — in fact, in simple non-associating fluids they are non-Gaussian.[3] For larger volumes, the central limit theorem predicts the fluctuations to become Gaussian in an uncorrelated ideal gas system. In bulk water, in large observation volumes, the fluctuations are roughly Gaussian near the mean, with non-Gaussian nature becoming apparent in the tails of the distribution.[4] Hydrophobic interfaces will further amplify this non-Gaussian behavior in the tails of the distribution.

1 Rahul Godawat, Sumanth N. Jamadagni and Shekhar Garde, *Proc. Natl. Acad. Sci. USA*, 2009, **106**(36), 15119–15124.
2 G. Hummer, S. Garde, A. E. García, A. Pohorille, and L. R. Pratt, *Proc. Natl. Acad. Sci. USA.*, 1996, **93**, 8951–8955.
3 D. M. Huang and D. Chandler, *Phys. Rev. E*, 2000, **61**, 1501–1506.
4 A. Patel, P. Varilly, and D. Chandler, *J. Phys. Chem. B*, 2010, **114**, 1632–1637.

**Professor Giovambattista** asked: This work clearly shows the relationship between the local compressibility of water next to different surfaces and the corresponding surface local hydrophobicity. It is also shown that this local compressibility can be used to define hydrophobicity maps for protein (heterogeneous and rough) surfaces. Since the hydrophobicity map is a time-averaged property that characterizes the protein surface, I wonder if you can comment on how the protein hydrophobicity maps change with time, as the protein atoms move due to thermal fluctuations?

**Professor Garde** answered: This is a very good question. The "hydrophobicity" of a given interface is affected by local chemistry and topography. In our calculations, the protein atoms were not constrained, and therefore, their dynamics over timescales of several nanoseconds are included in the results presented here. Your question points to two interesting future directions for investigation: (1) How do the flexibility and dynamics of a given surface affect the local water density fluctuations, and therefore, interface hydrophobicity? (2) Conformation dependent hydrophobicity maps: For proteins which display large conformational motions over slow timescales (*e.g.*, opening/closing of a large binding pocket), it would be necessary to map hydrophobicity of different conformations separately.

**Professor Debenedetti** addressed Dr Hummer: The polynomial fits to the simulation data shown in Figure 3 (equations of state) include inflection points for both the normal and parallel pressure. Such inflection points suggest the presence of critical points. Have you investigated this?

**Dr Hummer** answered: This is an interesting point that we have not yet investigated further. As noted in the paper, the cubic fits to the $P$–$H$ data have multiple roots for $P = 1$ atm for large roughness. This transition to metastability appears to coincide with the contact angle approaching 180°. For an open (grand canonical) system, we would thus expect either fluctuations between wet and dry states or "phase coexistence" with a dry layer at the rough wall, depending on the system size. For our system, the inflection points with $dP/dH = d^2P/dH^2 = 0$ occur at negative pressures (for fixed ambient temperature and varied roughness amplitude $a$). Nevertheless, it would be interesting to explore the behavior in this region, possibly with the help of your theoretical framework.[1]

1 T. M. Truskett, P. G. Debenedetti, and S. Torquato, *J. Chem. Phys.*, 2001, **114**, 2401–2418.

**Professor Debenedetti** asked Professor Garde: Your interesting calculations regarding the asymmetric effects of hydrophobic sites on hydrophilic substrates *vs.* hydrophilic sites on hydrophobic substrates are consistent with our earlier observations of hydration of patchy surfaces.[1] In that paper, we showed that the ability of a hydrophobic patch to repel water is sensitively dependent on its size and surroundings (*i.e.*, context matters). We also found that a single hydrophilic site located at the center of nanoscale plates has a clear effect on the progress of drying in the confined space, and that a single layer of hydrophilic sites on the border of otherwise hydrophobic nanoscale plates can prevent bulk cavitation.

1 N. Giovambattista, P. G. Debenedetti, and P. J. Rossky, *J. Phys. Chem. C*, 2007, **111**, 1323.

**Professor Garde** responded: Thank you for this comment and pointing to your interesting work. Indeed, our results on the asymmetric effects are consistent with your previous work. Our recent unpublished work on the hydration of heterogeneous surfaces under confinement indicates that confinements can amplify the context dependence significantly. Thus, effects of a single hydrophilic site in the hydrophobic background are likely more prominent in a confined system. Our present work shows that even in the case of hydration of isolated (non-confined) interfaces, the asymmetry is present and can be quantified.

**Professor Errington** asked Dr Hummer: The substrate–fluid potential that you adopted in your work is fairly coarse-grained. Essentially, each particle sees a structureless wall placed at an elevation that depends upon the particle's lateral position. It follows that the strength of the substrate–fluid potential is underestimated for a particle that sits deep within an interstice, and, conversely, is overestimated for a particle that sits atop the contour. Do you have a feel for how your results would change if a more detailed substrate–fluid potential was adopted?

**Dr Hummer** answered: We deliberately chose our potential to separate out the effects of roughness as much as possible. The effects you describe will depend on the detailed molecular interactions, the wall–particle sizes, wall–particle crystal lattice, *etc.* Purely based on perturbative arguments I would expect that enhancing the wall–water attractive interactions in the valleys, and weakening them at the tops, will pull the interface closer to the solid. I would expect the effect to be the strongest in the valleys, where it should favor "wet" over "dry" states; in contrast, at the tops the density tends to be high even for weak interactions, and the effect should be small. So overall I would expect a less hydrophobic surface if the valleys were more attractive.

Let me point out, however, that there is a serious practical issue in simulating the solid with explicit wall particles. Typically, only a few layers of wall particles can be included in a simulation and the results may also depend on the number of wall particle layers included. To answer these questions quantitatively, we are currently working in collaboration with Dr Vincent Shen (NIST) utilizing a Lennard-Jones fluid system to see how sensitive the results are if a more detailed substrate–fluid potential is adopted.

**Dr Daub** asked: Most (perhaps all) real surfaces show fluctuations in the structure at these length- and time-scales. Aside from the interesting possibility of being able to produce real surfaces with rigid molecular scale roughness as new technology is developed, it strikes me in retrospect that your simple functional form for the surface corrugation would be very easy to extend to incorporate time-dependent fluctuations, say by multiplying your Eqn. (2) by $\sin(\omega t)$. It would be interesting to see what happens as $\omega$ is varied above and below the typical timescale for hydrogen bonds and water reorientation times.

**Dr Hummer** answered: This is an excellent suggestion that we have not yet explored.

**Dr Angus** commented: Your work was done at static normal and parallel pressure conditions. Could you please comment on the behavior of the system under dynamic flow conditions?

**Dr Hummer** responded: At the moment we do not have concrete data for flow. It will certainly be interesting to see if the increased hydrophobicity caused by roughness results in an increased slip length and thus reduced friction, or if the roughness enhances the momentum transfer with the wall, thus reducing slip and enhancing friction. Note that for our ideal flat wall, we would have perfect slip boundary conditions for flow.

**Mr Chaimovich** asked: Can you explain your finding of a negative surface tension (a hydrophilic phenomenon on the observable level), considering your wall is only comprised of a Lennard-Jones force field (a hydrophobic attribute on the molecular level)? Might this inconsistency between the two scales stem from the fact that the chosen interaction parameters do not correspond with a realistic nonpolar substance?

**Dr Hummer** replied: The wall–water attractive interactions are sufficiently strong to result in a negative interfacial free energy. I should point out, though, that our rough estimate of a contact angle for a flat interface is not far below 90°. So the surface is only weakly hydrophilic.

**Mr Chaimovich** asked Professor Garde: Can you explain the asymmetry that you observe for nanoscale heterogeneity in terms of the tetrahedral network of hydrogen bonds? Might this asymmetry correspond with the fact that tetrahedral structure usually forms around small molecules (whether hydrophilic or hydrophobic in nature), which is markedly unlike the case for large surfaces?

**Professor Garde** answered: We have studied the hydration of extended interfaces in our simulations. We think that the asymmetry arises from the very different properties of water near hydrophobic and hydrophilic interfaces. Near extended hydrophobic interfaces hydration shells are soft, compressible, and accommodate significant density fluctuations. Naturally, water responds sensitively to inserting one hydrophilic site in such an interface. In contrast, we find that introduction of one hydrophobic site in a hydrophilic surface has little effect. The hydrogen bonding preferences of water in bulk, near small solutes, near extended interfaces, and near hydrophobic *vs.* hydrophilic interfaces will be different, and quantifying them will provide additional interesting information.

**Dr Christenson** addressed Dr Hummer: I was at first somewhat confused by the negative interfacial free-energy terms in your paper. Am I right in assuming that what you are referring to is what I would call the wetting tension, *i.e.* the difference between the solid–vapour interfacial free energy and solid–liquid interfacial free energy?

**Dr Hummer** responded: The negative interfacial free energies of the flat walls reflect favorable wall–water attractive interactions and are consistent with contact angles smaller than 90°. For a more detailed definition, I refer to the reply to Professor Henderson's question below.

**Professor Evans** addressed Professor Garde, Dr Hummer and Professor Debenedetti:

(i) Given that the large majority of simulation studies of confined water make use of molecular dynamics packages it is not surprising that these often correspond to a fixed number of molecules with either fixed reservoir ($P_p$) or fixed normal ($P_N$) pressure. My question is whether these are the most appropriate ensembles for performing detailed quantitative studies of confined water, especially those relating to phase separation?

One of the most significant properties of real water is that its chemical potential $\mu$ under ambient conditions lies very close to its value at saturation $\mu_{sat}(T)$. Thus simulation studies should attempt to mimic, as best they can, the proximity of a real reservoir of water near to saturation. The grand canonical ensemble is the natural choice for performing such simulations as one can monitor the 'distance' from bulk saturation/coexistence and, by calculating the grand potential, monitor phase transitions such as capillary evaporation.

Moreover, a big advantage of working in an open system (grand canonically) is that the solvation force, as is measured in the surface force apparatus, is given directly as a derivative of the excess grand potential of the fluid w.r.t wall separation $H$ at fixed $\mu$ and $T$, and can be evaluated easily.

(ii) Are there distinct advantages of working at fixed $P_N$? I note this route was used by Dominguez *et al.* in an early Monte Carlo study of capillary freezing and capillary evaporation of a Lennard-Jones fluid confined between planar walls. This ensemble was used because of insertion and sampling problems encountered using grand canonical methods at high fluid packings. But surely these difficulties can be overcome using modern sampling techniques? This paper also describes some of the issues that arise in relating phase boundaries calculated in the constant normal pressure ensemble to those that would be obtained for a fluid in contact with a reservoir. Are these taken on board by the water community?

(iii) Is working at fixed $P_p$ strictly equivalent to working at fixed $\mu$? Clearly for an infinite ($H \to \infty$) system this should be the case as $P_p \to P$, the bulk pressure of the reservoir and Gibbs–Duhem then relates $P$ to $\mu$. But are there finite size effects?

(iv) The observation of a large local compressibility for a fluid near a hydrophobic wall is very interesting. I regret that I have not followed the literature from my American cousins on this subject. But, once again, I would say that it is easier to interpret such measurements, *i.e.* of the local compressibility or the work of insertion of a model hydrophobe at locations close to the surface, when a grand-canonical approach is adopted. In the grand canonical ensemble the local compressibility, proportional to the derivative of the local density profile w.r.t. $\mu$, is given directly as an integral over the two-body distribution function of the inhomogeneous fluid. In other ensembles it is not clear to me that such direct relations continue to apply unless fixing $P_p$ is strictly equivalent to fixing the bulk (reservoir) pressure and therefore to fixing $\mu$.

1 H. Dominguez, M. P. Allen and R. Evans, *Mol. Phys.*, 1999, **96**, 209.

**Dr Hummer** responded: I agree that working in a grand-canonical ensemble would be advantageous. For our finite system under confinement, I would expect the ensembles to be different, in particular as we approach and enter metastability. In a grand canonical ensemble, if properly sampled, one could more directly explore possible metastabilities, in particular drying. As discussed in the response to Professor Debenedetti's question, our equations of state suggest such metastabilities at larger roughness amplitudes. As implied by your question, I would also expect effects on the compressibility and the magnitude of density fluctuations in a grand-canonical system. However, far from metastability those effects should be modest, and I would expect our main conclusions to hold also in the grand-canonical ensemble, at least qualitatively. But grand-canonical simulations of water remain challenging, with straight particle insertion and removal both having very low acceptance probabilities. One possibility with standard simulation codes is instead to combine canonical simulations with different particle numbers $N$ to construct the grand partition function.[1] An attractive alternative is to use the grand canonical transition matrix Monte Carlo[2] with advanced sampling techniques such as multicanonical sampling[3] to obtain interfacial properties directly from the grand potential.[4] We also refer to the comment by Professor Bratko below, and references therein.

1 J. C. Rasaiah, S. Garde and G. Hummer, *Annu. Rev. Phys. Chem.*, 2008, **59**, 713–740.
2 J. R. Errington, *J. Chem. Phys.*, 2003, **118**, 9915–9925.
3 B. A. Berg and T. Neuhaus, *Phys. Rev. Lett.*, 1992, **68**, 9.
4 E. M. Grzelak and J. R. Errington, *J. Chem. Phys.*, 2008, **128**, 014710.

**Professor Garde** replied: Thank you for these excellent comments. Short of doing direct comparisons with results of grand canonical ensemble simulations, your comment can not be quantitatively addressed. However, there are a number of factors that lend support to the robustness of our results obtained from MD simulations in NPT ensemble (see the responses by D. Bratko and G. Hummer).

The drying transition near idealized extended hydrophobic surfaces has now been well characterized and appreciated. Our focus is not so much on studying drying or phase transition in idealized systems, but on identifying molecular scale quantities that characterize hydrophobicity for applications to interesting and typically "wet" systems such as protein interfaces. To this end, we have shown that near realistic surfaces that interact with water with weak van der Waals attractions, local water density can be bulk-like and does not serve as a good quantifier of the hydrophobicity of those interfaces. In contrast, fluctuations of density, free energy of cavity formation, transverse water–water correlations correlate quantitatively with hydrophobicity or philicity of a given interface.[1] Fluctuations are enhanced and correlations are longer ranged near hydrophobic surfaces and both are gradually quenched as hydrophilicity of the surface increases. For these systems, we expect our findings to be robust, independent of the ensemble used to quantify them.

We also note that with availability of modern computing platforms, simulations of large systems (with many thousands of water molecules) are now routinely performed. For our protein simulations include ~10,000 water molecules. Use of such large systems will help minimize finite size effects.

1 Rahul Godawat, Sumanth N. Jamadagni, and Shekhar Garde, *Proc. Natl. Acad. Sci. USA*, 2009, **106**(36), 15119–15124.

**Professor Debenedetti** replied: The importance of choosing the "right" ensemble to simulate confined fluids depends on what one wants to calculate. For example, if one wants to simulate capillary drying, imposing a reservoir pressure rather than fixing the total volume will greatly facilitate drying. In simulations aimed at interpreting experiments on liquids in nano-scale confinement, such as the experiments of Chen, Mallamace and coworkers on confined supercooled water (*e.g.*, Mallamace *et al.*[1]) use of the grand canonical ensemble establishes a natural connection with laboratory protocols and measurements. On the other hand, if one wants to study dynamics in confinement, the results should not depend on whether one fixes the pressure or the volume.

1 Francesco Mallamace, Caterina Branca, Matteo Broccio, Carmelo Corsaro, Chung-Yuan Mou, and Sow-Hsin Chen, *Proc. Natl. Acad. Sci. USA*, 2007, **104**, 18387.

**Professor Evans** replied: I thank Professor Debenedetti for his helpful comment.

**Professor Bratko** commented: Dr Evans questioned the suitability of closed ensemble simulations in studies of properties of, and phenomena occurring in open systems that are naturally described by Grand Canonical statistics. Our experience with Grand Canonical Monte Carlo (GCMC) modeling lends guarded support to closed ensemble alternatives, notably the constant pressure (NPT) ensemble popular in Molecular Dynamics (MD) applications. When the pressure of an unperturbed solvent bath can be controlled, the approach will often secure an adequate control of chemical potential.[1–4] In a canonical setup, similar control can be obtained by equilibration with a saturated vapor phase.[5] In studies of (de)wetting and interfacial density fluctuations in water, predictions obtained by careful MD implementations of these techniques[1–5] are often consistent with GCMC studies[6–8], and provide viable alternatives to open ensemble methods when using Molecular Dynamics simulations.

1 Hari Acharya, Srivathsan Vembanur, Sumanth N. Jamadagni and Shekhar Garde, *Faraday Discuss.*, 2010, **46**, DOI:10.1039/b927019a.
2 S. Sarupria and S. Garde, *Phys. Rev. Lett.*, 2009, **103**, 037803.
3 N. Giovambattista, P. J. Rossky, P. Debenedetti, *Phys. Rev. E*, 2006, **73**, 041604.
4 N. Giovambattista, P. J. Rossky, P. Debenedetti, *J. Phys. Chem. B*, 2009, **113**, 13723.
5 A. J. Patel, P. Varilly, D. Chandler, *J. Phys. Chem. B*, 2010, **114**, 1632.
6 D. Bratko, R. Curtis, H. W. Blanch, J. M. Prausnitz, *J. Chem. Phys.*, 2001, **115**, 3873.
7 D. Bratko, C. D. Daub, A. Luzar, *J. Am. Chem. Soc.*, 2007, **129**, 2504.
8 D. Bratko, C. D. Daub, A. Luzar, *Faraday Discuss.*, 2009, **141**, 55.

**Professor Evans** replied: I thank Professor Bratko for his helpful comment.

**Dr Hummer** responded: Thank you for your helpful comment and the references.

**Professor Garde** responded: Thank you for your excellent comment and the references.

**Dr Henderson** addressed Dr Hummer: Sum rules involving integrals over pressure-tensor components are obtained by scaling the dimensions of the system at fixed temperature and chemical potentials (grand ensemble). If the volume is kept invariant while the interfacial area is altered, then this virial route yields an expression for the rate-of-change of surface-excess grand potential with change in surface area. For an un-patterned interface, this is the interfacial free energy per unit area, known as surface tension (although this historical name is not always helpful). However, if the substrate is patterned, either chemically or structurally, then deforming the area $A$ also typically alters at least one pattern wavelength, which means that the surface tension $\gamma$ will vary significantly with $A$ even though the adsorbed molecules are fluid. Hence, for patterned surfaces the virial route will yield a sum rule for the surface stress $\sigma \equiv \partial(\gamma A)/\partial A = \gamma + A\,\partial\gamma/\partial A$ instead of surface tension alone. This complication, first understood by Shuttleworth,[1] is well-known in the study of solid surfaces and solid–melt interfaces.[2] Less well-known is the equal relevance to the adsorption of fluid on chemically and structurally patterned substrates.[3,4] In the special case of stripe geometry it is possible to obtain sum rules for $\gamma$ and $A\partial\gamma/\partial A$, separately. For other patterns it appears to be impossible to use the virial route to separate surface tension from surface stress, requiring a laborious calculation of $A\partial\gamma/\partial A$ by some other route. It is not clear how relevant this problem is to the paper under discussion, but one always needs to be aware that a pressure tensor integral applied to patterned inhomogeneous fluid phenomena will seldom be a direct route to the calculation of interfacial free-energy, only surface stress. Numerical estimates imply that one cannot ignore the difference in typical cases.

1 R. Shuttleworth, *Proc. Phys. Soc. A*, 1950, **63**, 444.
2 J. Q. Broughton and G. H. Gilmer, *J. Chem. Phys.* 1986, **84**, 5759.
3 J. R. Henderson, *J. Phys.: Condens. Matter*, 1999, **11** 629.
4 J. R. Henderson, *Mol. Phys.*, 2003, **101**, 397.

**Dr Hummer** responded: We thank you for bringing up this interesting point. Indeed, for atomic and molecular systems, certain quantities are inherently discrete. This discreteness poses some challenges in relating thermodynamics to the statistical mechanics of the microscopic system. One such quantity is the chemical potential, with the particle number as a discrete conjugate variable. Another quantity is the interfacial free energy of a fluid and a solid, where the solid area comes in multiples of unit cells of the solid surface. To add to your comment, and for the sake of completeness, it may be instructive to rederive some basic (and well known) thermodynamic relations for the interfacial free energy. Consistent with our canonical simulation setup, let us consider a rectangular simulation box of dimensions $L_x$, $L_y$, and $L_z$. The interfaces at the top ($z = L_z$) and bottom ($z = 0$) each have an area $A = L_x L_y$. The system has a Helmholtz free energy $F = kT \ln Z = kT \ln \int \exp(-H/kT)\mathrm{d}^{3N}\mathbf{x}$, where $H$ is the Hamiltonian. The system is symmetric and infinite periodic in the $x$ and $y$ direction, but bounded in the $z$ direction. We define the interfacial free energy as

$$\gamma = \left.\frac{\partial F}{\partial A}\right|_{N,V,T} \quad (1)$$

To evaluate $\gamma$, we consider a particular deformation of the simulation box that preserves its volume $V = L_x L_y L_z$:

$$L_x \mapsto L_x(1+\varepsilon)$$
$$L_y \mapsto L_y \qquad (2)$$
$$L_z \mapsto L_z(1+\varepsilon)^{-1} = L_z(1-\varepsilon) + \mathcal{O}(\varepsilon^2)$$

For this deformation, we find that the area scales as $A \to A(1 + \varepsilon)$, and that

$$\gamma = \frac{1}{A}\frac{\partial F}{\partial \varepsilon}\bigg|_{\varepsilon=0} = \frac{1}{A}\left(L_x \frac{\partial F}{\partial L_x} - L_z \frac{\partial F}{\partial L_z}\right) \qquad (3)$$

The normal pressure, $P_{zz} = P_\perp$, and the in-plane (lateral) pressure, $P_{xx} = P_\parallel$, are defined as

$$P_{zz} = -\frac{1}{L_x L_y}\frac{\partial F}{\partial L_z}\bigg|_{N, L_x, L_y, T} \qquad (4)$$

$$P_{xx} = -\frac{1}{L_y L_z}\frac{\partial F}{\partial L_x}\bigg|_{N, L_y, L_z, T} \qquad (5)$$

We can thus rewrite $\gamma$ in terms of differences in normal and lateral pressures:
$$\gamma = L_z(P_{zz} - P_{xx}) \qquad (6)$$

This expression corresponds to Eq. (3) in our paper. In a second step, we relate thermodynamics to microscopic properties of the system. It is here where the difficulties arise that you point out. The transformation of Eq. 3 can be realized by rescaling the particle positions. For simplicity, we consider here only atomic coordinates (with additional terms arising for rigid molecules[1]). For Cartesian coordinates $x_i$, $y_i$, and $z_i$, their mapping according to Eq. 3 becomes:

$$x_i \mapsto x_i(1+\varepsilon)$$
$$y_i \mapsto y_i \qquad (7)$$
$$z_i \mapsto z_i(1+\varepsilon)^{-1} = z_i(1-\varepsilon) + \mathcal{O}(\varepsilon^2)$$

For simplicity we assume that the potential energy $U$ entering the Hamiltonian consists of the usual pair interactions $u$, plus boundary potentials $v_-$ and $v_+$ at the lower and upper boundary, respectively, that are periodic in $x$ and $y$:

$$U = \sum_{i<j} u(r_{ij}) + \sum_i v_-\left[z_i - z_-\left(\frac{x_i}{L_x}, \frac{y_i}{L_y}\right)\right] + \sum_i v_+\left[z_+\left(\frac{x_i}{L_x}, \frac{y_i}{L_i}\right) - z_i\right] \qquad (8)$$

Note the use of scaled $x$ and $y$ coordinates in the boundary potential. For this form, and with $z_\pm \mapsto z_i(1+\varepsilon)^{-1} z_\pm$ one can evaluate $\partial H/\partial \varepsilon$, and in turn obtain expressions for the diagonal components of the pressure tensor,

$$P_{zz} = \frac{NkT}{V} + \frac{1}{V}\left\langle \sum_{i<j} F_{ij,z} z_{ij} + \sum_i f_i - (z_i - z_-) - \sum_i f_{i,+}(z_+ - z_i)\right\rangle = \frac{\langle F_+ \rangle}{L_x L_y}$$
$$= -\frac{\langle F_- \rangle}{L_x L_y} \qquad (9)$$

$$P_{xx} = \frac{NkT}{V} + \frac{1}{V}\left\langle \sum_{i<j} F_{ij,x} x_{ij}\right\rangle \qquad (10)$$

Here $\langle F_+ \rangle = \langle F_- \rangle$ are the "piston" forces acting on the walls.

$P_{xx}$ does not contain $x$ and $y$ contributions from the boundary forces. This is a direct consequence of the assumption of periodic boundary potentials. Under the infinitesimal transformation considered here, the wavelength of the roughness itself changes, $\lambda \to \lambda(1 + \varepsilon)$. As you point out, to avoid this effect under periodic boundary conditions, and calculate $\gamma$ for a fixed roughness wavelength, is not trivial. One possible route would be to calculate the free energy change associated with a finite deformation of the box. By using, say, thermodynamic integration one could consider the following path: in the first step, remove the roughness; then deform the box with at boundaries such that exactly one more (or one less) wavelength fits into the $x$-dimension, and with conserved volume; finally, regrow the roughness. Clearly, such a path is not only tedious computationally, but also likely subject to pronounced finite-size effects. Here we instead consider the interfacial free energy associated with deformations that infinitesimally perturb the roughness wavelength in one dimension. It would be interesting to explore the magnitude of the difference in interfacial free energy for fixed and scaled $\lambda$. This could most easily be done by calculating the difference between $P_{xx}$ and $P_{yy}$ in a system with roughness only in the $x$ direction.

1 G. Hummer, N. Grønbech-Jensen, M. Neumann, *J. Chem. Phys.*, 1998, **109**, 2791–2797.

**Mr Kang** asked: It was very interesting to know that there can be a transition from hydrophilic to hydrophobic as roughness changes.

Could you comment on what parameters determine the periodicity and the amplitude of the critical roughness for transition?

**Dr Hummer** answered: For a roughness wavelength of about 11.6 Å, the transition from "hydrophilic" to "hydrophobic" occurs at roughness amplitudes $a$ of 2–3 Å. For a fixed amplitude of $a = 3$ Å, the transition occurs at wavelengths below 15 Å.

**Professor J. Klein** communicated: In your interesting paper you write that "Water confined between a "Janus interface" of adjoining hydrophobic and hydrophilic surfaces was found to fluctuate significantly during shear deformations.[19]", where ref. 19 is by Zhang, Zhu and Granick.[1] This is in fact not correct: It was shown explicitly by de Gennes[2] in a paper analyzing the results of ref. 19 that the results of ref. 19 were due to a solid protrusion bridging the two surfaces rather than to any liquid between them. This suggestion of a possible protrusion was subsequently 'confirmed' by Lin and Granick,[3] which demonstrated that mica surfaces prepared in the Granick lab prior to 2003 (including those in ref. 19 by Zhang, Zhu and Granick, 2002) were quite heavily contaminated with solid Pt nanoparticles arising from their method of preparation. The height of these contaminant Pt nanoparticles was *ca.* 3–5 nm, just the surface separation where ref. 19 claimed fluctuations during shear. In view of this (*i.e.* that the fluctuations[19] under shear that you relate to were in fact due to a solid contaminant rather than liquid water), does this change any of the conclusions of your manuscript?

1 X. Zhang, Y. Zhu and S. Granick, *Science*, 2002, **295**, 663–666.
2 P. G. de Gennes, *Adv. Colloid Interface Sci.*, 2003, **100–102**, 129–135.
3 Z. Lin and S. Granick, *Langmuir*, 2003, **19**, 7061–7070.

**Dr Hummer** communicated in reply: Thank you for bringing these alternative interpretations to our attention. In response to your question, in our highly controlled, very simple system such issues do not arise, and so our conclusions are not affected.

**Mr Stirnemann** addressed Professor Garde by communicating: You highlight the limitations of context-independent hydropathy scales. From your study, what would be the respective influences on hydropathy of the exposed surface local topology *vs.* the chemical nature of the proximal groups? Could this help us understand why out of the large hydrophobic patch in Fig. 8c (according to the Kyte–Dolittle scale), your binding frequency map shows that one specific site is markedly more hydrophobic (Fig. 8e), while its proximal environment does not seem much different than that of the others.

**Professor Garde** communicated in reply: This is a challenging question to answer. Indeed, both the local chemistry and topology affect the local density fluctuations, and therefore, hydrophobicity. Separating the effects of one from the other, will however, be challenging. One might perform systematic studies of increasingly rough surfaces of a given chemistry to try and de-convolute the two effects, but how one combines them to develop hydrophobicity maps of complex protein interfaces is not clear to us. Indeed, if such de-convolution were possible we could address the question of why a specific site appears as a hydrophobic hot spot. Perhaps, a combination of our approach with that of Prof. Peter Rossky, who presented preliminary work on developing an atom-based hydropathy scale in his introductory lecture, is a way toward developing a computationally efficient (first pass) predictive scheme that performs better than the existing hydropathy scales.

# Concluding remarks for FD 146: Answers and questions

Frank H. Stillinger

Received 26th May 2010, Accepted 26th May 2010
DOI: 10.1039/c005398h

## Introduction

In view of the attractiveness of the basic science involved, and the significance of its applications, it was quite natural and appropriate that a *Faraday Discussion* in 2010 would be devoted to the Wetting Dynamics of Hydrophobic and Structured Surfaces. Powerful motivation for continued scientific and engineering advances in this area arise from sources as diverse as the fundamental role of hydrophobic interactions in molecular biology, the importance of natural and artificial self-cleaning surfaces, the production of water-repellant textiles, and the technological requirements of microfluidics. The high rate of current related research activity is hard to miss; even as the nearly two dozen invited papers for this *Faraday Discussion* were being prepared by their authors, collected by the conference organizers, and distributed to participants, the scientific literature exhibited a continuing stream of publications in the same scientific area covered by this *Faraday Discussion*, but from yet other research groups.[1] As further evidence demonstrating widespread creative involvement in this pervasive subject, one can cite two other recent and closely connected *Faraday Discussions*, namely FD129 "The Dynamics and Structure of the Liquid–Liquid Interface", and FD141 "Water—From Interfaces to the Bulk".

This *Faraday Discussion* 146 arrived with an important distinction in comparison with its forerunners. The formal meeting was immediately preceded by a three-day Graduate Research Seminar (April 9–11, 2010), held locally at the Virginia Commonwealth University School of Engineering. This introductory activity was arranged for the benefit of involved undergraduate, graduate, and postdoctoral students, numbering roughly 40, who would also be part of the following FD 146. Beyond its social amenities, this Graduate Research Seminar offered five formal lectures by established experts (Widom, Yeomans, Evans, Klein, and Quéré) who would also be involved as invited speakers in the *Faraday Discussion* sessions to follow. In addition, this Graduate Research Seminar included student oral presentations, and two poster sessions. The organizers' intentions behind arranging this introductory meeting were to familiarize the young participants with major aspects of the general scientific field, with its terminology, and with at least some of its more seasoned practitioners. Evidently this strategy was successful in bringing the student group into more comfortable interaction with ongoing research activities; during the subsequent three-day *Faraday Discussion* itself, many of the technical questions and comments directed at the invited speakers were posed confidently and with insight by the younger population of students and committed scientists, an encouraging sign for the intellectual future of the field. A broader implication is that at least some future *Faraday Discussions* might benefit similarly by arranging analogous preliminary meetings for their own groups of devoted students and early-career scientists.

In addition to the formal invited lectures presented at *Faraday Discussion* 146 (April 12–14, 2010), the program also included its own pair of poster sessions.

Department of Chemistry, Princeton University, Princeton, NJ, 08544, USA

Among those poster presentations, three were selected for awards to recognize their excellence. Those awards were announced publicly at the close of Session 6, Tuesday afternoon.

Viewed in a conceptually coarse-grained way, the contributed papers FD146: 1–23 can be crudely classified into two categories, "experimental", and "theoretical/simulational". In this naively simplified view there were 12 of the former and 11 of the latter. This near-unity ratio may reflect a little something about the prejudices of research funding sources, but perhaps more importantly it is roughly consistent with the distribution of opportunities for substantial scientific advances in the immediate future. The listing of abstracts for the poster session that was distributed to participants presented a somewhat different ratio; with a nearly 50% incremental dominance of the former over the latter. Considering the fact that poster presenters demographically tend to be significantly younger on average than invited speakers, this discrepancy may be a harbinger of a future trend in research activity surrounding this Faraday Discussion's chosen subject.

A successful scientific meeting is a positively memorable event, and this FD146 qualifies for that description. For those participants who traveled from Europe to attend and contribute to the meeting, expecting to return on schedule, there was an additional unpleasant memory that arose unexpectedly. On the last day of the meeting, April 14, the Icelandic volcano Eyjafjallajökull began to emit a seriously disruptive amount of gas and ash, with the unfortunate result that airline travel across the Atlantic Ocean was interrupted for several days.

Remarks below have been divided into three parts. The first involves a brief view of some aspects of the collection of invited papers, as well as the discussions that they stimulated, and the poster presentations. These can fairly be taken as representing the current scientific and technological status of the field. This is followed by two lists of suggestions for possible future research, one for experiment and one for theory/computation. These lists include, but extend well beyond, the Faraday Discussion's selected topic of Wetting Dynamics of Hydrophobic and Structured Surfaces. Nevertheless they represent only a small fraction of the opportunities for future advances. The reason for adopting this broader viewpoint is the generally acknowledged necessity of establishing as many intellectual ties as possible to other areas of the physical and biological sciences, with the expectation that those connections and the conceptual feedbacks that they can generate will yield fundamental and useful insights for aqueous surface chemistry, physics, and biology.

## Present status

The formal program began with a comprehensive introductory lecture delivered by Prof. P. J. Rossky. Its emphasis was on the molecular level interactions and cooperative phenomena that underlie the macroscopic observations central to this *Faraday Discussion*. Ultimately these regimes at quite different length scales need to be much more deductively connected than they are at present. This introduction was followed by two sessions (five invited papers) devoted primarily to superhydrophobic surfaces, two subsequent sessions (seven invited papers) for dynamic transitions at various surfaces, one session each for liquid–vapor interfaces and nanobubbles (three invited papers) and for hydrophobic surfaces (two invited papers), and two final sessions (five invited papers) focusing on heterogeneous surfaces. These presentations and the audience questions that they generated can be described overall as indicative of a vigorous and imaginative scientific field that is in its intellectual adolescence. The experimental and theoretical advances reported (and this is true of the poster presentations too) involve clever experimental and theoretical techniques, but much has yet to be accomplished. The scientific areas covered by *Faraday Discussion* 146 will remain lively and productive; they are unlikely to "dry up" intellectually in the foreseeable future.

It is worth mentioning that six of the invited talks presented reported classical molecular dynamics simulations results that utilized either the simple point-charge model "SPC"[2] to describe water intermolecular interactions (146/17[3]), or its extension "SPC/E"[4] that incorporates polarization self energy (146/06,[5] 146/13,[6] 146/18,[7] 146/22,[8] and 146/23[9]). A significant number of distinctively different alternative model interactions have been proposed to describe water.[10,11] Therefore it is at least temporarily beneficial for the subject under consideration here to have simulation efforts focused on essentially a single model. This allows consistent comparison of results that examine different physical situations to produce a more comprehensive view of the molecular phenomena that water and its solutions produce. In the future it will inevitably become advantageous to graduate to simulations consistently based on a different and more physically accurate interaction model for water.

As a minor side issue arising both in one of the public presentations as well as during informal conversations, the semantic appropriateness of the adjectives "hydrophobic" and "superhydrophobic" was raised for the situations in which these words are normally invoked. It is generally acknowledged that however weak it might be, there is always a net attraction between a droplet of pure water or other liquid and a nearby solid (uncharged) substrate, due at least to van der Waals interactions. So one can argue that "phobia" is really always "philia", but with highly variable magnitude. No serious alternative terminology has been proposed, however.

The fact that water, aqueous solutions, and their related surface phenomena were thematically central for FD146 invites a bit of historical perspective. Dispersed within the vast inventory of valid scientific contributions concerning water that have accumulated over many years are a few bizarre claims. And indeed the final refutations of these bizarre claims speak well for the normal operation of the scientific method. Notorious specific examples are "polywater",[6] "cold fusion",[13] and "dilution memory".[14] By contrast the invited and contributed research presentations at FD146 were consistently representative of "normal" scientific inquiry, constituting valid and useful progress.

## Prospects for experiment

Many of the distinguishing properties that pure liquid water exhibits in thermodynamic equilibrium are magnified as it is supercooled below the freezing point. These magnifying properties include negative thermal expansion, high isothermal compressibility and isobaric heat capacity, and isothermal reduction in shear viscosity with increasing pressure. These metastable bulk-phase properties have been widely explored and well documented.[15] However there has been relatively little attention devoted to the water surface properties in the supercooled regime that would bear on the subject of this *Faraday Discussion*. Specifically, the surface tension $\gamma(T)$ needs careful determination for supercooling toward $-35\,°C$, the range where several bulk properties hint at an impending divergence,[15] to establish if it exhibits its own singularity. The obvious corollary issue is how the static contact angle $\theta_c(T)$ for supercooled water droplets on various hydrophobic and superhydrophobic surfaces behaves as temperature declines. The water droplets typically examined experimentally for investigations of the types covered in this *Faraday Discussion* have the advantage of relatively low ice nucleation rates because of their small volumes, a major advantage in attempting to probe the deep supercooling regime.

The simple structure of the isolated water molecule (nominally exhibiting symmetry $C_{2v}$) implies that it is achiral (geometrically unchanged under mirror imaging). Consequently its liquid behavior on a patterned surface of low symmetry that itself displays a definite handedness will not change if that substrate's pattern handedness is reversed (*i.e.*, mirror imaged). However many substances have optically active chiral molecules with stable liquid ranges that are convenient for experiment. An example is 3-methyl hexane (m.p. $-119.4\,°C$, b.p. $92\,°C$). The pure *D* and *L* enantiomers of such substances would exhibit identical liquid–vapor surface tensions as a function of temperature, and would behave identically on an achiral substrate (*e.g.*, contact angle as a function of temperature). However mixtures of the two mirror-image enantiomers could be expected to show a composition dependence of the equilibrium liquid–vapor and liquid–substrate interfacial free energies. This in turn should lead to measurable $D$, $L$ mixing effects on droplet contact angles.

At least equally instructive would be how these pure enantiomorphs and their racemic mixtures would behave next to chiral solvophobic substrates. In particular chiral ligands could be grafted onto a suitable flat substrate such as silicon or gold, against which $D$ and $L$ enantiomers would then in principle behave differently. In the case of the just-mentioned optically active alkane liquid 3-methyl hexane, perfluorinated optically active ligands should produce non-wetting surfaces. Several distinct possibilities arise when these chirally-distinct surfaces are brought into near contact, depending on whether they display the same or opposite chirality, and even whether the intervening liquid is chiral, a racemic mixture, or even non-chiral.

It should not escape attention that some polyatomic substances exhibit two-dimensional crystal structures in their liquid–vapor interfaces over a non-trivial temperature range above their bulk thermodynamic melting points. This unusual phenomenon has been established for the normal alkanes $C_nH_{2n+2}$ in the chain-length range $16 \leq n \leq 50$,[16,17] and as would be expected it affects the static and dynamical properties of substantially planar interfaces.[18] The influence of this surface crystallization on static and dynamical properties of small droplets, specifically their contact angles at various lyophobic substrates, is largely unknown and so deserves experimental investigation.

The aqueous surface phenomena receiving attention at this *Faraday Discussion* would benefit from being viewed in the wider scientific context of other classes of liquids and their own characteristic surface phenomena. An extreme related area might be identified as involving liquid metals; these typically display much larger surface tensions than those for aqueous fluids, and have virtually no tendency to wet organic substrates. The most obvious candidates for "temperature-convenient" measurement are mercury [m.p. $-38.87$ °C, b.p. 356.58 °C, $\gamma(20$ °C$) \cong 475$ dyn/cm] and gallium [m.p. 29.78 °C, b.p. 2403° C, $\gamma$ (m.p.) $\cong 718$ dyn/cm]. However the former has a well-deserved bad reputation on account of the toxic nature of its vapor around room temperature. But by contrast the very high boiling point of the latter indicates that it has a far lower tendency to vaporize until very strongly heated. It would be edifying to classify various ionic-crystal, semiconductor, and metallic substrates as "mercurophobic" and "galliophobic" (alternatively "mercurophilic" and "galliophilic") *via* contact angle observations, as well as to determine quantitative details of mean forces as a function of distance acting between pairs of solid objects embedded in these metallic liquids.

On account of their historical role in this subject and their frequent mention in the literature, the leaves of the Lotus (*Nelumbo nucifera*, *Nelumbo lutea*) have become the "poster child" of superhydrophobicity. Consequently this phenomenon has traditionally been called the "Lotus Effect". But in addition to the Lotus many other plant species also possess this beneficial characteristic.[19] Considering the fact that the superhydrophobicity leads to leaf self-cleaning as well as to reduction in opportunities for bacterial and fungal infections, it is natural to ask if this property might not also be useful for other plant species that do not currently exhibit it. In particular ornamental flowers, shrubs, and trees forced to survive in dirty urban environments might benefit from this characteristic. More specifically, genetic engineering (a.k.a. recombinant DNA[20]) has become a major contributor to optimizing the productivity of a wide variety of important crops. It would therefore be potentially valuable to identify and isolate the gene(s) that generate superhydrophobic leaf surfaces in various species, and to see if gene insertion could be exploited for additional agricultural advantage.

## Prospects for theory/simulation

The geometric patterns of water-molecule hydrogen bonds present at a hydrophobic or superhydrophobic interface, and how they compare with patterns inside the bulk liquid at the same temperature and pressure have not been aggressively investigated, although they could be an important source of scientific insight. In particular, the occurrence frequency with which successive hydrogen bonds are arranged to form closed polygons of different sizes is a fundamental characteristic. Although some attention has been expended in this direction for solvation of small hydrophobic solutes such as methane,[21] more detailed simulational examination of extended hydrophobic/superhydrophobic surfaces is warranted. Of course this hinges on the definition of "hydrogen bond", a concept that does not enjoy a unique definition, in spite of the consensus opinion of its basic relevance. But whether a geometric or a potential energy definition is chosen, it should be consistent with the agreed-upon presence and arrangements of hydrogen bonds between neighboring molecules in the crystal polymorphs of ice.[22]

Beyond hydrogen bond topology issues, the precise way that the distribution functions of various orders for water vary with distance from a hydrophobic or superhydrophobic surface has not yet been theoretically specified in sufficient detail. This concerns specifically the spatial rate at which surface perturbations of molecular distribution functions die away to be replaced by their bulk characteristics as the normal distance from the surface increases. That may indeed involve hydrogen-bond polygon size distribution variations, but it also simply involves number density and polarization density variations. These spatially decaying perturbations would be involved in the long-range tail of the water-mediated interaction between a pair of hydrophobes, such as a pair of flat surfaces, where the prevailing temperature, pressure, and distance were such that the drying phenomenon was avoided. In addition to the conventional descriptions provided by statistical mechanics of condensed matter systems, it may also be useful to examine the predictions of Casimir–Lifshitz theory for the effect of electromagnetic field fluctuations which (depending on the complex dielectric functions of the materials involved) can yield long-range attractions or repulsions.[23]

Within the macroscopic description regime, droplet contact angles can indeed be well defined and measured. This is true not only for static equilibrium contact angles $\theta_c$, but also for the advancing and receding contact angles $\theta_a$ and $\theta_r$ under droplet transport circumstances. But because water static and dynamic phenomena at hydrophobic and structured surfaces span length scales from the molecular to the macroscopic regimes, a basic issue is how, or even whether, contact angles can be precisely and uniquely defined for very small droplet systems. Computer simulations typically incorporate $10^3$ to $10^4$ molecules, which for liquid water involve droplet diameters of roughly 10 nm or less. Simply estimating contact angles in these simulations roughly from graphical images of instantaneous configurations is ultimately not a very satisfying procedure. If instead an objective and precise algorithm for extracting contact angles from molecular distribution functions were to become available, the effects of variable droplet size and of substrate type on those contact angles would become accessible and would yield useful insights into the physics and chemistry involved.

The capillary length $l_c$ for a given liquid roughly represents the outcome of the competition between the liquid–vapor surface tension $\gamma$ and the strength $g$ of the gravitational field. It is defined by the elementary relation:

$$l_c = (\gamma/g\rho)^{1/2}.$$

Here $\rho$ is the mass density of the liquid under consideration. For pure liquid water at the earth's surface and at room temperature this length is approximately 2.7 mm. This is the order of magnitude of the vertical thickness of a puddle of water sitting atop a flat and horizontal superhydrophobic surface, in other words an essentially macroscopic length. However it is an engaging fact that the effective value of the gravitational acceleration $g$ can be increased by up to a factor of $10^6$ in an ultracentrifuge.[24] This would formally reduce $l_c$ for room-temperature water to approximately 2.7 μm. It would be a challenging theoretical problem then to deduce the structure of the water hydrogen bond network geometry in such extreme $g$-flattened "puddles", as well as to derive the $g$ dependence of the contact angle. Nevertheless, water models currently in use for numerical simulation could probably supply a qualitatively accurate description.

The development of more accurate models (*i.e.*, approximations) for water molecule interactions and their utilization in simulations remains a major theoretical challenge. As stressed earlier there are advantages for various research groups at any given period to employ a common model interaction, however improvements eventually must be aggressively sought. In the long run some variant of the *ab initio* quantum mechanical method for calculating the potential energy in a group of interacting water molecules "on the run" may be the most advantageous strategy.[25] However at present that approach is not yet capable of dealing with the typical numbers of molecules in simulations at the accuracy required to contribute effectively to the subjects under consideration in *Faraday Discussion* 146. Instead, more precise and descriptive "semi-empirical" potentials should be devised. Specifically it would be useful to include intramolecular vibrational degrees of freedom, and the dependence of their frequencies on hydrogen bonding. Furthermore, it is important to recognize that the dipole moments of water molecules depend on their environment, so that moment magnitudes and directions will not be the same in bulk phases as in interfaces. Further research will be required to determine how best to represent these attributes in terms of minimally complex mathematical functions whose sums combine to produce the necessary $N$-molecule interaction potential energy function.

Presuming that a future *Faraday Discussion* devoted to the same or similar subjects as this one will occur in the foreseeable future, it will be instructive then to see if the kinds of qualitative suggestions listed here for both experiment and for theory/simulation will have generated substantive advances.

## Acknowledgements

The author thanks the organizers for the opportunity to attend and contribute to this *Faraday Discussion* 146. He is grateful specifically to Professors Alenka Luzar and Peter Rossky for providing technical information about the subjects covered at this meeting and about their proponents.

## References

1 Examples just from the physics literature are the following: (*a*) P. J. Feibelman, *Phys. Today*, 2010, **63**(2), 34; (*b*) N. Savva, S. Kalliadasis and G. A. Pavliotis, *Phys. Rev. Lett.*, 2010, **104**, 084501; (*c*) T. Hofmann, M. Tasinkevych, A. Checco, E. Dobisz, S. Dietrich and B. M. Ocko, *Phys. Rev. Lett.*, 2010, **104**, 106102; (*d*) P. Tsai, R. G. H. Lammertink, M. Wesling and D. Lohse, *Phys. Rev. Lett.*, 2010, **104**, 116102; (*e*) J. C. Fogarty, H. M. Aktulga, A. Y. Grama, A. C. T. van Duin and S. A. Pandit, *J. Chem. Phys.*, 2010, **132**, 174704.
2 H. Berendsen, J. Postma, W. van Gunsteren, and J. Hermans, in *Intermolecular Forces*, B. Pullman, editor, Reidel: Dordrecht, 1981, pp. 331–342.
3 Lingle Wang, Richard A. Friesner and Bruce J. Berne, *Faraday Discuss.*, 2010, **146**, DOI: 10.1039/b925521b.
4 H. J. C. Berendsen, J. R. Grigera and T. P. Straatsma, *J. Phys. Chem.*, 1987, **91**, 6269.
5 Christopher D. Daub, Jihang Wang, Shobbit Kudesia, Dusan Bratko and Alenka Luzar, *Faraday Discuss.*, 2010, **146**, DOI: 10.1039/b927061m.
6 Takahiro Koishi, Kenji Yasuoka, Xiao Cheng Zeng and Shigenori Fujikawa, *Faraday Discuss.*, 2010, **146**, DOI: 10.1039/b926919c.

7 Guillaume Stirnemann, Peter J. Rossky, James T. Hynes and Damien Laage, *Faraday Discuss.*, 2010, DOI: 10.1039/b925673c.
8 Jeetain Mittal and Gerhard Hummer, *Faraday Discuss.*, 2010, **146**, DOI: 10.1039/b925913a.
9 Hari Acharya, Srivathsan Vembanur, Sumanth N. Jamadagni and Shekhar Garde, *Faraday Discuss.*, 2010, **146**, DOI: 10.1039/b927019a.
10 B. Guillot, *J. Mol. Liq.*, 2002, **101**, 219.
11 C. Vega, J. L. F. Abascal, M. M. Conde and J. L. Aragones, *Faraday Discuss.*, 2009, **141**, 251.
12 F. Franks, *Polywater*, MIT Press: Cambridge, Massachusetts, 1981.
13 J. R. Huizenga, *Cold Fusion, The Scientific Fiasco of the Century*, University of Rochester Press: Rochester, New York, 1992.
14 E. Davenas, F. Beauvais, J. Amara, M. Oberbaum, B. Robinzon, A. Miadonna, A. Tedeschi, B. Pomeranz, P. Fortner, P. Belon, J. Sainte-Laudy, B. Poitevin and J. Benveniste, *Nature*, 1988, **333**, 816.
15 C. A. Angell, in *Water, A Comprehensive Treatise*, Vol. 7, F. Franks, editor (Plenum Press, New York, 1982), pp. 1–81.
16 X. Z. Wu, B. M. Ocko, E. B. Sirota, S. K. Sinha, M. Deutsch, B. H. Cao and M. W. Kim, *Science*, 1993, **261**, 1018.
17 B. M. Ocko, X. Z. Wu, E. B. Sirota, S. K. Sinha, O. Gang and M. Deutsch, *Phys. Rev. E*, 1997, **55**, 3164.
18 S. Prasad and A. Dhinojwala, *Phys. Rev. Lett.*, 2005, **95**, 117801.
19 C. Neinhuis and W. Barthlott, *Ann. Bot.*, 1997, **79**, 667.
20 J. D. Watson, R. M. Meyers, A. A. Caudy, and J. A. Witkowski, *Recombinant DNA: Genes and Genomes: A Short Course*, 3rd edition (W. H. Freeman and Company), 2007.
21 D. Chandler, *Nature*, 2005, **437**, 640, see Fig. 1.
22 D. Eisenberg and W. Kauzmann, *The Structure and Properties of Water*, Oxford University Press: New York, 1969, Chap. 3.
23 J. N. Munday, F. Capasso and V. A. Parsegian, *Nature*, 2009, **457**, 170.
24 T. Svedberg and K. O. Pedersen, *The Ultracentrifuge*, Clarendon, Oxford, 1940.
25 (*a*) R. Car and M. Parrinello, *Phys. Rev. Lett.*, 1985, **55**, 2471; (*b*) T. D. Kuhne, M. Krack, F. R. Mohamed and M. Parrinello, *Phys. Rev. Lett.*, 2007, **98**, 066401.

# Poster titles

Aerosol assisted chemical vapour deposition of elastomers – A route to self-cleaning thin films, **Colin R. Crick and Ivan P. Parkin**, *University College London, UK*

Water droplet and menisci profiles: A comparison study between macroscopic predictions and results from nanoscale computer simulations, **Nicolas Giovambattista, Sergey V. Buldyrev and Pablo G. Debenedetti**, *Brooklyn College, USA*

Anisotropic superhydrophobic surface inspired by rice leaf, **Zhiguang Guo and Bao-lian Su**, *Namur University, Belgium*

Water vapor adsorption and microscopic wetting of heterogeneous surfaces – A molecular simulation study, **Milán Szőri, Pál Jedlovszky and Martina Roeselová**, *Academy of Sciences of the Czech Republic, Czech Republic*

Fabrication of superhydrophobic surfaces with regular pillar-like and hierarchical structures, **Kuan-Hung Cho, Kuan-Yu Yeh, Wei-Fan Kuan and Li-Jen Chen**, *National Taiwan University, Taiwan*

Femtosecond laser-patterned pure substrates: Cassie–Baxter to Wenzel wetting transitions, **Anne-Marie Kietzig, Mehr Negar Mirvakili, Savvas G. Hatzikiriakos and Peter Englezos**, *The University of British Columbia, Canada*

Superhydrophobic polyimide coatings on aluminium plate and their anti-frosting properties, **Zhengyu Shi, Mei Li and Qinghua Lu**, *Shanghai Jiao Tong University, China*

Modifying wettability of surfaces by nanoparticles: Experiments and modelling using the Wenzel law, **Mikko Alava, Tiina Nypelö, Monika Österberg, Janne Laine and Lei Dong**, *Aalto University, Finland*

An improved determination of the spring constant and slip length using large colloidal probe atomic force microscopy, **Sean P. McBride and Bruce M. Law**, *Kansas State University, USA*

Molecular dynamics simulations of droplets wetting a polymer surface, **Valentina Marcon and Nico van der Vegt**, *Technische Universität Darmstadt, Germany*

Sequence dependent dynamics of water molecules in the DNA minor grooves revealed by MD simulations, **Yoshiteru Yonetani and Hidetoshi Kono**, *Japan Atomic Energy Agency, Japan*

Dynamic intrusion and extrusion of water in hydrophobic mesoporous materials, **Ludivine Guillemot, Elisabeth Charlaix, Gérard Vigier and Anne Galarneau**, *Université Claude Bernard Lyon 1, France*

Effect of 2,2,2-triflouroethanol (TFE) on micellar properties of Triton X-165. Similarities with proteins/TFE interactions, **Pedro A. Galera-Gómez, Maria A. Elorza, Concepción Arias, Begoña Elorza, Concepción Civera and Francisco Garcia-Blanco**, *Universidad Complutense, Spain*

Equilibrium shapes of coalescing drops on hysteric surfaces, **Bharadwaj Prabhala, Mahesh Panchagnula and Srikanth Vedantam**, *Tennessee Technological University, USA*

Droplet impingement on superhydrophobic surfaces: Compressibility and water hammer induced dynamic wetting transitions, **Kripa K. Varanasi and Tao Deng**, *Massachusetts Institute of Technology, USA*

Controlling nucleation and growth of water using heterogeneous hydrophobic–hydrophilic surfaces, **Kripa K. Varanasi and Tao Deng**, *Massachusetts Institute of Technology, USA*

Computer simulation of a nematic nanodroplet wetting a flat surface, **Davide Vanzo, Roberto Berardi, Matteo Ricci and Claudio Zannoni**, *Università di Bologna, Italy*

Effect of porosity and thickness of the carbon layer on the hydrophobicity of carbon nanofiber (CNF) layers, **Sergio Pacheco Benito and Leon Lefferts**, *University of Twente, The Netherlands*

Engineering artificial omniphobic surfaces at different scales: From nanostructures on silicon to microstructures on PDMS, **Thi Phuong Nhung Nguyen, Renaud Dufour, Yannick Coffinier, Maxime Harnois, Florian Lapierre, Vincent Thomy, Rabah Boukherroub and Vincent Senez**, *Université Lille Nord de France, France*

The energetics of hysteresis, **Y. Kwon, S. Choi, N. Anantharaju, J. Lee, M. V. Panchagnula and N. A. Patankar**, *Northwestern University, USA*

Deceleration-driven wetting transition during 'gentle' drop deposition on textured surfaces, **Adam T. Paxson, Hyuk-Min Kwon, Kripa K. Varanasi and Neelesh A. Patankar**, *Northwestern University, USA*

Graphical approach to structure in the first (and second) solvent shell of hydrophobes, **Barbara Logan Mooney and L. René Corrales**, *University of Arizona, USA*

Vapor stabiizing surfaces, **N. A. Patankar**, *Northwestern University, USA*

Self-propelled dropwise condensate on superhydrophobic surfaces, **Jonathan B. Boreyko and Chuan-Hua Chen**, *Duke University, USA*

Restoring superhydrophobicity of lotus leaves with vibration-induced dewetting, **Jonathan B. Boreyko and Chuan-Hua Chen**, *Duke University, USA*

Influence of breath figures on growth of NaCl saturated water, **J. Guadarrama, R. D. Narhe, W. González-Viñas and D. Beysens**, *University of Navarra, Spain*

The impact of the contact angle on the performance of multiphase microreactors, **Kristin Hecht, Georg Fröhlich, Furkan Özkan, Peter Pfeifer, Roland Dittmeyer and Bettina Kraushaar-Czarnetzki**, *Karlsruhe Institute of Technology (KIT), Germany*

Temperature induced hydrophilic to hydrophobic switch in water confining organic pores, **Natalia Pérez-Hernández, Eduardo H. L. Falcao, Quan Trung-Luong, Cirilio Pérez, Juergen Eckert, Martina Havenith and Julio D. Martín**, *Instituto de Investigaciones Químicas (CSIS), Spain*

Using molecular simulation to understand wetting at rough surfaces, **Eric M. Grzelak, Vaibhaw Kumar and Jeffrey R. Errington**, *The State University of New York Buffalo, USA*

Drying of colloidal droplets on superhydrophobic surfaces, **Wei Xu, Rajesh Leeladhar and Chang-Hwan Choi**, *Stevens Institute of Technology, USA*

Competition of electrostatic and hydrophobic interactions between small hydrophobes and model enclosures, **Lingle Wang, Richard A. Friesner and B. J. Berne**, *Columbia University, USA*

Oil drop hydration and coalescence, **Robin C. Underwood, Jill Tomlinson-Phillips and Dor Ben-Amotz**, *Purdue University, USA*

Quantifying (1) temperature-dependent structure and thermodynamics and (2) thermal transport at surfactant interfaces using non-equilibrium molecular dynamics simulations, **Hari Acharya and Sherkhar Garde**, *Rensselaer Polytechnic Institute, USA*

Polyurethanes with alkyl ammonium co-polyoxetanes soft blocks: Effect of time, thermal processing and soft block molecular weight on surface morphology, **Kennard Brunson, Asima Chakravorty and Kenneth J. Wynne**, *Virginia Commonwealth University, USA*

Anomalous waterlike behaviour in spherically-symmetric water models optimized with $S_{rel}$, **Aviel Chaimovich and M. Scott Shell**, *University of California, USA*

Wetting on functionalized nanostructured materials, **Anindarupa Chunder and Lei Zhai**, *University of Central Florida, USA*

Salty nanodrops and their interactions with electric fields, **Christopher D. Daub, Dusan Bratko and Alenka Luzar**, *Virginia Commonwealth University, USA*

Insights into the hydrophobic interaction by direct measurement of light-modulated interactions and bilayer rupture, **Stephen H. Donaldson Jr., Bradley F. Chmelka and Jacob N. Israelachvili**, *University of California, USA*

Dynamic mean field theory for fluids in porous materials: Comparison with higher order approximations and molecular simulations, **John R. Edison and Peter A. Monson**, *University of Massachusetts, USA*

Bacterial biofilm omniphobicity, **Alexander K. Epstein and Joanna Aizenberg**, *Harvard University, USA*

Near superhydrophobic surfaces *via* cold crystallization, **Ying Zhang, Wei Zhang, Murari L. Gupta and Kenneth J. Wynne**, *Virginia Commonwealth University, USA*

Lubrication forces in air, **Chris Honig, John Sader and William Ducker**, *Virginia Tech, USA*

Engineering of a microfluidic device for co-culturing of cells with dynamic interaction, **Neha Jaina, Bryan Pfister and Raquel Perez Castilliejos**, *New Jersey Institute of Technology, USA*

Self-assembly of TMAO at hydrophobic interfaces and its effect on protein adsorption: Insights from experiments and simulations, **Gaurav Anand, Sumanth N. Jamadagni, Shekhar Garde and Georges Belfort**, *Rensselaer Polytechnic Institute, USA*

Dynamic wetting of fibrous surface, **Sung Hoon Kang, Boaz Pokroy, L. Mahadevan and Joanna Aizenberg**, *Harvard University, USA*

Monte Carlo simulations to study effect of roughness on wetting behaviour, **Vaibhaw Kumar and Jeffrey R. Errington**, *The State University of New York Buffalo, USA*

Quantum MD simulation of nano-sized water droplets on hydrophobic surfaces, **Hui Li and Xiao Cheng Zeng**, *University of Nebraska Lincoln, USA*

Effect of temperature on the surface forces measured between hydrophobic silica surfaces, **Zuoli Li, Jianli Wang and Roe-Hoan Yoon**, *Virginia Tech, USA*

Measurement of forces between TMCS-coated colloidal surfaces in aqueous solutions, **Dean Mastropietro and William Ducker**, *Virginia Tech, USA*

Structure and dynamics of water confined in amorphous silica nanopores, **Anatoli Milischuk and Branka M. Ladanyi**, *Colorado State University, USA*

Kinetics of thinning of the wetting films formed on quartz in $C_{18}$ TACl solutions, **Lei Pan and Roe-Hoan Yoon**, *Virginia Tech, USA*

The role of attractive forces at aqueous interfaces: Insight from Local Molecular Field theory, **Richard C. Remsing, Jocelyn M. Rodgers and John D. Weeks**, *University of Maryland, USA*

What determines contact angle of a nanodrop on a heterogeneous surface? **John Ritchie, Jihang Wang, Dusan Bratko and Alenka Luzar**, *Virginia Commonwealth University, USA*

Hydration shell of hydrophobic solutes and proteins – Exploring the pressure dimension, **Sapna Sarupia and Shekhar Garde**, *Princeton University, USA*

Switchable nano-wetting dynamics, **Jamileh Seyed-Yazdi, Jihang Wang, Dusan Bratko and Alenka Luzar**, *Virginia Commonwealth University, USA*

The origin of the attraction between heterogeneously charged surfaces, **Gilad Silbert, Yael Dror, Susan Perkin, Nir Kampf and Jacob Klein**, *Weizmann Institute of Science, Israel*

Characterization of hexane–water Pickering emulsions stabilized by hydrophobically-modified poly(N-isopropylacrylamide)-based microgel particles, **J. B. Thorne, L. S. Benée and M. J. Snowden**, *University of Greenwich at Medway, UK*

Water properties at contact with amphiphilic surfaces, **Naga Rajesh Tummula, Shi Liu, Dimitrios Argyris, Tuan A. Ho and Alberto Striolo**, *University of Oklahoma, USA*

Dynamics at a Janus interface, **Michael von Domaros, Jihang Wang, Dusan Bratko and Alenka Luzar**, *Virginia Commonwealth University, USA*

Surface-bound proteins with preserved functionality, **Jiandi Wan, Marlon S. Thomas, Sean Guthrie and Valentine I. Vullev,** Princeton University, USA

Nanoscale *vs* microscale gelation on surface of crosslinked polyurethane, **Chenyu Wang, Wei Zhang and Kenneth J. Wynne**, *Virginia Commonwealth University, USA*

Wetting free energy on heterogeneous surfaces: From synthetic to biological, **Jihang Wang, Dusan Bratko and Alenka Luzar**, *Virginia Commonwealth University, USA*

A bug's life – On superhydrophobic surfaces, **Alex H. F. Wu, K. L. Cho, Irving I. Liaw, Hua Zhang and Robert N. Lamb**, *University of New South Wales, Australia*

Water-mediated mid-range interaction of antiparallel β-sheets, **Soohaeng Yoo and Sotiris S. Xantheas**, *Pacific Northwest National Laboratory, USA*

Network constrained surface phase separation, **Wei Zhang, Chenyu Wang and Kenneth J. Wynne**, *Virginia Commonwealth University, USA*

Temperature-induced dynamic pinning on structured superhydrophobic surfaces, **Lidiya Mishchenko, Benjamin Hatton, J. Ashley Taylor, Vaibhav Bahadur, Tom Krupenkin and Joanna Aizenberg**, *Harvard University, USA*

Directional collecting water drops, **Lei Jiang**, *Institute of Chemistry, CAS, China*

The Skinner Poster Prize for the best poster was awarded to Ludivine Guillemot of Université Claude Bernard Lyon 1 University, France, for her poster on Dynamic intrusion and extrusion of water in hydrophobic mesoporous materials. The second and third awards were sponsored by the American Chemical Society (Division of Colloid and Surface Chemistry) and were awarded to Jonathan Boreyko (Duke University) for his poster on Self-propelled dropwise condensate on superhydrophobic surfaces and to Davide Vanzo (Università di Bologna) for his work on Computer simulation of a nematic nanodroplet wetting a flat surface.

# List of participants

Mr Hari Acharya, *Rensselaer Polytechnic Institute, USA*
Mr Travers Anderson, *University of California, Santa Barbara, USA*
Dr Rick Angus, *Hollingsworth and Vose, USA*
Professor Bruce Berne, *Columbia University, USA*
Mr Jonathan Boreyko, *Duke University, USA*
Mr Dusan Bratko, *Virginia Commonwealth University, USA*
Mr Kennard Brunson, *Virginia Commonwealth University, USA*
Professor Hans-Juergen Butt, *Max Planck Institute for Polymer Research, Germany*
Mr Kyler Carroll, *Virginia Commonwealth University, USA*
Mr Aviel Chaimovich, *UCSB, USA*
Ms Asima Chakravorty, *Virginia Commonwealth University, France*
Professor Elisabeth Charlaix, *University of Lyon, USA*
Professor Chuan-Hua Chen, *Duke University, Taiwan*
Professor Li-Jen Chen, *National Taiwan University, Taiwan*
Dr Kuan-Hung Cho, *National Taiwan University, USA*
Professor Chang-Hwan Choi, *Stevens Institute of Technology, United Kingdom*
Dr Hugo Christenson, *University of Leeds, USA*
Dr Anindarupa Chunder, *Nanoscience Technology Center, USA*
Professor Rene Corrales, *The University of Arizona, USA*
Professor Vincent Craig, *Australian National University, Australia*
Dr Colin Crick, *University College London, United Kingdom*
Mrs Maria D'Acunzi, *MPI for Polymer Research, USA*
Dr Chris Daub, *Virginia Commonwealth University, USA*
Dr Debnath De, *Johnson Matthey ECT, USA*
Professor Pablo Debenedetti, *Princeton University, USA*
Dr Stephen Donaldson, *UCSB, Finland*
Dr Lei Dong, *Aalto University, USA*
Professor William Ducker, *Virginia Tech, USA*
Mr John Edison, *University of Massachusetts, USA*
Mr Alexander Epstein, *Harvard University, USA*
Professor Jeffery Errington, *University at Buffalo, United Kingdom*
Professor Bob Evans, *University of Bristol, USA*
Professor John Fenn, *Virginia Commonwealth University, USA*
Professor Tomoko Fujiwara, *University of Memphis, USA*
Professor Di Gao, *University of Pittsburgh, USA*
Professor Shekhar Garde, *Rensselaer Polytechnic Institute, USA*
Ms Morwenna Gilbert, *Royal Society of Chemistry, United Kingdom*
Professor Nicolas Giovambattista, *Brooklyn College of the City University of New York, USA*
Dr Jose Guadarrama, *University of Navarra, Spain*
Ms Ludivine Guillemot, *Laboratoire PMCN, France*
Dr Zhiguang Guo, *University of Namur, Belgium*
Dr Murari Gupta, *Virginia Commonwealth University, USA*
Dr Malte Hammer, *University of California, Santa Barbara, USA*
Dr Benjamin Hatton, *Harvard University, USA*
Professor Savvas Hatzikiriakos, *University of British Columbia, Canada*
Ms Kristin Hecht, *Karlsruhe Insitute of Technology, Germany*
Dr Jim Henderson, *University of Leeds, United Kingdom*
Dr Chris Honig, *Virginia Tech, USA*
Dr Gerhard Hummer, *National Institutes of Health, USA*

Professor Jacob Israelachvili, *University of California, Santa Barbara, USA*
Dr Neha Jain, *New Jersey Institute of Technology, USA*
Dr Sumanth Jamadagni, *Rensselaer Polytechnic Institute, USA*
Dr Lei Jiang, *Institute of Chemistry, CAS, China*
Mr Sung Kang, *Harvard University, USA*
Ms Anne-Marie Kietzig, *University of British Columbia, Canada*
Professor Jacob Klein, *Weizmann Institute of Science, Israel*
Professor Michael Klein, *University of Pennsylvania, USA*
Dr Takahiro Koishi, *University of Fukui, Japan*
Dr Vaibhaw Kumar, *University at Buffalo, USA*
Dr Yongjoo Kwon, *Seoul National University, South Korea*
Professor Robert Lamb, *The University of Melbourne, Australia*
Dr Florian Lapierre, *BIOMEMS Group, France*
Professor Bruce Law, *Kansas State University, USA*
Dr Mei Li, *Shanghai Jiao Tong University, USA*
Dr Hui Li, *University of Nebraska Lincoln, USA*
Dr Zuoli Li, *Virginia Tech, USA*
Mr Fernando Luna-Vera, *Virginia Commonwealth University, USA*
Dr Helen Lunn, *Royal Society of Chemistry, United Kingdom*
Professor Alenka Luzar, *Virginia Commonwealth University, USA*
Dr Valentina Marcon, *Centers of Smart Interfaces - TU Darmstadt, Germany*
Mr Dean Mastropietro, *Virginia Tech, USA*
Mr Sean McBride, *Kansas State University, USA*
Professor Thomas McCarthy, *University of Massachusetts Amherst, USA*
Dr Anatoli Milischuk, *Colorado State University, USA*
Ms Lidiya Mischenko, *Harvard University, USA*
Professor Jeetain Mittal, *Lehigh University, USA*
Professor Peter Monson, *University of Massachusetts, USA*
Ms Barbara Mooney, *The University of Arizona, USA*
Professor Frieder Mugele, *University of Twente, The Netherlands*
Ms Deborah Ortiz, *Georgia Tech, USA*
Mr Sergio Pacheco Benito, *University of Twente, The Netherlands*
Dr Lei Pan, *Virginia Tech, USA*
Professor Mahesh Panchagnula, *Tennessee Tech University, USA*
Dr Neelesh Patankar, *Northwestern University, USA*
Mr Amish Patel, *Rensselaer Polytechnic Institute, USA*
Dr Natalia Pérez-Hernández, *Instituto de Investigaciones Quimicas, CSIC, Spain*
Dr Susan Perkin, *University College London, United Kingdom*
Mr Daniel Peter, *Lam Research AG, Austria*
Professor David Quéré, *ESPCI, France*
Professor Jay Rasaiah, *University of Maine, USA*
Mr Richard Remsing, *University of Maryland, USA*
Mr John Ritchie, *Virginia Commonwealth University, USA*
Dr Martina Roeselova, *Czech Academy of Sciences, Czech Republic*
Miss Anna Roffey, *Royal Society of Chemistry, United Kingdom*
Professor Peter Rossky, *The University of Texas at Austin, USA*
Ms Sapna Sarupria, *Princeton University, USA*
Dr Rossen Sedev, *University of South Australia, Australia*
Mrs Jamileh Seyed-Yazdi, *Virginia Commonwealth University, USA*
Mr David Sherwood, *Exosect Limited, United Kingdom*
Mr Gilad Silbert, *Weizmann Institute of Science, Rehovot, Israel*
Dr Christos Stamboulides, *University of British Columbia, Canada*
Dr Frank Stillinger, *Princeton University, USA*

Mr Guillaume Stirnemann, *École Normale Supérieure, France*
Dr Vincent Thomy, *BioMEMS Group IEMN, France*
Mrs Joanna Thorne, *University of Greenwich, United Kingdom*
Dr Naga Rajesh Tummala, *University of Oklahoma, USA*
Ms Robin Underwood, *Purdue University, USA*
Mr Davide Vanzo, *Universita di Bologna, Italy*
Dr Kripa Varanasi, *MIT, USA*
Dr Doris Vollmer, *MPI for Polymer Research, Germany*
Mr Michael von Domaros, *Virginia Commonwealth University, USA*
Dr Jiandi Wan, *Princeton University, USA*
Mr Chenyu Wang, *Virginia Commonwealth University, USA*
Mr Jihang Wang, *Virginia Commonwealth University, USA*
Mr Lingle Wang, *Columbia University, USA*
Professor John Weeks, *University of Maryland, USA*
Professor Benjamin Widom, *Cornell University, USA*
Dr Geoff Willmott, *Industrial Research Limited, New Zealand*
Mr Alex Wu, *The University of New South Wales, Australia*
Professor Kenneth Wynne, *Virginia Commonwealth University, USA*
Professor Julia Yeomans, *University of Oxford, United Kingdom*
Dr Yoshiteru Yonetani, *Japan Atomic Energy Agency, Japan*
Dr Soohaeng Yoo, *Pacific Northwest National Laboratory, USA*
Professor Roe-Hoon Yoon, *Virginia Tech, USA*
Dr Xiao Cheng Zeng, *University of Nebraska Lincoln, USA*
Mr Wei Zhang, *Virginia Commonwealth University, USA*

# Index of contributors*

Acharya, H., 195, 283, **353**
Anderson, T. H., **299**
Angus, R., 367
Audry, M.-C., **113**
Auerhammer, G. K., **35**
Berne, B. J., **247**, 283
Blossey, R., **125**
Boreyko, J., 79, 195, 283
Boukherroub, R., **125**
Bratko, D., **67**, 79, 195, 283, 367
Brunet, P., **125**
Butt, H.-J., **35**, 79, 195, 367
Cao, L., **57**
Chaimovich, A., **299**, 367
Charlaix, E., **113**, 195
Chen, C.-H., 79, 195, 367
Cho, K.-H., **223**
Choi, C.-H., 79
Christenson, H. K., 79, 195, 367
Chunder, A., 79
Clanet, C., **19**
Coffinier, Y., **125**
Corrales, R., 79, 195
Craig, V. S. J., **141**, 195, 283, 367
D'Acunzi, M., **35**
Daub, C., **67**, 79, 367
Debenedetti, P., 79, 195, 283, 367
Ducker, W., 195
Edison, J., **167**
Epstein, A., 79
Errington, J., 79
Evans, B., 79, 195, 283, 367
Friesner, R. A., **247**
Fujikawa, S., **185**
Gao, D., **57**, 79
Garde, S., 79, **353**, 367
Giovambattista, N., 79, 195, 367
Hecht, K., 283
Henderson, J., 79, 283, 367
Hendy, S. C., **233**
Honig, C., 195
Hummer, G., 79, 195, 283, **341**, 367
Hynes, J. T., **263**
Indekeu, J. O., **217**
Israelachvili, J., **299**, 367
Jamadagni, S. N., 79, 283, **353**
Jiang, L., 79
Joseph, P., **113**
Kang, S., 79, 283, 367
Kirby, N., **223**

Klein, J., 79, 195, **309**, 367
Koga, K., **217**
Koishi, T., **185**, 195
Krumpfer, J. W., **103**
Kudesia, S., **67**
Kusumaatmaja, H., **153**
Laage, D., **263**
Lamb, R., 79, **223**
Lapierre, F., **125**
Law, B., 79, 195, 283, 367
Liaw, I. I., **223**
Liu, G., **141**
Luzar, A., **67**, 79, 195, 283, 367
Mammen, L., **35**
McCarthy, T., 79, **103**, 195, 283
Mischenko, L., 79, 195, 283, 367
Mittal, J., **341**
Mognetti, B. M., **153**
Monson, P., **167**, 195
Mooney, B., 283
Moran, G., **223**
Mugele, F., **49**, 79, 195
Neto, C., **233**
Pacheco Benito, S., 79, 195, 283
Pan, L., **325**
Panchagnula, M., 195
Patankar, N., 79, 195
Patel, A., 79, 195, 283, 367
Perkin, S., 195, **309**, 367
Peter, D., 195, 283
Piednoir, A., **113**
Quéré, D., **19**, 79, 195, 283
Rathgen, H., **49**
Reyssat, M., **19**
Richard, D., **19**
Ritchie, J., 283
Rossky, P. J., **13**, 79, **263**
Sarupria, S., 79, 367
Scott Shell, M., **299**
Sedev, R., 195
Seyed-Yazdi, J., 79, 195, 283, 367
Sherwood, D., 195
Silbert, G., **309**, 367
Singh, M., **35**
Stillinger, F., 283, **395**
Stirnemann, G., 79, 195, **263**, 283, 367
Thomy, V., **125**, 195
Vanzo, D., 367
Varanasi, K., 79
Vembanur, S., **353**

Vollmer, D., **35**, 79, 195, 367
Wan, J., **67**, 195
Wang, J., 195, 367
Wang, L., **247**
Weeks, J., 79, 195
Widom, B., **217**, 283

Willmott, G., 195, **233**, 283
Wu, A. H. F., 79, 195, **223**, 283
Yasuoka, K., **185**
Yeomans, J., 79, **153**, 195
Yoon, R.-H., **325**, 367
Zeng, X. C., **185**

* The page numbers in **bold** type indicate papers submitted for discussions.